AF007224

Series on Probability and Statistics Vol. 1

The Third Nagoya Lévy Seminar

Gaussian Random Fields

Series on Probability and Statistics Vol. 1

The Third Nagoya Lévy Seminar

Gaussian Random Fields

Nagoya, Japan *15 – 20 Aug. 1990*

Editors:

K. ITÔ

Kyoto University

T. HIDA

Meijo University

World Scientific

Singapore • New Jersey • London • Hong Kong

Published by

World Scientific Publishing Co. Pte. Ltd.
5 Toh Tuck Link, Singapore 596224
USA office: 27 Warren Street, Suite 401-402, Hackensack, NJ 07601
UK office: 57 Shelton Street, Covent Garden, London WC2H 9HE

British Library Cataloguing-in-Publication Data
A catalogue record for this book is available from the British Library.

GAUSSIAN RANDOM FIELDS

Copyright © 1991 by World Scientific Publishing Co. Pte. Ltd.

All rights reserved. This book, or parts thereof, may not be reproduced in any form or by any means, electronic or mechanical, including photocopying, recording or any information storage and retrieval system now known or to be invented, without written permission from the publisher.

For photocopying of material in this volume, please pay a copying fee through the Copyright Clearance Center, Inc., 222 Rosewood Drive, Danvers, MA 01923, USA. In this case permission to photocopy is not required from the publisher.

ISBN-13 978-981-02-0505-8
ISBN-10 981-02-0505-8
ISBN-13 978-981-02-3804-9 (pbk)
ISBN-10 981-02-3804-5 (pbk)

PREFACE

These are notes of lectures presented at the International Conference on Gaussian Random Fields held at Nagoya for the period August 15–20, 1990. Part I involves papers by the speakers of the plenary session on the opening day while the notes of other invited papers are in Part II.

The conference was originally organized as the third Nagoya Lévy Seminar on the anniversary of the Lévy Brownian motion. Later it was suggested to us to organize it also as one of the satellite meetings of the 1990 International Congress of Mathematics (ICM 90) held at Kyoto. The scope of the conference, therefore, has been considerably widened so as to include other topics in addition to Gaussian random fields.

The organizers are grateful to the invited speakers, primarily of course for their contributions, but also for the efforts they made to stay within the space limitations. Thanks are also due to other invited speakers who did not send us their papers; we list their talks by title after the contents. Finally, we thank Dr. N. Obata and Dr. K. Saitô who helped us in typing final versions of the manuscripts.

May 15, 1991

Kiyosi Itô Kyoto University
and
Takeyuki Hida Meijo University

Academic Advisors

Louis H.Y. Chen

National University of Singapore

G. Kallianpur

University of North Carolina

Yu.A. Rozanov

Steklov Mathematical Institute

CONTENTS

PLENARY LECTURES

INVITED LECTURES

The following lectures were delivered at the conference, but are not included in this volume.

S.Albeverio and Z.M.Ma

Dirichlet forms associated with perfect processes, diffusion processes and Hunt processes.

L.H.Y.Chen

Some recent developments in Poisson approximation.

G.W.Johnson

Homogeneous chaos, p-forms, scaling and the Feynman integral.

H.Kaneta

Maximum distance separable codes.

P.Krée and E.Carlen

On sharp L^p estimates for iterate stochastic integrals and some related topics.

J.-H.Lou

On the best possible constants of Poincaré inequalities.

H.Ogura

Stochastic spherical harmonics and their applications.

I.Shigekawa

Sobolev spaces over the Wiener space.

M.Yamazato

The class of hitting time distributions of 1-dimensional generalized diffusion processes and related subclasses of infinitely divisible distributions.

and D.Q.Luu.

Part I Plenary Session Papers

LOOP SPACES AND LOGARITHMIC SOBOLEV INEQUALITIES

LEONARD GROSS
Department of Mathematics, Cornell University
Ithaca, NY 14853

ABSTRACT

A logarithmic Sobolev inequality is proved for a perturbation of the natural Dirichlet form operator over the Brownian bridge measure space of a compact Lie group.

1. **Introduction.**

Let M be a finite dimensional Riemannian manifold. Pick a point m in M and denote by WM the set of continuous functions x from $[0,1]$ into M such that $x(0) = m$. Then Brownian motion in M, beginning at m, induces a probability measure P on WM as is well known [IW]. In case M is the real line and $m = 0$ then (WM,P) is just the usual Wiener space with Wiener measure. For general Riemannian manifolds and a function $f:WM \to R$ there is a notion of gradient which we shall describe shortly when M is a compact Lie group. Having both a measure on WM as well as a gradient operation ∇ for functions on WM there are a large number of interesting questions in analysis that one may ask about the Dirichlet form operator $\nabla^*\nabla$ and its perturbations, as operators on $L^2(P)$.

The loop space L consisting of those paths x which return to m at time one supports a natural probability measure μ, namely the conditioned Brownian measure:

$$\mu = P(\ |x(1) = m). \tag{1.1}$$

There is also a natural gradient operator ∇_0 for functions on L as well, which we shall describe below when M is a compact Lie group. Thus in $L^2(\mu)$ one also has a natural Dirichlet form operator $\nabla_0^*\nabla_0$ which is of great interest along with its perturbations. If M is the real line or R^n then this kind of operator has been studied intensively both in the quantum field theory literature, where it appears as the number operator or as a Hamiltonian, and in the mathematics literature, where it is an infinite dimensional Ornstein-Uhlenbeck operator. Several lectures at this conference will be devoted to variants of these operators. If M is nonlinear, however, the study of these "Laplacians" on these infinite dimensional Riemannian manifolds is in its infancy. For some of the recent work we refer the reader to [A,AM1,2,AV, Ge,Gr3,5,K1,2,M,MM]. In this lecture I will describe some recent results on the role of these operators in infinite dimensional Sobolev type inequalities when M is a compact Lie group. But first let us consider the definition of gradient for this easy case.

Denote by $\mathcal{G}$ the Lie algebra of G. We choose an inner product on $\mathcal{G}$ which is invariant under the adjoint action of G. We shall keep this inner product fixed throughout and denote the associated norm on $\mathcal{G}$ by $|\ |$. Let H denote the Hilbert space consisting of absolutely continuous functions from $[0,1]$ into $\mathcal{G}$ which vanish at zero and have square integrable derivative. Put

$$|h|^2 = \int_0^1 |\dot{h}(s)|^2 ds \tag{1.2}$$

wherein a dot denotes the derivative. If g is in WG and h is in H we write

$$(e^h g)(s) = \exp(h(s))g(s). \tag{1.3}$$

Then we define the directional derivative of a function $f:WG \to R$ by

$$\partial_h f(g) = df(e^{th}g)/dt \quad \text{at} \quad t = 0 \tag{1.4}$$

if the derivative exists. Moreover if $\partial_h f(g)$ is a continuous linear function of h then the gradient of f may be defined by the equation

$$(\nabla f(g), h) = \partial_h f(g) \qquad h \text{ in } H. \tag{1.5}$$

Now let

$$H_0 = \{h \text{ in } H : h(1) = 0\} \tag{1.6}$$

and observe that if h is in H_0 and g is in the space L of based loops then so is $e^h g$. Hence if f is a function on L then (1.4) is well defined and (1.5), for h in H_0, defines a tangential gradient of f, which we denote by $\nabla_0 f$. Of course $\nabla_0 f$ lies in H_0. We have now two examples of infinite dimensional manifolds, each with a natural probability measure and gradient operation. In the next section we shall use them to describe some Sobolev type inequalities. Let us mention however that the gradient operations can be extended to differential forms and the Lie group G replaced by an arbitrary compact Riemannian manifold. An intriguing long range goal of this infinite dimensional analysis would be to prove some infinite dimensional analog of the Hodge DeRham decomposition. Much of the machinery required for this has been developed in recent years [S1,2,3,Su and the preceding references]. In particular S. Kusuoka has already proved a DeRham theorem [K2].

2. Logarithmic Sobolev inequalities.

The standard Sobolev inequalities on R^n are highly dimension dependent and do not appear to have a meaningful formulation in infinite dimensions. We refer the reader to [Gr6] for further discussion of this difficulty. A part of the problem is the nonexistence of a useful form of infinite dimensional Lebesgue measure. But there are analogous inequalities for Gauss measure which are dimension independent and have simple extensions to infinite dimensions. Let us denote by ν the Gauss measure on R^n:

$$d\nu(x) = (2\pi)^{-n/2}\exp(-|x|^2/2)dx. \qquad (2.1)$$

Then [Gr2] for a real function f whose weak gradient is in $L^2(\nu)$ one has

$$\text{(LS)} \int_{R^n} f(x)^2 \log|f(x)|d\nu(x) \le \int_{R^n} |\mathrm{grad} f(x)|^2 d\nu(x) + \|f\|^2_{L^2(\nu)} \log\|f\|_{L^2(\nu)}. \qquad (2.2)$$

The inequality (LS) is a good example of the stimulus provided to mathematics by mathematical physics. In order to prove the semiboundedness of certain Hamiltonians in quantum field theory E. Nelson proved his famous hypercontractivity inequalities [N1,2] which were shown by Paul Federbush [Fe] to imply an inequality such as (LS). The present author [Gr2] later showed the reverse implication. See [DGS] for a historical review and bibliography and [Gr4] for an updated bibliography on this topic. In this report we will be concerned only with (LS) and its infinite dimensional versions. Let us first consider some of the general properties of inequalities such as (LS).

If we regard the gradient operator as a densely defined closed operator from $L^2(R^n,\nu)$ into $L(R^n,\nu) \otimes R^n$ then the operator $N = \mathrm{grad}^*\mathrm{grad}$ is a positive self-adjoint operator on $L^2(R^n,\nu)$ and (LS) reads

$$\int_{R^n} f(x)^2 \log|f(x)|d\nu(x) \le (Nf,f)_{L^2(\nu)} + \|f\|^2_{L^2(\nu)} \log\|f\|_{L^2(\nu)}. \qquad (2.3)$$

Definition 1. A self-adjoint operator A on a probability measure space (X,μ) is called a logarithmic Sobolev generator (LS generator) if

$$\int_X f(x)^2 \log|f(x)|d\mu \le (Af,f) + \|f\|^2 \log\|f\| \qquad (2.4)$$

for all f in the domain of A. The norms are $L^2(\mu)$ norms.

<u>Properties of logarithmic Sobolev generators</u>.

a. An LS generator is a nonnegative operator.

b. (semi boundedness of perturbations) If A is an LS generator on $L^2(X,\mu)$ and V is a real valued measurable function on X such that $\|e^{-V}\|_{L^2} < \infty$ then $A+V$ is bounded below with

$$A+V \geq -\left[\log\|e^{-V}\|_{L^2}\right] I. \tag{2.5}$$

Moreover (2.4) and (2.5) are equivalent in the sense that (2.4) holds for all f if and only if (2.5) holds for all V.

c. (additivity) If A is an LS generator on $L^2(X_i,\mu_i)$ for $i = 1$ and 2 then $A \otimes I + I \otimes A$ is a LS generator on $L^2(X_1 \times X_2,\mu_1 \times \mu_2)$.

d. (mass gap) If A is an LS generator on $L^2(R^n,\mu)$ for some probability measure μ and whose quadratic form is the Dirichlet form for μ, that is

$$(Af,f) = \int |\mathrm{grad}\ f(x)|^2 d\mu(x) \tag{2.6}$$

then the spectrum of A consists of zero (with multiplicity one) together with a subset of $[1,\infty)$. There is thus a spectral gap (so called mass gap because of its interpretation in quantum field theory) of at least one between the lowest eigenvalue zero and the rest of the spectrum.

Property a. follows immediately from Jensen's inequality and the convexity of $t\log t$ (put $t = f(x)^2$). For the most efficient proofs of b. and c. see [F]. For a simple direct proof of d. see [R2]. Now one can deduce (LS) by induction with the help of c. if one knew (LS) for the case $n = 1$. For the simplest proof of (LS) in the case $n = 1$ see [R1]. There are many other proofs of (LS) using a wide variety of techniques.

It would be fair to credit the additivity property c. for the fact that logarithmic Sobolev inequalities are well adapted to infinite dimensions. Let us see now how one can formulate (LS) in infinite dimensions. On the countable product R^{∞} denote by μ the product of Gaussian measures with mean zero and variance one. If t_j is the jth coordinate function on R^{∞} and f is a cylinder function on R^{∞} which depends only on the first n coordinates then (LS) can clearly be written as

$$\int_{R^{\infty}} f(x)^2 \log|f(x)|d\mu(x)$$

$$< \int_{R^{\infty}} \sum_{j=1}^{\infty} (\partial f(x)/\partial t_j)^2 d\mu(x) + \|f\|^2 \log\|f\| \qquad (2.7)$$

since the marginal distribution of μ with respect to the σ-field generated by the first n coordinates is υ. For such a function one defines

$$\text{grad } f(x) = (\partial f(x)/\partial t_1, \ldots, \partial f(x)/\partial t_j, \ldots)$$

as an element in the Hilbert space ℓ^2 for each x in R^{∞}. The inequality (2.7) extends to the form closure of this domain and we thus have an infinite dimensional version of (LS). It is easy to rewrite this in a way that makes it look more interesting. Denote by C the usual Wiener space consisting of the continuous functions on [0,1] which vanish at zero and by H the Carmeron-Martin Hilbert space defined by (1.2) with $\mathscr{G} = R$ and $G = R$. For this group the product (1.3) reduces to $g(s)+h(s)$ and the gradient (1.5) is therefore just the usual H derivative [Gr1];

$$(\text{grad } f(g), h) = df(g+th)/dt \quad \text{at} \quad t = 0. \qquad (2.8)$$

It is well understood how, by choosing any orthonormal basis of H, the product measure μ of (2.7) maps to Wiener measure w on C while at the same time the gradient operation in (2.7) maps to the H derivative. Thus (2.7) may be written as an inequality over C:

$$\int_C f(x)^2 \log|f(x)|dw(x)$$

$$\leq \int_C |\text{grad } f(x)|^2 d\mu(x) + \|f\|^2 \log\|f\|. \qquad (2.9)$$

Of course the same argument shows that the same inequality holds over any abstract Wiener space. In particular (2.9) holds if w is the path space measure for n dimensional Brownian motion. We shall make use of this inequality in the next section where we shall choose the n dimensional state space to be the Lie algebra $\mathcal{G}$.

3. Group valued Wiener space.

So far we have stated logarithmic Sobolev inequalities only for Gaussian measures, both finite and infinite dimensional. In this section we shall rewrite these same Gaussian inequalities in a mildly disguised form and in the next section we shall derive some non-Gaussian logarithmic Sobolev inequalities. Let G be a compact Lie group with Lie algebra $\mathcal{G}$. We choose an AdG invariant inner product on $\mathcal{G}$ and use the notation of Section 1 for WG, P, H and $\nabla f(g)$. Since $\mathcal{G}$ is an n dimensional Euclidean space there is a standard $\mathcal{G}$ valued Brownian motion $B(s)$, $0 < s < 1$ with $B(0) = 0$. It is well known that the measure P on WG is induced from the Brownian motion B by the solution to the stochastic differential equation

$$dg(s) = g(s)\circ dB(s) \qquad g(0) = \text{identity in } G. \tag{3.1}$$

Here the stochastic integral is to be interpreted as a Fisk-Stratonowich integral. Let us write the solution to (3.1) as $g = I(B)$. In the $\mathcal{G}$ valued Wiener space we have the usual H derivative grad whereas for functions on WG we have a gradient operation defined by (1.5). The map I from almost all of $\mathcal{G}$ valued Wiener space into WG not only induces the measure P from Wiener measure but also relates the two gradients as follows.

Lemma 3.1. If $f:WG \to R$ is reasonable and $F(B) = f(I(B))$ then

$$|\text{grad } F(B)| = |(\nabla f)(I(B))|. \tag{3.2}$$

For a proof we refer the reader to [Gr5, Lemma 4.15].

We may therefore rewrite (2.9) entirely in terms of the group valued path space instead of the Lie algebra valued path space thus

$$\int_{WG} f(g)^2 \log|f(g)|dP(g)$$

$$\leq \int_{WG} |\nabla f(g)|^2 dP(g) + \|f\|^2 \log\|f\|. \quad (3.3)$$

The point of this rewriting is that it now allows us to pass naturally to an interesting cofinite dimensional submanifold of WG, the space of based loops.

4. The space of based loops.

The space L of paths in G which return to the identity element e at time $s = 1$ supports the conditional Brownian measure μ defined in (1.1). The process $s \to g(s)$ associated to μ is usually referred to as pinned Brownian motion or Brownian Bridge and μ is Brownian Bridge measure. We have already defined the tangential gradient of a function on L in (1.5) with H replaced by H_0. Under pointwise multiplication WG is a group and L is a subgroup. By virtue of (1.4) H can be identified with a space of right invariant vector fields on WG and we shall regard it in this manner as the right invariant Lie algebra of WG. The discussion in the introduction shows that H_0 may similarly be interpreted as the right invariant Lie algebra of the based loop group L.

In order to state our main theorem it will be necessary to describe another Brownian motion. Let

$$b(t) = \int_0^t Adg(s)dB(s) \quad \text{(Ito integral)}. \quad (4.1)$$

It is easy to verify by the Ito calculus that $b(\ )$ is indeed a $\mathcal{G}$ valued Brownian motion and that it is related to the G valued process by means of the Fisk-Stratonowich stochastic differential equation

$$dg(s) = db(s) \circ g(s). \quad (4.2)$$

B and b are respectively the right and left $\mathcal{G}$ valued Brownian motions associated to P. Let

$$V(g) = |b(1)|^2 + 1. \quad (4.3)$$

<u>MAIN THEOREM</u>. There is a constant c such that the operator

$$A = c(\nabla_0^* \nabla_0 + V) \tag{4.4}$$

is a logarithmic Sobolev generator in $L^2(L,\mu)$ where μ is Brownian Bridge measure.

Remarks. 4.1. If G is not commutative then the Brownian Bridge measure μ is not Gaussian.

4.2. If one eliminates the potential in (4.4) then A is not a logarithmic Sobolev generator when G is the circle group. The unperturbed Dirichlet form operator $\nabla_0^* \nabla_0$ cannot by itself be an LS operator if G is not simply connected because any function which is constant on each homotopy class belongs to the eigenvalue zero, whereas property d. in Section 2, properly generalized to infinite dimensions, shows that this eigenvalue has multiplicity one.

4.3. The restriction of the Dirichlet form operator $\nabla_0^* \nabla_0$ to the functions supported on a single homotopy class may have a mass gap even if it is not an LS generator on this space. A technique that could prove the existence of a mass gap would very likely yield a method for proving a Hodge theorem over a loop space. E. Getzler's [Ge] motive in studying logarithmic Sobolev inequalities over loop space was in fact to prove a mass gap and eventually a Hodge theorem. Unfortunately the powerful method of Bakry and Emery for proving logarithmic Sobolev inequalities fails to be applicable in the present situation as Getzler showed.

4.4. For the general case of a compact manifold M instead of merely a compact Lie group nothing is known about logarithmic Sobolev inequalities over WM even for unrestricted endpoint, when one uses the intrinsic gradient for path spaces. The curvature of M seems to play a significant role already for the unpinned process.

The idea behind the proof of the main theorem is very simple. We wish to prove the inequality (2.4) for a smooth function f defined on the loop space L. We simply extend the function f to a smooth function on WG in a controlled way and apply the already known Gaussian inequality (3.3) to the extended function. One must disintegrate the measure P in such a way as to express it as a product of μ with Lebesgue measure on $\mathcal{G}$ together with a density. Unfortunately the density is unbounded and forces one to add the potential V in (4.4). See [Gr5] for details.

References

[A] H. Airault, *Projection of the infinitesimal generator of a diffusion*, J. of Funct. Anal. 85 (1989), 353-391.

[AM1] H. Airault and P. Malliavin, *Integration geometrique sur l'espace de Wiener*, Bull. Sc. math. 2^e serie, 112 (1988), 3-52.

[AM2] ______________________, *Le processus D'Ornstein- Uhlenbeck sur une sous variete*, Paris preprint 1990.

[AV] H. Airault and J. Van Biesen, *Geometrie riemannienne en codimension finie sur l'espace de Wiener*, C. R. Acad. Sci. Paris, v. 311 (1990), 125-130.

[DGS] E. B. Davies, L. Gross and B. Simon, *Hypercontractivity: a bibliographic review*, Proc. Hoegh-krohn, Memorial Conference, Ed. S. Albeverio, to appear.

[F] W. Faris, *Product spaces and Nelson's inequality*, Helv. Phys. Acta 48 (1975), 721-730.

[Fe] P. Federbush, *A partially alternate derivation of a result of Nelson*, J. Math. Phys. 10 (1969), 50-52.

[Ge] E. Getzler, *Dirichlet forms on loop space*, Bull. Sc. math. 2^e serie, 113 (1989), 151-174.

[Gr1] L. Gross, *Potential theory on Hilbert space*, J. of Funct. Anal. 1 (1967), 123-181.

[Gr2] ________, *Logarithmic Sobolev inequalities*, Amer. J. Math. 97 (1975), 1061-1083.

[Gr3] ________, *Logarithmic Sobolev inequalities on Lie groups*, to appear.

[Gr4] ________, *Logarithmic Sobolev inequalities for the heat kernel on a Lie group*, Bielefeld Conf. on White Noise Analysis (1989), 108-130, Eds. T. Hida, H. H. Kuo, J. Potthoff, L. Streit, World Scientific.

[Gr5] ________, *Logarithmic Sobolev inequalities on loop groups*, J. of Funct. Anal., to appear.

[Gr6] ________, *Logarithmic Sobolev inequalities over some infinite dimensional manifolds*, Proceedings of the Colorado Springs Conference, May 1989, World Scientific, to appear.

[IW] N. Ikeda and S. Watanabe, *Stochastic Differential Equations and Diffusion Processes*, North-Holland, New York 1981.

[K1] S. Kusuoka, *Analysis on Wiener spaces II. Differential forms*, RIMS report 705, July 1990.

[K2] __________, *DeRham cohomology of Wiener-Riemannian manifolds*, preprint August 1990, RIMS, Kyoto.

[M] P. Malliavin, *Hypoellipticity in infinite dimension*, Proceedings of the 1989 North-Western Conference, M. Pinsky, Editor, to appear 1990.

[MM] M-P. Malliavin and P. Malliavin, *Quasi invariant integration on loop groups*, J. of Funct. Anal., to appear.

[N1] E. Nelson, *A quartic interaction in two dimensions*, in Mathematical Theory of Elementary Particles, 69-73 (R. Goodman and I. Segal, eds.) M.I.T. Press, Cambridge, MA 1966.

[N2] _________, *The free Markoff field*, J. of Funct. Anal. 12 (1973), 211-227.

[R1] O. S. Rothaus, *Lower bounds for eigenvalues of regular Sturm-Liouville operators and the logarithmic Sobolev inequality*, Duke Math. J. 45 (1978), 351-362.

[R2] _____________, *Diffusion on compact Riemannian manifolds and logarithmic Sobolev inequalities*, J. of Funct. Anal. 42 (1981), 102-109. See 2nd Lemma p.105.

[S1] I. Shigekawa, *Absolute continuity of probability laws of Wiener functionals*, Proc. of the Japan Academy 54 (1978), 230-233.

[S2] ____________, *Derivatives of Wiener functionals and absolute continuity of induced measures*, J. Math. Kyoto Univ. 20 (1980), 263-289.

[S3] ____________, *DeRham-Hodge-Kodaira's decomposition on an abstract Wiener space*, J. Math. Kyoto Univ. 26 (1986), 191-202.

[Su] H. Sugita, *Sobolev spaces of Wiener functionals and Malliavin's calculus*, J. Math. Kyoto Univ. 25 (1985), 31-48.

Traces, Natural Extensions and Feynman Distributions

G. Kallianpur
Department of Statistics, University of North Carolina
Chapel Hill, NC 27599-3260 USA

Abstract

A subspace of Wiener functionals with chaos expansions that have natural extensions is introduced as a test function space on which a suitable nuclear topology is defined. Feynman distributions are defined (and evaluated) as continuous linear functionals on this space. The existence and properties of k-traces of Hilbert space valued kernels play an important part in the work. The theory applies to L^2 functionals defined on conventional and abstract Wiener spaces as well as the Gaussian space used in Hida's white noise analysis.

1 Introduction

In a paper containing many insightful ideas, Hu and Meyer have considered extensions of multiple Wiener integrals in an attempt to obtain a satisfactory definition of the Feynman integral at least for a wide class of functionals on Wiener space [2]. In some recent work, Johnson and Kallianpur have provided a rigorous justification for the Hu-Meyer approach [4,5]. This is done by introducing the notion of scale invariant liftings as well as different types of Hilbert space-valued traces of symmetric kernels of multiple Wiener integrals.

In this paper we first summarize the two main results of [5] leading to the definition of the natural extension of the multiple Wiener-Itô integral. These results derived in [5] for functionals on the conventional Wiener space are seen to hold also for any abstract Wiener space or for the Gaussian white noise probability space $(\mathcal{S}(\mathbf{R})', \mathbf{P})$. The main concern of this paper is to investigate the notions of k-trace and natural extensions of chaos expansions and to show how they can be used to define a natural topology on a subspace Φ^c of $\mathbf{L}^2$, the space of square integrable Wiener functionals. This topology is nuclear and the generalized Wiener functionals which are elements of the dual of Φ^c are defined to be the Feynman distributions with complex parameter z ranging over a domain that includes $\pm\sqrt{-1}$. The test functional space Φ^c seems to be quite reasonable in view of the fact that the Feynman integrals evaluated in

[2] and [5] are given explicitly in terms of traces. It would be interesting to compare the approach adopted here with the Hida distributions discussed in the recent papers [8,9,10]. The aim of the paper is similar to that of A.S. Ustunel and the author [7] but the approach here is different. More comments are made at the end of Section 3.

For reasons of space, discussion of concrete examples is deferred to a later paper.

2 Scale-Invariant Measurability and Scale-Invariant Liftings.

Johnson and Skoug introduced this concept with reference to the conventional Wiener space in which $\Omega = C_0(\mathbf{R}_+)$, the space of real-valued continuous functions on $\mathbf{R}_+$ vanishing at 0 and P_σ is Wiener measure with variance parameter $\sigma > 0$ [3]. Then there exist supports Ω_σ for P_σ such that $\Omega_{\sigma_2} \cap \Omega_{\sigma_2} = 0$ if $\sigma_1 \neq \sigma_2$ (of course, P_{σ_1} and P_{σ_2} are mutually singular) and $\Omega_\sigma = \sigma\Omega_1$. Denote by $\mathbf{B}_\sigma$, the P_σ completion of $\mathbf{B}$, the σ-field of Borel sets of Ω.

A subset A of Ω is scale-invariant (or s-invariant) measurable if $\sigma A \in \mathbf{B}_1$ for all $\sigma > 0$. An s-invariant measurable set N is called an s-invariant null set if $P_1(\sigma N) = 0\ \forall \sigma > 0$. The family of s-invariant measurable sets is denoted by $\mathbf{B}_s = \bigcap_{\sigma>0} \mathbf{B}_\sigma$. A real-valued function F on Ω is said to be s-invariant measurable if it is measurable w.r.t. $\mathbf{B}_s$.

If F has domain D, we say that F is s-a.s. defined and s-invariant measurable iff the set where F is not defined is s-invariant null and for each $\sigma > 0$, the restriction of F to Ω_σ is P_σ-a.s. defined and P_σ-measurable.

Let $I_n^\sigma(f_n)$ for $\sigma > 0$ and $f_n \in \mathbf{L}^2(\mathbf{R}_+^n)$ be the multiple Wiener-Itô integral of order n. Then

$$(2.1) \qquad I_n^\sigma(f_n)(\sigma x) = \sigma^n I_n(f_n)(x), P_1 - a.s.$$

Scale-invariant lifting. We present here a very brief outline of the notion of lifting from [see 6]. This idea goes back to I.E. Segal. It is convenient to consider a real separable Hilbert space H (otherwise arbitrary) and μ, the canonical Gauss measure on the finite dimensional Borel cylinder sets of H, i.e. the finitely additive measure with characteristic functional $e^{-\frac{1}{2}\|h\|^2}, (h \in H)$.

A representation of μ is defined to be a pair $(L, \tilde{P})$ where $\tilde{P}$ is a probability measure on a measure space $(\tilde{\Omega}, \tilde{A})$ and L is a mapping (more precisely, an equivalence class of mappings) from H into the space of real valued random variables on $(\tilde{\Omega}, \tilde{A}, \tilde{P})$ such that L is linear in the sense that

$$(2.2) \qquad L(a_1, h_1 + a_2 h_2)(\tilde{\omega}) = a_1 L(h_1)(\tilde{\omega}) + a_2 L(h)2)(\tilde{\omega})\tilde{P} - \text{a.s.}$$

for $h_1, h_2 \in H, a_1, a_2 \in \mathbf{R}$. Further, for all cylinder sets

$$C = \{h \in H : ((h, h_1), \cdots, (h, h_k) \in B\}, (B Borel),$$

$$\mu(C) \quad = \quad \tilde{P}\{\tilde{\omega} \in \tilde{\Omega} : (L(h_1(\tilde{\omega}), \cdots, L(h_k)(\tilde{\omega})) \in B\}.$$

$\tilde{\Omega}$ is called a representation space. It is known that a representation space always exists. In fact, among the possible choices of $\tilde{\Omega}$ are (1) $\Omega = C_0(\mathbf{R}_+)$ (which has already been chosen above), (2) Ω is a separable Banach space with $\Omega^* \hookrightarrow H \hookrightarrow \Omega$ and where (Ω, H, P) is an abstract Wiener space and (3) $\Omega = \mathcal{S}(\mathbf{R})'$, the space with Gaussian white noise measure P. In this case we take $H = L^2(\mathbf{R})$. The supports Ω_σ of P_σ are defined somewhat differently than in [5]. If $\Omega = \mathcal{S}(\mathbf{R})'$ let $(\phi_j) \subset \mathcal{S}(\mathbf{R})$ be a complete orthonormal sequence (CONS) in H. Denoting by $<,>$ the duality relation between $S(\mathbf{R})'$ and $S(\mathbf{R})$, the random variable $< x, \phi_j >$ $(j = 1, 2, \cdots)$ are i.i.d. standard Gaussian random variables under P_1. For $h \in H, L(h)$ is defined by

$$L(h)(x) = \sum_{j=1}^{\infty}(h, \phi_j) < x, \phi_j > \; forall x \in \Omega$$

for which the series converges and $= 0$ otherwise. Then, under $P_\sigma, L(h)$ has a Gaussian distribution with mean 0 and variance $\sigma^2\|h\|^2$. In the more familiar notation, $L(h) = I_1(h)$. We may take the support of P_σ to be

$$\Omega_\sigma = \left\{ x \in \Omega : \lim_{n\to\infty} \frac{1}{n} \sum_{j=1}^{n} < x, \phi_j >^2 = \sigma^2 \right\}.$$

The procedure is entirely similar in the case when (Ω, P) is an abstract Wiener space.

We turn next to the definition of s-invariant liftings. To fix ideas as well as for a clearer exposition we shall, from now on, simply refer to $(\Omega, \mathcal{B}_\sigma(\Omega), P_\sigma)$ and H without specifying which of the models (1), (2) or (3) is meant.

A function $f : H \to \mathbf{R}$ is a Borel cylinder function if and only if it can be written as

$$f(h) = g((h, h_1, \cdots, (h, h_k)) \tag{2.3}$$

for some $k \geq 1$ and $h_1, \cdots, h_k$ in H where $g : \mathbf{R^k} \to \mathbf{R}$ is Borel measurable. We define Rf, the lifting of f, to be

$$R(f)(\cdot) := g(I_1(h_1)(\cdot), \cdots, I_1(h_k)(\cdot)). \tag{2.4}$$

Let $\sigma > 0$. $\mathcal{L}^0_\sigma(H, \mu)$ will denote the class of functions $f : H \to \mathbf{R}$ with the following properties: For all $\pi \in \mathcal{P}$, (the class of orthogonal projections on H with finite dimensional ranges), the function $(f \circ \pi)(h)$ is a Borel cylinder function and for all sequences $\{\pi_N\}$ from $\mathcal{P}$ converging strongly to the identity $(\pi_N \to I)$, the sequence $\{R(f \circ \pi_N)\}$ is Cauchy in P_σ-probability. Under these circumstances, one can show that all these sequences converge in P_σ-probability to the same limit $R_\sigma(f)$, called the σ–lifting of f. $R_\sigma(f)$ is defined P_σ-a.s. The lifting usually discussed is, in our present terminology, the 1-lifting.

If f has a σ-lifting for all $\sigma > 0$, we let $Rf = R_\sigma f$ on Ω_σ and we call Rf the scale-invariant lifting (or s-lifting) of f. In this case, for every $\sigma > 0, Rf$ is defined P_σ-a.s. Thus Rf is s-a.s. defined and scale-invariant measurable.

For any $\sigma > 0$, we let $\mathcal{L}^2_\sigma(H,\mu)$ denote the set of all $f \in \mathcal{L}^0_\sigma(H,\mu)$ such that for all sequences $\{\Pi_N\}$ from $\mathcal{P}$ with $\Pi_N \uparrow I$,

$$\int_{\mathcal{C}_0} |R(f \circ \Pi_N) - R(f \circ \Pi_{N'})|^2 dP_\sigma \to 0 \tag{2.5}$$

as $N, N' \to \infty$. Note that if $f \in \mathcal{L}^2_\sigma(H,\mu)$, then

$$\int_{\mathcal{L}_0} |R_\sigma(f)|^2 dP_\sigma < \infty. \tag{2.6}$$

When $f \in \mathcal{L}^2_\sigma(H,\mu)$, we call $R_\sigma(f)$ a $\sigma - \mathcal{L}^2$–lifting. If f belongs to $\mathcal{L}^2_\sigma(H,\mu)$ for all $\sigma > 0$, we call $Rf := R_\sigma f$ on Ω_σ, a scale-invariant$\mathcal{L}^2$ – lifting. If Rf is a scale-invariant $\mathcal{L}^2$-lifting then, for every $\sigma > 0, Rf$ is defined P_σ-a.s. and belongs to the space $L^2(\Omega, P_\sigma)$ which can be identified with $L^2(\Omega_\sigma, P_\sigma)$.

Let $\mathbf{L}^2 = L^2(\Omega, P)$ and $I_n(f_n)$, the nth order multiple Wiener integral with $f_n \in H^{\hat{\otimes} n}$($n^{th}$ symmetric tensor product). When $H = L^2(\mathbf{R}), f_n(t_1, \cdots, t_n)$ is symmetric and belongs to $L^2(\mathbf{R}^n)$.

Hilbert space–valued traces. Let $f_n \in H^{\hat{\otimes} n}$ and k an integer such that $0 \leq k \leq [\frac{n}{2}]$, ($[a]$ denotes, as usual, the integer part in a). The k-traces of f_n are defined as follows: For $k = 0$ define $Tr^\circ f_n = f_n$. If (e_j) is any CONS in $H^{\hat{\otimes} k} (k \geq 1)$, denote by $f_n[e_j \otimes e_j]$ the unique element in $H^{\hat{\otimes} n-2k}$ such that

$$(f_n[e_j \otimes e_j], g) = (f_n, e_j \otimes e_j \otimes g)$$

for all $g \in H^{\hat{\otimes} n-2k}$.

Definition. For $k \geq 1$ we say that $Tr^k f_n$ exists and equals $h \in H^{\hat{\otimes} n-2k}$ iff for every CONS (e_j) in $H^{\hat{\otimes} k}$

$$\sum_{j=1}^{\infty} f_n[e_j \otimes e_j]$$

converges in $H^{\hat{\otimes} n-2k}$ to h.

When $H = L^2(\mathbf{R}_+)$ the above definition may be equivalently stated in the following way. For $f \in \hat{L}^2(\mathbf{R}^n)$, there exists $h \in \hat{L}^2(\mathbf{R}_+^{n-2k})$ such that for every CONS (e_j) in $\hat{L}^2(\mathbf{R}^k)$ the series

$$\sum_{j=1}^{\infty} \int_{\mathbf{R}_+^k \times \mathbf{R}_+^k} f_n(s_1, \cdots, s_k, s_{k+1}, \cdots, s_{2k}, \cdots, s_n) e_j(s_1, \cdots, s_k) e_j(s_{k+1}, \cdots, s_{2k}) \cdot ds_1 \cdots ds_k ds_{k+1} \cdots ds_{2k}$$

converges to h in the norm of $L^2(\mathbf{R}_+^{n-2k})$. Then $Tr^k f_n \equiv h$. For $n = 2$ and $k = 1, Tr^1 f_2$ is the familiar numerical trace $\sum_{j=1}^{\infty} \int_{\mathbf{R}_+ \times \mathbf{R}_+} f_2(s_1, s_2) e_j(s_1) e_j(s_2) ds_1 ds_2$.

We are now in a position to state the results that lead to the definition of the Hu-Meyer natural extension. These are stated in terms of scaled liftings of n-linear forms. For $f_n \in H^{\hat{\otimes} n}, \psi_n(f_n)(h) = (f_n, h^{\otimes n})$ will be called the n-form associated with f_n. The proofs of the following results are given in detail in [5] for the case $H = L^2(\mathbf{R}_+)$. The extension of the notions of s-invariant measurability to abstract Wiener spaces and the Gaussian white noise space $(\mathcal{S}(\mathbf{R})', \mathbf{P})$ make the proofs go through for these probability spaces as well so that the theorems may be of independent interest for white noise analysis.

In the case $\Omega = \mathcal{S}(\mathbf{R})'$ and $H = L^2(\mathbf{R})$,

$$\psi_n(f_n)(h) = \int \cdots \int_{\mathbf{R}^n} f_n(t_1, \cdots, t_n) h(t_1) \cdots h(t_n) dt_1 \cdots dt_n, \ h \in H.$$

The most general form of Theorem 2.1 requires a slightly weaker definition of trace which is called a limiting k-trace and is denoted by $\vec{Tr}^k f_n$. The definition given above is modified by taking the CONS (e_j) to be of the form $\{\phi_{i_1} \otimes \cdots \otimes \phi_{i_n}\}$ where (ϕ_i) is an abitrary CONS in H and furthermore, the infinite series (2.6) is to be understood as the limit of partial sums $\sum_{i_1, \cdots i_n = 1}^{N} f_n[\phi_{i_1} \otimes \cdots \otimes \phi_{i_n}]$ as $N \to \infty$.

Theorem 2.1 $\psi_n(f_n)$ has a scaled $\mathcal{L}^2$-lifting $R[\psi_n]$ if and only if $\vec{Tr}^k f_n$ exists for $k = 0, 1, \cdots [\frac{n}{2}]$. In this case, for every $\sigma > 0, P_\sigma$-a.s. on Ω_σ (i.e., s-a.s.)

$$R[\psi_n] = \sum_{k=0}^{[\frac{n}{2}]} \sigma^{2k} C_{n,k} I_{n-2k}^{\sigma}(\vec{Tr} f_n). \tag{2.7}$$

Here and in what follows $C_{n,k}$ is the constant $\frac{n!}{(n-2k)! 2^k k!}$

Theorem 2.2 Assume that $Tr^k f_n$ exists for $k = 0, \cdots, [\frac{n}{2}]$. Then for every $\sigma > 0, P_\sigma$-a.s. on Ω_σ,

$$I_n^{\sigma}(f_n) = \sum_{k=0}^{[\frac{n}{2}]} (-1)^k \sigma^{2k} C_{n,k} R\left[\psi_{n-2k}(Tr_k f_n)\right]. \tag{2.8}$$

Definition. Under the assumption of Theorem 2.2 the natural extension of $I_n(f_n)$ is defined by

$$N\left[I_n(f_n)\right] = \sum_{k=0}^{[\frac{n}{2}]} (-1)^k C_{n,k} R\left[\psi_{n-2k}(Tr^k f_n)\right]. \tag{2.9}$$

An application of Theorem 2.2 yields the following useful expression for $N[I_n(f_n)]$.

Theorem 2.3 Let $Tr^k f_n$ exist for $0 \le k \le [\frac{n}{2}]$. Then for any $\sigma > 0$ we have P_σ-a.s. on Ω_σ,

$$N[I_n(f_n)] = \sum_{k=0}^{[n2]} (\sigma^2 - 1)^k C_{n,k} I^\sigma_{n-2k}(Tr^k f_n). \tag{2.10}$$

The following subclass of $\mathbf{L}^2$ is of particular interest from the point of view of this paper.

$$\begin{aligned}(\mathbf{L}^2)_{Tr} &:= \\ \{\phi \in \mathbf{L}^2 : \phi &= E(\phi) + \sum_{n\geq 1} I_n(f_n), \text{ where for each } n, Tr^k f_n \text{ exists for } 0 \leq k \leq \frac{n}{2}\}.\end{aligned}$$

The following proposition gives a sufficient condition for a functional $\phi \in \mathbf{L}^2$ to belong to $(\mathbf{L}^2)_{Tr}$. The proof given here is based on the proof of Proposition 9.5 of [5] where H is taken to be $L^2(\mathbf{R}_+)$ and is modified to apply to kernels of multiple Wiener-Itô integrals in abstract Wiener spaces.

Proposition 2.1. Let $f_n \in H^{\hat{\otimes} n}$ and suppose that there exists a CONS (ϕ_i) in H such that the coefficients $(a_{i_1 \cdots i_n})$ in the expansion for f_n,

$$f_n = \sum_{i_1,\cdots,i_n=1}^{\infty} a_{i_1,\cdots,i_n} \phi_{i_1} \otimes \cdots \otimes \phi_{i_n} \tag{2.11}$$

belong to ℓ_1.
Then for $0 \leq k \leq [\frac{n}{2}], Tr^k f_n$ exists and is given by

$$Tr^k f_n = \sum_{i_{2k+1},\cdots,i_n=1}^{\infty} \left(\sum_{j_1,\cdots,j_k=1}^{\infty} a_{j_1,j_1,j_2,j_2,\cdots,j_k,j_k,i_{2k+1},\cdots,i_n} \right) \phi_{i_{2k+1}} \otimes \cdots \otimes \phi_{i_n}. \tag{2.12}$$

Proof. It suffices to show that, for any CONS (e_j) in $H^{\otimes k}$, the series $\sum_{j=1}^{\infty} f_n[e_j \otimes e_j]$ converges in $H^{\otimes n-2k}$ and

$$\sum_{j=1}^{\infty} f_n[e_j \otimes e_j] = \sum_{i_{2k+1},\cdots,i_n=1}^{\infty} \left(\sum_{j_1,\cdots,j_k=1}^{\infty} a_{j_1,j_1,\cdots,j_k,j_k,i_{2k+1},\cdots,i_n} \right) \phi_{i_{2k+1}} \otimes \cdots \otimes \phi_{i_n}. \tag{2.13}$$

For $h \in H^{\otimes 2k}$, the element $f_n[h]$ is defined by the relation

$$(f_n, h \otimes g)_{H^{\otimes n}} = (f_n[h], g)_{H^{\otimes n-2k}}$$

for each $g \in H^{\otimes n-2k}$. Let

$$f_n^N = \sum_{i_1,\cdots,i_n=1}^{N_1,\cdots,N_n} a_{i_1,\cdots,i_n} \phi_{i_1} \otimes \cdots \otimes \phi_{i_n} \tag{2.14}$$

where $N_1, \cdots N_n$ are integers. It is easy to see that $\|f_n - f_n^N\|_{H^{\otimes n}} \to 0$ as $(N) \to \infty$, i.e. $N_1, \cdots, N_n \to \infty$ and therefore

$$(f_n[h], g) = lim_{(N)\to\infty}(f_n^N[h], g).$$

Letting $h = e_j \otimes e_j$, substituting from (2.11) in the above relation, we have

$$\begin{aligned}&(f_n[e_j \otimes e_j], g)\\ =& \sum_{i_1,\cdots i_n=1}^{\infty} a_{i_1,\cdots,i_n}(\phi_{i_1} \otimes \cdots \otimes \phi_{i_k}, e_j)(\phi_{i_{k+1}} \otimes \cdots \otimes \phi_{i_{2k}}, e_j)(\phi_{i_{2k+1}} \otimes \cdots \otimes \phi_{i_p}, g)\end{aligned}$$

for all $g \in H^{\otimes n-2k}$ from which it follows that

$$\begin{aligned}&f_n[e_j \otimes e_j] =\\ (2.15)\ \sum_{i_1,\cdots,i_n=1}^{\infty} \quad & a_{i_1,\cdots,i_n}(\phi_{i_1} \otimes \cdots \otimes \phi_{i_k}, e_j)(\phi_{i_{k+1}} \otimes \cdots \otimes \phi_{i_{2k}}, e_j)\phi_{i_{2k+1}} \otimes \cdots \otimes \phi_{i_n}.\end{aligned}$$

Hence

$$\begin{aligned}(2.16)\qquad & \sum_{j=1}^{\infty} f_n[e_j \otimes e_j] = \sum_{j=1}^{\infty} [\text{the series in } (2.15)]\\ =& \sum_{i_1,\cdots,i_n=1}^{\infty} a_{i_1,\cdots,i_n}\left\{\sum_{j=1}^{\infty}(\phi_{i_1} \otimes \cdots \otimes \phi_{i_k}, e_j)(\phi_{i_{k+1}} \otimes \cdots \otimes \phi_{i_{2k}}, e_j)\right\}\cdot\\ & \phi_{i_{2k+1}} \otimes \cdots \otimes \phi_{i_n}.\end{aligned}$$

We justify the interchange in summation as follows. The sequences

$$\{(\phi_{i_1} \otimes \cdots \otimes \phi_{i_k}, e_j)\}$$

and

$$\{(\phi_{i_k+1} \otimes \cdots \otimes \phi_{i_{2k}}, e_j)\}$$

are in ℓ_2 as functions of j. Hence by the Schwarz inequality

$$\begin{aligned}(2.17)\qquad & \sum_{j=1}^{\infty} |(\phi_{i_1} \otimes \cdots \otimes \phi_{i_k}, e_j)| \left|(\phi_{i_{k+1}} \otimes \cdots \otimes \phi_{i_{2k}}, e_j)\right|\\ \leq & \left[\sum_j (\phi_{i_1} \otimes \cdots \otimes \phi_{i_k}, e_j)^2\right]^{\frac{1}{2}} \left[\sum_j (\phi_{i_{k+1}} \otimes \cdots \otimes \phi_{i_{2k}}, e_j)^2\right]^{\frac{1}{2}} = 1.\end{aligned}$$

Using (2.17) and the ℓ_1-assumption on the coefficients $a_{i_1,\cdots,i_n}$ and regarding the sums in the two series in (2.16) as Bochner integrals we can now apply the Fubini theorem for Bochner integrals to obtain the desired interchange of summation. Now using the second equality in (2.16) we obtain

$$\begin{aligned}\sum_{j=1}^{\infty} f_n[e_j \otimes e_j] &= \sum_{i_1,\cdots,i_n=1}^{\infty} a_{i_1,\cdots,i_n} \left(\phi_{i_1} \otimes \cdots \otimes \phi_{i_k}, \phi_{i_{k+1}} \otimes \cdots \otimes \phi_{i_{2k}}\right) \phi_{i_{2k+1}} \otimes \cdots \otimes \phi_{i_n} \\ &= \sum_{i_{2k+1},\cdots,i_n=1}^{\infty} \left(\sum_{j_1,\cdots,j_k=1}^{\infty} a_{j_1,j_1,\cdots,j_k,j_k,i_{2k+1},\cdots,i_n} \right) \phi_{i_{2k+1}} \otimes \cdots \otimes \phi_{i_n},\end{aligned}$$

the last equality following from the fact that coefficients $a_{i_1,\cdots,i_n}$ are symmetric in the indices and in ℓ_1. □

3 The space Φ^c and Feynman distributions.

In this section we shall introduce one example of a space of test functionals which is particularly suited for the definition of Feynman distributions via the concept of the natural extension of a multiple Wiener integral.

Let A be a positive self-adjoint operator on H such that $H_\infty = \bigcap_{p=1}^{\infty} Dom A^p$ is dense in H. We assume that A^{-p_0} is Hilbert-Schmidt for some $p_0 \geq 1$. It is convenient to assume that $\|A^{-p_0}\|_2 < 1$ where $\|\cdot\|_2$ denotes Hilbert-Schmidt norm. Let $(e_j) \subset H_\infty$ be the CONS in H with $Ae_j = \lambda_j e_j$. Fixing a constant $c > 1$ we define

$$\begin{aligned}\Phi^c := \{\phi \in \mathbf{L}^2 : \phi = E(\phi) + \sum_{n\geq 1} I_n(f_n), \\ \sum_{n\geq 1}^{\infty} n! c^{2n} \|(A^p)^{\otimes n} f_n\|_{H^{\otimes n}}^2 < \infty \text{ for all } p \in \mathbf{N}\}\end{aligned}$$

For each positive integer p, define

$$\begin{aligned}\|\phi\|_p^2 &= E(\phi)^2 + \sum_{n=1}^{\infty} n! c^{2n} \|(A^p)^{\otimes n} f_n\|^2, \quad (\phi \in \Phi^c) \\ &= E(\phi)^2 + \sum_{n\geq 1} n! c^{2n} \sum_{j_1,\cdots,j_n} (\lambda_{j_1} \cdots \lambda_{j_n})^{2p} (f_n, e_{j_1} \otimes \cdots \otimes e_{j_n})^2.\end{aligned}$$

Then $\Phi^c = \bigcap_p \Phi_p^c$ where Φ_p^c is the $\|\cdot\|_p$ - completion of Φ^c in $\mathbf{L}^2$. The Hilbert-Schmidt condition on A^{-p_0} implies that Φ^c is a countable Hilbertian nuclear space. For $\phi \in \Phi^c$ as above, we have for each n,

$$f_n = \sum_{j_1,\cdots,j_n} a_{j_1 \cdots j_n} e_{j_1} \otimes \cdots \otimes e_{j_n}$$

and

$$(A^p)^{\otimes n} f_n = \sum_{j_1,\cdots,j_n} (\lambda_{j_1} \cdots \lambda_{j_n})^p a_{j_1,\ldots,j_n} e_{j_1} \otimes \cdots \otimes e_{j_n}$$

where

$$\sum_n n! c^{2n} \sum_{j_1,\ldots,j_n} (\lambda_{j_1} \cdots \lambda_{j_n})^{2p} a^2_{j_1\cdots j_n} < \infty \text{ for every p .}$$

Hence for every n and $j_1, \cdots, j_n$, for some positive constant α_p,

$$|a_{j_1,\cdots,j_n}| \leq \frac{\alpha_p}{|\lambda_{j_1}, \cdots, \lambda_{j_n}|^p}$$

so that

$$\sum_{j_1,\cdots,j_n} \left| a_{j_1,\cdots,j_n} \right| \leq \alpha_p \left(\sum_{j=1}^{\infty} \frac{1}{\lambda_j^p} \right)^n < \infty$$

if $p \geq 2p_0$. It follows from Proposition 2.1 that $Tr^k f_n$ exists for $0 \leq k \leq \left[\frac{n}{2}\right]$.

We have thus proved the following result.

Proposition 3.1
(a) Φ^c is a countably Hilbertian nuclear space
and
(b) $\Phi^c \subset (\mathbf{L}^2)_{Tr}$.

The space of distributions on Wiener space is given by $(\Phi^c)'$ the strong dual of Φ^c. The following result shows that the test function space Φ^c is sufficiently rich. For $h \in H$, let

$$\rho_h = \exp\{I_1(h) - \frac{1}{2}\|h\|^2\}$$

Proposition 3.2

$$\{\rho_h, h \in H_\infty\} \subset \Phi^c.$$

Proof. For any h in H it is easy to see that $Tr^k h^{\otimes n}$ exists and equals $\|h\|^{2k} h^{\otimes n-2k}$ for $0 \leq k \leq \left[\frac{n}{2}\right]$. Indeed, if (e_j) is any CONS in $H^{\otimes k}$ and $g \in H^{\otimes n-2k}$,

$$(h^{\otimes n}, e_j \otimes e_j \otimes g) = (h^{\otimes 2k}, e_j \otimes e_j)(h^{\otimes n-2k}, g).$$

Hence, by definition of k-trace (see [5])

$$\sum_{i=1}^{\infty} h^{\otimes n} \left[e_j \otimes e_j \right] = \sum_{i=1}^{\infty} (h^{\otimes k}, e_j)^2 h^{\otimes n-2k}$$

converges in $H^{\otimes n-2k}$ to $\|h\|^{2k} h^{\otimes n-2k}$. Next,

$$\rho_h = \sum_{n \geq 0} \frac{1}{n!} I_n(h^{\otimes n})$$

and if $h \in H_\infty$, then for p $\in$ N,

$$\|\rho_h\|_p^2 = \sum_{n\geq 0} \frac{c^{2n}}{n!} \|(A^p h)^{\otimes n}\|_{H^{\otimes n}}^2 \tag{3.1}$$

$$= \sum_{n\geq 0} \frac{c^{2n}}{n!} \|A^p h\|^2 = \exp(c^2 \|A^p h\|^2) < \infty. \tag{3.2}$$

We shall now introduce the Feynman distribution as an element of $(\Phi^c)'$ and point out its link with the natural extension. The following lemma is needed in our calculations.

Lemma 3.1. For $0 \leq k \leq \left[\frac{n}{2}\right]$, and A^{-p} Hilbert-Schmidt,

$$\|(A^p)^{\otimes n-2k} Tr^k f_n\| \leq \|(A^p)^{\otimes n} f_n\| \|A^{-p}\|_2^{2k} \tag{3.3}$$

where f_n's are the kernels in the chaos expansion of $\phi \in \Phi^c$.

Proof. From the definition of $Tr^k f_n$ and our assumption on A,

$$Tr^k f_n = \sum_{j_1,\cdots,j_k=1}^{\infty} \sum_{i_{2k+1},\cdots,i_n=1}^{\infty} \left(f_n, e_{j_1}^{\otimes 2} \otimes \cdots \otimes e_{j_k}^{\otimes 2} \otimes e_{i_{2k+1}} \otimes \cdots \otimes e_{i_n}\right) e_{i_{2k+1}} \otimes \cdots \otimes e_{i_n}$$

and

$$(A^p)^{\otimes n-2k} Tr^k f_n = \sum_{j_1,\cdots,j_k=1}^{\infty} \sum_{i_{2k+1}\cdots i_n=1}^{\infty} \left(f_n, e_{j_1}^{\otimes 2} \otimes \cdots \otimes e_{j_k}^{\otimes 2} \otimes e_{i_{2k+1}} \otimes \cdots \otimes e_{i_n}\right) \left(A^p e_{i_{2k+1}}\right) \otimes \cdots \otimes (A^p e_{i_n}).$$

Recalling that $A^p e_j = \lambda_j^p e_j$ we have

$$\begin{aligned}(A^p)^{\otimes n-2k} Tr^k f_n &= \sum_{j_1,\cdots,j_k} \sum_{i_{2k+1},\cdots,i_n} \left(f_n, e_{j_1}^{\otimes 2} \otimes \cdots \otimes e_{j_k}^{\otimes 2} \otimes A^p e_{i_{2k+1}} \otimes \cdots \otimes A^p e_{i_n}\right) e_{i_{2k+1}} \otimes \cdots \otimes e_{i_n} \\ &= \sum_{j_1,\cdots,j_k} \sum_{i_{2k+1},\cdots,i_n} ((A^p)^{\otimes n} f_n, (A^{-p} e_{j_1})^{\otimes 2} \otimes \cdots \otimes (A^{-p} e_{j_k})^{\otimes 2} \\ &\quad \otimes e_{i_{2k+1}} \otimes \cdots \otimes e_{i_n}) e_{i_{2k+1}} \otimes \cdots \otimes e_{i_n}.\end{aligned}$$

Hence

$$\left\|(A^p)^{\otimes n-2k}Tr^k f_n\right\|^2 = \sum_{j_1,\cdots,j_k}\left((A^p)^{\otimes n} f_n, (A^{-p}e_{j_1})^{\otimes 2}\otimes\cdots\otimes(A^{-p}e_{j_k})^{\otimes 2}\right)^2$$
$$\leq \|(A^p)^{\otimes n} f_n\|^2 \|A^{-p}\|_2^{4k} \qquad \square$$

For $z \in \mathbf{C}$ define

$$K_z\phi = \sum_{n\geq 0}\sum_{2k\leq n} C_{n,k}(z-1)^k z^{\frac{n}{2}-k} I_{n-2k}(Tr^k f_n) \tag{3.4}$$

where $\phi = E\phi + \sum_{n\geq 1} I_n(f_n)$ and, as before $C_{n,k} = \frac{n!}{(n-2k)!2^k k!}$.
Theorem 3.1. Let $a > 0$ and the integer p chosen such that $\|A^{-p}\|_2^2 < c^2 a^{-1}$. Then for values of z satisfying

$$|z-1| < a \text{ and } |z| + a\|A^{-p}\|_2^2 < c^2 \tag{3.5}$$

we have

$$E_{P_1}|K_z\phi| \leq B\|\phi\|_p \text{ for all } \phi\in\Phi^c, \tag{3.6}$$

where B is a positive constant involving a, c and p.
Proof.

$$E_{P_1}|K_z\phi| \leq \sum_{n\geq 0}\sum_{2k\leq n} C_{n,k}|z-1|^k |z|^{\frac{n}{2}-k}\left((n-2k)!\right)^{\frac{1}{2}} \|Tr^k f_n\|$$
$$\leq \sum_{n\geq 0}\sum_{2k\leq n} C_{n,k}|z-1|^k |z|^{\frac{n}{2}-k}\left((n-2k)!\right)^{\frac{1}{2}} \left\|(A^p)^{\otimes n} f_n\right\| \left\|A^{-p}\right\|_2^{2k}$$
$$\leq \sum_{n\geq 0}(n!)^{\frac{1}{2}} c^n \left\|(A^p)^{\otimes n} f_n\right\| \theta_n(z), \text{ say}$$

where we have used Lemma 3.1 and

$$\theta_n(z) = \frac{1}{c^n}\sum_{2k\leq n}\left\{\frac{n!}{(n-2k)!(2k)!}\right\}^{\frac{1}{2}}\left(\frac{|z-1|}{a}\right)^k |z|^{\frac{n}{2}-k} a^k \left\|A^{-p}\right\|_2^{2k}\frac{(2k!)^{\frac{1}{2}}}{2^k k!}$$
$$\leq \frac{1}{c^n}\left[\sum_{2k\leq n}\frac{n!}{(n-2k)!(2k)!}|z|^{n-2k}\left(a\|A^{-p}\|_2^2\right)^{2k}\right]^{\frac{1}{2}}$$
$$\left[\sum_{2k\leq n}\frac{(2k)!}{2^{2k}(k!)^2}\left(\frac{|z-1|}{a}\right)^{2k}\right]^{\frac{1}{2}}$$
$$\leq \frac{1}{c^n}\left(|z| + a\|A^{-p}\|_2^2\right)^{\frac{n}{2}}\left[\sum_{k=0}^{\infty}\frac{(2k)!}{2^{2k}(k!)^2}\left(\frac{|z-1|}{a}\right)^{2k}\right]^{\frac{1}{2}}.$$

The infinite series converges for $|z-1|<a$. Denoting the sum by $B_1(z,a)^2$ we obtain

$$\theta_n(z) \leq \frac{1}{c^n}\left(|z|+a\|A^{-p}\|_2^2\right)^{\frac{n}{2}} B_1(z,a). \tag{3.7}$$

Hence from the Schwarz inequality we have

$$\begin{aligned} E_{P_1}|K_z\phi| &\leq \left\{\sum_{n\geq 0} n!c^{2n}\|(A^p)^{\otimes n}f_n\|^2\right\}^{\frac{1}{2}}\left\{\sum_{n\geq 0}\theta_n(z)^2\right\}^{\frac{1}{2}} \\ &\leq B_1(z,a)\|\phi\|_p\left\{\sum_{n\geq 0} c^{-2n}\left[|z|+a\|A^{-p}\|_2^2\right]^n\right\}^{\frac{1}{2}}. \end{aligned}$$

The quantity in the curly parenthesis is finite for $|z|+a\|A^{-p}\|_2^2<c^2$ and will be denoted by $B_2(z,a,c,p)$. Writing $B=B_1B_2$ we have

$$E_{P_1}|K_z\phi| \leq B(z,a,c,p)\|\phi\|_p \tag{3.8}$$

for all $\phi\in\Phi^c$ for values of z satisfying (3.5).

Definition. For z as in Theorem 3.1, define the distribution $F_z\in(\Phi^c)'$ by

$$< F_z,\phi >= E_{P_1}(K_z\phi),\ \phi\in\Phi.$$

Clearly F_z is a linear functional on Φ^c and continuous in view of (3.6).

F_z is called the Feynman distribution with complex parameter z.

From (3.4) it is easy to calculate $< F_z,\phi >$. Noting that the terms corresponding to odd values of n vanish in the expression for $E_{P_1}(K_z\phi)$ we get after some simplification

$$< F_z,\phi >= E(\phi)+\sum_{m=1}^{\infty}\frac{(2m)!}{m!}\left(\frac{z-1}{2}\right)^m Tr^m f_{2m}. \tag{3.9}$$

The "true" Feynman distribution is given by $F_i,(i=\sqrt{-1})$ provided we can show that i is in the domain of convergence given by (3.5). We can achieve this by choosing $a=2$ and $\|A^{-p}\|_2$ to satisfy the inequality $\|A^{-p}\|_2^2<\frac{1}{2}(c^2-1)$. We then have

$$< F_i,\phi >= E(\phi)+\sum_{m=1}^{\infty}\frac{(2m)!}{m!}\left(\frac{i-1}{2}\right)^m Tr^m f_{2m}. \tag{3.10}$$

Feynman distribution and the natural extension. Our choice of $K_z\phi$ is based on the natural extension of the multiple Wiener integral. When $z=\sigma^2$ (real and positive) $K_z\phi$ becomes

$$\begin{aligned}(K_{\sigma^2}\phi)(x) &= \sum_{n\geq 0}\sum_{2k\leq n} C_{n,k}\left(\sigma^2-1\right)\sigma^{n-2k}I_{n-2k}\left(Tr^k f_n\right)(x),\\ &= \sum_{n\geq 0}\sum_{2k\leq n} C_{n,k}\left(\sigma^2-1\right)^k I^{\sigma}_{n-2k}\left(Tr^k f_n\right)(\sigma x)\\ &= \sum_{n\geq 0} N\left[I_n(f_n)\right](\sigma x),\ P_1 - \text{ a.s. for } x\in\Omega_1,\end{aligned}$$

where $N[I_n(f_n)]$ is the natural extension of $I_n(f_n)$.

A somewhat different definition of F_z and F_i leading, however, to the 'same' distribution has been given in [7]. The natural extension is not involved directly in that definition. Moreover, the test function spaces are not the same. In the cited paper the Hilbertian norm $\|\|_p$ is defined directly in terms of the second quantization operator $\Gamma(A^p)$ whereas here it is necessary to use a weighting. [7] also uses some results on the Malliavin-Watanabe calculus.

Remark. Suppose $H = L^2(\mathrm{R})$ and let A be the operator $A = -\frac{d^2}{du^2}+1+u^2$ with domain $\mathcal{D}A \subset L^2(\mathrm{R})$ and $\{e_j, j\in \mathrm{N}_0\}$, the CONS in $\mathrm{L}^2(\mathrm{R})$ consisting of the eigenfunctions of A (the e_j's are Hermite functions). The corresponding eignevalues are $\{2j+2, j\in\mathrm{N}_0\}$. Let us take $c=1$ in the definition of Φ^c and write Φ for Φ^1. Then it has been shown in [9] that, using the notation of [9], $\Phi = (\mathcal{S})$, i.e.

$$\Phi = \left\{\phi\in \mathrm{L}^2 : \phi = \sum_n I_n(f_n) \text{ with } f_n\in\mathcal{S}(\mathrm{R}^n)\right\}.$$

We also have the following corollary which may be a known result though we have not been able to find a reference to it. It shows a link between traces and smoothness properties of kernels.

Corollary. Let $f_n \in \mathcal{S}(\mathrm{R}^n)$ be a symmetric function. Then $Tr^k f_n$ exists for all $k, 0\leq k\leq [\frac{n}{2}]$.

4 Acknowledgement.

Research funded by the Air Force Office of Scientific Research Contract No. F49620 85C 0144.

References

[1] D.M. Chung, Scale invariant measurability in abstract Wiener space, *Pacific J. of Mathematics*, **130**, (1987), 27-40.

[2] Y.Z. Hu and P.A. Meyer, Chaos de Wiener et integrale de Feynman, Séminaire de Probabilités XXII, Université de Strasbourg, (1987) 51-71, Lecture Notes in Math, **1321**, Springer-Verlag, Berlin 1988.

[3] G.W. Johnson and D.L. Skoug, Scale invariant measurability in Wiener space, *Pacific J. of Math*, 83, (1979), 157-176.

[4] G.W. Johnson and G. Kallianpur, Some remarks on Hu and Meyer's paper and infinite dimensional calculus on finitely additive canonical Hilbert space, *Theory of Probab. and Its Applications*, vol. XXXIV, **4** (1989), 742-752.

[5] G.W. Johnson and G. Kallianpur, Homogeneous chaos, p-Forms, Scaling and the Feynman Integral, Tech Report 274, Univ. of North Carolina, Center for Stochastic Processes (1989), To appear.

[6] G. Kallianpur and R.L. Karandikar, White Noise Theory of Prediction, Filtering and Smoothing, Stochastic Monographs 3, Gordon and Breach, New York, (1988).

[7] G. Kallianpur and A.S. Ustunel, Distributions on abstract Wiener space and applications to the Feynman integral (in preparation).

[8] H.H. Kuo, Lectures on white noise calculus, (1991), (Preprint)

[9] Yuh-Jia Lee, Analytic version of test functionals, Fourier transform and a characterization of measures in white noise calculus, (1989), To appear in *J. of Functional Analysis*.

[10] J. Potthoff and L. Streit, A characterization of Hida distributions (1990) (Preprint)

CENTRAL LIMIT THEOREMS ON RANDOM MEASURES AND STOCHASTIC DIFFERENCE EQUATIONS

HIROSHI KUNITA

Department of applied science, Kyushu University 36, Fukuoka 812, Japan

0. Introduction

Let $\{\xi_j^n, j \in \mathbf{N}\}, n = 1, 2, \ldots$ be a sequence of stationary stochastic processes with values in $\mathbf{R}^m$. The sequence is assumed to converge in law to a stationary process $\{\xi_j, j \in \mathbf{N}\}$. Define a family of random measures by

$$B^n(t, E) = \frac{1}{\sqrt{n}} \sum_{j=1}^{[nt]} (\chi_E(\xi_j^n) - \pi^n(E)), \tag{1}$$

where E ia a Borel set in $\mathbf{R}^m$, χ_E is the indicator function of the set E and π^n is the common distribution of the random variables $\xi_j^n, j \in \mathbf{N}$. Then for each n, $(B^n(t, .)$, $t \in [0, \infty))$ may be regarded as a stochastic process with values in the space of signed measures on $\mathbf{R}^m$, cadlag (right continuous with left hand limits) with respect to time t. In the previous paper[6], the author has shown that the sequence $\{B^n(t, .)\}_n$ converges in law in the space $\mathbf{D}$ with respect to Skorohod's J_1-topology, provided that $\xi_j^n, j \in \mathbf{N}$ are independent random variables for any n. Furthermore, the limit $B(t, E)$ is characterized as a Brownian random measure, i.e., it is a Gaussian random measure for each fixed t and is a Brownian motion for each fixed E.

In this paper we will extend the above limit theorem to the case where $\xi_j^n, j \in \mathbf{N}$ are not necessarily independsent but have some mixing property. Another object will be to discuss the weak convergence of the solutions of stochastic difference equations represented by

$$\psi_j = \psi_{j-1} + \frac{1}{\sqrt{n}} f^n(\psi_{j-1}, \xi_j^n) + \frac{1}{n} g^n(\psi_{j-1}), \qquad j = 1, 2, \ldots \tag{2}$$

where $f^n(x, \lambda)$ and $g^n(x)$; $x \in \mathbf{R}^d$, $\lambda \in \mathbf{R}^m$ are continuous functions with values in $\mathbf{R}^d$ converging to $f(x, \lambda)$ and $g(x)$ respectively as $n \to \infty$. Let ψ_j^n be the solution of Eq.2 with the initial condition $\psi_0 = x_0$ and set $\varphi_t^n = \psi_{[nt]}^n$. We will show that the sequence $\{\varphi_t^n\}_n$ converges in law to a diffusion process. Our goal is to prove that the limit φ_t satisfies a stochastic differential equation driven by the Brownian random measure $B(t, E)$. Indeed for any n the process φ_t^n satisfies

$$\begin{aligned} \varphi_t^n &= x_0 + \int_0^t \int_{\mathbf{R}^m} f^n(\varphi_{s-}^n, \lambda) B^n(d\lambda, ds) \\ &\quad + \int_0^{[nt]/n} \{g^n(\varphi_{s-}^n) + b^n(\varphi_{s-}^n)\} ds, \end{aligned} \tag{3}$$

where

$$b^n(x) = \sqrt{n} \int_0^{[nt]/n} \int_{\mathbf{R}^m} f^n(\varphi_{s-}^n, \lambda) \pi^n(d\lambda) ds. \tag{4}$$

A question is that if the limit φ_t exists, it satisfies

$$\varphi_t(x) = x_0 + \int_0^t \int_{\mathbf{R}^m} f(\varphi_{s-}, \lambda) B(ds, d\lambda) + \int_0^t (g + b)(\varphi_{s-}) ds, \tag{5}$$

where $b(x)$ is the limit of the sequence $\{b^n(x)\}_n$. We show that the answer is affirmative if $\xi_j^n, j \in \mathrm{N}$ are independent random variables for any n. However if $\xi_j^n, j \in \mathrm{N}$ are not independent, a correction term $\int_0^t c(\varphi_{s-}(x))ds$ has to be added to the right hand side of Eq.5.

Our discussion is based on the author's previous work Fujiwara-Kunita[2], where we established limit theorems for stochastic difference-differential equations. In the same paper, we obtained several limit theorems for stochastic ordinary differential equations. It might be interesting to compare our results with them, since these two are parallel in many aspects.

Limit theorems for stochastic difference equations driven by a mixing array of random variables has been studied by several authors under various assumptions. See Kushner-Hai Huang[8], Iizuka-Matsuda[3] and Watanabe[10]. In these works , the limit process is characterized as a diffusion process associated with a certain second order differential operator. A feature of our result is that the limit process is charactrerized directly as a solution of a stochastic differential equation driven by the Brownian random measre. Another feature is that conditions required for the limit theorems are relaxed considerably.

The case where the limit process may have jumps is also interesting. Recently, Fujiwara[1] obtained such limit theorems for stochastic difference equations. We will discuss the similar problems for Eq.2 separately[7].

1. Uniform mixing case.

Let N be the set of all positive integers. Let $\{\xi_j^n; j \in \mathrm{N}\}, n = 1, 2, \ldots$ and $\{\xi_j; j \in \mathrm{N}\}$ be stationary stochastic processes with values in the common space $\mathbf{R}^m$. We set

$$\pi^n(E) = P(\xi_j^n \in E), \qquad \pi(E) = P(\xi_j \in E). \tag{6}$$

These measures do not depend on j since they are stationary. We will introduce two conditions to $\{\xi_j^n\}_n$ and $\{\xi_j\}$.

Condition (ξ.1) *Any finite dimensional distributions of* $\{\xi_j^n\}$, $n = 1, 2, \ldots$ *converge weakly to the corresponding finite dimensional distribution of* $\{\xi_j\}$.

Then the sequence $\{\pi^n\}$ converges weakly to π.

The second condition is concerned with the mixing property. Let h, k $(h < k)$ be elements of N. We denote by $\mathcal{F}_{h,k}^n$ the least σ-field for which ξ_j^n, $h \leq j \leq k$ are measureble. We often write $\mathcal{F}_{1,k}^n$ by $\mathcal{F}_k^n$ and $\mathcal{F}_{h,\infty}^n = \sigma(\cup_{k=h+1}^{\infty} \mathcal{F}_{h,k}^n)$. The *uniform mixing rate* (ϕ *mixing rate*) of the process $\{\xi_j^n; j \in \mathrm{N}\}$ is a sequence $\{\phi_k^n, k \in \mathrm{N}\}$ of nonnegative numbers defined by

$$\phi_k^n = \sup_{h \in \mathrm{N}} \sup\{|P(F|E) - P(F)|; E \in \mathcal{F}_h^n, P(E) > 0, F \in \mathcal{F}_{h+k,\infty}^n\}. \tag{7}$$

Condition (ξ.2) *The sequence* $\{\phi_k^n\}$ *satisfies* $\sum_{k=1}^{\infty}(\sup_n \phi_k^n)^{1/2} < \infty$.

Let $\{\phi_k\}$ be the uniform mixing rate of the process $\{\xi_j\}$. Then the inequality $\phi_k \le \sup_n \phi_k^n$ is easily verified for any $k \in \mathbf{N}$. Therefore $\sum_{k=1}^{\infty}(\phi_k)^{1/2} < \infty$ is satisfied.

Now let E_1 and E_2 be Borel sets in $\mathbf{R}^m$. By a well known mixing inequality (Jacod-Shiryaev[4], Chapter VIII), we have

$$|E[\chi_{E_1}(\xi_1)\chi_{E_2}(\xi_j)] - \pi(E_1)\pi(E_2)| \le 2(\phi_{j-1})^{\frac{1}{2}}\pi(E_1)\pi(E_2)^{\frac{1}{2}}. \tag{8}$$

Therefore the infinite sum

$$V(E_1, E_2) = \sum_{j=2}^{\infty}\{E[\chi_{E_1}(\xi_1)\chi_{E_2}(\xi_j)] - \pi(E_1)\pi(E_2)\} \tag{9}$$

converges and satisfies

$$|V(E_1, E_2)| \le 2\left(\sum_{j=1}^{\infty}(\phi_j)^{\frac{1}{2}}\right)\pi(E_1)\pi(E_2)^{\frac{1}{2}}. \tag{10}$$

Similarly the bilinear form V

$$V(u, v) = \int\int_{\mathbf{R}^m \times \mathbf{R}^m} u(\lambda)v(\lambda')V(d\lambda, d\lambda') \tag{11}$$

is well defined for $u, v \in L^2(\pi)$ and satisfies

$$|V(u, v)| \le 2\left(\sum_{j=1}^{\infty}(\phi_j)^{\frac{1}{2}}\right)\pi(|u|)\pi(v^2)^{\frac{1}{2}}, \tag{12}$$

where $\pi(u) = \int u(\lambda)\pi(d\lambda)$. The bilinear form V is identically 0 if and only if $\xi_j, j \in \mathbf{N}$ are independent. A stationary process $\{\xi_j\}$ is called *reversible* (or *symmetric*) if $P(\xi_1 \in E_1, \xi_j \in E_2) = P(\xi_1 \in E_2, \xi_j \in E_1)$ holds for any j and E_1, E_2. If the process $\{\xi_j\}$ is reversible, then the bilinear form V of Eq.11 is symmetric, i.e. $V(u, v) = V(v, u)$ holds for any bounded continuous functions u and v.

Now let $B^n(t, E)$ be random measures defined by Eq.1. We shall define the weak convergence of the sequence $\{(B^n(t)\}_n$ rigorously. Let $\mathcal{G}(\mathbf{R}^m)$ be a ring of subsets of $\mathbf{R}^m$ consisting of countable sets such that the σ-field generated by the ring $\mathcal{G}(\mathbf{R}^m)$ coincides with the topological Borel field $\mathcal{B}(\mathbf{R}^m)$ of $\mathbf{R}^m$. We assume that any element of $\mathcal{G}(\mathbf{R}^m)$ satisfy $\lim_{n\to\infty} \pi^n(E) = \pi(E)$. The set of all finitely additive set functions on $\mathcal{G}(\mathbf{R}^m)$ is denoted by $\mathcal{M}(\mathbf{R}^m)$. It is a complete separable metric space by the metric

$$d(\mu, \nu) = \sum_{k=1}^{\infty}\frac{1}{2^k}\frac{|\mu(E_k) - \nu(E_k)|}{1 + |\mu(E_k) - \nu(E_k)|} \quad \text{where} \quad \{E_k\} = \mathcal{G}(\mathbf{R}^m). \tag{13}$$

Let $\mathbf{D} = \mathbf{D}([0,\infty)\colon \mathcal{M}(\mathbf{R}^m))$ be the set of all cadlag maps from $[0,\infty)$ into $\mathcal{M}(\mathbf{R}^m)$. It is a complete metric space by Skorohod's J_1-metric. Typical elements of $\mathbf{D}$ is denoted by $B = B(t, E)$. The topological Borel field of $\mathbf{D}$ is denoted by $\mathcal{B}(\mathbf{D})$. Since $(B^n(t), t \in [0,\infty))$ can be regarded as a cadlag process with values in $\mathcal{M}(\mathbf{R}^m)$, the law of $B^n(t)$ is defined on $(\mathbf{D}, \mathcal{B}(\mathbf{D}))$ by

$$P^n(A) = P(\{\omega; B^n(\omega) \in A\}), \qquad A \in \mathcal{B}(\mathbf{D}). \tag{14}$$

If the sequence of measures $\{P^n\}_n$ converges weakly, then the sequence $\{B^n(t)\}_n$ is said to *converge in law*.

Theorem 1.1. *Assume Conditions* (ξ.1) *and* (ξ.2). *Then the sequence* $\{B^n(t)\}_n$ *converges in law. Let* $(B(t,E), P^\infty)$ *be its weak limit. Then, for any* $E_1, ..., E_q$ *of* $\mathcal{G}(\mathbf{R}^m)$, $(B(t,E_1), ..., B(t,E_q))$ *is a q-dimensional Brownian motion with mean 0 and covariance*

$$E[B(t,E_i)B(t,E_j)] = t(\pi(E_i \cap E_j) + V(E_i, E_j) + V(E_j, E_i) - \pi(E_i)\pi(E_j)). \quad (15)$$

Further for any t, $B(t,E)$ *can be extended continuously to any* $E \in \mathcal{B}(\mathbf{R}^m)$, *which is a Gaussian random measure.*

The above $B(t,E)$ is called a *Brownian random measure.*

We shall define stochastic integrals based on the Brownian random measure $B(t,E)$. Set

$$\mathcal{B}_t = \bigcap_{\varepsilon>0} \sigma(B(s,\cdot); 0 \le s \le t+\varepsilon). \quad (16)$$

Then $\{\mathcal{B}_t\colon\ t \ge 0\}$ is a filtration of σ-fields of $\mathbf{D}$. If $u \in L^2(\pi)$, the integral $\int_{\mathbf{R}^m} u(\lambda)B(t,d\lambda)$ is well defined. It is a Brownian motion with mean 0 and variance

$$E[|\int_{\mathbf{R}^m} u(\lambda)B(t,d\lambda)|^2] = t\left(\pi(u^2) + 2V(u,u) - \pi(u)^2\right). \quad (17)$$

Now let $u(x,\lambda)$, $x \in \mathbf{R}^d, \lambda \in \mathbf{R}^m$ be a measurable function, continuous in x such that $u(x,\cdot) \in L^2(\pi)$ for any x. Let ψ_t be an $\mathbf{R}^d$-valued $(\mathcal{B}_t)$ adapted continuous process. Itô's stochastic integral of $u(\psi_s, \lambda)$ based on $B(ds, d\lambda)$ is defined by

$$\int_0^t \int_{\mathbf{R}^m} u(\psi_s, \lambda)B(ds, d\lambda)$$
$$= \lim_{|\Delta|\to 0} \sum_{k=1}^n \int_{\mathbf{R}^m} u(\psi_{t_{k-1}}, \lambda)\Big(B(t_k, d\lambda) - B(t_{k-1}, d\lambda)\Big), \quad (18)$$

where $\Delta = \{0 = t_0 < \cdots < t_n = t\}$ and $|\Delta| = \max t_k - t_{k-1}$. It is a continuous local martingale. Next let $v(x,\lambda)$ be another function with the same property as $u(x,\lambda)$ and let η_t be a continuous process adapted to $(\mathcal{B}_t)$. We use the notations

$$\pi(u(x)v(y)) = \int_{\mathbf{R}^m} u(x,\lambda)v(y,\lambda)\pi(d\lambda), \quad (19)$$

$$V(u(x), v(y)) = \int\int_{\mathbf{R}^m \times \mathbf{R}^m} u(x,\lambda)v(y,\lambda')V(d\lambda, d\lambda'). \quad (20)$$

Then the Itô integral $\int_0^t \int_{\mathbf{R}^m} v(\eta_s, \lambda)B(ds, d\lambda)$ is well defined as a continuous local martingale. It is easily verified that their joint quadratic variation is equal to

$$< \int_0^t \int_{\mathbf{R}^m} u(\psi_s, \lambda)B(ds, d\lambda), \int_0^t \int_{\mathbf{R}^m} v(\eta_s, \lambda)B(ds, d\lambda) >$$
$$= \int_0^t \Big(\pi(u(\psi_s)v(\eta_s)) + V(u(\psi_s), v(\eta_s)) + V(v(\eta_s), u(\psi_s))\Big) ds. \quad (21)$$

The Stratonovich integral of $u(\psi_t, \lambda)$ based on the Brownian random measure $B(t,E)$ is defined by

$$\int_0^t \int_{\mathbf{R}^m} u(\psi_s, \lambda)B(\circ ds, d\lambda)$$
$$= \lim_{|\Delta|\to 0} \sum_{k=1}^n \int_{\mathbf{R}^m} \frac{1}{2}\{u(\psi_{t_k}, \lambda) + u(\psi_{t_{k-1}}, \lambda)\}\Big(B(t_k, d\lambda) - B(t_{k-1}, d\lambda)\Big), \quad (22)$$

if the limit exists. It is easily verified that the integral exists if $u(x,\lambda)$ is continuous and is a C^2-function with respect to x and ψ_t is a continuous semimartingale. The Stratonovich integral and Itô integral are related by

$$\int_0^t \int_{\mathbf{R}^m} u(\psi_s,\lambda)B(\circ ds,d\lambda)$$
$$= \int_0^t \int_{\mathbf{R}^m} u(\psi_s,\lambda)B(ds,d\lambda) + \frac{1}{2}\sum_i < \psi_t^i, \int_0^t \int_{\mathbf{R}^m} \frac{\partial u}{\partial x_i}(\psi_s,\lambda)B(ds,d\lambda) >, \quad (23)$$

where $\psi_t = (\psi_t^1, ..., \psi_t^d)$.

Now let us return to the sequence of Eq.2. We will introduce assumptions to the sequences of coefficients $\{f^n(x,\lambda)\}_n$ and $\{g^n(x)\}_n$. Let $\mathbf{C}^k = \mathbf{C}^k(\mathbf{R}^d : \mathbf{R}^d)$ be the space of k times continuously differentiable maps from $\mathbf{R}^d$ into itself. For $k = 1, 2$ and $N > 0$ we define the seminorms $\|u\|_{k,N}$ by

$$\|u\|_{k,N} = \sup_{|x|\le N} \frac{|u(x)|}{1+|x|} + \sum_{1\le|\alpha|\le k} \sup_{|x|\le N} |D^\alpha u(x)|, \quad (24)$$

and the norms $\|u\|_k^*$ by $\lim_{N\to\infty} \|u\|_{k,N}$. We denote by $\mathbf{C}_{b*}^k$ the set of all $u \in \mathbf{C}^k$ such that $\|u\|_k^* < \infty$. We assume that $f^n(\lambda) \equiv f^n(x,\lambda)$ in Eq.2 is a piecewise continuous $\mathbf{C}_{b*}^2$-valued function.

Condition (f.1) *The sequence $\{f^n(\lambda)\}_n$ satisfies*

$$\lim_{c\to\infty} \sup_n \int_{c<|\lambda|<\sqrt{n}M} (\|f^n(\lambda)\|_{2,N})^2 \pi^n(d\lambda) = 0 \quad (25)$$

for any $M > 0$ and $N > 0$.

Condition (f.2) *(1) There exists a piecewise continuous $\mathbf{C}_{b*}^2$-valued function $f(\lambda) = f(x,\lambda)$ such that $\pi(f(x)) = 0$ holds for any x of $\mathbf{R}^d$ and for any $N > 0$, $\|f^n(\lambda) - f(\lambda)\|_{1,N}$ converges to 0 uniformly on compact sets with respect to λ. (2) The sequence $\{b^n(x)\}_n$ defined by (0.4) is in $\mathbf{C}_{b*}^2$, $\sup_n \|b^n\|_{2,N} < \infty$ for any $N > 0$. Further there exists $b(x)$ of $\mathbf{C}_{b*}^1$ such that for any $N > 0$, $\|b^n - b\|_{1,N}$ converges to 0.*

Condition (g.1) *There exists $g(x)$ of $\mathbf{C}_{b*}^1$ such that for any $N > 0$ $\|g^n - g\|_{1,N}$ converges to 0.*

The law of the pair $(B^n(t), \varphi_t^n)$ can be defined on the Skorohod space $\mathbf{D} = \mathbf{D}([0,\infty) : \mathcal{M}(\mathbf{R}^m) \times \mathbf{R}^d)$. We denote it by $\bar{P}^n$. If the sequence $\{\bar{P}^n\}_n$ converges weakly, the sequence $\{(B^n(t), \varphi_t^n)\}_n$ is said to *converge in law*.

Theorem 1.2. *Assume Conditions (ξ.1), (ξ.2), (f.1), (f.2) and (g.1). Then the sequence $\{(B^n(t), \varphi_t^n)\}_n$ converges in law. Let $(B(t), \varphi_t, \bar{P}^\infty)$ be its limit law. Then φ_t is represented by*

$$\varphi_t = x_0 + \int_0^t \int_{\mathbf{R}^m} f(\varphi_s,\lambda)B(ds,d\lambda) + \int_0^t (g+b+c)(\varphi_s)ds, \quad (26)$$

where $c(x) = (c_1(x), ..., c_d(x))$ is defined by

$$c_i(x) = \sum_j V(f_j(x), \frac{\partial}{\partial x_j} f_i(x)). \quad (27)$$

In the next section we will give the proofs of Theorems 1.1 and 1.2, since they are long.

Now define a second order differential operator $\mathcal{L}$ by

$$\begin{aligned}\mathcal{L}u(x) &= \frac{1}{2}\sum_{i,j}\Big(\pi(f_i(x)f_j(x)) + V(f_i(x), f_j(x)) + V(f_j(x), f_i(x))\Big)\frac{\partial^2 u}{\partial x_i \partial x_j} \\ &\quad + \sum_i (b_i + c_i + g_i)(x)\frac{\partial u}{\partial x_i}. \end{aligned} \tag{28}$$

Introducing the first order differential operators (vector fields) with parameter λ by

$$L(\lambda)u(x) = \sum_i f_i(x,\lambda)\frac{\partial u}{\partial x_i}(x), \tag{29}$$

the above $\mathcal{L}$ is represented by

$$\begin{aligned}\mathcal{L} &= \frac{1}{2}\sum_{i,j}\pi(f_i(x)f_j(x))\frac{\partial^2}{\partial x_i \partial x_j} \\ &\quad + \int\!\!\int_{\mathbf{R}^m\times\mathbf{R}^m} L(\lambda)L(\lambda')V(d\lambda, d\lambda') + \sum_i (b_i + g_i)\frac{\partial}{\partial x_i}. \end{aligned} \tag{30}$$

The following is immediate from Eq.26 and Itô's formula.

Corollary 1.3. *Assume the same condition as in Teorem 1.2. For any $\mathbf{C}^2$-function u,*

$$u(\varphi_t) - \int_0^t \mathcal{L}u(\varphi_s)ds \tag{31}$$

is a local martingale with respect to $\bar{P}$.

In the sequel we consider the case where the stationary process $\{\xi_j^n, j \in \mathbf{N}\}$ does not depend on n. We denote it by $\{\xi_j, j \in \mathbf{N}\}$. Let $\{\phi_k\}$ be the uniform mixing rate of $\{\xi_j, j \in \mathbf{N}\}$. The following corollary is immediate from Theorems 1.1-1.2.

Corollary 1.4. *Assume $\sum_{k=1}^{\infty}(\phi_k)^{1/2} < \infty$.*
(1) Then the assertion of Theorem 1.1 is valid.
(2) Consider stochastic difference equations:

$$\psi_j = \psi_{j-1} + \frac{1}{\sqrt{n}}f(\psi_{j-1}, \xi_j) + \frac{1}{n}g(\psi_{j-1}), \qquad j = 1, 2, \ldots \tag{32}$$

where $f(x,\lambda)$ is a piecewise continuous $\mathbf{C}^2_{b*}$-valued function satisfying $\pi(f(x)) = 0$ for all x and

$$\int_{\mathbf{R}^m}(\|f(\lambda)\|_{2,N})^2\pi(d\lambda) < \infty, \qquad \forall N. \tag{33}$$

Further $g(x)$ is a $\mathbf{C}^1_{b*}$-function. Then the conclusion of Theorems 1.2 is valid.

Finally we will apply the above limit theorem to limits of stoschastic flow of diffeomorphisms studied by Matsumoto-Shigekawa[8]. Let $\{\xi_j, j \in \mathbf{N}\}$ be a sequence of stationary process with values in $\mathbf{R}^m$ satisfying the uniform mixing conditions of Corollary 1.4. Let π be the distribution function of ξ_1. Let $f(\lambda) = f(x,\lambda)$ be a continuous $\mathbf{C}^2_{b*}$-valued function satisfying $\pi(f(x)) = 0$ for all x and Eq.33. Consider the ordinary differential equation with parameter λ:

$$\frac{d\eta_t}{dt} = f(\eta_t, \lambda). \tag{34}$$

Let $\eta_t^\lambda(x)$ be the solution with the initial condition $\eta_0 = x$. Then it defines a flow of diffeomorphisms. Set

$$f^n(x,\lambda) = \sqrt{n}\int_0^{1/\sqrt{n}} f(\eta_s^\lambda(x),\lambda)ds, \qquad \text{and} \qquad g^n(x) = 0. \tag{35}$$

Consider Eq.2, where $\xi_j^n = \xi_j$. The solution with the initial condition $\psi_0 = x$ is represented by

$$\psi_k^n(x) = \eta_{1/\sqrt{n}}^{\xi_k} \circ \ldots \circ \eta_{1/\sqrt{n}}^{\xi_1}(x), \tag{36}$$

where the symbol $\circ$ denotes the composition of the maps. Hence it defines a diffeomorphism for each fixed n, k and ω.

We shall consider the limit of $\varphi_t^n(x) \equiv \psi_{[nt]}^n(x)$. We can examine directly that the sequence $\{f^n(x,\lambda)\}_n$ of Eq.35 satisfies Condition (f.1) and converges to $f(x,\lambda)$ in the sense of Condition (f.2). Consider $b^n(x)$ of Eq.4. It is written as

$$b_i^n(x) = n\int_{\mathbf{R}^m}\left(\int_0^{1/\sqrt{n}} f_i(\eta_s^\lambda(x),\lambda)ds\right)\pi(d\lambda). \tag{37}$$

By the mean value theorem, we have

$$b_i^n(x) = n\int_0^{1/\sqrt{n}} ds\int_0^s du\left(\sum_j\int_{\mathbf{R}^m}\frac{\partial}{\partial x_j}f_i(\eta_{\theta s}^\lambda(x),\lambda)f_j(\eta_u^\lambda(x),\lambda)\pi(d\lambda)\right).$$

It converges to

$$b_i(x) = \frac{1}{2}\sum_j \pi\left(\frac{\partial}{\partial x_j}f_i(x)f_j(x)\right) \tag{38}$$

in the sense of Condition (f.2). Then the sequence $\{(B^n(t),\varphi_t^n)\}_n$ converges in law and the limit $(B(t),\varphi_t)$ satisfies Eq.26 with $g = 0$ and c of Eq.27. The infinitesimal generator of Eq.30 is written by

$$\mathcal{L}u(x) = \frac{1}{2}\int_{\mathbf{R}^m} L(\lambda)^2 u(x)\pi(d\lambda) + \int\int_{\mathbf{R}^m\times\mathbf{R}^m} L(\lambda)L(\lambda')u(x)V(d\lambda,d\lambda'). \tag{39}$$

Suppose that $\{\xi_j\}$ is reversible. Then noticing Eq.23, it is easily verified that the stochastic differential equation is represented by

$$\varphi_t = x_0 + \int_0^t\int_{\mathbf{R}^m} f(\varphi_s,\lambda)B(\circ ds, d\lambda), \tag{40}$$

using the Stratonovich integral. In particular if $\xi_j, j \in \mathbf{N}$ are independent identically distributed random variables, $\{\xi_j\}$ is reversible and Eq.40 holds. Furthermore, the kernel $V(d\lambda, d\lambda')$ is identically 0 so that the generator is written by

$$\mathcal{L} = \int_{\mathbf{R}^m} L(\lambda)^2\pi(d\lambda). \tag{41}$$

2. Proof of Theorems

Let us first observe that Theorems 1.1 and 1.2 can be proved in a unified way. Indeed, set for arbitrarily fixed $E_1, ..., E_q \in \mathcal{G}(\mathbf{R}^m)$,

$$\chi^n(\lambda) = (\chi_{E_1}(\lambda) - \pi^n(E_1), ..., \chi_{E_q}(\lambda) - \pi^n(E_q)), \tag{42}$$

and define for $y \in \mathbf{R}^q$, $F^n(y,\lambda) = \chi^n(\lambda)$ and $G^n(y) = 0$. Consider the stochastic difference equation

$$\Psi_k = \Psi_{k-1} + \frac{1}{\sqrt{n}} F_n(\Psi_{k-1}, \xi_k^n) + \frac{1}{n} G^n(\Psi_{k-1}). \tag{43}$$

Let Ψ_k^n be the solution with the initial condition $\Psi_0 = 0$. Set $\Phi_t^n = \Psi_{[nt]}^n$. Then $\Phi_t^n = \bar{B}^n(t)$ holds where $\bar{B}^n(t) = (B^n(t,E_1), ..., B^n(t,E_q))$ is defined by Eq.1. On the other hand for $y = (z,x) \in \mathbf{R}^{q+d}$ set

$$F^n(y,\lambda) = (\chi^n(\lambda), f^n(x,\lambda)), \qquad G^n(y) = (0, g^n(x)). \tag{44}$$

Consider a stochastic difference equation based on the above F^n and G^n: Let Ψ_k^n be the solution with the initial condition $\Psi_0 = (0, x_0)$. Set $\Phi_t^n = \Psi_{[nt]}^n$. Then Φ_t^n is represented by $\Phi_t^n = (\bar{B}^n(t), \varphi_t^n)$ where φ_t^n is defined by Eq.3. Therefore we consider Eq.43 and discuss the weak convergence of $\{\Phi_t^n\}_n$ rather than the weak vonvergence of $\{\bar{B}^n(t)\}_n$ or $\{\bar{B}^n(t), \varphi_t^n\}_n$.

Consider the sequence of Eq.43, where $F^n(y,\lambda), y \in \mathbf{R}^e, \lambda \in \mathbf{R}^m$ are $\mathbf{R}^e$-valued functions, twice continuously differentiable with respect to y and piecewise continuous with respect to λ. Further $G^n(y)$ are continuously differentiable $\mathbf{R}^e$-valued functions. We assume that the sequence $\{F^n(y,\lambda)\}_n$ and $\{G^n(y)\}_n$ satisfy Conditions (f.1), (f.2) and (g.1). Let $F(y,\lambda)$ and $G(\lambda)$ be their limits, respectively. For a C^2-function $u(y)$, we define

$$\mathcal{L}u(y) = \frac{1}{2}\sum_{i,j=1}^{e} A_{ij}(y)\frac{\partial^2 u}{\partial y^i \partial y^j}(y) + \sum_{i=1}^{e}\left(B_i(y) + C_i(y) + G_i(y)\right)\frac{\partial u}{\partial y^i}(y), \tag{45}$$

where

$$A_{ij}(y) = \pi(F_i(y)F_j(y)) + V(F_i(y), F_j(y)) + V(F_j(y), F_i(y)), \tag{46}$$

$$C_i(y) = \sum_j V\left(F_j(y), \frac{\partial}{\partial y_j}F_i(y)\right). \tag{47}$$

Theorem 2.1. *Assume Conditions* (ξ.1), (ξ.2), *(f.1),(f.2) and (g.1) for Eq. 43. Let* Ψ_k^n *be the solution with the initial condition* $\Psi_0 = y$. *Set* $\Phi_t^n \equiv \Psi_{[nt]}^n$. *Let* Q^n *be the law of* Φ_t^n *on the space* $\mathbf{D}([0,\infty)\colon \mathbf{R}^e)$. *Then the sequnece* $\{Q^n\}_n$ *converges weakly. Let* (Φ_t, Q^∞) *be its limit. Then for any* $\mathbf{C}^2$*-function* u *on* $\mathbf{R}^e$,

$$u(\Phi_t) - \int_0^t \mathcal{L}u(\Phi_s)ds, \tag{48}$$

is a locally square integrable martingale with respect to Q^∞.

For the proof of the theorem, we will apply a criterion of the weak convergence obtained by Fujiwara-Kunita[2]. It tells us that Theorem 2.1 is valid if the following Conditions (A.I)-(A.IV) are satisfied. For a positive constant M, set

$$\bar{F}^{n,M}(y) = E[F^n(y,\xi_j^{n,M})], \quad \tilde{F}^{n,M}(y,\lambda) = F^{n,M}(y,\lambda) - \bar{F}^{n,M}(y). \tag{49}$$

Condition (A.I) (1) For every $M > 0$ and $N > 0$ there exists a sequence of non-decreasing $\mathcal{F}^n_{[nt]}$ adapted cadlag processes $\{D^n_t\}_n$ such that

$$\frac{1}{n}\sum_{k=[ns]+1}^{[nt]}\sum_{\ell=k+1}^{[nt]} \|E[\tilde{F}^{n,M}(\xi_\ell^{n,M})|\mathcal{F}^n_k]\|_{2,N}\|\tilde{F}^{n,M}(\xi_k^{n,M})\|_{1,N} + \frac{1}{n}\sum_{k=[ns]+1}^{[nt]}(\|F^n(\xi_k^{n,M})\|_{1,N})^2 \le D^n_t - D^n_s, \tag{50}$$

for any $0 \le s \le t$ and that $\{D^n_t\}_n$ is C-tight (tight and any weak limit law is supported by continuous functions) and uniformly integrable for any t.

(2) $\{\|\bar{F}^{n,M}\|_{1,N}\}_n$ is bounded for every $M > 0$ and $N > 0$.

Condition (A.II) (1) For any bounded continuous function h, $N > 0$, $0 < \delta < M$ and $0 \le s < t$,

$$\lim_{n\to\infty} E\left[\sup_{|y|\le N}\left|E\left[\sum_{k=[ns]+1}^{[nt]} h\Big(\frac{1}{\sqrt{n}}F^n(y,\xi_k^n)\Big)\chi_{(\delta,M]}\Big(\|\frac{1}{\sqrt{n}}F^n(\xi_k^n)\|_{1,N}\Big)\Big|\mathcal{F}^n_{[ns]}\right]\right|\right] = 0.$$

(2) For any $N > 0$, $0 \le s < t$ and i, j,

$$\lim_{M\to 0}\limsup_{n\to\infty} E\left[\sup_{|y|\le N}\left|\frac{1}{n}E\left[\sum_{k=[ns]+1}^{[nt]} L^{n,M}_{k,k}(y)\Big|\mathcal{F}^n_{[ns]}\right] - (t-s)\pi(F_i(y)F_j(y))\right|\right] = 0,$$

where $L^{n,M}_{k,\ell}(y) = \tilde{F}^{n,M}_i(y,\xi_k^{n,M})\tilde{F}^{n,M}_j(y,\xi_\ell^{n,M})$.

(3) For any $N > 0$, $0 \le s < t, M > 0$ and i, j,

$$\lim_{n\to\infty} E\left[\sup_{|y|\le N}\left|\frac{1}{n}E\left[\sum_{k=[ns]+1}^{[nt]}\sum_{\ell=k+1}^{[nt]} L^{n,M}_{k,\ell}(y)\Big|\mathcal{F}^n_{[ns]}\right] - (t-s)V(F_i(y),F_j(y))\right|\right] = 0.$$

(4) For any $N > 0$, $0 \le s < t, M > 0$ and i, j,

$$\lim_{n\to\infty} E\left[\sup_{|y|\le N}\left|E\left[\frac{1}{n}\sum_{k=[ns]+1}^{[nt]}\sum_{\ell=k+1}^{[nt]} \hat{L}^{n,M}_{k,\ell}(y)\Big|\mathcal{F}^n_{[ns]}\right] - (t-s)C_i(y)\right|\right] = 0,$$

where $\hat{L}^{n,M}_{k,\ell}(y) = \sum_j \tilde{F}^{n,M}_j(y,\xi_k^{n,M})(\partial\tilde{F}^{n,M}_i/\partial y_j)(y,\xi_\ell^{n,M})$.

(5) $\lim_{n\to\infty} B^n(y) + G^n(y) = B(y) + G(y)$ uniformly on compact sets.

Condition (A.III) For any α with $|\alpha| \le 1$, $M > 0$, $N > 0$ and $0 \le s < t$,

$$\lim_{n\to\infty}\sup_{s\in[0,t]} E\left[\sup_{|y|\le N}\left|\frac{1}{\sqrt{n}}\sum_{k=[ns]+1}^{[nt]} E\Big[D^\alpha\tilde{F}^{n,M}(y,\xi_k^{n,M})\Big|\mathcal{F}^n_{[ns]}\Big]\right|\right] = 0,$$

$$\lim_{n\to\infty} E\Big[\sup_{|y|\le N}\Big(\frac{1}{\sqrt{n}}\Big)^3\sum_{k=[ns]+1}^{[nt]}\sum_{\ell=\kappa+1}^{[nt]}\Big|E\Big[D^\alpha\tilde{F}^{n,M}(y,\xi_\ell^{n,M})\Big|\mathcal{F}^n_k\Big]\Big| \times|\tilde{F}^{n,M}(y,\xi_k^{n,M})|^2\Big] = 0.$$

Condition (A.IV) For any $t > 0$ and $N > 0$,

$$\lim_{M\to\infty} \limsup_{n\to\infty} P\Big(\sup_{k\le[nt]} \|\frac{1}{\sqrt{n}} F^n(\xi_k^n)\|_{1,N} > M \Big) = 0.$$

In the sequel, we will examine these conditions one by one.

Lemma 2.2. *Condition (A.I) is satisfied.*

Proof. Our discussion is similar to that of Theroem 4.2 in Fujiwara-Kunita[2]. We will fix $M > 0$ and $N > 0$, and set

$$K_k^n = \sum_{\ell=k+1}^{[nt]} \Big\| E\Big[\tilde{F}^{n,M}(\xi_\ell^{n,M})\Big|\mathcal{F}_k^n\Big]\Big\|_{2,N} \|\tilde{F}^{n,M}(\xi_k^{n,M})\|_{1,N} + \Big(\|F^{n,M}(\xi_k^{n,M})\|_{1,N}\Big)^2.$$

We can prove similarly as in the proof of Lemma 4.3 in Fujiwara-Kunita[2] that $\{K_k^n\}_{n,k}$ is also uniformly integrable. In the following, we will prove the tightness of $D_t^n \equiv (1/n)\Sigma_{k=1}^{[nt]} K_k^n$, $n \in \mathbf{N}$. We have

$$\frac{1}{n}\sum_{k=[n\sigma]}^{[n(\sigma+\theta)]} K_k^n \le \frac{1}{n}\sum_{k=[n\sigma]}^{[n(\sigma+\theta)]} K_k^n \chi_{(0,c)}(K_k^n) + \frac{1}{n}\sum_{k=1}^{[nt]} K_k^n \chi_{[c,\infty)}(K_k^n)$$

for any stopping time σ and positive constant θ. Since $\{K_k^n\}_{n,k}$ is uniformly integrable,

$$\lim_{c\to\infty} \sup_n \frac{1}{n}\sum_{k=1}^{[nt]} E[K_k^n \chi_{[c,\infty)}(K_k^n)] = 0.$$

Consequently,

$$\limsup_{\theta\to 0} \sup_n \sup_\sigma E\Big[\frac{1}{n}\sum_{k=[n\sigma]}^{[n(\sigma+\theta)]} K_k^n\Big] = 0.$$

Therefore the sequence $\{D_t^n\}_n$ is tight by Aldous's theorem.

For the C-tightness of $\{D_t^n\}_n$, we have to show that

$$\lim_{n\to\infty} P\Big(\sup_{k\le[nt]} \frac{1}{n} K_k^n \ge \delta \Big) = 0$$

holds for any t and $\delta > 0$. Note that

$$P\Big(\sup_{k\le[nt]} \frac{1}{n} K_k^n \ge \delta \Big) \le \sum_{k=1}^{[nt]} P(K_k^n \ge n\delta) \le \frac{[nt]}{n\delta} \sup_{k\le[nt]} E[K_k^n : K_k^n \ge n\delta].$$

The last term converges to 0 as n tends to infinity since $\{K_k^n\}_{n,k}$ is uniformly integrable. Therefore $\{D_t^n\}_n$ is C-tight. The uniform integrability of $\{D_t^n\}_n$ is obvious. qed

Lemma 2.3. *Condition (A.II) is satisfied.*

Proof. We have

$$E\left[\sum_{k=[ns]+1}^{[nt]} \chi_{(\delta,M]}\left(\|\frac{1}{\sqrt{n}}F^n(\xi_k^n)\|_{1,N}\right)\right]$$
$$\le ([nt]-[ns])P(\sqrt{n}\delta \le \|F^n(\xi_k^n)\|_{1,N} \le \sqrt{n}M)$$
$$\le ([nt]-[ns])(\sqrt{n}\delta)^{-2}E[(\|F^n(\xi_1^n)\|_{1,N})^2 : \sqrt{n}\delta \le \|F^n(\xi_1^n)\|_{1,N} \le \sqrt{n}M].$$

The right hand side converges to 0 as n tend to infinity by Condition (f.1). Then (A.II)(1) is satisfied.

We next consider (A.II)(2),(3) and (4). Since the proofs of these are similar, we only prove (3). We can prove similarly as in the proof of Theorem 4.2 in Fujiwara-Kunita[2] that

$$\lim_{n\to\infty}\sup_{|y|\le N}\frac{1}{n}\left|\sum_{k=[ns]+1}^{[nt]}\sum_{\ell=k+1}^{[nt]}\left(E[L_{k,\ell}^{n,M}(y)|\mathcal{F}_{[ns]}^n]-\bar{L}_{k,\ell}^{n,M}(y)\right)\right|=0. \tag{51}$$

We wish to prove

$$\lim_{n\to\infty}\sup_{|y|\le N}\left|\frac{1}{n}\sum_{k=[ns]+1}^{[nt]}\sum_{\ell=k+1}^{[nt]}\bar{L}_{k,\ell}^{n,M}(y)-(t-s)V(F_i(y),F_j(y))\right|=0, \tag{52}$$

where $\bar{L}_{k,\ell}^{n,M}(y)=E[L_{k,\ell}^{n,M}(y)]$. Then (A.II)(3) will follow. Now for any fixed $k_0\in\mathrm{N}$, we have the mixing inequality

$$\sup_{|y|\le N}\frac{1}{n}\left|\sum_{k=[ns]+1}^{[nt]}\sum_{\ell=k+k_0+1}^{[nt]}\bar{L}_{k,\ell}^{n,M}(y)\right|$$
$$\le\frac{2}{n}\sum_{k=[ns]+1}^{[nt]}\sum_{\ell=k+k_0+1}^{[nt]}(\phi_{\ell-k}^n)^{\frac{1}{2}}\sup_{|y|\le N}E\left[\tilde{F}_i^{n,M}(y,\xi_k^{n,M})^2\right]^{\frac{1}{2}}E\left[\tilde{F}_j^{n,M}(y,\xi_\ell^{n,M})^2\right]^{\frac{1}{2}}.$$

Since $E[\tilde{F}_i^{n,M}(y,\xi_k^{n,M})^2]$ and $E[\tilde{F}_j^{n,M}(y,\xi_\ell^{n,M})^2]$ are bounded in n, there exists a positive constant C such that the above is bounded by $C\Sigma_{k=k_0}^{\infty}(\phi_k^n)^{1/2}$. Now for any given $\delta>0$ choose $k_0\in\mathrm{N}$ such that $C\sum_{k=k_0}^{\infty}(\phi_k^n)^{1/2}<\delta$ holds for all n. Then

$$\sup_{|y|\le N}\left|\frac{1}{n}\sum_{k=[ns]+1}^{[nt]}\sum_{\ell=k+k_0+1}^{[nt]}\bar{L}_{k,\ell}^{n,M}(y)\right|<\delta \tag{53}$$

holds for any n. Similarly we can choose the above $k_0\in\mathrm{N}$ such that

$$\sup_{|y|\le N}\left|\frac{1}{n}\sum_{k=[ns]+1}^{[nt]}\sum_{\ell=k+k_0+1}^{[nt]}E[F_i(y,\xi_k)F_j(y,\xi_\ell)]\right|<\delta \tag{54}$$

for all n. Furthermore there exists n_0 such that

$$\sup_{|y|\le N}\left|\frac{1}{n}\sum_{k=[ns]+1}^{[nt]}\sum_{\ell=k+1}^{[nt]}E[F_i(y,\xi_k)F_j(y,\xi_\ell)]-(t-s)V(F_i(y),F_j(y))\right|<\delta \tag{55}$$

holds for all $n\ge n_0$.

Let $\psi_c(\lambda)$ be a smooth function such that $0 \le \psi_c(\lambda) \le 1$, $\psi_c(\lambda) = 1$ if $|\lambda| \le c$, $= 0$ if $|\lambda| \ge c+1$. Since $\{\|F^n(\xi_k^{n,M})\|_{2,N}^2\}_n$ is uniformly integrable for any k by Condition (f.2), there exists $c > 0$ such that

$$\sup_{|y|\le N}\left|E\left[L_{k,\ell}^{n,M}(y)(1-\psi_c(\xi_k^{n,M})\psi_c(\xi_\ell^{n,M}))\right]\right| < \frac{\delta}{k_0(t-s)} \tag{56}$$

hold for all n. Then we have

$$\sup_{|y|\le N}\left|\frac{1}{n}\sum_{k=[ns]+1}^{[nt]}\sum_{\ell=k+1}^{k+k_0} E\left[L_{k,\ell}^{n,M}(y)(1-\psi_c(\xi_k^{n,M})\psi_c(\xi_\ell^{n,M}))\right]\right| < \delta. \tag{57}$$

Also for all n we have

$$\sup_{|y|\le N}\left|\frac{1}{n}\sum_{k=[ns]+1}^{[nt]}\sum_{\ell=k+1}^{k+k_0} E\left[F_i(y,\xi_k)F_j(y,\xi_\ell)(1-\psi_c(\xi_k)\psi_c(\xi_\ell))\right]\right| < \delta. \tag{58}$$

On the other hand, the finite dimensional distributions of $\{\xi_j^n\}$ converges to the corresponding distribution of $\{\xi_j\}$. Since $\tilde{F}_i^{n,\epsilon}(\lambda)$ converges to $F_i(\lambda)$ with respect to $\|\ \|_{1,N}$ norm uniformly on compact sets of λ, there exists $n_1 \in \mathrm{N}$ such that

$$\begin{aligned}\sup_{|y|\le N}\Bigg|&\frac{1}{n}\sum_{k=[ns]+1}^{[nt]}\sum_{\ell=k+1}^{k+k_0} E\left[L_{k,\ell}^{n,M}(y)\psi_c(\xi_k^{n,M})\psi_c(\xi_\ell^{n,M})\right]\\ &-\frac{1}{n}\sum_{k=[ns]+1}^{[nt]}\sum_{\ell=k+1}^{k+k_0} E\left[F_i(y,\xi_k)F_j(y,\xi_\ell)\psi_c(\xi_k)\psi_c(\xi_\ell)\right]\Bigg| < \delta\end{aligned} \tag{59}$$

holds for all $n \ge n_1$. These inequalities Eqs.53-59 imply

$$\sup_{|y|\le N}\left|\frac{1}{n}\sum_{k=[ns]+1}^{[nt]}\sum_{\ell=k+1}^{[nt]} \bar{L}_{k,\ell}^{n,M}(y) - (t-s)V(F_i(y),F_j(y))\right| < 6\delta, \tag{60}$$

if $n \ge \max(n_0, n_1)$. This proves Eq.52.

The proof of (A.II)(5) is immediate. It is omitted. qed.

Lemma 2.4. *Conditions (A.III) and (A.IV) are satisfied.*

Proof. Note that

$$\begin{aligned}&\sum_{k=[ns]+1}^{[nt]} E\left[\left|E\left[D^\alpha \tilde{F}_i^{n,M}(y,\xi_k^{n,M})\middle|\mathcal{F}_{[ns]}^n\right]\right|\right]\\ &\qquad\le \sum_{k=[ns]+1}^{[nt]} 2(\varphi_{k-[ns]}^n)^{\frac{1}{2}} E\left[D^\alpha \tilde{F}_i^{n,M}(y,\xi_k^{n,M})^2\right]^{\frac{1}{2}}.\end{aligned}$$

Since $\{E[D^\alpha \tilde{F}_i^{n,M}(y,\xi_k^{n,M})^2]\}_n$ is a bounded sequence, the last term of the above divided by $\sqrt{n}$ converges to 0 as $n \to \infty$. This proves the first equality of (A.III). The second equality can be proved similarly.

Finally, we have

$$P\left(\sup_{k\le[nt]}\|\frac{1}{\sqrt{n}}F^n(\xi_k^n)\|_{1,N} > M\right) \le [nt]P\left(\|F^n(\xi_k^n)\|_{1,N} > \sqrt{n}M\right).$$

The last term converges to 0 as n tends to infinity. Therefore Condition (A.IV) is satisfied. qed.

Proof of Theorem 1.1. Consider Eq.43 with $F^n(y,\lambda) = (\chi^n(\lambda))$ and $G^n(y) = 0$. These functions satisfies Conditions (f.1), (f.2) and (g.1) obviously. Let Ψ_k^n be the solution such that $\Psi_0 = 0$ and let Q^n be its law on $\mathbf{D} = \mathbf{D}([0,\infty)\colon \mathbf{R}^q)$. Then the sequence $\{Q^n\}_n$ converges weakly to Q^∞ by virtue of Theorem 2.1. Since Φ_t^n coincides with $(B^n(t,E_1),...,B^n(t,E_q))$, the limit (Φ_t, Q^∞) can be identified with $((B(t,E_1),...,B(t,E_q)), Q^\infty)$. Note that the operator $\mathcal{L}$ of Eq.45 is represented by

$$\mathcal{L}u(y) = \frac{1}{2}\sum_{i,j}\Big(\pi(E_i\cap E_j) + V(E_i,E_j) + V(E_j,E_i) - \pi(E_i)\pi(E_j)\Big)\frac{\partial^2 u}{\partial y_i \partial y_j}(y). \quad (61)$$

Then the limit $(B(t,E_1),...,B(t,E_q))$ is a Brownian motion with the covariance function Eq.15. See Kunita[6].

Finally let P^n be the law of $B^n(t)$ on the space $\mathbf{D}([0,\infty) : \mathcal{M}(\mathbf{R}^m))$. Then the tightness of the laws $\{Q^n\}_n$ for any $(E_1,...,E_q)$ implies the tightness of $\{P^n\}_n$. Hence a subsequence of $\{P^n\}_n$ converges weakly. Let P^∞ be any limit law. Then the law of $(B(t,E_1),...,B(t,E_q))$ with respect to P^∞ coincides with Q^∞. Since this is valid for any $(E_1,...E_q)$, the measure P^∞ is unique. This proves Theoem 1.1. qed

Proof of Theorem 1.2. Set $y = (z,x)$ where $z \in \mathbf{R}^q, x \in \mathbf{R}^d$ and define $F^n(y,\lambda)$ and $G^n(y)$ by Eq.44. Then Conditions (f.1), (f.2) and (g.1) are satisfied. Consider Eqs.43 based on these F^n and G^n. Then we have $\Phi_t^n = (\bar{B}^n(t), \varphi_t^n)$. Therefore by Theorem 2.1, the sequence $\{(\bar{B}^n(t),\varphi_t^n)\}_n$ converges in law. Let $(\bar{B}(t),\varphi_t,\bar{Q}^\infty)$ be its limit. We shall consider the first component $\bar{B}(t)$. If u is a function of z only, then $\mathcal{L}u$ is represented as Eq.61. Therefore $\bar{B}(t)$ is a Brownian motion with covariance Eq.15 by Theorem 1.1. Then we can show similarly as in the proof of Theorem 1.1 that the sequence of the laws $\{\bar{P}^n\}$ on the space $\mathbf{D}([0,\infty) : \mathcal{M}(\mathbf{R}^m)\times\mathbf{R}^d))$ converges weakly. Let $(B(t),\varphi_t,\bar{P}^\infty)$ be its limit law.

We will show Eq.26. We assume $d = 1$ for simplicity. Then Theorem 2.1 tells us that

$$M_t \equiv \varphi_t - x_0 - \int_0^t (b+c+g)(\varphi_s)ds \quad (62)$$

is a locally square intgrable martingale with respect to $\bar{P}^\infty$. Its bracket process is computed from coefficients of $\partial^2/\partial x^2$ in Eq.45. Indeed,

$$< M_t > \; = \int_0^t \Big(\pi(f(\varphi_s)^2) + 2V(f(\varphi_s), f(\varphi_s))\Big)\, ds. \quad (63)$$

Similarly the joint quadratic variation of M_t and $B(t,E)$ is obtained from the coefficient of $\partial^2/\partial z\partial x$, i.e.,

$$< M_t, B(t,E) >= \int_0^t A_2(\varphi_s, E)ds, \quad (64)$$

where $A_2(x,E) = \pi(\chi_E f(x)) + V(\chi_E, f(x)) + V(f(x),\chi_E)$. Now define $\tilde{M}_t$ by

$$\tilde{M}_t = \int_0^t \int_{\mathbf{R}^m} f(\varphi_s,\lambda)B(ds,d\lambda). \quad (65)$$

We can compute $<\tilde{M}_t>$ by Eq.21. Then we get $<M>_t=<\tilde{M}>_t$. Further from Eq.64, we have

$$\begin{aligned}
&< M_t, \int_0^t \int_{\mathbf{R}^m} f(\varphi_s,\lambda)B(ds,d\lambda) > \\
&= \int_0^t \int_{\mathbf{R}^m} f(\varphi_s,\lambda)A_2(\varphi_s,d\lambda)ds \\
&= \int_0^t \left(\pi(f(\varphi_s)^2) + 2V(f(\varphi_s), f(\varphi_s))\right) ds.
\end{aligned}$$

Consequently we have $< M_t, \tilde{M}_t >=< M_t >$. This implies $< M_t - \tilde{M}_t >=< M_t > -2 < M_t, \tilde{M}_t > + < \tilde{M}_t >= 0$, proving $M_t = \tilde{M}_t$. Then Eq.62 and Eq.65 imply Eq. 26. qed.

3. Strong mixing case.

Let $\{\xi_j^n, j \in \mathbf{N}\}, n = 1,2,...$ and $\{\xi_j, j \in \mathbf{N}\}$ be stationary processes satisfying Condition (ξ.1) as in Section 1. Let π^n and π be the distributions of ξ_1^n and ξ_1 respectively. The *strong mixing rate* $\{\alpha_k^n,\ k \in \mathbf{N}\}$ of the process $\{\xi_j^n; j \in \mathbf{N}\}$ is defined by

$$\alpha_k^n = \sup_{h\in\mathbf{N}} \sup\{|P(E\cap F) - P(E)P(F)| : E \in \mathcal{F}_h^n, F \in \mathcal{F}_{h+k,\infty}^n\}, \tag{66}$$

where $\mathcal{F}_{h,k}^n = \sigma(\xi_j^n\colon h \le j \le k)$. We denote that of $\{\xi_j, j \in \mathbf{N}\}$ by $\{\alpha_k, k \in \mathbf{N}\}$. For $\beta > 0$ we introduce a condition:

Condition (ξ.2)$_\beta$. *The sequence $\{\alpha_k^n\}$ satisfies $\sum_{k=1}^\infty (\sup_n \alpha_k^n)^\beta < \infty$.*

If $\beta < \beta'$, then $(\xi.2)_\beta$ implies $(\xi.2)_{\beta'}$ since $\alpha_k^n \le 1$ holds for any n, k. Let ϕ_k^n be the uniform mixing rate of $\{\xi_j^n\}$. Then the inequality $\alpha_k^n \le \phi_k^n$ holds for any n, k. Therefore Condition (ξ.2) implies Condition $(\xi.2)_{1/2}$.

From (ξ.1) and $(\xi.2)_\beta$, the inequality $\alpha_k \le \sup_n \alpha_k^n$ is easily verified for any $k \in \mathbf{N}$. Therefore $\sum_{k=1}^\infty (\alpha_k)^\beta < \infty$ is satisfied.

Let χ_E be the indicator function of the set E. By a well known mixing inequality, we have

$$|E[\chi_{E_1}(\xi_1)\chi_{E_2}(\xi_j)] - \pi(E_1)\pi(E_2)| \le C(\alpha_{j-1})^\beta \pi(E_1)^{1/r}\pi(E_2)^{1/r}$$

if $\beta + 2/r = 1$. Therefore if $u, v \in L^r(\pi)$, then $V(u,v)$ is well defined by Eq.12 and satisfies

$$|V(u,v)| \le C(\sum_{j=1}^\infty (\alpha_j)^\beta)\|u\|_r \|v\|_r.$$

The following theorem is stronger than that of Theorem 1.1.

Theorem 3.1. *Suppose (ξ.1) and $(\xi.2)_{1/2}$. Then the assertion of Theorem 1.1 is valid.*

Now consider Eq. 4. We introduce an assumption for $\{f^n(x,\lambda)\}$.

Condition (f.1)$_\delta$. *The sequence $\{f^n(\lambda)\}_n$ is in $\mathbf{C}_{b*}^2$ and satisfies*

$$\sup_n \int_{|\lambda|\le\sqrt{n}M} (\|f^n(\lambda)\|_{2,N})^{2+\delta}\pi^n(d\lambda) < \infty \tag{67}$$

for any $M > 0$ and $N > 0$.

Theorem 3.2. *(c.f.* Fujiwara[1]) *Assume Conditions* $(\xi.1), (\xi.2)_\beta$ *and* $(f.1)_\delta$, *(f.2), (g.1) for some* $\delta \in (0,\infty)$ *and* $\beta \in (0, \delta/(2+\delta)(1+d))$. *Then the assertion of Theorem 1.2 is valid.*

Proofs of these theorems are again reduced to checking Conditions (A.I)-(A.IV) introduced in the previous section. They can be carried out similarly as in the proof of Theorem 4.2 in Fujiwara-Kunita[2]. So it is omitted.

Corollary 3.3. *Assume that the sequence* $\{\xi_j^n; j \in \mathrm{N}\}$ *does not depend on n.*

(1) Assume (a) $\sum_k (\alpha_k)^\beta < \infty$ *for* $\beta = 1/2$. *Then the assertion of Theorem 1.1 is valid.*

(2) Consider Eq.27, where $f^n(x,\lambda)$ *and* $g^n(x)$ *satisfy* $(f.1)_\delta$, *(f.2) and (g.1) for some* $\delta > 0$. *Assume (a) for* $\beta \in (0, \delta/(2+\delta)(1+d))$. *Then the assertion of Theorem 1.2 is valid.*

References

1. T. Fujiwara, *Limit theorems for random difference equations driven by mixing processes*, to appear.

2. T. Fujiwara and H. Kunita, *Limit theorems for stochastic difference-differential equations*, Nagoya Math. J., submitted

3. M. Iizuka and H. Matsuda, *Weak convergence of discrete time non-Markovian processes related to selection models in population genetics*, J. Math. Biology, **15**(1982), 107-127.

4. J. Jacod and A.N. Shiryaev, *Limit theorems for stochastic processes*, Springer-Verlag, 1987.

5. H. Kunita, *Stochastic flows and stochastic differential equations*, Cambridge Univ. Press, 1990.

6. H. Kunita, *Limits of random measures induced by an array of independent random variables*, to appear.

7. H. Kunita, *Limits on random measures and stochastic difference equations induced by mixing array of random vsariables*, to appear.

8. H. J. Kushner and Hai-Huang, *On the weak convergence of a sequence of general stochastic difference equations*, SIAM J. Appl. Math., **40**(1981),528-541.

9. H. Matsumoto and I. Shigekawa, *Limit theorems for stochastic flows of diffeomorphisms of jump type*, Z. Wahrscheinlichkeitstheorie verw. Gebiete, **69**(1965), 507-540.

10. H. Watanabe, *Diffusion approximations of some stochastic difference equations II*, Hiroshima Math. J., **14**(1984), 15-34.

Stochastic Partial Differential Equations : White Noise Approach

YU.A. ROZANOV

Abstract. Generalized test functions and Boundary conditions. Correlation operator and Reproducing Kernel space. Forecast boundary problems and Markov property. Probability model identification.

1. INTRODUCTION.

We wish to start with observation that white noise approach to the corresponding stochastic disturbance term in the most classical differential equations of Mathematical Physics brings us the most well known Probability models in Random Fields theory. This remarkable the phenomenon in particular happens for the *String Equation*

$$\frac{\partial^2 \xi}{\partial t_1^2} - \frac{\partial^2 \xi}{\partial t_2^2} = \eta$$

in a rectangular $T \subseteq R^2$ formed by its characteristics (see a fig.1), the *Poisson Equation*

$$\triangle \xi = \eta$$

with the Laplacian $\triangle$ in a region $T \subseteq R^3$ formed by the whole R^3-space but its origin O and the *Heat Equation*

$$\frac{\partial \xi}{\partial t} - A\xi = \eta$$

in $T = R^n \times (0, \infty)$ where $A = \triangle$ represents the Laplacian on R^n, say, and t runs a time-interval $(0, \infty)$. It turns out that with the standard while noise approach to the generalized disturbance term

$$\eta = (g, \eta) \in L_2(T),$$

these equation brings us correspondingly the well known *Brownian Sheet*, Lévy Brownian Motion and *Stochastic Itô Equation*; namely in each case there is the unique solution $\xi \in W$ (in a proper functional class W),

$$\xi = (x, \xi), x \in D = C_0^\infty(T),$$

and this generalized field ξ in the case of the String / Poisson equations has a trace $\xi(t)$ at every point $t \in T$ which represents the Brownian Sheet

/ the Lévy Brownian Motion, and in the case of the Heat equation its generalized solution ξ has a generalized trace

$$\xi_t = (\varphi, \xi_t), \varphi \in C_0^\infty(R^n),$$

at every section $R^n \times \{t\}$ of $T = R^n \times (0, \infty)$ which represents the solution of the functional Itô equation

$$d(\varphi, \xi_t)dt + d(\varphi \times 1_{(0,t)}, \eta),$$

where $1_{(0,t)}$ is an indicator of a time interval $(0, t), 0 < t < \infty$.

These examples represent a linear partial differential equation

$$L\xi = \eta, \tag{1}$$

(with infinitely differential coefficients, say) in a region $T \subseteq R^d$ treated in a generalized sense as

$$(L^*\varphi, \xi) = (\varphi, \eta), \varphi \in D = C_0^\infty(T), \tag{1'}$$

with a formally conjugate differential operator L^* such that the conjugate equation

$$L^* f = x$$

for any $x \in D = C_0^\infty(T)$ has the unique solution $f \in L_2(T)$, and the very equation (1) has the unique solution $\xi \in W$ which can be described by means of the test functions $x \in D$ as

$$(x, \xi) = (L^* f, \xi) = (f, \eta); \tag{2}$$

the corresponding functional space $W \ni \xi$ can be characterized by a continuity (x, ξ) over $x \in D$ with respect to a norm

$$\|x\|_X = \|f\|_{L_2}, x = L^* g, f \in L_2(T) \tag{3}$$

- of course with our white noise approach the continuity of ξ, η is to be understood as for the generalized random functions in a Hilbert space $L_2(\Omega)$ on a probability space Ω.

2. Generalized test functions and boundary conditions.

According to the continuity in (1') over $\varphi \in D = C_0^\infty(T)$ in $L_2(T)$ -space the generalized random fields ξ described by the formula (2) can be extended to

$$\xi = (x, \xi), x \in X, \tag{4}$$

over a closure $X = [L^* D]$ with respect to the norm (3) which represents the certain Hilbert space

$$X = L^* L_2(T) \subseteq D' \tag{5}$$

of Schwartz distributions $x \in D'$, namely with a limit

$$\lim \varphi = f \in L_2(T),$$

we have

$$\lim(L^* \varphi, \xi) = (L^* f, \xi) = (x, \xi),$$

where

$$x = L^* f = \lim L^* \varphi \in D';$$

more over as it is actually indicated in (2) we have

$$D = C_0^\infty(T) \subseteq L^* L_2(T) = X,$$

and in particular it shows that

$$X = [D] \tag{5'}$$

is a closure of $D = C_0^\infty(T)$ with respect to the Hilbert norms (3); the corresponding inner product in X is translated from $L_2(T)$ - space by the one-to-one conjugate differential operator

$$L^* : L_2(T) \to X,$$

i.e.

$$(L^* f, L^* g) = (g, f)_{L_2}. \tag{6}$$

One can treat

$$(x, \xi) = (\xi, x)$$

in (4) as a result of testing the random field ξ by generalized test functions

$$x = (\varphi, x), \varphi \in D = C_0^\infty(T),$$

and consider ξ as a function of the functional variable $x \in X \subseteq D'$ in the region $T \subseteq R^d$.

With this approach considering the generalized random function $\xi \in W$ in any region

$$S \subseteq T,$$

One can determine *boundary values*

$$(x,\xi) = (\xi, x),\ supp\ \ x \subseteq \Gamma, \tag{7}$$

on a boundary $\Gamma = \partial S$ of S which represent a result of testing ξ by all Schwartz distributions $x \in X$ with supports on Γ. According to (5) the *boundary test functions* in (7) are all

$$x = L^* f, \tag{8}$$

where $f \in L_2(T)$ satisfy the differential equation

$$L^* f = 0$$

outside of the boundary Γ. And one can observe that some boundary values in (7) are determined by the very equation (1) in the region S,

$$(L^*, \xi) = (f, \eta), f \in L_2(S) = [C_0^\infty(S)]$$

namely all of them represent a result of testing ξ by the boundary test functions (8) with the corresponding $f \in L_2(S)$, $f = 0$ outside of S. Of course considering the only region $S \subseteq T$ we have to deal with a proper functional class $W(S)$ represented by a restriction of all functions from W in S. Let us take any *collection* $X^+(\Gamma)$ of the boundary test functions with

$$x = L^* f, f \in L_2(S)$$

in (8) from a *complete system* generated in our Hilbert space X *all* boundary test functions; it turns out that the generalized random fields ξ can be identified in any region $S \subseteq T$ as the unique solution $\xi \in W(S)$ of the equation (1) in S *conditioned* on the boundary $\Gamma = \partial S$ by the boundary values

$$(x, \xi), x \in X^+(\Gamma), \tag{9}$$

where the collection $X^+(\Gamma)$ of the boundary test functions determines a type *of* the corresponding *boundary conditions.*

For our examples (we start with) all boundary test functions on a regular boundary Γ are generated in X by Schwartz distributions

$$x = (\varphi, x), \varphi \in D = C_0^\infty(T),$$

which represent the very φ and normal derivatives $\frac{\partial \varphi}{\partial n}$ on Γ distributed there with the corresponding weight functions $x(s), s \in \Gamma$, as

$$(\varphi, x) = \int_\Gamma \varphi x(s) ds, \int_\Gamma \frac{\partial \varphi}{\partial n} x(s) ds,$$

and by these boundary test functions one can determine some kind of a *trace* of the generalized random field ξ and its generalized derivative $\frac{\partial \xi}{\partial n}$ on the boundary Γ

$$(10) \qquad \int_\Gamma \xi x(s)ds, \int_\Gamma \frac{\partial \xi}{\partial n} x(s)ds \equiv (\xi, x) \equiv (x, \xi)$$

- it is worthy to note here, that in a usual sence the random fields considered are *non-differentiable* anywhere.

One can observe that the complete collection (10) depends on the boundary Γ type. For the *Brownian Sheet* ξ in a case of the boundary Γformed by characteristic of the String operator L (see fig.2), we have in (10) the collection of *only*

$$(x, \xi) = \int_\Gamma \xi x(s)ds$$

which represents the very ξ on Γ, and in a case of the regular Γ with no parts formed by the characterics (see fig.3)we have in (10) the collection which represent in addition to the very ξ also its generalized derivative $\frac{\partial \xi}{\partial n}$ trace on the whole Γ, for the *Lévy Brownian Motion* ξ in a case of any regular Γ we have in (10) the collection which represents the trace of both ξ, $\frac{\partial \xi}{\partial n}$ on Γ, and for the generalized random field ξ associated with the stochastic Itô equation in a cylinder $S = G \times (t_1, t_2)$ with a regular base $G \subseteq R^n$, say, we have in (10) the collection which represents the very ξ trace on the whole boundary $\Gamma = \partial S$ plus the generalized derivative $\frac{\partial \xi}{\partial n}$ trace on the side-boundary of the cylinder S (see fig.4).

Going back to the boundary condition (9) for the general equation (1) in the region $S \subseteq T$ on the boundary $\Gamma = \partial S$ we indicate a few types $X^+(\Gamma)$ of the conditions which are associated with the most well known boundary problems in Partial Differential Equations. The stochastic *String equation* in the rectangular S formed by the characteristics has the unique solution $\xi \in W(S)$ conditioned by the trace

$$\xi(s), s \in \gamma_1 \cup \gamma_2,$$

on the only part $\gamma_1 \cup \gamma_2$ of the boundary Γ (see the fig.2), and in the general regular region S in the case of $\Gamma = \partial S$ with no parts formed by the characteristics the unique solution $\xi \in W(S)$ does exist *conditioned* by the trace of ξ and its generalized derivative $\frac{\partial \xi}{\partial n}$ on the part γ_0 plus the trace of the very ξ / or plus the trace of $\frac{\partial \xi}{\partial n}$ on the only part $\gamma_1 \cup \gamma_2$ of the boundary Γ (see the fig.3); these boundary conditions are associated correspondingly with the well known *Gursa problem* and so called *the first / second mixed boundary problems.* The stochastic *Poisson equation* in any regular region $S \subseteq T$ has the unique solution $\xi \in W(S)$ conditioned by the trace of the very ξ on $\Gamma = \partial S$ / or by the trace of its generalized derivative $\frac{\partial \xi}{\partial n}$ plus the boundary value

$$(x, \xi) \equiv (\xi, x) = \int \xi(s) ds$$

with the boundary test function

$$x = (\varphi, x) = \int_{\Gamma} \varphi(s) ds, \varphi \in D = C_0^{\infty}(T);$$

these boundary conditions are associated with the well known *Dirichret / Neumann boundary problems.* The stochastic *Heat equation* in the cylinder $S = G \times (t_1, t_2)$, say, has the unique solution $\xi \in W(S)$ *conditioned* by the trace ξ on the base boundary $G \times \{t_1\}$ plus the trace of the very ξ / or the trace of its generalized derivative $\frac{\partial \xi}{\partial n}$ on the side-boundary of the cylinder S (see the fig.4); these boundary conditions are associated with the well known *first / second mixed boundary problems.*

3. Correlation Operator and Reproducing Kernel Space.

For any generalized random fields $\xi = (x, \xi)$ with

$$E(x, \xi) = 0, x \in D = C_0^{\infty}(T),$$

we suggest to determine a *correlation operator B* as

$$B : D \in x \to u = Bx \in D'$$

where Schwartz distributions $u = Bx$ are

$$Bx = (\varphi, Bx) = E(\varphi, \xi)(x, \xi), \varphi \in D.$$

The corresponding *reproducing Kernel space* (we denote it as W) can be determined as a closure

$$W = [BD] \subseteq D'$$

with respect to a Hilbert norm

$$\|u\|_W = (E|(x, \xi)|)^{1/2} = \|x\|_X.$$

In our case represented in (1),(2) the reproducing Kernel space in a closure

$$W = D \subseteq D'$$

of $D = C_0^\infty(T)$ with the norm $\|u\|_W$ determined as

$$\|u\|_W = (u, Pu)^{1/2} \tag{11}$$

by means of a differential operator

$$P = L^* L \geq 0$$

which brings us the Schwartz distributions $x = Pu \in X$ well defined on the whole $W = [D]$; moreover the differential operator $P = L^* L$ (considered on Schwartz distributions $u \in W \subseteq D'$) is the *unitary operator*

$$P : W \longrightarrow X = PW \tag{12}$$

and the correlation operator B is represented by

$$B = P^{-1} : X \longrightarrow W = BW.$$

One can observe that for $\xi \in W$ the original equation (1) is equivalent to differential equation

$$P\xi = \eta \tag{13}$$

where $\eta^* = L^* \eta$ represents a white noise on the reproducing Kernel space $W = [D]$.

4. Forecast boundary problems and Markov property

The differential equation (1)/(13) give a powerful instrumentation to deal with a general *Forecast problem* which requires to determine the *best forecast* $\hat{\xi} = E(\xi/\mathbf{B})$ of the random field ξ based on some data $\mathbf{B}$,

$$\hat{\xi} = (x, \hat{\xi}) = E[(x, \xi)/\mathbf{B}], x \in X;$$

namely the solution of this problem can be given by means of the unique solution $\hat{\xi} \in W$ of the equations

$$L\hat{\xi} = \hat{\eta}/P\hat{\xi} = L^* \hat{\eta} \tag{14}$$

with

$$\hat{\eta} = E(\eta/\mathbf{B})/L^*\hat{\eta} = E(L^*\eta/\mathbf{B}) = L^*\hat{\eta}.$$

In particular one can determine $\hat{\xi}$ in some region $S \subseteq T$ with respect to $\mathbf{B}$ representing complete data on η outside of S / or complete data on the very ξ *outside* of S. In the first case $\hat{\xi} \in W(S)$ can be given as the unique solution of

$$L\hat{\xi} = 0$$

with the *boundary conditions*

$$(x, \hat{\xi}) = (f, \eta), x \in X^+(\Gamma),$$

which are set on the boundary $\Gamma = \partial S$ by mean of the collection $X^+(\Gamma)$ of the boundary test function (8) with $f \in L_2(S^c)$ in a complementary S^c to S, i.e. $f = 0$ in the region S and $L^* f = 0$ in the complementary region $T - (S \cup \Gamma)$, and in the second case $\hat{\xi} \in W(S)$ can be given as the unique solution of

$$P\hat{\xi} = 0 \tag{16}$$

with the *boundary conditions*

$$(x, \hat{\xi}) = (x, \xi), x \in X^+(\Gamma),$$

set by means of the collection $X^+(\Gamma)$ which generates in X *all* boundary test functions (8). The equations (15)/(16) are a result of application of the general equations (14), any explanation where (15) does not need any explanations and (16) can be explained by a fact that the right side $L^*\eta$ of (14) with the *Gaussian* white noise η on $L_2(T)$–space in S does *not depend* on ξ *outside* of the region S.

Actually the fact indicatred above and applied also to the complementary region $T - (S \cup \Gamma)$ shows us that the independent parts $L^*\eta$ in S and in $T - (S \cup \Gamma)$ are *conditionally* independent with respect to *all* boundary values (7) and this phenomenon brings us the following *Markov property*: the random field ξ in *any* region S conditioned by all boundary values on the boundary $\Gamma = \partial S$ is conditionally independent on its part outside of S. (It is worthy to note here that $\xi \in W$ represents in S and in the complementary region $T - (S \cup \bar{\Gamma})$ the unique solution of the equation

$$P\xi = \eta^*(= L^*\eta),$$

with the same boundary conditions as for (16)).

5. Stochastic elliptic equations: identification

BY A SOLUTION TRAJECTORY

In our approach to the random field ξ by means of the corresponding stochastic partial differential equation we could start with the very equation (13) representing $\xi \in W$ as its unique solution in the region $T \subseteq R^d$ which can be described as

$$(x, \xi) = (Bx, \eta^*), B = P^{-1}, x \in X,$$

by introduction of the Hilbert space $W = [D] \subseteq D'$ with the norm

$$\|u\|_W = (u, Pu)^{1/2}, u \in D,$$

and the corresponding

$$X = PW = [D] \subseteq D'$$

appeared as a result of the *unitary mapping*

$$W \underset{B = P^{-1}}{\overset{\xrightarrow{P}}{\longleftarrow}} X$$

see (11),(12). This scheme can be applied to a fairly wide class of stochastic differential equations

$$P\xi = \eta^* \tag{17}$$

with a differential operator $P \geq 0$ and a white noise η^* on the Hilbert space $W = [D]$ which appears to be the *reproducing Kernel space* for the unique solution $\xi \in W$ in the region $T \subseteq R^d$ where the functional class W can be characterized by a continuity of (x, ξ) over $x \in D = C_0^\infty(T)$ with respect to the norm

$$\|u\|_X = \|Bx\|_W$$

with $B = P^{-1}$ as the *correlation operator* of $\xi \in W$.

One can imagine that dealing with the equation (17) we might not know coefficients of the operator

$$P = \sum_{|k| \leq 2p} a_k \partial^k \tag{18}$$

(which are supposed to be infinitely differential say) and a problem is to identify them by a *trajectory* of the generalized field ξ.

Let us consider the elliptic operator (18) such that

$$\|u\|_W = (u, Pu)^{1/2} \asymp \|u\|_p, u \in C_0^\infty(T_{loc}),$$

is equivalent to a *Sobolev norm*

$$\|u\|_p = (\sum_{|k| \leq p} \|\partial^k u\|_{L_2}^2)^{1/2}$$

in any bounded region $T_{loc} \subset T$. For example this kind of operator P appears in a stationary *Schrödinger equation* of a form

$$-\Delta\xi + U\xi = \eta^*$$

with a potential $U(= a_0)$ which in case of $U \equiv 1$ brings us in $T = R^d$ the *stationary Markov Free Field* with a spectral density

$$\frac{1}{|\lambda|^2 + 1}, \lambda \in R^d,$$

or it appears in a case of a general *stationary random field* on the R^d–space with a *spectral density*

$$\frac{1}{P(i\lambda)}, \lambda \in R^d,$$

where

$$P(i\lambda) = \sum_{|k| \leq 2p} a_k (i\lambda)^k \geq 0$$

is a polynomial such that

$$P(i\lambda) \asymp |\lambda|^{2p}, |\lambda| \to \infty,$$

which corresponds to the operator

$$P = P(\partial) = \sum_{|k| \leq 2p} a_k \partial^k$$

with the constant coefficients $a_k, |k| \leq 2p$.

The following general result holds true for the $2p$–order elliptic operator P in the region $T \subseteq R^d$: any its coefficient a_k,

$$|k| > 2p - \frac{d}{2}, \tag{19}$$

can be determined at each point $t \in T$ by means of some quadratic functional of a form

$$S_n = \frac{1}{B_n} \sum [(x_{mn}, Q\xi)^2 - A_n]$$

where for a given

$$k = (k_1, ..., k_d), |k| = k_1 + \cdots + k_d,$$

one has to take correspondingly some differential operator Q and a proper collection of numbers A_n, B_n and test functions $x_{mn} \in C_0^\infty(T_{loc})$ in a neighbourhood T_{loc} of the point $t \in T$ considered. It shows in particular that in a case of

$$2p < \frac{d}{2} \tag{20}$$

all coefficient $a_k, |k| \geq 2p$ can be completely determined by the trajectory of ξ. And it turns out that the conditions (19),(20) can not be improved.

6. References

The results presented are selected from various topics of the Yu.A. Rozanov book"Random Fields and Stochastic Partial Differential Equations". - Nauka, Moskow, 1990 (to appear) where one can find all kinds of references.

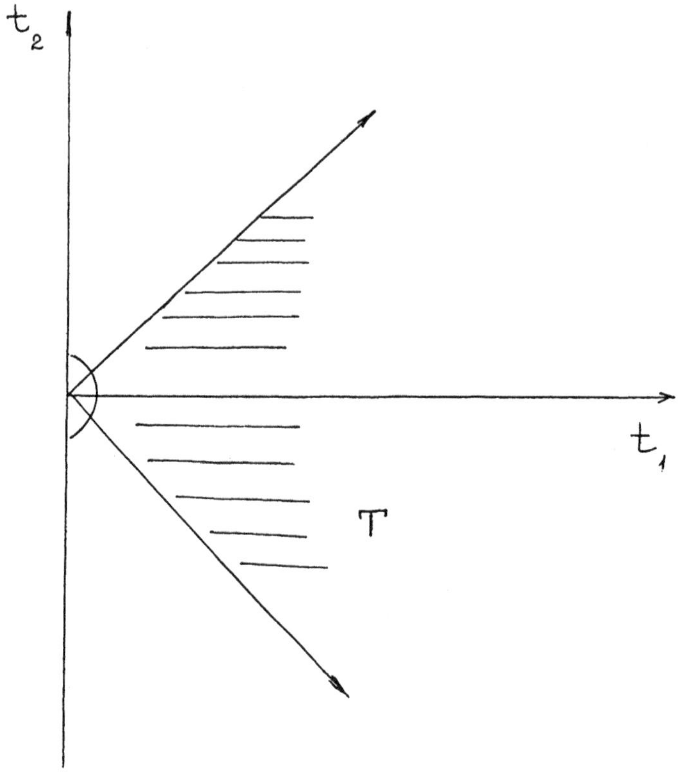

fig. 1

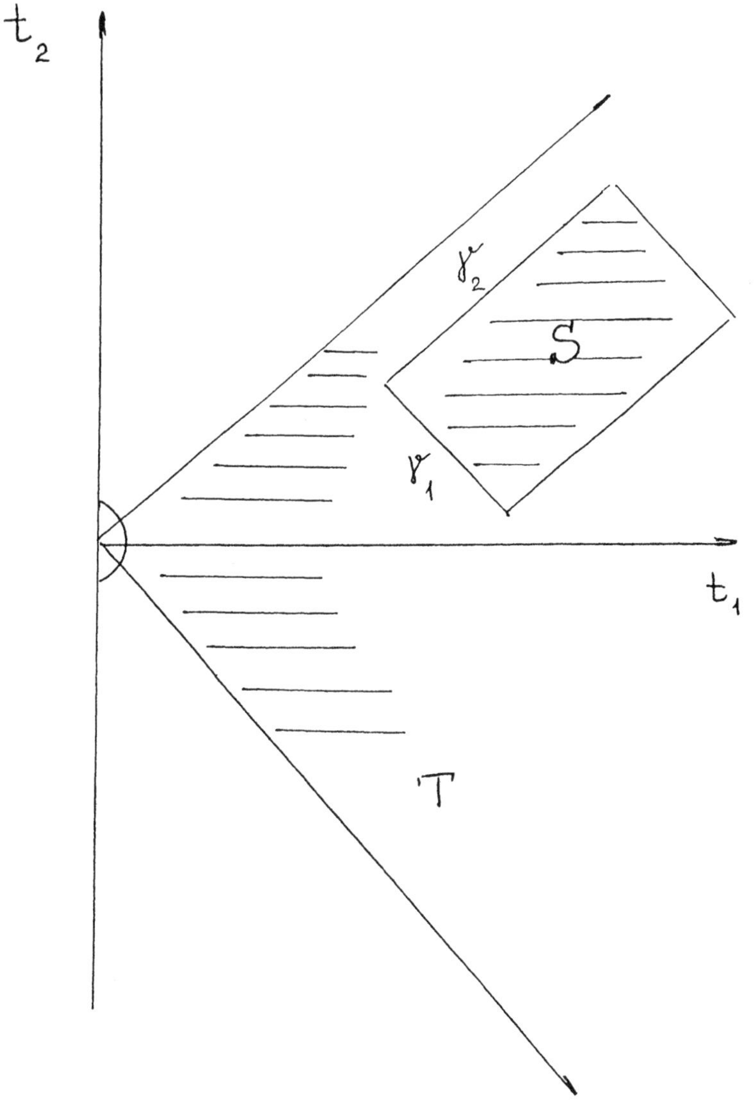

Fig. 2

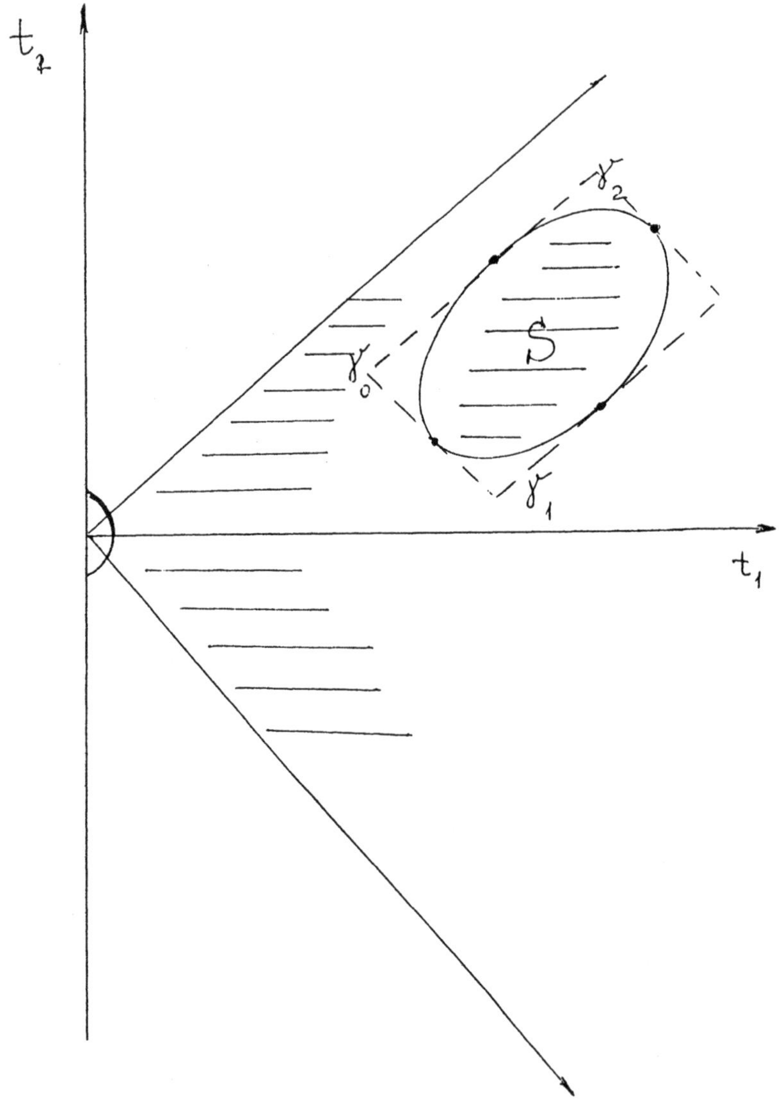

Fig. 3

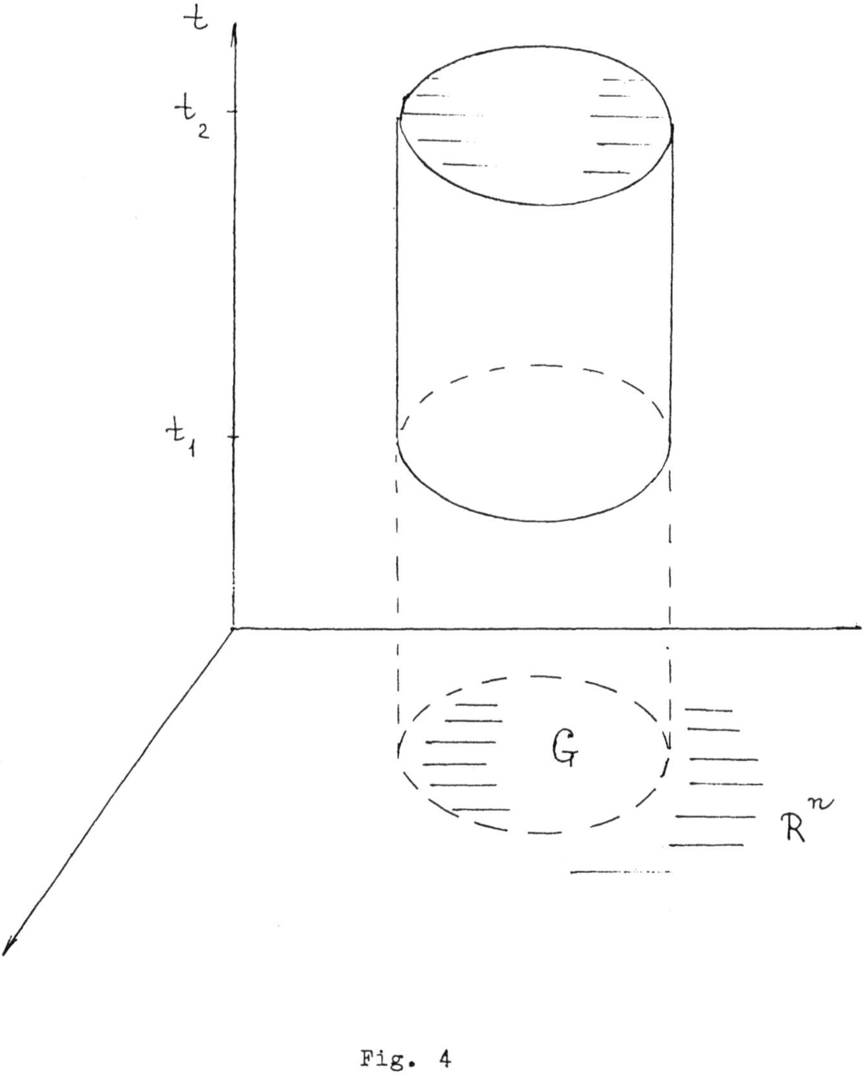

Fig. 4

WHITE NOISE ANALYSIS AND WHAT IT CAN DO FOR PHYSICS

Dedicated to Professor Takeyuki Hida at the occasion of his retirement from Nagoya University

J. Potthoff
Dept. of Mathematics, L S U, Baton Rouge LA, USA
and
L. Streit
BiBoS, Univ. Bielefeld, 4800 Bielefeld

I. Introduction.

Through the centuries, mathematics and physics have been a highly interactive business. White Noise of course first belonged to the physicists. In 1928 Johnson at Bell Labs measured voltage fluctuations across a resistor and observed that their spectrum was flat [Jo]. But Nyquist, in a companion paper [Ny], noted immediately that this noise was not, in fact could not be, pure white. Really white noise, he showed, is so unreal that we must ask the mathematicians to give it to us. This they did and with it, over the years, came a rich toolbox of stochastic and infinite dimensional analysis which now is ready for use in various branches of theoretical physics. - Consequently, we shall introduce the mathematical theory of (Gaussian) White Noise in section II, in III we shall construct Hida distributions and test functions, IV will review some of their properties, and finally, V and VI will illustrate applications of White Noise Analysis in the study of Feynman integrals and of canonical quantum field theories.

II. White Noise.

White Noise is a generalized Gaussian random field X with mean zero and covariance

$$E\left(\langle X,f\rangle \langle X,g\rangle \right) = (f,g)_{L^2(\mathbb{R}^d)} \qquad f,g \in S(\mathbb{R}^d).$$

Hence its characteristic functional C is

$$C(f) = E\left(e^{i\langle X,f\rangle} \right) = \int_{S^*} e^{i\langle X,f\rangle} \, d\mu[X\}$$

For d=1 - the case we shall mostly focus on below - it is intimately related to the Wiener process B:

$$B(t) = \langle X, 1_{[0,t]} \rangle$$

is a version of it [Hi].

It is customary to abbreviate

$$L^2(S^*,d\mu) = (L^2)$$

and it is well known (see e.g. [Hi]) that the algebra of polynomials P is dense in (L^2). Denoting the projection on the span of nth order polynomials by P_n we can introduce "normal" or "Wick ordered" products through

$$:\prod_{i=1}^{n} \langle X,f_i\rangle : = (1 - P_{n-1}) \prod_{i=1}^{n} \langle X,f_i\rangle.$$

In terms of the distribution valued fields this leads to

$$: X(t_1).......X(t_n) : \in S^*(\mathbb{R}^n)$$

and (L^2) orthogonality implies

$$E(: X(s_1).......X(s_m) : : X(t_1).......X(t_n) :) = n! \; \delta_{mn}\delta(s,t).$$

The polynomial algebra P is spanned by elements $\langle :X^{\otimes n}:, \otimes f_i \rangle$ and, in view of this orthogonality relation, its closure (L^2) is obtained by extending the kernel functions $\otimes f_i$ to

$$(F_n) \in \bigoplus_n \mathrm{Sy}\; L^2(\mathbb{R}^n, n!\, d^n t).$$

It is in this sense, and in this sense only, that we may write

$$\Phi[X] = \sum_{n=0}^{\infty} \int d^n t \; F_n(t_1,.......t_n) : X(t_1).......X(t_n) : .$$

Other very useful characterizations of White Noise functionals $\Phi \in (L^2)$ are given through the "T-" and "S-transform", linear maps defined by

$$T: \Phi \to E(\Phi\, e^{i<\cdot,f>}) \equiv (T\Phi)(f)$$

and

$$S: \Phi \to E(\Phi\,[\cdot + f]) \equiv (S\Phi)(f).$$

III. Hida Distributions.

For our present purpose, the main interest of the above constructions is to go beyond them, to consider non-L^2 functionals of White Noise such as e.g.

$$\int :X^2(t): dt$$

or Donsker's δ-function

$$\delta(B(t) - y).$$

In the first case the kernel functions F_n are

$$F_n(t_1,\ldots\ldots t_n) = \delta_{n2}\, \delta(t_1 - t_2) \in \mathrm{Sy}\, L^2(\mathbf{R}^n,\, n!\, d^n t)$$

while in the second example all the kernel functions F_n are square integrable, but

$$\sum n!\, \|F_n\|^2 = \infty.$$

Thus, a good generalization will be one that relaxes both the L^2 and the l^2 properties.

A standard procedure in finite dimensional analysis is to enlarge spaces through the construction of Gelfand triples; recall

$$S(\mathbf{R}) \subset L^2(\mathbf{R},dx) \subset S^*(\mathbf{R}),$$

obtainable e.g. by putting

$$S(\mathbf{R}) = \bigcap_n D(H_0{}^n),$$

where

$$H_o = 1 + x^2 - \frac{d^2}{dx^2}.$$

Here we proceed similarly [K] [KT], replacing H_o by its second quantization.

$$\Gamma(H_0):\ :\prod_{i=1}^{n} <X,f_i>: \ \to\ :\prod_{i=1}^{n} <X,H_0 f_i>:$$

extends to a positive self adjoint operator on (L^2) and we obtain a nuclear countable Hilbert space by introducing the Hilbert spaces

$$(S)_p \equiv D(\Gamma(H_0{}^p)) \qquad p=1,2,3,\ldots$$

and their projective limit

$$(S) = \lim_{\overleftarrow{p}} (S)_p.$$

In the sequel it is the triple

$$(S) \subset (L^2) \subset (S)^*$$

that we shall consider, the names of Hida test function(al)s and Hida distributions have become customary for (S) and $(S)^*$. - Of course this rigging is not unique, as an alternative we mention the dual pair D, D^* studied by Meyer, Shigekawa, Sugita, Watanabe and others [M1][P][Sh][Su][Wa]. Evidently the choice, as in finite dimensional distribution theory, will depend on the problems one wants to solve; for the purposes of quantum physics $(S)^*$ has the decisive advantage of being larger than D^*:

$$(S) \subset D \subset (L^2) \subset D^* \subset (S)^*.$$

An even larger space of "enlarged Hida distributions" has recently been proposed by P. A. Meyer [M2].

IV. - Properties of Test Functionals and Hida Distributions.

We give a short list of useful features; for demonstrations and more we refer to the literature, e.g. [WN] and references therein, as well as the specific references given below.

* (S) is nuclear
* (S) is an algebra under pointwise multiplication
* (S) has Gateaux derivatives D_h in all directions $h \epsilon (S)^*$:
  $$D_h : (S) \to (S)$$
  is continuous for all h in $(S)^*$ [PY].
* In particular for the Dirac distribution δ_t we use the notation $\partial_t \equiv D_{\delta_t}$ ("Hida derivative")
* This Hida derivative acts on (S) as Frechet derivative [KPY]
  $\partial \cdot : (S) \to S(\mathbb{R}) \otimes (L^2)$, and $(\partial F, \partial F) \epsilon (S)$ for all $F \epsilon (S)$ [HPS].
* For any $F \epsilon (S)$ there is a version $\widetilde{F}$ with the kernel functions $\widetilde{F}_n \ \epsilon \ S(\mathbb{R}^n)$, for which $\widetilde{F}[X] = \sum_{n=o}^{\infty} < : X^{\otimes n} : , \widetilde{F}_n >$, an expression which extends

  continuously to all $X \epsilon S^*(\mathbb{R})$ [KY].

Regarding the space of Hida distributions we note here

* All Hida distributions are of finite order: $(S)^* = \bigcup_p (S)_{-p}$.

* T- and S-transform extend to $(S)^*$.
* Any positive Hida distribution Φ is a measure, i.e. there is a measure ν_Φ such

  that $\Phi \geq 0 \Rightarrow \langle\Phi,F\rangle = \int \tilde{F} d\nu_\Phi$ [K],[Yo].
* There is the following characterization of $(S)^*$ in terms of "U-functionals":

  Let V: $S(\mathbf{R}) \to \mathbf{C}$ be such that

  1) $V(\lambda f)$, as a function of λ, extends to an entire analytic function $v(z,f)$

  2) for some p,a,b, v obeys the bound $| v(z,f) | \leq a \exp\left(b \| H_0{}^p f \|\right)$.

  Then we call V a U-functional, and the following are equivalent [PS1,3]:

  (1) V is a U-functional,

  (2) V is the S-transform of a Hida distribution,

  (3) V is the T-transform of a Hida distribution.

  (for related results see [K], [M2], [Y], [L]).
* As a direct consequence, pointwise multiplication of U-functionals induces two algebraic structures on the space of Hida distributions: convolutions and Wick products are allowed in $(S)^*$ [PS1].

This characterization offers e.g. the possibility to often decide immediately whether a given measure may be expressed in terms of a positive Hida distribution: its characteristic function would then be the T-transform of the latter, i.e. a U-functional. This serves to exclude Poisson processes, but also to include the Gibbs measures of various Euclidean quantum field theories, they all have a representation in terms of a positive Hida distribution [AHP].

V. - The Feynman Integral.

Quantum theory, since its inception, has owed much of its progress to very recent developments in mathematics. but the converse is also true: various more or less heuristic approaches of quantum physics have remained so or have stagnated for years and even decades until the appropriate mathematical tools have become available; infinite dimensional and stochastic analysis are a case in point.

A most remarkable construct of quantum physics is the Feynman integral

$$\int_{\substack{y(o)=y_1\\ y(t)=y_2}} \prod_{\tau} dy(\tau)\, e^{\frac{i}{\hbar} S[y]}$$

with the "action functional" $S[y] = \frac{m}{2} \int_0^t \dot{y}^2 d\tau - \int_0^t V(y) d\tau$.

For the physicist it has become an invaluable conceptual and calculational tool, yet for the mathematician its proper definition and treatment continue to be a challenge, witness the contributions of D. M. Chung, G. W. Johnson, G. Kallianpur, D. Storvick to this conference. Of course many mathematically rigorous definitions have been given, but typically they function only for potentials V such as

$$V(y) = gy^2 + \int dm(\alpha)\, e^{i\alpha y}, \qquad \text{see e.g. [AH2].}$$

In particular the Feynman integral is not one with respect to a measure, but as it turns out we can indeed express it as a weighted average over Brownian trajectories $y(\tau)=y_1+B(\tau)$ where the weight is a Hida distribution $I\epsilon(S)^*$:

$$\int_{\substack{y(o)=y_1\\ y(t)=y_2}} \prod_{\tau} dy(\tau)\, e^{\frac{i}{\hbar} S[y]} F[y] = < I, F[y] >,$$

where the functional I is a Hida distribution $I\epsilon(S)^*$, given in terms of its T-transform

$$(TI)(f) = e^{-\frac{1}{2} \|f\|^2_{L^2([0,t])}} <\alpha|\, e^{i\lambda Qf(t)} U^{(\lambda \dot{f})}(t,0)\, e^{-i\lambda Qf(0)} |\,\beta>$$

with $U^{(g)}$ the unitary propagator obeying

$$\frac{\partial}{\partial t} U^{(g)}(t,s) = -i(H+g(t)Q)\, U^{(g)}(t,s) \text{ and the initial condition } U^{(g)}(s,s) = Id,$$

whenever the matrix element $<\alpha|\,\beta>$ is a U-functional [FPS].

We note that in earlier approaches the restrictions on the potential V were limitations of analytic perturbation theory whereas here analyticity is required only with respect to the "small" perturbation $\lambda\dot{f}(\tau)Q$; for quantum models in a bounded domain this perturbation is even bounded and we obtain matrix elements that are entire of order one.

VI. - Quantum Field Theory through Dirichlet Forms.

This is another subject that has been on the physicists' back burner for a long time for lack of good mathematical tools. Its aim is to give a non-perturbative description of quantum (field) dynamics, i.e. of the self-adjoint generator H of time translations, through the knowledge of its ground state. In a more or less heuristic fashion the history of this attempt goes back thirty years:

1960	Coester and Haag [CH], Araki [A]
1975	Albeverio and Hoegh-Krohn [AH1]
1988ff.	Hida, Potthoff and Streit [HPS], Albeverio, Roeckner [AHP][AR], Razafimanantena[R],.......

The idea is best illustrated by taking a look at it in non-relativistic quantum mechanics [AHS], in particular by noting the form that the usual Schroedinger Hamiltonian takes in the following two equivalent representations

	Schroedinger Representation	Ground State Representation
Hilbert Space	$L^2(\mathbf{R}^n, d^n x)$	$L^2(\mathbf{R}^n, \psi_0{}^2(x)d^n x)$
Hamiltonian	$H = - \Delta_x + V(x)$	$\widetilde{H} = \nabla^* \nabla.$

Here the "ground state" ψ_0 is the eigenfunction associated with the minimal eigenvalue (zero without loss of generality) of H, and the remarkable feature of the ground state representation is that the Hamiltonian $\widetilde{H}$, contrary to its counterpart H is not defined through a perturbation ansatz, the dynamics is encoded in the measure $d\mu=\psi_0{}^2(x)d^n x$. Instead of the perturbative ansatz $H=\Delta+V$, one thus defines $\widetilde{H}$ through an energy form

$$(f,\widetilde{H}f) \equiv \epsilon(f) = \int |\nabla f|^2 \, d\mu,$$

admitting such measures that the form ϵ is closable so that there is a unique self adjoint operator $\widetilde{H}$ associated with it. As an extra bonus we are assured that $\widetilde{H}$ is the generator of a diffusion process which, under mild conditions, solves a stochastic differential equation [Fu], see [AR] [AHP] for the infinite dimensional case.

Now for quantum field theory the idea evidently is to try the same kind of ansatz

$$\epsilon(F) = \int |\nabla F|^2 \, d\nu.$$

However, here, in the infinite dimensional case there is no flat reference measure

$$d\nu \neq \rho(x) d^\infty x.$$

Even worse, Haag's theorem tells us that for any two field theories the corresponding measures will be singular with respect to each other, in short there is no Radon-Nikodym derivative ρ.

That is unless we resolve both quandaries at the same time by

(a) using the White Noise measure μ as reference measure and

(b) admitting (positive) Hida distributions Φ as generalized Radon-Nikodym derivatives.

This then leads to

$$\epsilon(F) = \int |\nabla F|^2 \, d\nu_\Phi = \langle \Phi, |\nabla F|^2 \rangle \quad \text{with } D(\epsilon)=(S).$$

As in the finite dimensional case we admit such $\Phi \epsilon (S)^*$ that make the form ϵ closable so that

$$\bar{\epsilon}(F) = || H^{\frac{1}{2}} F ||_{L^2(d\nu)}$$

for a unique self adjoint H.

Such energy forms were first constructed in [HPS], but of course the question remains

Q: Which of the known quantum field theories can we describe in this way?

Partial answers were given in [AHP], along with more on the forms and the associated processes. More recently, a general criterion was found which we shall now sketch:

If the quantum field ϕ obeys a so-called "ϕ-bound"

$$\pm\phi(f) \leq \alpha H + \beta \, |H_o{}^p f|^2 + \gamma$$

in the sense of quadratic forms for some real valued constants α,β,p, then there is a positive Hida distribution $\Phi\epsilon(S)^*$ such that $\nu=\nu_\Phi$ and the vacuum expectation values of the field are given by

$$E(\phi(f_1)....\phi(f_n)) = <\Phi, <X,f_1>.....<X,f_1> >. \qquad [PS2]$$

It is worth pointing out that such ϕ-bounds arise naturally in quantum field theory, see e.g. [GJ], so much so indeed that virtually all Bose quantum fields so far constructed may be expressed in terms of White Noise.

*

Clearly, the above does not exhaust the actual or potential applications of White Noise analysis in theoretical physics. As examples only let us mention the work of P. L. Chow on hydrodynamics, the modelling of fluid flow in porous media by Lindstroem, Oksendal and Uboe [LOU], studies of selfinteracting Brownian paths by H. Watanabe, and the investigations of Yoshizawa, Hida, Okamoto, Obata on the group $O(\infty)$ acting in (L^2). Together with the variational questions studied by Hida and SiSi this might well mark one of the future growth directions along the interface between White Noise Analysis and physics.

References:

[A] H. Araki: Hamiltonian Formalism and the Canonical Commutation Relations in Quantum Field Theory. J. Math. Phys. 1, 492 (1960)

[AH1]S. Albeverio, R. Hoegh-Krohn: Dirichlet Forms and Diffusion Processes on Rigged Hilbert Spaces. Z. Wahrsch. verw. Gebiete 40, 1 (1975).

[AH2]S. Albeverio, R. Hoegh-Krohn: "Mathematical Theory of Feynman Path Integrals". Springer LNM no. 523, Berlin, 1976.

[AHPS. Albeverio, T. Hida, J. Potthoff, L. Streit: The Vacuum of the Hoegh-Krohn Model as a Generalized White Noise Functional. Phys. Lett.B 217, 511 (1989)

S. Albeverio, T. Hida, J. Potthoff, M. Roeckner, L. Streit: Dirichlet Forms in Terms of White Noise Analysis I: Construction and QFT Examples. Rev. Math.Phys.1, 291 (1990).

S. Albeverio, T. Hida, J. Potthoff, M. Roeckner, L. Streit: Dirichlet Forms in Terms of White Noise Analysis II: Closability and Diffusion Processes. Rev.Math. Phys.1, 313 (1990).

[AHS]S. Albeverio, R. Hoegh-Krohn, L. Streit: J. Math. Phys. 18, 907 (1977).

[AR] S. Albeverio, M. Roeckner: Classicasl Dirichlet forms on topological Vector spaces - Construction of an Associated Diffusion Process. Prob. Th. Rel. Fields 83, 405 (1989).

[CH] F. Coester, R. Haag: Phys. Rev. 117, 1137 (1960).

[FPS] M. de Faria, J. Potthoff, L. Streit: The Feynman Integrand as a Hida Distribution, BiBoS preprint no.432.

[Fu] M. Fukushima: *Dirichlet Forms and Markov Processes.* Kodansha - North-Holland, 1980.

[GJ] J. Glimm, A. Jaffe: *Quantum Physics. A Functional Integral Point of View.* Springer, Berlin, 1981.

[Hi] T. Hida: *Brownian Motion.* Springer, Berlin, 1980.

[HPS]T. Hida, J. Potthoff, L. Streit: Dirichlet Forms and White Noise Analysis. Comm.math. Phys. 166, 235 (1988).

[Jo] J. B. Johnson: Phys. Rev. 32, 97 (1928).

[KPY]H. H. Kuo, J. Potthoff, J.-a. Yan: Continuity of Affine Transformations of White Noise Test Functionals and Applications. Preprint 1990.

[K] Yu. G. Kondratiev: Nuclear Spaces of Entire Functions in Problems of Infinite Dimensional Analysis. Soviet Math. Dokl. 22, 588 (1980).

[KT] I. Kubo, S. Takenaka: Calculus on Gaussian White Noise I-IV. Proc. Japan Acad. 56, 376,411 (1980); 57, 433 (1981); 58, 186 (1982).

[KY] I. Kubo Y.Yokoi: A Remark on the Space of Testing Random Variables in the White Noise Calculus. Nagoya Math. J. 115, 139 (1989).

[L] Y. J. Lee: Generalized functions on infinite dimensional spaces and its application to white noise calculus. J. Funct. Anal. 82, 429 (1989)

[LOU]T. Lindstrom, B. Oksendal, J. Uboe: Dynamical Systems in Random Media: A White Noise Functional Approach. Dept. of Math. Oslo, 1990.

[M1] P. A. Meyer: Quelques resultats sur le semigroupe d'Ornstein-Uhlenbeck en dimension infinie. In "Theory and Application of Random Fields", G. Kallianpur, ed., Springer, Berlin, 1983.

[M2] P. A. Meyer: Les "fonctions caractéristiques" des distributions sur l'espace de Wiener. Preprint, Sém. de Probabilité, Univ. de Strasbourg.

[Ny] H. Nyquist: Phys. Rev. 32, 110 (1928).

[P] J. Potthoff: On Meyer's Equivalence. Nagoya Math. J. 111, 99 (1988).

[PS1] J. Potthoff, L. Streit: A Characterization of Hida Distributions. To appear in J. Funct. Anal.

[PS2] J. Potthoff, L. Streit: Invariant States on Random and Quantum Fields: ϕ-Bounds and White Noise Analysis. BiBoS preprint no. 422, 1990.

[PS3] J. Potthoff, L. Streit: Generalized Radon-Nikodym Derivatives and Cameron-Martin-Theory, these proceedings.

[PY] J. Potthoff, J.-a.Yan: Some Results about Test and Generalized Functionals of White Noise. To appear in Proc. Singapore Prob. Conf. 1989, L. Y. Chen ed.

[R] E.A. Razafimanantena: Construction of Energy Forms in Terms of White Noise Analysis. Submitted to Stoch. Proc. Appl.

[Sh] I. Shigekawa: Derivatives of Wiener Functionals and Absolute Continuity of Induced Measures. J. Math. Kyoto Univ. 20, 263 (1980).

[Su] H. Sugita: Positive Generalized Wiener Functionals and Potential Theory over Abstract Wiener Spaces. Osaka J. Math. 25, 665 (1988).

[Wa] S. Watanabe: Malliavin's Calculus in Terms of Generalized Wiener Functionals. In "Theory and Application of Random Fields", G. Kallianpur, ed., Springer, Berlin, 1983.

[WN] T. Hida, H. H. Kuo, J. Potthoff, L. Streit (eds.): *White Noise Analysis.* World Scientific, Singapore 1990.

[Y] J.-a. Yan: A Characterization of White Noise Functionals. Preprint, Beijing 1990.

[Yo] Y. Yokoi: Hiroshima Math. J. 20, 137 (1990).

Part II Invited Lectures

The Low Density Limit in the Finite Temperature Case (I)

L.Accardi Y.G. Lu*

Centro Matematico V.Volterra
Dipartimento di Matematica
Universita' di Roma II

ABSTRACT

In this paper, the low density limit for Bose particles in the finite temperature case is studied. We prove that the wave operator at time t/z^2 (z^2 the fugacity) tends, as the density of the initial state goes to zero, and in the sense of the matrix elements with respect to some appropriately chosen collective vectors, to the solution of a stochastic differential equation whose explicit form is given. Here we give a detailed outline of the basic steps of the proof. The full details are contained in [12].

§1. Introduction

One of the basic problems in the theory of dissipative quantum phenomena is to explain the mechanism of dissipation. Usually one assumes, in analogy with the classical situation, that the dissipation in a system is related to its interaction with an outer system (a field, a gas, ...). The problem with these models however is that, while dissipation is in most cases a macroscopic phenomenon, described by few macroscopically measurable quantities (temperature, relaxation times, ...), these models are based on a microscopic description of the reservoir. Therefore there is an excess of information and an insufficient separation between the scales of magnitude relevant for the dissipation phenomena and those which are not. As a consequence of this, even in the simplest models it is practically impossible to read the physics from these models. To overcome these difficulties, one frenquently introduces some limiting situations which evidentiate the scales of magnitudes in which one is interested and eliminate the remaining ones. Several such limiting situations

*On leave of absence from Beijing Normal University

are known: weak coupling, singular coupling , low density, $\cdots$. In the present paper we describe some recent advances in the last problem.

The low density limit in the Boson Fock case has been investigated in [1] where also the physical meaning and the motivations have been explained (cf. also [0]). From these papers one knows how the number processes can be obtained from a quantum Hamiltonian model via a certain limit procedure.

But the physically more meaningful case is the one in which the initial state of the reservoir is a finite temperature rather than the Fock state. The present paper is devoted to this case.

In [0] we have explained the meaning of the, apparently self-contradictory, statement *low density limit in the Fock case*: the fact is that the low density interaction breaks in five pieces, one of which lives entirely in the Fock space (cf. (2.11) and (2.11a) below), and in [1], [0] only this piece of the full interaction was studied. In the presnet paper we confront ourselves with the full low density interaction and we show that new qualitative phenomena arise. In particular, we are able also in this case to deduce a limiting equation, however in general there is no reason to expect the solution of this equation to be unitary, contrarily to what we proved in the Fock case. We strongly suspect that this phenomenon is not intrinsic to the physical model, but depends on our choice of the collective vector (i.e. the collective coherent vectors obtained by applying the collective Weyl operators (2.6) to the cyclic vector (2.7)–c.f. below). We refer to [0] for a motivation of this conjecture.

As in our other papers on the approximation problem, the result follows from a very detailed analysis of the iterated series. Contrarily to the weak coupling case, in the low density limit, this series includes some disconnected diagrams, and this makes its analysis more difficult.

A single algorithm to approximate quantum fields by quantum noises and the Schrödinger (resp. Heisenberg) equation by quantum stochastic diffrential equations, does not exist. This is because the approximation technique strongly depends on:

i) The nature of the field (Boson, Fermion).

ii) The state of the field.

iii) The Hamitonian of the system interacting with the field.

iv) The interaction between the system and the field.

v) Some special properties, like the rotating approximation, ...

The experience accumulated on several models however, suggests the following general strategy (cf. [0]):

1) One writes the Schrödinger equation in interaction representation ((1.3) below).

2) One introduces the collective vectors appropriate for the given interaction (in the present paper–the collective coherent vectors described below).

3) One expands the solution of the Schrödinger equation in the iterated series and considers the matix element of each term with respect to a pair of collective vectors. The technique works when the series converges on a sufficiently rich set of collective vectors (or at least each term of the series is well defined).

4) One tries to separate those terms which are negligible (i.e. give zero contribution) in the limit from the relevent terms.

5) One finds the limit of the relevent terms.

6) The limit (1.13) is usually given as a complicated series. This series has to be identified to the solution of a quantum stochastic differential equation.

7) One has to obtain a uniform estimate for the terms of the series in order to be able to take the limit term by term.

In the present paper, our physical model and essential assumptions are the same as in [1] (except for the replacement of the Fock state with a finite temperature state). We consider a "System+Reservoir" model described by: –the system Hilbert space H_0;

–the one particle reservoir Hilbert space H_1;

–the system Hamiltonian H_S;

–the one particle reservoir Hamiltonian H_R and associated one particle reservoir evolution

$$S_t^0 := \exp(itH_R) \tag{1.0}$$

which is a one parameter unitary group on $B(H_1)$.

Let be given, on the CCR C^*-algebra over H_1 (cf. [11]), a gauge invariant quasi - free state $\varphi^{(z)}$ with fugacity z^2, i.e. for each $f \in H_1$,

$$\varphi^{(z)}(W(f)) \; = \; \exp\Big(-\frac{1}{2} < f, (1+z^2e^{-\frac{1}{2}H\beta})(1-z^2e^{-\frac{1}{2}H\beta})^{-1}f >\Big) \tag{1.1}$$

where H is a self - adjoint operator which in this paper will be supposed to commute with H_R, and $W(f)$ is the Weyl operator with test function f. Moreover up to GNS-construction one can write the left hand side of (1.1) as

$$< \Phi_z, \; W(f) \; \Phi_z >$$

where the Weyl operators act on a Hilbert space $\mathcal{H}_z$ and Φ_z is a cyclic vector for the Weyl algebra. For any Hilbert space $\mathcal{H}$, we shall denote $\Gamma(\mathcal{H})$ the Fock space over $\mathcal{H}$.

As in [1], the interaction between System and Reservoir is of the form

$$V := i(D \otimes A^+(g_0)A(g_1) - D^+ \otimes A(g_1)^+ A(g_0)) \tag{1.2}$$

and the time evolution in the interaction picture is defined by

$$U_t := e^{itH_{\rm fr}} \cdot e^{-itH_{\rm tot}} \tag{1.3}$$

where

$$H_{\rm fr} := H_0 \otimes 1 + 1 \otimes d\Gamma(H_R) \tag{1.4a}$$

is called the free Hamiltonian and

$$H_{\rm tot} := H_{\rm fr} + V \tag{1.4b}$$

the total Hamiltonian.

A simple computation shows that the time evolution U_t is the solution of the following ordinary differential equation

$$\frac{d}{dt} U_t = -iV(t) \cdot U_t \tag{1.5}$$

where $V(t)$ is defined by

$$V(t) := e^{itH_{fr}}\, V\, e^{-itH_{fr}} \tag{1.6}$$

In [1] (also [0]) we have shown that, if
1) g_0 and g_1 have disjoint energy spectra sets, i.e.

$$< g_0,\ S_t^0\, g_1 > \, = \, 0, \qquad\qquad \forall t \in \mathbf{R} \tag{1.7}$$

2) the rotating wave approximation holds:

$$e^{itH_0}\, D\, e^{-itH_0} = e^{-it(\omega_0-\omega_1)}\, D\,, \qquad\qquad \text{for some } \omega_0 \neq \omega_1 \tag{1.8}$$

then, (1.6) can be rewritten as

$$V(t) = i(D \otimes A^+(S_t g_0) A(S_t g_1) - D^+ \otimes A(S_t g_1)^+ A(S_t g_0)) \tag{1.9}$$

where S_t is a unitary group satisfying

$$S_t g_0 = S_t^0\, e^{-it\omega_0} g_0; \qquad\qquad S_t g_1 = S_t^0\, e^{it\omega_1} g_1 \tag{1.10}$$

The physical meaning of these two conditions have been explained at length in [0].

The equation (1.5), (1.9), (1.10) will be our starting point in the present paper independently of the specific examples provided by the conditions (1.7) and (1.8). Moreover, we shall suppose that the unitary group $\{S_t\}_{t\in\mathbf{R}}$ commutes with H_R and that there exists a non-zero subset K of H_1 such that $g_0, g_1 \in K$,

$$\int_{\mathbf{R}} | < f, S_t g > | dt < \infty\,, \quad \forall f, g \in K \tag{1.11}$$

and for each $f, g \in K$, the series

$$\sum_{n=1}^{\infty} z^n \cdot \int_{\mathbf{R}} | < f, S_t \exp(-\beta n H) g > | dt \tag{1.11a}$$

has a positive convergence radius.

Given the condition (1.11), one can define the Hilbert space $\{K, (\cdot|\cdot)\}$ in the same way as [1], i.e. by completing K with the scalar product

$$(f|g) := \int_{\mathbf{R}} < f, S_t g > dt \quad , \forall\, f, g \in K \tag{1.12}$$

In analogy with [1], in this paper we shall investigate the limit

$$\lim_{z\to 0} < u \otimes W\Big(z \int_{S/z^2}^{T/z^2} S_u f du\Big)\Phi_z, U_{t/z^2} v \otimes W\Big(z \int_{S'/z^2}^{T'/z^2} S_u f' du\Big)\Phi_z > \tag{1.13}$$

It will be proved that the limit (1.13) exists and is equal to

$$< u \otimes W(\chi_{[S,T]} \otimes f)\Psi, U(t)v \otimes W(\chi_{[S',T']} \otimes f')\Psi > \tag{1.14}$$

where the scalar product in (1.14) is meant in the space of a Fock quantum Brownian motion (cf. [0], [1]) and $U(t)$ is a unitary Markovian cocycle satisfying a quantum stochastic differential equation (q.s.d.e.). More precisely, the main result of the present paper is the following:

$THEOREM(1.1)$ For each $f, f', g_0, g_1 \in K$, $u, v \in H_0$, $D \in B(H_0)$, $S, T, S', T' \in \mathbf{R}$, $t \geq 0$, if t, g_0, g_1 satisfying the condition

$$t < \frac{1}{16\,\|D\|}\,; \qquad \|g\|_S^2 := \Big(\max_{g,g' \in \{g_0, g_1, e^{-\frac{1}{2}\beta H} g_0, e^{-\frac{1}{2}\beta H} g_1\}} \int_{-\infty}^{\infty} | < g, S_t g' > | dt \Big)^2 < \frac{1}{16\,\|D\|} \tag{1.15}$$

the low density limit (1.13) exsits and is equal to (1.14) where, $U(t)$ is the solution of the q.s.d.e.

$$U(t) = 1 + \sum_{\varepsilon \in \{0,1\}} \int_0^t \Big[D_1(\varepsilon) \otimes dN_s(g_\varepsilon, g_{1-\varepsilon}) +$$

$$+ D_2(\varepsilon) \otimes dN_s(g_\varepsilon, g_\varepsilon) + \big(D_3(\varepsilon) + D_\varepsilon < g_{1-\varepsilon}, e^{-\frac{1}{2}\beta H} g_\varepsilon > \big) \otimes 1 ds \Big] U(s) \tag{1.16}$$

on $H_0 \otimes \Gamma\Big(L^2(\mathbf{R}) \otimes \{K, (\cdot|\cdot)\}\Big)$ and

(i) Ψ is the vacuum of $\Gamma\Big(L^2(\mathbf{R}) \otimes \{K, (\cdot|\cdot)\}\Big)$;

(ii) $N_s(g, g') := N(\chi_{[0,s]} \otimes |g >< g'|)$ is a number process

(iii) For any operator T on $L^2(\mathbf{R}) \otimes K$, $N(T)$ denotes the number operator generated by T and characterized, for T self-adjoint, by the condition

$$e^{it\, N(T)}\, W(\xi)\Psi \;=\; W(e^{itT}\xi)\Psi\,, \qquad \forall\, \xi \in L^2(\mathbf{R}) \otimes K$$

and extended by complex linearity to arbitrary T.

(iv) $D_1(\varepsilon)$, $D_2(\varepsilon)$ and $D_3(\varepsilon)$ are given by (4.15), (4.16) and (4.7) respectively.

REMARK Since in the Hilbert space $\Gamma(L^2(\mathbf{R}) \otimes K)$, the set $\{W(\chi \otimes f)\Psi;\ \chi \in L^2(\mathbf{R}),\ f \in K\}$ is a total subset, we know that the linear functions

$$X \mapsto < u \otimes W(\chi_{[S,T]} \otimes f)\Psi, Xv \otimes W(\chi_{[S',T']} \otimes f')\Psi >$$

can separate (bounded) operators. Therefore Theroem (1.1) gives a stochastic process as the low density limit of the time evolution of the original Hamiltonian model.

§2. Preliminaries

A simple and important consequence of condition (1.11) is that (cf. Lemma (2.1) in [1]): for each $f, f' \in K$, $S, S', T, T' \in \mathbf{R}$, one has

$$\lim_{z\to 0} < z\int_{S/z^2}^{T/z^2} S_u f du, z\int_{S'/z^2}^{T'/z^2} S_u f' du > = < \chi_{[S,T]}, \chi_{[S',T']} >_{L^2(\mathbf{R})} \cdot \int_{\mathbf{R}} < f, S_t f' > dt \tag{2.1}$$

Using the CCR and the Gaussianity of the state, (2.1) implies that (cf. [1]) for each $n \in \mathbf{N}$, $\{f_k\}_{k=1}^n \subset K$, $\{S_k, T_k\}_{k=1}^n \subset \mathbf{R}$, $\{x_k\}_{k=1}^n \subset \mathbf{R}$, the expression

$$< \Phi_z, W\Big(x_1 z\int_{S_1/z^2}^{T_1/z^2} S_u f_1 du\Big)\cdots W\Big(x_n z\int_{S_n/z^2}^{T_n/z^2} S_u f_n du\Big)\Phi_z >$$

as $z \to 0$, converges uniformly for $\{x_k, S_k, T_k\}_{k=1}^n$ in a bounded set of $\mathbf{R}$,

$$< \Psi, W\Big(x_1\chi_{[S_1,T_1]} \otimes f_1\Big)\cdots W\Big(x_n\chi_{[S_n,T_n]} \otimes f_n\Big)\Psi > \tag{2.2}$$

where, Ψ and $W\Big(\chi_{[S,T]} \otimes f\Big)$ are the vacuum and the Weyl operators of the Fock Browian motion over $L^2(\mathbf{R}) \otimes K$, where K is the Hilbert space described by (1.12).

Denote H_1^ι the conjugate Hilbert space of H_1, i.e.

$$\iota : H_1 \longrightarrow H_1 \quad \iota(\lambda f) := \bar{\lambda}\iota(f) \tag{2.3}$$

$$< \iota(f), \iota(g) >_\iota := < g, f > \tag{2.4}$$

It is well known that up to unitary isomorphism the following identifications take place

$$\Gamma(H_1) = \Gamma(H_1) \otimes \Gamma(H_1^\iota) \tag{2.5}$$

$$W(f) = W(Q_+ f) \otimes W(Q_- f) ; \qquad \forall f \in H_1 \tag{2.6}$$

$$\Phi_z = \Phi \otimes \Phi^\iota \tag{2.7}$$

where Φ, Φ^ι are the vacuum vectors in $\Gamma(H_1)$ and $\Gamma(H_1^\iota)$ respectively and

$$Q_+ := \sqrt{\frac{Q+1}{2}} = \sqrt{\frac{1}{2}\cdot\frac{2}{1 - z^2\exp(-\frac{1}{2}H\beta)}} = \frac{1}{\sqrt{1 - z^2\exp(-\frac{1}{2}H\beta)}} \tag{2.8}$$

$$Q_- := \iota\sqrt{\frac{Q-1}{2}} = \iota\sqrt{\frac{1}{2}\cdot\frac{2\exp(-\frac{1}{2}H\beta)}{1 - z^2\exp(-\frac{1}{2}H\beta)}} = z\iota\sqrt{\frac{\exp(-\frac{1}{2}H\beta)}{1 - z^2\exp(-\frac{1}{2}H\beta)}} =: zQ^- \tag{2.9}$$

Moreover, we have

$$A(f) = A(Q_+ f) \otimes 1 + 1 \otimes A^+(Q_- f) \tag{2.10}$$

So up to unitary isomorphism:

$$A^+(f)A(g) = \Big(A^+(Q_+f)\otimes 1 + 1\otimes A(Q_-f)\Big)\cdot\Big(A(Q_+g)\otimes 1 + 1\otimes A^+(Q_-g)\Big) =$$

$$= A^+(Q_+f)A(Q_+g)\otimes 1 + z\Big(A^+(Q_+f)\otimes A^+(Q^-g) + A(Q_+g)\otimes A(Q^-f)\Big) +$$

$$+z^2 1\otimes A(Q^-f)A^+(Q^-g) \tag{2.11}$$

By the CCR, the last term of the right hand side of (2.11) is equal to

$$A^+(Q^-g)A(Q^-f) + < Q^-f, Q^-g >_\iota$$

therefore the right hand side of (2.11) can be rewritten as

$$A^+(Q_+f)A(Q_+g)\otimes 1 + z\Big(A^+(Q_+f)\otimes A^+(Q^-g) + A(Q_+g)\otimes A(Q^-f)\Big) +$$

$$+z^2\Big(1\otimes A^+(Q^-g)A(Q^-f) + < Q^-f, Q^-g >_\iota\Big)$$

$$=: A_0(f,g) + z\Big(A_{-1}(f,g) + A_1(f,g)\Big) + z^2 A_2(f,g) + z^2 A_{-2}(f,g) \tag{2.11a}$$

where the right hand side of (2.11a) defines the expressions $A_\sigma(f,g)$, $\sigma = 0, \pm 1, \pm 2$. Our starting point in this paper is, as in [1], the iterated solution of (1.5):

$$U_t = \sum_{n=0}^{\infty}(-i)^n \int_0^t dt_1 \cdots \int_0^{t_{n-1}} dt_n V(t_1)\cdots V(t_n) \tag{2.12}$$

Let us now briefly comment upon the difference between the Fock and the finite temperature cases.

In (2.11a), the term $A_0(f,g)$ is the interaction term in the Fock case and the term $z^2 A_{\pm 2}(f,g)$ will not play an important role. In the term $z\Big(A_{-1}(f,g)+A_1(f,g)\Big)$, if there exists some operator acting on the coherent vector $W\Big(z\int_{S/z^2}^{T/z^2} S_u f du\Big)\Phi_z$ or $W\Big(z\int_{S'/z^2}^{T'/z^2} S_u f' du\Big)\Phi_z$, we shall get so many z that the term will go to zero. But the new situation is not the same as that in the Fock case: there it is impossible that all the operators are used to produce scalar products (in the sense explained after formula (3.6) below), but here is possible. For example, in the case $n = 2$, from the term

$$A(S_{t_1}Q_+g_0)\otimes A(Q_-S_{t_1}g_1)\cdot A^+(S_{t_2}Q_+g_0)\otimes A^+(Q_-S_{t_2}g_1) \tag{2.13}$$

one can get the following term

$$< Q_+g_0, S_{t_2-t_1}Q_+g_0 > \cdot < Q_-g_1, Q_-S_{t_2-t_1}g_1 >_\iota \tag{2.14}$$

In the low density limit, this term will give rise to the contribution:

$$\lim_{z\to 0}\int_0^{t/z^2} dt_1 \int_0^{t_1} dt_2 < Q_+g_0, S_{t_2-t_1}Q_+g_0 > \cdot < Q_-g_1, Q_-S_{t_2-t_1}g_1 >_\iota$$

$$= \lim_{z\to 0}\int_0^{t/z^2} dt_1 \int_0^{t_1} dt_2 z^2 < Q_+g_0, S_{t_2-t_1}Q_+g_0 > \cdot < Q^-g_1, Q^-S_{t_2-t_1}g_1 >_\iota$$

$$= \lim_{z\to 0}\int_0^{t} dt_1 \int_{-t_1/z^2}^{0} dt_2 < Q_+g_0, S_{t_2}Q_+g_0 > \cdot < Q^-g_1, Q^-S_{t_2}g_1 >_\iota$$

$$= \int_0^{t} dt_1 \int_{-\infty}^{0} du < g_0, S_u g_0 > \cdot \overline{< e^{-\frac{1}{4}\beta H}g_1, S_u e^{-\frac{1}{4}\beta H}g_1 >} \tag{2.15}$$

which has no counterpart in the purely Fock case. In the case $n \geq 2$, one gets similar terms.

§3. The Negligible Terms and the Uniform Estimate

The section is devoted to
i) isolate those terms which are negligible in the limit;
ii) discuss briefly the uniform estimate.
First of all for each $n \in \mathbf{N}$ and $\varepsilon \in \{0,1\}^n$ we replace, in the product

$$A^+(S_{t_1}g_{\varepsilon(1)})A(S_{t_1}g_{1-\varepsilon(1)})\cdots A^+(S_{t_n}g_{\varepsilon(n)})A(S_{t_n}g_{1-\varepsilon(n)}) \tag{3.1}$$

each product of A^+A by its expression (2.11a). Thus we obtain a sum of 5^n factors:

$$\sum_{\sigma\in\{0,\pm1,\pm2\}^n} A_{\sigma(1)}(S_{t_1}g_{\varepsilon(1)}, S_{t_1}g_{1-\varepsilon(1)})\cdots A_{\sigma(n)}(S_{t_n}g_{\varepsilon(n)}, S_{t_n}g_{1-\varepsilon(n)})z^{\sum_{k=1}^n |\sigma(k)|} \tag{3.2}$$

In order to deal with the limit (1.13) we shall start by investigating the limit of the n–th term of the iterated series (2.12), i.e.

$$\lim_{z\to 0} < u\otimes W\Big(z\int_{S/z^2}^{T/z^2} S_u f du\Big)\Phi_z, (-i)^n \int_0^{t/z^2} dt_1 \cdots \int_0^{t_{n-1}} dt_n V(t_1)\cdots V(t_n)$$

$$v\otimes W\Big(z\int_{S'/z^2}^{T'/z^2} S_u f' du\Big)\Phi_z > \tag{3.3}$$

Because of (1.9), the system part in (3.3) does not depend on t, hence the control of the limit (3.3) is equivalent to the control of its reservoir part, i.e. the limit

$$\lim_{z\to 0} < W\Big(z\int_{S/z^2}^{T/z^2} S_u f du\Big)\Phi_z, \sum_{\sigma\in\{0,\pm1,\pm2\}^n} \int_0^{t/z^2} dt_1 \cdots \int_0^{t_{n-1}} dt_n$$

$$A_{\sigma(1)}(S_{t_1}g_{\varepsilon(1)}, S_{t_1}g_{1-\varepsilon(1)}) \cdots A_{\sigma(n)}(S_{t_n}g_{\varepsilon(n)}, S_{t_n}g_{1-\varepsilon(n)}) z^{\sum_{k=1}^n |\sigma(k)|} W\Big(z\int_{S'/z^2}^{T'/z^2} S_u f' du\Big)\Phi_z > \tag{3.4}$$

As usual, our first step is the change of variables

$$z^2 t_j \hookrightarrow t_j, \quad j = 1, 2, \cdots, n \tag{3.5}$$

under which (3.4) becomes a sum (over $\sigma \in \{0, \pm1, \pm2\}^n$) of

$$\lim_{z\to 0} < W\Big(z\int_{S/z^2}^{T/z^2} S_u f du\Big)\Phi_z, z^{-2}\int_0^{t} dt_1 \cdots z^{-2}\int_0^{t_{n-1}} dt_n$$

$$A_{\sigma(1)}(S_{t_1/z^2}g_{\varepsilon(1)}, S_{t_1/z^2}g_{1-\varepsilon(1)}) \cdots A_{\sigma(n)}(S_{t_n/z^2}g_{\varepsilon(n)}, S_{t_n/z^2}g_{1-\varepsilon(n)}) z^{\sum_{k=1}^n |\sigma(k)|}$$

$$v \otimes W\Big(z\int_{S'/z^2}^{T'/z^2} S_u f' du\Big)\Phi_z > \tag{3.6}$$

Following the general strategy explained in Section 0 of [0] we want to separate, in (3.6), the terms which are negliglible in the limit from those which, in the limit, give a nontrival contribution. To this goal first notice that, if a term of (3.6) contains a factor $A_{\sigma(j)}$ with $\sigma(j) = 2$, i.e. a term of the form $z^2(1 \otimes A^+A)$ (cf. (2.11a)) then this term goes to zero in the limit. The qualitative reason for this is that for the product A^+A there are only 3 possibilities:

(i) both A^+ and A act on the collective coherent vectors;

(ii) both A^+ and A are **used to produce a scalar product** (recall that this means that, in the procedure of normal order, each of them is commuted with some other operator hence, by the CCR, it gives rise to a scalar factor);

(iii) one of the two operators acts on a coherent vector and the other one is used to produce a scalar product.

In the case (i), the actions on the coherent vector give rise to a factor z^3; in the case (ii) and (iii), the arguments of Section 7 in [0] imply that each scalar product **produces a factor** z^2 (notice that if the creator (resp. the annihilator) is used to produce a scalar product with an annihilator (resp. a creator) then they have different time indices). Therefore in all cases a factor z^4 (including the original z^2 factor) will arise. Since the change of variables (3.5) in the corresponding integrals absorbs only a factor z^2 (due to the scaling t/z^2), the resulting term tends to zero as $z \to 0$. Thus denoting II^n_ε the sum of all terms in (3.2) for which some of the indices $\sigma(j)$ is equal to 2, one has $II^n_\varepsilon \to 0$ as $z \to 0$. The remaining sum, denoted I^n_ε, contains only 4^n terms.

Also in this sum several terms will go to zero in the limit because of the following 5 situations:

1) the terms, corresponding to $\sigma(j) = -1$, i.e. those of the form $z(A^+ \otimes A^+)$ may give a non trival contribution only if they are both used to produce a scalar product. In fact: if both act on a coherent vector, then a factor z^4 arises; if $1 \otimes A^+$ acts on a coherent vector, then due to (2.6) and (2.9), a factor z^3 arises; if $A^+ \otimes 1$ acts on a coherent vector, a factor z^2 arises (because Q_- is of order z). For the factors z^4 and z^3 one can apply the same arguments as before. In the third case, i.e. action of $A^+ \otimes 1$, and commutation of $1 \otimes A^+$, the action gives a factor z and the commutation, a factor z. Thus we have again a factor z^3 (including the original z factor) and the preceeding arguments apply. The same conclusion holds for the terms with $\sigma(j) = +1$, i.e. the terms of the form $z(A \otimes A)$.

2) If, in a factor of the form $z(A^+ \otimes A^+)$, one A^+ is used to produce a scalar product with an annihilator taken from a factor of the type $A \otimes A$ at time t_k and the other one with an annihilator from the type of $A \otimes A$ at time t_h, then the arguments of [4a] can be used to show that the resulting scalars will give zero contribution in the limit, unless $h = k$. The same result holds for the terms of the form $z(A \otimes A)$.

3) Some term of the form $z(A^+ \otimes A^+)$ might arise in the processes of normal ordering through the following mechanism: consider product

$$(A^+_{t_p/z^2} A_{t_p/z^2} \otimes 1)(A^+_{t_{p+1}/z^2} A_{t_{p+1}/z^2} \otimes 1) \cdots (A^+_{t_{q-1}/z^2} A_{t_{q-1}/z^2} \otimes 1) z (A^+_{t_q/z^2} \otimes A^+_{t_q/z^2}) \tag{3.7}$$

If in the product (3.7) the annihilator A_{t_p/z^2} is used to produce a scalar product with the creator $A^+_{t_{p+1}/z^2}$, the annihilator A_{t_{p+1}/z^2} is used to produce a scalar product with the creator $A^+_{t_{p+1}/z^2}$, ..., the annihilator A_{t_{q-1}/z^2} is used to produce a scalar product with the creator $A^+_{t_q/z^2}$, then, corresponding to this term, one has the following integral

$$\cdots z^{-2} \int_0^{t_p} dt_{p+1} \cdots z^{-2} \int_0^{t_{q-1}} dt_q \int_0^{t_q} dt_{q+1} < S_{t_p/z^2} g, S_{t_{p+1}/z^2} g' >$$

$$< S_{t_{p+1}/z^2} g, S_{t_{p+2}/z^2} g' > \cdots\cdots < S_{t_{q-1}/z^2} g, S_{t_q/z^2} g' > z A^+_{t_p/z^2} \otimes A^+_{t_q/z^2} \tag{3.8}$$

where, $g, g' \in \{Q_+ g_0, Q_+ g_1\}$. Such term will be called on **approximative** $z(A^+ \otimes A^+)$ term. With the change of variable

$$(t_{j+1} - t_j)/z^2 \hookrightarrow t_{j+1}, \qquad j = p, \cdots, q-1 \tag{3.9}$$

(3.8) becomes

$$\cdots \int_{-t_p/z^2}^0 dt_{p+1} \int_{-t_p/z^2 - t_{p+1}}^0 dt_{p+2} \cdots \int_{-t_p/z^2 - t_{p+1} - \cdots - t_{q-1}}^0 dt_q \int_0^{t_p + z^2(t_{p+1} + \cdots + t_q)} dt_{q+1}$$

$$< g, S_{t_{p+1}} g' >< g, S_{t_{p+2}} g' > \cdots < g, S_{t_q} g' > z A^+_{t_p/z^2} \otimes A^+_{t_p/z^2 + (t_{p+1} + \cdots + t_{q-1})} \tag{3.10}$$

This is essentially (i.e. up to a term of order $o(1)$) the same as the constant

$$\prod_{h=p+1}^{q} \int_{-\infty}^{0} dt_h < g, S_{t_h} g' >$$

times $z(A^+_{t_p/z^2} \otimes A^+_{t_p/z^2})$. Symmetrically one can consider the **approximative** $z(A \otimes A)$ terms.

4) The above considerations imply that, in the normal ordered form of the sum (3.2), the only terms which may survive in the limit are those of the form:

$$C(g,g') \prod_{\alpha} (A^+_{t_\alpha} A_{t_\alpha} \otimes 1) \prod_{h} < Q_+ S_{t_{p_h}} g, Q_+ S_{t_{q_h}} g >< Q^- S_{t_{p_h}} g', Q^- S_{t_{q_h}} g' >_{\iota} \prod_{\beta} < Q^- g', Q^- g' > \tag{3.11}$$

where, $g, g' \in \{g_0, g_1\}$ and $C(g,g')$ is a constant. Notice that the pair of indices (p_h, q_h) is the same in the two scalar products in the second product in (3.11). But among these terms all those such that, for some index h, $q_h \geq p_h + 2$, are negligible because they are not time consecutive in the sense of [1], [2].

5) Finally, the products of the factors of the form $A^+ A \otimes 1$ in (3.11) are of pure Fock type, and therefore they are controlled by the techniques developed in [1].

Having discussed the limit term by term, we need a uniform estimate in order to exchange the order of the limit $z \to 0$ and the sum over $n = 0, 1, \cdots$ in the series expansion of U_{t/z^2} in (1.13). In order to do this one can combine the arguments about the uniform estimate in [1] and [4] since the expression (3.2) is a sum of terms of the types which have been considered (separately) in [1] and [4]. Therefore we know that if the time variable t is in a certain small intival $[0, t_0]$ and $\|g\|_S^2 \leq \frac{1}{16\|D\|}$ then we can change the order of the limit $z \to 0$ and the sum over $n = 0, 1, \cdots$.

§4. The Quantum Stochastic Differential Equantion

In this section we investigate the quantum stochastic differential equation satisfied by the limit of U_{t/z^2}.

In the following, the limit (1.13) will be denoted $< u, G(s) >$.

With the change of variable $z^2 t_1 \hookrightarrow t_1$, the n–th term (1.13) which corresponds to the n–th term of the iterated series of U_{t/z^2}. becomes

$$< u \otimes W\Big(z \int_{S/z^2}^{T/z^2} S_u f du\Big) \Phi_z, z^{-2} \int_0^t dt_1 (-i)^n V(t_1/z^2) \int_0^{t_1/z^2} dt_2 \int_0^{t_2} dt_3 \cdots \int_0^{t_{n-1}} dt_n$$

$$V(t_2) \cdots V(t_n) v \otimes W\Big(z \int_{S'/z^2}^{T'/z^2} S_u f' du\Big) \Phi_z > \tag{4.1}$$

Applying (2.11) to expand $V(t_1/z^2)$ as

$$i\Big(D\otimes A^+(S_{t_2/z^2}Q_+g_0)A(S_{t_2/z^2}Q_+g_1)\otimes 1 + zD\otimes\big(A^+(S_{t_2/z^2}Q_+g_0)\otimes A^+(S_{t_2/z^2}Q^-g_1)+$$
$$+A(S_{t_2/z^2}Q_+g_1)\otimes A(S_{t_2/z^2}Q^-g_0)\big)+$$
$$+z^2D\otimes 1\otimes A^+(S_{t_2/z^2}Q^-g_1)A(S_{t_2/z^2}Q^-g_0)+z^2D\otimes 1\otimes 1 < Q^-g_0, Q^-g_1 >_\iota - \text{c.c.}\Big) \quad (4.2)$$

we notice the following:

1) Replacing $V(t_1/z^2)$ in (4.1) by $z^2D\otimes 1\otimes 1 < Q^-g_0, Q^-g_1 >_\iota$, one has the limit

$$< u\otimes W(\chi_{[S,T]}\otimes f)\Psi, D\otimes 1\otimes 1 < g_0, e^{-\frac{1}{2}H}g_1 >_\iota \int_0^t dt_1 U(t_1)v\otimes W(\chi_{[S',T']}\otimes f')\Psi > \quad (4.3)$$

2) Replacing $V(t_1/z^2)$ in (4.1) by $z^2D\otimes 1\otimes A^+(S_{t_2/z^2}Q^-g_1)A(S_{t_2/z^2}Q^-g_0)$ and expanding the product $V(t_2)\cdots V(t_n)$ one finds that it is a type II term so it tends to 0.

3) Replacing $V(t_1/z^2)$ in (4.1) by $zD\otimes A^+(S_{t_2/z^2}Q_+g_0)\otimes A^+(S_{t_2/z^2}Q^-g_1)$, then the operator $A^+\otimes A^+$ surely acts on a collective coherent vector. Therefore the arguments in Section 3 show that it tends to zero.

4) Replacing $V(t_1/z^2)$ in (4.1) by $zD\otimes A(S_{t_2/z^2}Q_+g_1)\otimes A(S_{t_2/z^2}Q^-g_0)$, if some annihilator acts on the coherent vector $W\big(z\int_{S'/z^2}^{T'/z^2} S_u f' du\big)\Phi_z$, then the arguments in Section 3 show that it tends to zero. Therefore the operator $A\otimes A$ (which can be in approximative form in the sense defined after (3.8)) should be used to produce a scalar product with two creators from the expansion of $V(t_2)\cdots V(t_n)$. If the two creators have different time indices then the limit, as $z\to 0$, is equal to zero from the arguments in Section 3. If the two creators have the same time indices, it is surely of the form of $A^+(S_{t_j/z^2}Q_+g)\otimes A^+(S_{t_j/z^2}Q^-g')$, where $j=2,\cdots,n$. In the case of $j=2$ a simple computation shows that the limit (4.1), in which $V(t_1/z^2)$ has been replaced by $zD\otimes A(S_{t_2/z^2}Q_+g_1)\otimes A(S_{t_2/z^2}Q^-g_0)$ is equal to

$$< u\otimes W(\chi_{[S,T]}\otimes f)\Psi, \sum_{\epsilon(1)\in\{0,1\}} DD_{\epsilon(1)}\otimes 1\otimes 1 \int_{-\infty}^0 du < g_1, S_t g_{\epsilon(1)} >< g_0, S_t e^{-\frac{1}{2}\beta H} g_{1-\epsilon(1)} >_\iota$$
$$\int_0^t dt_1 U(t_1)v\otimes W(\chi_{[S',T']}\otimes f')\Psi > \quad (4.4)$$

If $j>2$, the arguments of Section 3 imply that this $A\otimes A$ term should be in the approximative form. Up to a constant this is essentially the same as the case of $j=2$. Summing up we know that the limit of

$$\sum_{n=1}^{\infty}\sum_{\epsilon\in\{0,1\}} < u\otimes W\Big(z\int_{S/z^2}^{T/z^2} S_u f du\Big)\Phi_z, z^{-1}\int_0^t dt_1 D_\epsilon\otimes A(S_{t_1/z^2}Q_+g_{1-\epsilon})\otimes A(S_{t_1/z^2}Q_-g_\epsilon)$$

$$(-i)^{n-1}\int_0^{t_1/z^2}dt_2\int_0^{t_2}dt_3\cdots\int_0^{t_{n-1}}dt_nV(t_2)\cdots V(t_n)v\otimes W\Big(z\int_{S'/z^2}^{T'/z^2}S_uf'du\Big)\Phi_z> \tag{4.5}$$

is equal to

$$\sum_{\varepsilon\in\{0,1\}}\int_0^t dt_1 <u,D_3(\varepsilon)G(t_1)> \tag{4.6}$$

where,

$$D_3(\varepsilon):=\sum_{k=1}^{\infty}\sum_{\sigma\in\{0,1\}^k}D_\varepsilon D_{\sigma(k)}\cdots D_{\sigma(1)}\int_{-\infty}^0 ds_1\cdots\int_{-\infty}^0 ds_k <g_{1-\varepsilon},S_{s_1}g_{\sigma(1)}>$$

$$<g_{1-\sigma(1)},S_{s_2}g_{\sigma(2)}>\cdots<g_{1-\sigma(k-1)},S_{s_k}g_{1-\sigma(k)}>\cdot\overline{<g_\varepsilon,S_{s_k+\cdots+s_1}g_{\sigma(k)}>} \tag{4.7}$$

5) Now we consider the term replacing $V(t_1/z^2)$ in (4.1) by

$$D\otimes A^+(S_{t_1/z^2}Q+g_0)A(S_{t_1/z^2}Q+g_1)\otimes 1 \tag{4.8}$$

i.e.

$$<u\otimes W\Big(z\int_{S/z^2}^{T/z^2}S_ufdu\Big)\Phi_z,z^{-2}\int_0^t dt_1D\otimes A^+(S_{t_1/z^2}Q+g_0)A(S_{t_1/z^2}Q+g_1)\otimes 1$$

$$(-i)^{n-1}\int_0^{t_1/z^2}dt_2\int_0^{t_2}dt_3\cdots\int_0^{t_{n-1}}dt_nV(t_2)\cdots V(t_n)v\otimes W\Big(z\int_{S'/z^2}^{T'/z^2}S_uf'du\Big)\Phi_z> \tag{4.9}$$

The treatment of this term is the same as that in the Section 6 of [1]. In fact the creator $A^+(S_{t_2/z^2}Q+g_0)$ acts surely on the collective coherent vector, hence (4.9) is equal to

$$<u\otimes W\Big(z\int_{S/z^2}^{T/z^2}S_ufdu\Big)\Phi_z,z^{-1}\int_0^t dt_1\int_{S/z^2}^{T/z^2}<S_uf,S_{t_1/z^2}Q+g_0>D\otimes A(S_{t_1/z^2}Q+g_1)\otimes 1\cdot$$

$$(-i)^{n-1}\int_0^{t_1/z^2}dt_2\int_0^{t_2}dt_3\cdots\int_0^{t_{n-1}}dt_nV(t_2)\cdots V(t_n)v\otimes W\Big(z\int_{S'/z^2}^{T'/z^2}S_uf'du\Big)\Phi_z> \tag{4.10}$$

The annihilator $A(S_{t_2/z^2}Q+g_1)$ plays a role in the following two ways:

(i) acting on coherent vector;

(ii) being used to produce a scalar product with a creator which comes from the $A^+A\otimes 1$ in the expansion of $V(t_2)$.

Thus we obtain a sum of two terms

$$<u\otimes W\Big(z\int_{S/z^2}^{T/z^2}S_ufdu\Big)\Phi_z,D\otimes 1\otimes 1\int_0^t dt_1\int_{S/z^2}^{T/z^2}<S_uf,S_{t_1/z^2}g_0>du$$

$$\int_{S'/z^2}^{T'/z^2} < S_{t_1/z^2}g_1, S_v f' > dv U_{t_2/z^2} v \otimes W\Big(z\int_{S'/z^2}^{T'/z^2} S_u f' du\Big)\Phi_z > \tag{4.11}$$

and

$$< u \otimes W\Big(z\int_{S/z^2}^{T/z^2} S_u f du\Big)\Phi_z, \int_0^t dt_1 \int_{S/z^2}^{T/z^2} < S_u f, S_{t_1/z^2} g_0 > du$$

$$z^{-2}\int_0^{t_1} dt_2 \sum_{\varepsilon(1)\in\{0,1\}} DD_{\varepsilon(1)} \otimes 1 \otimes 1 < S_{t_1/z^2}g_1, S_{t_2/z^2}g_{\varepsilon(1)} > dv \cdot \frac{1}{z} 1 \otimes A(S_{t_2/z^2}g_{1-\varepsilon(1)}) \otimes 1$$

$$(-i)^{n-2}\int_0^{t_2/z^2} dt_3 \int_0^{t_3} dt_4 \cdots \int_0^{t_{n-1}} dt_n V(t_3)\cdots V(t_n) v \otimes W\Big(z\int_{S'/z^2}^{T'/z^2} S_u f' du\Big)\Phi_z > \tag{4.12}$$

Notice that up to the factor

$$z^{-2}\int_0^{t_1} dt_2 \sum_{\varepsilon(1)\in\{0,1\}} D_{\varepsilon(1)} \otimes 1 \otimes 1 < S_{t_1/z^2}g_1, S_{t_2/z^2}g_{\varepsilon(1)} > dv \tag{4.13}$$

(4.12) is of the same type as (4.10) and (4.13) tends, as $z \to 0$, to a certain limit. Thus we can repeat above arguments and prove that the summation of (4.9) when n runs over $1, 2, \cdots$ and its symmetric part (the c.c. part in (4.2)) converges to

$$\sum_{\varepsilon\in\{0,1\}} \sum_{n=0}^{\infty} \sum_{\sigma\in\{0,1\}^n} < D_\varepsilon^+ D_{\sigma(n)}^+ \cdot \ldots \cdot D_{\sigma(1)}^+ D_\varepsilon^+ u, G(s) >$$

$$(g_{1-\varepsilon}|g_{\sigma(1)})_- \cdot (g_{1-\sigma(1)}|g_{\sigma(2)})_- \cdot \ldots (g_{1-\sigma(n)}|g_\varepsilon)_-$$

$$\chi_{[S,T]}(s) \cdot (f|g_\varepsilon) \cdot \chi_{[S',T']}(s) \cdot (g_{1-\varepsilon}|f')$$

$$+ \sum_{\varepsilon\in\{0,1\}} \sum_{n=0}^{\infty} \sum_{\sigma\in\{0,1\}^n} < D_{1-\varepsilon}^+ D_{\sigma(n)}^+ \cdot \ldots \cdot D_{\sigma(1)}^+ D_\varepsilon^+ u, G(s) >$$

$$(g_{1-\varepsilon}|g_{\sigma(1)})_- \cdot (g_{1-\sigma(1)}|g_{\sigma(2)})_- \cdot \ldots (g_{1-\sigma(n)}|g_{1-\varepsilon})_-$$

$$\chi_{[S,T]}(s) \cdot (f|g_\varepsilon) \cdot \chi_{[S',T']}(s) \cdot (g_\varepsilon|f') \tag{4.14}$$

where, $\sigma(0) := \varepsilon$.

Now denote

$$D_1(\varepsilon) := D_\varepsilon + \sum_{n=0}^{\infty} \sum_{\sigma\in\{0,1\}^n} D_\varepsilon D_{\sigma(1)} \cdot \ldots \cdot D_{\sigma(n)} D_{1-\varepsilon}$$

$$(g_{1-\varepsilon}|g_{\sigma(1)})_- \cdot (g_{1-\sigma(1)}|g_{\sigma(2)})_- \cdot \ldots \cdot (g_{1-\sigma(n)}|g_{1-\varepsilon})_- \tag{4.15}$$

$$D_2(\varepsilon) := \sum_{n=0}^{\infty} \sum_{\sigma \in \{0,1\}^n} D_\varepsilon D_{\sigma(1)} \cdot \ldots \cdot D_{\sigma(n)} D_\varepsilon$$

$$(g_{1-\varepsilon}|g_{\sigma(1)})_- \cdot (g_{1-\sigma(1)}|g_{\sigma(2)})_- \cdot \ldots (g_{1-\sigma(n)}|g_\varepsilon)_- \tag{4.16}$$

and summing up above discussion, one knows that $< u, G(t) >$ satisfies the following integral equation:

$$< u, G(t) > = < u, G(0) > + \sum_{\varepsilon \in \{0,1\}} \int_0^t ds$$

$$\Big(\chi_{[S,T]}(s) \cdot \chi_{[S',T']}(s) \cdot (g_\varepsilon|f') \cdot (f|g_\varepsilon) < D_1^+(\varepsilon)u, G(s) > +$$

$$+ \chi_{[S,T]}(s) \cdot \chi_{[S',T']}(s) \cdot (f|g_\varepsilon)_- \cdot (g_{1-\varepsilon}|f') \cdot < D_2^+(\varepsilon)u, G(s) > +$$

$$+ < \big(D_3(\varepsilon) + D_\varepsilon < g_{1-\varepsilon}, e^{-\frac{1}{2}\beta H} g_\varepsilon >\big)^+ u, G(s) > \Big) \tag{4.17}$$

On the other hand, if we define

$$< u, F(t) > := < u \otimes W(\chi_{[S,T]} \otimes f)\Psi, U(t)v \otimes W(\chi_{[S',T']} \otimes f')\Psi > \tag{4.18}$$

where $U(t)$ is the (unique) solution of the quantum stochastic differential equation (1.16), then a simple computation shows that $< u, F(t) >$ satisfies the integral equation (4.17). The uniqueness of the solution of the quantum stochastic differential equation (1.16) implies that (4.17) has a unique solution. Therefore we obtain our conclusion.

REFERENCES

[0]. L. Accardi, Y.G.Lu, A. Frigerio and R. Alicki: An inviatation to the weak coupling limit and the low density limit, to appear in: Quantum probability and related fields VI. World Scientific.

[1]. L. Accardi, Y.G.Lu: The number process as low density limit of Hamiltonian models, to appear in Commun. Math. Phys.

[2]. L. Accardi, A. Frigerio, Y.G.Lu: The weak coupling limit as a functinal central limit theorem, Commun. Math. Phys. (**131**) 537–570.

[3]. L. Accardi, Y.G.Lu: On the low density limit of Boson Models. Lect. Notes in Math. (**1442**), 17–53.

[4]. L. Accardi, Y.G.Lu: The weak coupling limit for nonlinear interactions. *Stochastic Processes, Physics and Geometry* 1–26. World Scientific (1990).

[4a]. Y.G.Lu: The weak coupling limit for quadratic interactions, submitted to J. Math. Phys.

[5]. R.Dümcke: The low density limit for n-level systems, Lect. Notes in Math (**1136**).

[6]. R.Dümcke: The low density limit for an N-level system interacting with a free Bose or fermi gas, Commun. Math. Phys. (**97**). 331–359.

[7]. A. Frigerio, H. Maassen: Quantum Poisson processes and dilations of dynamical semigroups. Prob. Th. Rel. Fields. (**83**), 489–508.

[8]. H. Grad: Principles of the kinietic theory of Gases. Handbuch der Physik, vol. 12, Springer (1958).

[9]. R. L. Hudson, K. R. Pathasarathy: Quantum Ito's formula and stochastic evolution. Commun. Math. Phys. (**93**), 301-323 (1984).

[10]. P. F. Palmer: Thesis, Oxford University.

[11]. D.Petz: An invitation to the C^*-algebra of the canonical commutation relation. Leuven Notes in Math. and Theor. Phys., eds. Fannes M. Verbeure A., Leuven Univ. Press 1990.

[12]. L. Accardi, Y.G.Lu: The Low Density Limit in the Finite Temperature case, submitted to *Nagoya Mathematical Journal.*

DIFFUSION ON p - ADIC NUMBERS

by

Sergio Albeverio* ** and Witold Karwowski ** #

Abstract

We give a direct construction of a random walk with continuous time and state space the field of p - adic numbers.

1. Introduction

The mathematics of p-adic numbers is a classical domain of number theory, originated with Hensel, at the end of last century, see e.g. [Ko], [Sch]. p-adic numbers form a complete metric commutative field Q_p, with respect to a suitable non-archimedean norm, which topologically is a locally compact totally discontinuous space with the cardinality of the continuum, containing the (uncomplete) field Q of rationals as a closed subset (the ring $\mathbb{Z}$ is totally bounded in Q_p). Very little has been undertaken to study specific probabilistic structures on Q_p. Our starting point was precisely to construct directly a diffusion process on Q_p (continuous time, state space Q_p), by giving a system of transition probabilities. As we shall show in this paper this is indeed possible. We start out by determining first suitable transition probabilities from a ball centered at the origin to equidistant balls of the same radius, using translation invariance; again using translation invariance we extend then these transition probabilities to transition probabilities from any starting ball. The final step to get a process running on points is to let the initial ball shrink to its center. After our work was finished we became aware of [BrO], in which the relation between Q_p and a certain regular tree is exploited to yield, by constructing a process on that tree, a process on Q_p. By a computation involving p-adic Fourier transforms the authors of [BrO] compute the absolute probabilities of the process, which are essentially identifiable with the ones of our process. We think however our method of construction and the way we determine the absolute probabilities maintain a certain interest being direct and avoiding use of p-adic Fourier transforms. Moreover our construction is the starting point for further studies, involving long time asymptotics [AK]. Before we start describing our work, let us mention some further relations of our construction with other studies. [Se], [Ca] (see also e.g. [FiT] and references therein) have studied discrete potential theory on certain special trees, exploiting algebraic - combinatorial means. It would be interesting to find the relations between the process constructed in our work and those

* Department of Mathematics, Ruhr-Universität, D4630 Bochum 1(FRG); BiBoS; CERFIM (Locarno)

** SFB 237 Essen - Bochum - Düsseldorf

Institute of Theoretical Physics, University of Wroclaw, Wroclaw (Poland)

of [Ca] (see also [Gr]). For further references on stochastic processes on trees see also e.g. [Ly1,2], [Pr], [Sa] and references therein. It must be underlined that in recent years motivations for studying random processes and fields on trees have been increased by the relevance ultrametric structures have received in physics, particularly solid state physics, see e.g. [DoS], [GrWH], [OgS], [Schr], [RaT], [Sa] and references therein. To quote just one particularly important example let us mention the study of ultrametric structures in spin glasses and hierarchical trees (with applications also to optimization theory, neural networks and biological structures). Of course there should be and there are relations between stochastic processes on p-adics and quantum theory on p-adics, although these relations have not yet been studied in details. For quantum theory on p-adics (and more generally physics on p-adics) see e.g. [VlV], [Pa], [EU], [BeC], [Me] and references therein. We hope our results will also stimulate further studies of these relations. Our report is a preliminary one – for more details and complements see [AK].

2. The construction of the diffusion

Let $p > 1$ stand for a prime number. A p-adic number x can be defined through the formal power series

$$x = \sum_{i=N}^{\infty} \gamma_i p^i \tag{1}$$

where N is an integer ($i.e.$, $N \in \mathbb{Z}$), and γ_i assume values

$$\gamma_i = 0, 1, ..., p-1.$$

With the appropriate definition of addition and multiplication Q_p , the set of all p-adic numbers, becomes a field [Ko], [Sch].
Let x be given by (1) and i_0 the smallest value of i such that $\gamma_{i_0} \neq 0$. Then we define

$$\|x\|_p = p^{-i_0}. \tag{2}$$

It is well known that the map $x \to \|x\|_p$ defines a norm in Q_p. This norm has in addition the non-Archimedian triangle property

$$\|x+y\|_p \leq \max\{\|x\|_p, \|y\|_p\} \tag{3}$$

Q_p with this norm is a complete separable metric locally compact totally discrete space (with the cardinality of the continuum). The series (1) is convergent to x in $\| \quad \|_p$- norm. Of course $Q \subset Q_p$ densely.
Let $a \in Q_p$ and $M \in \mathbb{Z}$. The set

$$K(a, p^M) = \{x \in Q_p; \|x - a\|_p \leq p^M\} \tag{4}$$

is called a sphere of radius p^M centered at a.
As an immediate consequence of the non-Archimedian property we have that if

$$x \in K(a,p^M) \text{ then } K(a,p^M) = K(x,p^M).$$

We also have that if

$$a = \sum_{j=-m}^{\infty} \alpha_{-M+j} p^{-M+j}$$

then the sphere $K(a,p^M)$ is completely determined by the numbers $\alpha_{-(M+m)}, \alpha_{-(M+m-1)}, \ldots, \alpha_{(-M+1)}$. This justifies the notation

$$K(a,p^M) \equiv \{\alpha_{-(M+m)}, \ldots \alpha_{-(M+1)}\} \tag{5}$$

Note that

$$\{\alpha_{-(M+m)}, \ldots, \alpha_{-(M+1)}\} = \{0, \alpha_{-(M+m)}, \ldots, \alpha_{-(M+1)}\} \tag{6}$$

and also

$$\{\alpha_{-(M+m)}, \ldots, \alpha_{-(M+2)}\} = \bigcup_{\alpha_{-(M+1)}=0}^{p-1} \{\alpha_{-(M+m)}, \ldots, \alpha_{-(M+1)}\} = K(a, p^{M+1}). \tag{7}$$

Iterating latter formula we conclude that Q_p is a countable union of disjoint spheres of radius p^M, and this is true for any $M \in \mathbb{Z}$.
Let $\mathcal{K}^M = \{K_i^M\}_{i=1}^{\infty}$ be the family of disjoint spheres of radius p^M such that $Q_p = \bigcup_{i=1}^{\infty} K_i^M$.
Our aim is to construct a stationary process with Q_p as state space. We begin by constructing a process with $\mathcal{K}^M$ as state space. This can be done by solving following system of forward and backward Kolmogorov equations

$$\dot{P}_{K_i K_j}(t) = -\tilde{a}(K_j) P_{K_i K_j}(t) + \sum_{\substack{l=1 \\ l \neq j}}^{\infty} \tilde{u}(K_l, K_j) P_{K_l K_j}(t), \tag{8a}$$

$$\dot{P}_{K_i K_j}(t) = -\tilde{a}(K_i) P_{K_i K_j}(t) + \sum_{l=1}^{\infty} \tilde{u}(K_l, K_j) P_{K_l K_j}(t) \tag{8b}$$

$t \in [0,\infty)$ $i,j \in \mathbb{N}$ with the initial condition $P_{K_i K_j}(0) = \delta_{ij}$. $\tilde{a}(K_j)$ is interpreted as intensity of the state K_j and $\tilde{u}(K_i, K_j)$ as the infinitesimal transition probability. To simplify the notation we dropped the index M in K_i^M.

Sometimes it is convenient to write equations (8) in the matrix form. If we put $A = (a_{ij})$ with

$$a_{ii} = -\tilde{a}(K_i), \quad a_{ij} = \tilde{u}(K_i, K_j) \quad i \neq j \tag{9}$$

and

$$P(t) = (P_{ij}(t)) \text{ with } P_{ij}(t) = P_{K_i K_j}(t) \quad i,j \in \mathbb{N} \tag{10}$$

then the equations (8) become

$$\dot{P}(t) = P(t)A, \quad \dot{P}(t) = AP(t) \tag{11}$$

Following facts concerning existence and uniqueness of the solution of (11) and hence of (8) can easily be established, see e.g. [Bha–Re]:

Theorem 1. If the elements $a_{ij}(t)$ are continuous functions of t with

$$a_{ij}(t) \geq 0 \quad i \neq j, \qquad a_{ii}(t) \leq 0 \quad i \in \mathbb{N}, \tag{12}$$

$$\sum_{j=1}^{\infty} a_{ij}(t) \leq 0 \text{ for all } i \in \mathbb{N}, \tag{13}$$

then there is a matrix $P(t)$ such that each $P_{ij}(t)$ is a continuous function with continuous first derivative and

$$\dot{P}_{ij}(t) = \sum_{k=1}^{\infty} P_{ik}(t)a_{kj}, \quad \dot{P}_{ij}(t) = \sum_{k=1}^{\infty} a_{ik}P_{kj}(t) \tag{14}$$

$$P_{ij}(t) \geq 0, \quad P_{ij}(0) = \delta_{ij} \tag{15}$$

$$\sum_{j=1}^{\infty} P_{ij}(t) \leq 1 \text{ for all } i \in \mathbb{N} \tag{16}$$

$$P_{ij}(t+\tau) = \sum_{k=1}^{\infty} P_{ik}(t)P_{kj}(\tau) \tag{17}$$

If moreover $\sum_{j=1}^{\infty} P_{ij}(t) = 1$ for some $i \in \mathbb{N}$ them $P_{ij}(t)$ is unique . □
To proceed further we have to determine the matrix A i.e. the coefficients $\tilde{a}(K_j)$ and $\tilde{u}(K_i, K_j)$. Define the natural p-adic distance between $K_i, K_j, i \neq j$ by $\text{dist}_p(K_i, K_j) = \|x_i - x_j\|_p$ where $x_i \in K_i$ and $x_j \in K_j$. Note that $\|x_i - x_j\|_p$ is independent of the choice of x_i, x_j. We shall assume that $\tilde{u}(K_i, K_j)$ depends only on the distance between K_i, K_j. We shall also require

$$\tilde{a}(K_j^M) = \sum_{\substack{i=1 \\ i \neq j}}^{\infty} \tilde{u}(K_i^M, K_j^M). \tag{18}$$

More precisely let $a(M), M \in \mathbb{Z}$ be a real number such that

a) $a(M) \geq a(M+1)$ (19)

b) $\lim_{M \to +\infty} a(M) = 0, \lim_{M \to -\infty} a(M) = W$, where W is a positive number or $+\infty$.

For any $m \in \mathbb{N}$ we put

$$U(M+m) \equiv a(M+m-1) - a(M+m) \tag{20}$$

and

$$u(M, m) \equiv (p-1)^{-1} p^{-m+1} U(M+m). \tag{21}$$

Then we get by direct computations

Lemma 2 Following relations hold

1) $a(M+m) = (p-1)\sum_{i=m+1}^{\infty} p^{i-1}u(M,i), m \in \mathbb{N}_0 \equiv \{0\} \cup \mathbb{N}$ (22)

2) $u(M+1, m-1) = pu(M,m), m \in \mathbb{N}, m > 1.$ □ (23)

Note that $\text{dist}_p(K_i^M, K_j^M) = p^{M+m}$, for some $m \in \mathbb{N}$ (depending on i, j). We now choose

$$\tilde{u}(K_i^M, K_j^M) \equiv u(M,m), \quad m \in \mathbb{N}. \tag{24}$$

We then have

Lemma 3

$$\tilde{a}(K_j^M) = a(M). \tag{25}$$

Proof. Given $j \in \mathbb{N}$ the number of spheres K_i^M such that $\text{dist}_p(K_i^M, K_j^M) = p^{M+m}$ is $(p-1)p^{m-1}$. Thus by (18) and (25) we have

$$\tilde{a}(K_j^M) = \sum_{\substack{i=1 \\ i \neq j}} \tilde{u}(K_i^M, K_j^M) = (p-1)\sum_{i=1}^{\infty} p^{i-1}u(M,i) = a(M).$$

□

Since $a(M)$ is decreasing it follows from (20), (21) and (18), (24), (25) that the elements a_{ij} defined by (9) satisfy conditions (12), (13). Thus by theorem 1 there exists a solution of (11) or equivalently (8) with properties (14)-(17). In fact we shall find the unique solution satisfying $\sum_{j=1}^{\infty} P_{ij}(t) = 1$.

We begin with the following observation. According to our choice the matrix A is symmetric, $A = A^T$. If $P(t)$ fulfills $\dot{P}(t) = P(t)A$ and $P(t) = P(t)^T$, (^{T} meaning transpose), then $\dot{P}(t) = \dot{P}(t)^T = (P(t)A)^T = A^T P(t)^T = AP(t)$ i.e. $P(t)$ satisfies both forward and backward Kolmogorov equations.

Thus we shall be looking for a symmetric solution of the forward Kolmogorov equations (8a). To simplify notations we shall use following reformulation. Consider

$$\dot{P}_{K_j}(t) = -\tilde{a}(K_j)P_{K_j}(t) + \sum_{\substack{l=1 \\ l \neq j}} \tilde{u}(K_l, K_j)P_{K_l}(t) \quad t \in [0, \infty), j \in \mathbb{N}. \tag{26}$$

If $P_{K_i K_j}(t)$ is a solution of (8a) with the initial condition $P_{K_i K_j}(0) = \delta_{ij}$ then $P_{K_j}(t) \equiv P_{K_i K_j}(t)$, $j \in \mathbb{N}$ is a solution of (26) with the initial condition $P_{K_j}(0) = \delta_{ij}$. And vice versa, a family $\mathcal{P}$ of solutions of (26) such that for any $i \in \mathbb{N}$ there is $P_{K_j}(t) \in \mathcal{P}$ with $P_{K_j}(0) = \delta_{ij}$ defines a solution of (8a) with the initial condition $P_{K_i K_j}(0) = \delta_{ij}$. Thus our original task to find a symmetric solution of (8a) is equivalent to finding the family $\mathcal{P}$ of solutions of (26) such that for any pair $i, j \in \mathbb{N}$ there are $P_{K_l}, Q_{K_k} \in \mathcal{P}$ such that $P_{K_l}(0) = \delta_{il}, Q_{K_k}(0) = \delta_{jk}$ and $P_{K_j}(t) = Q_{K_i}(t)$.

Using the notation given by (5) and taking into account (24) and (25) we can write the system of equations (26) in the following form

$$\dot{P}_{\{\alpha_{-(M+m)},\dots,\alpha_{-(M+1)}\}} = -a(M)P_{\{\alpha_{-(M+m)},\dots,\alpha_{-(M+1)}\}}$$

$$+u(M,1) \sum_{\alpha'_{-(M+1)}\neq\alpha_{-(M+1)}} P_{\{\alpha_{-(M+m)},\dots,\alpha'_{-(M+1)}\}}$$

$$+u(M,2) \sum_{\substack{\alpha'_{-(M+1)} \\ \alpha'_{-(M+2)}\neq\alpha_{-(M+2)}}} P_{\{\alpha_{-(M+m)},\dots,\alpha'_{-(M+2)},\alpha'_{-(M+1)}\}}$$

$$+\dots + u(M,m) \sum_{\substack{\alpha'_{-(M+1)},\dots,\alpha'_{(M+m-1)} \\ \alpha'_{-(M+m)}\neq\alpha_{-(M+m)}}} P_{\{\alpha'_{-(M+m)},\dots,\alpha'_{-(M+1)}\}}$$

$$+\sum_{i=1} u(M,m+i) \sum_{\alpha'_{-(M+m+i)}\neq 0} P_{\{\alpha'_{-(M+m+i)}\}}, m \in I\!N_0.$$

We remark that in the case $m = 0$ the equation simply reads

$$\dot{P}_{K(0,p^M)} = -a(M)P_{K(0,p^M)} + \sum_{i=1} u(M,i) \sum_{\alpha'_{-(M+i)}\neq 0} P_{\{\alpha'_{-(M+i)}\}}.$$

In the last sum we used (7) and the fact that

$$P_{K_i \cup K_j} = P_{K_i} + P_{K_j}.$$

By exactly the same reason we have

$$\dot{P}_{\{\alpha_{-(M+m)},\dots,\alpha_{-(M+1)}\}} = -(a(M) + u(M,1))P_{\{\alpha_{-(M+m)},\dots,\alpha_{-(M+1)}\}}$$

$$+ \sum_{j=1}^{m-1} (u(M,j) - u(M,j+1))P_{\{\alpha_{-(M+m)},\dots,\alpha_{-(M+j+1)}\}}$$

$$+\sum_{i=0}^{\infty} (u(M,m+i) - u(M,m+i+1))P_{M,m+i}, \quad m \in I\!N \tag{27}$$

where

$$P_{M,m+i} \equiv \sum_{\alpha'_{-(M+1)},\dots,\alpha'_{-(M+m+i)}} P_{\{\alpha'_{-(M+1)},\dots,\alpha'_{-(M+m+i)}\}} \tag{27'}$$

is the probability to find the process at time t in the sphere of radius p^{M+m+i} centered at zero. Moreover we have the equation

$$\dot{P}_{K(0,p^M)} = -(a(M) + u(M,1))P_{K(0,p^M)} + \sum_{i=1}^{\infty} (u(M,i) - u(M,i+1))P_{M,i}.$$

At this point we make following observation.
Let $a, b \in Q_p$. If $\|b\|_p \leq p^M$ then $K(a, p^M) + b = K(a - b, p^M) = K(a, p^M)$. If $\|b\|_p < \max\{p^M, \|a\|_p\}$ then $\text{dist}_p(0, K(a, p^M)) = \text{dist}_p(0, K(a, p^M) + b)$. For any $b \in Q_p$ and any two spheres K_1, K_2 one has dist $(K_1, K_2) =$ dist $(K_1 + b, K_2 + b)$. This together with our assumption that the infinitesimal transition probabilities depend only on the distances between elementary spheres implies

Theorem 4. The system (27) is invariant under p-adic translations in the sense that if the spheres $\{\alpha_{-(M+m)}, ..., \alpha_{-(M+j)}\}$ $, m \in \mathbb{N}, j = 1, ...m$ are substituted by $\{\alpha_{-(M+m)}, ..., \alpha_{-(M+j)}\} + b, b \in Q_p$ then the resulting system coincides with (27). □

Let $P_{\{\alpha_{-(M+m)}, ..., \alpha_{-(M+1)}\}}(t), m \in \mathbb{N}$ be the solution of the system (16) corresponding to the initial conditions

$$P_{\{\alpha_{-(M+1)}\}}(0) = 1 \quad \text{if } \alpha_{-(M+1)} = 0$$

$$P_{\{\alpha_{-(M+1)}\}}(0) = 0 \quad \text{if } \alpha_{-(M+1)} \neq 0,$$

$$P_{\{\alpha_{-(M+m)}, ..., \alpha_{-(M+1)}\}}(0) = 0, \quad m \in \mathbb{N}, m > 1.$$

This defines transition probabilities

$$P_t\left(K(0, p^M), \{\alpha_{-(M+m)}, ..., \alpha_{-(M+1)}\}\right) \equiv P_{\{\alpha_{-(M+m)}, ..., \alpha_{-(M+1)}\}}(t),$$

$m \in \mathbb{N}$, for state $K(0, p^M)$ at zero and $\{\alpha_{-(M+m)}, ..., \alpha_{-(M+1)}\}$ at time t.
As a direct consequence of Theorem 4 we have

Corollary 5. If K_i, K_j are disjoint spheres of radius p^M then the transition probabilities

$$P_t(K_i, K_j)$$

depend only on $\text{dist}_p(K_i, K_j)$. In particular we have

$$P_t\left(K(0, p^M), \{\alpha_{-(M+m)}, ..., \alpha_{-(M+1)}\}\right)$$

$$= P_t\left(K(b, p^M), \{\alpha_{-(M+m)}, ..., \alpha_{-(M+1)}\} - b\right)$$

□

If we sum up (27) over $\alpha_{-(M+1)}$ we get

$$\dot{P}_{\{\alpha_{-(M+m)}, ..., \alpha_{-(M+2)}\}} = -\left(a(M) - (p-1)u(M,1) + pu(M,2)\right) P_{\{\alpha_{-(M+m)}, ..., \alpha_{-(M+2)}\}}$$

$$+ \sum_{j=2}^{n-1} p\left(u(M,j) - u(M,j+1)\right) P_{\{\alpha_{-(M+m)}, ..., \alpha_{-(M+j+1)}\}}$$

$$+ p\sum_{i=0}^{\infty}\left(u(M, m+i) - u(M, m+i+1)\right) P_{M, m+i}, m \in \mathbb{N}, m \geq 2.$$

For $m = 1$ we have the equation

$$\dot{P}_{K(0,p^{M+1})} = -(a(M) - (p-1)u(M,1) + pu(M,2))\, P_{K(0,p^{M+1})}$$

$$+p\sum_{i=0}^{\infty}(u(M,2+i) - u(M,3+i))\, P_{M,2+i}.$$

By Lemma 2, (26) can be written as

$$\dot{P}_{\{\alpha_{-(M+m)},\ldots,\alpha_{-(M+2)}\}} = -(a(M+1) + u(M+1,1))\, P_{\{\alpha_{-(M+m)},\ldots,\alpha_{-(M+2)}\}}$$

$$+\sum_{j=1}^{m-2}(u(M+1,j) - u(M+1.j+1))\, P_{\{\alpha_{-(M+m)},\ldots,\alpha_{-(M+j+2)}\}} +$$

$$\sum_{i=0}^{\infty}(u(M+1,m+i-1) - u(M+1,m+i))\, P_{M+1,m-1+i}, m \in I\!N, m \geq 2$$

or

$$\dot{P}_{\{\alpha_{-(M+1+m)},\ldots,\alpha_{-(M+1+1)}\}} = -(a(M+1) + u(M+1,1))\, P_{\{\alpha_{-(M+1+m)},\ldots,\alpha_{-(M+1+1)}\}}$$

$$+\sum_{j=1}^{m-1}(u(M+1,j) - u(M+1.j+1))\, P_{\{\alpha_{-(M+1+m)},\ldots,\alpha_{-(M+j+2)}\}} \tag{28}$$

$$+\sum_{i=0}^{\infty}(u(M+1,m+i) - u(M+1,m+i+1))\, P_{M+1,m+i}, m \in I\!N.$$

Moreover we have the equation

$$\dot{P}_{K(0,p^{M+1})} = -(a(M+1) + u(M+1,1))\, P_{K(0,p^{M+1})}$$

$$+\sum_{i=0}^{\infty}(u(M+1,i) - u((M+1,i+1))\, P_{M+1,i}.$$

Thus we have

Theorem 6. If the sequence $(a(M), M \in \mathbb{Z})$ satisfies conditions (19) a) and b) and $u(M,m), m \in \mathbb{N}$ are given by (20) and (21) the system (28) of equations obtained from the system (27) by summing over $\alpha_{-(M+1)}$ is identical with the system (27) with M substituted by $M+1$. □

Let $P_{\alpha'_{-(M+1)}}, \alpha'_{-(M+1)} = 0,...,p-1$ denote the solution of the system (27) corresponding to the initial conditions

$$P_{\{\alpha'_{-(M+1)}\}}(0) = 1$$
$$P_{\{\alpha_{-(M+1)}\}}(0) = 0, \quad \alpha_{-(M+1)} \neq \alpha'_{-(M+1)}$$
$$P_{\{\alpha_{-(M+m)},...,\alpha_{-(M+1)}\}}(0) = 0 \quad m \in \mathbb{N}, m > 1.$$

By Corollary 5 the functions $P_{\{\alpha_{-(M+m)},...,\alpha_{-(M+1)}\}}(t), m > 1$ are the same in all cases. Observing moreover that the system (27) is invariant under permutation of spheres $\{\alpha_{-(M+1)}\}$, $\alpha_{-(M+1)} = 0,...,p-1$, we conclude that $\sum_{\alpha_{-(M+1)}=0}^{p-1} P_{\{\alpha_{-(M+1)}\}}(t)$ is the same in all cases.
These considerations prove following corollary to Theorem 6.

Corollary 7. Given $M \in \mathbb{Z}$, any of the solutions of the system (28) corresponding to one of the boundary conditions

$$P_{\{\alpha'_{-(M+1)}\}}(0) = 1, \alpha'_{-(M+1)} = 0,1,..,p-1$$

$$P_{\{\alpha_{-(M+1)}\}}(0) = 0 \text{ for } \alpha_{-(M+1)} \neq \alpha'_{-(M+1)},$$

$$P_{\{\alpha_{-(M+m)},...\alpha_{-(M+1)}\}}(0) = 0 \ m \in \mathbb{N}, m > 1$$

defines the same solution for the system (28) with M substituted by $M+1$, namely that satisfying initial conditions

$$P_{K(0,p^{M+1})}(0) = 0$$
$$P_{\{\alpha_{-(M+2)}\}}(0) = 0, \quad \alpha_{-(M+2)} \neq 0$$
$$P_{\{\alpha_{-(M+1+m)},...,\alpha_{-(M+2)}\}}(0) = 0, \quad m \in \mathbb{N}, m > 2.$$

□

To solve the system (27) we apply following procedure :
For any $m \in \mathbb{N}$ we sum up the corresponding equations over the $\alpha_{-(M+m)},...,\alpha_{-(M+1)}$, recalling that $\alpha_{-(M+j)}$ takes values $\{0,...,p-1\}$ for all $j = 1,...,m$.
This together with the equation

$$\dot{P}_{M,0} = -(a(M) + u(M,1))\, P_{M,0} + \sum_{i=1}^{\infty} (u(M,i) - u(M,i+1))\, P_{M,i}$$

where $P_{M,0} \equiv P_{K(0,p^M)} = P_{\{\alpha_{-(M+1)}=0\}}$ yields following system of equations

$$\dot{P}_{M,m} = -\left(a(M+m) + p^m u(M,m+1)\right) P_{M,m}$$

$$+p^m \sum_{i=1}^{\infty} \left(u(M,m+i) - u(M,m+i+1)\right) P_{M,m+i}, m \in \mathbb{N}_0. \quad (29)$$

Direct computations show that the following equality holds

$$p\dot{P}_{M,m} - \dot{P}_{M,m+1} = \left(a(M+m) + p^m u(M,m+1)\right)(pP_{M,m} - P_{M,m+1}).$$

Taking into account our initial condition we obtain

$$pP_{M,m}(t) - P_{M,m+1}(t) = (p-1)e^{-[a(M+m)+p^m u(M,m+1)]t}.$$

Observe that, for arbitrary $i \in \mathbb{N}$:

$$p^{-i}P_{M,m+i}(t) - p^{-(i+1)}P_{M,m+i+1}(t)$$

$$= \frac{p-1}{p^{i+1}} e^{-[a(M+m+i)+p^{m+i}u(M,m+i+1)]t},$$

Summing up over i from $i=0$ to $i=k-1, k \in \mathbb{N}$ we get

$$P_{M,m}(t) - p^{-k}P_{M,m+k}(t) = \frac{(p-1)}{p} \sum_{i=0}^{k-1} p^{-i} e^{-[a(M+m+i)+p^{m+i}u(M,m+i+1)]t}$$

which for $k \to \infty$ converges absolutely and uniformly on $0 \le t < \infty$ to

$$P_{M,m}(t) = \frac{p-1}{p} \sum_{i=0}^{\infty} p^{-i} e^{-[a(M+m+i)+p^{m+i}u(M,m+i+1)]t}.$$

The formulae (20) and (21) imply

$$a(M+m+i) + p^{(m+i)}u(M,m+i+1)$$

$$= (p-1)^{-1}[pa(M+m+i) - a(M+m+i+1)].$$

Hence

$$P_{M,m}(t) = \frac{p-1}{p} \sum_{i=0}^{\infty} p^{-i} e^{-(p-1)^{-1}[pa(M+m+i)-a(M+m+i+1)]t}. \quad (30)$$

(Note that the series is convergent, since a is monotone decreasing and $p > 1$). Thus the probability to find the process at time t in the sphere $K(0, p^{M+m})$ provided $P_{K(0,p^M)}(0) = 1$, is independent of M, i.e. of the radius of the elementary sphere, but depends on $M+m$, i.e. on the radius of the sphere in question. This justifies the new notation

$$P_{M+m}(t) \equiv P_{M,m}(t). \tag{31}$$

Let us consider $P_N(t)$. This could have been obtained as the solution of the system (27) with $M = N$ but also for instance with $M = N-1$. If in the first case the initial condition is

$$P_N(0) \equiv P_{\{\alpha_{-(N+1)}=0\}}(0) = 1,$$

then by Corollary 7 in the second case the process starting from any sphere $\{0, \alpha_{-N}\}, \alpha_{-N} = 0, ..., p-1$ yields the same $P_N(t)$.

Let $x \in K(0, p^N)$, for some $N \in \mathbb{Z}$. For any $M < N$ define $K^M = \{y \in Q_p; \|x-y\|_p \leq p^M\}$. Then $\{x\} = \bigcap_{M<N} K^M$. By multiple application of Corollary 7 we get

$$P_N(t) = P_t\left(K^M; K(0, p^M)\right).$$

This motivates us to define

$$P_t\left(x; K(0, p^N)\right) \equiv P_N(t) \tag{32}$$

for all $x \in K(0, p^N)$. By Corollary 5 we have

$$P_t\left(x; K(a, p^N)\right) = P_N(t) \tag{33}$$

for all $x \in K(a, p^N)$.
To find $P_t\left(x; K(a, p^N)\right)$ for $x \notin K(a, p^N)$ we begin by solving following problem : Compute

$$P_{\{\alpha_{-(N+m)}, ..., \alpha_{-(N+1)}\}}(t) \tag{34}$$

with $\alpha_{-(N+m)} \neq 0$, the process starting at $K(0, p^N)$. Observe that the spheres $\{\alpha_{-(M+m)}, ..., \alpha_{-(M+1)}\}$ with $\alpha_{-(M+m)} \neq 0$ fill the set $K(0, p^{N+m}) \setminus K(0, p^{N+m-1})$ and since they are all equidistant from $K(0, p^M)$ the functions (34) are all equal, by Corollary 5. The number of such spheres is $(p-1)p^{m-1}$ thus

$$P_{\{\alpha_{-(N+m)}, ..., \alpha_{-(N+1)}\}}(t) = (p-1)^{-1} p^{1-m} \left(P_{N+m}(t) - P_{N+m-1}(t)\right)$$

Since

$$P_t\left(K(0, p^N), \{\alpha_{-(N+m)}, ..., \alpha_{-(N+1)}\}\right) \equiv P_{\{\alpha_{-(N+m)}, ..., \alpha_{-(N+1)}\}}(t)$$

the Corollaries 5 and 7 imply

$$P_t\left(x, K(a, p^N)\right) = (p-1)^{-1} p^{1-m} \left(P_{N+m}(t) - P_{N+m-1}(t)\right) \tag{35}$$

for $x \in Q_p$ such that $\text{dist}_p\left(x, K(a, p^N)\right) = p^{N+m}$.
For $A = \bigcup_{i=1}^{\infty} K_i$ where K_i are disjoint spheres of not necessarily equal radiuses, we have

$$P_t(x, A) = \sum_{i=1}^{\infty} P_t(x, K_i), \tag{36}$$

where $P_t(x, K_i)$ is given by formula (33) if $x \in K_i$ and by (35) when $x \notin K_i$. The transition probability is determined by the unique solution of the Kolmogorov equations (8). The uniqueness follows from the fact that the condition $\sum_{j=1}^{\infty} P_{ij}(t) = 1$ holds. Indeed

$$\sum_{j=1}^{\infty} P_{K(0,p^M),K_j^M} = \lim_{m\to\infty} P_{M+m}(t) =$$

$$= \lim_{m\to\infty} \frac{p-1}{p} \sum_{i=0}^{\infty} p^{-i} \exp\{[pa(M+m+i) - a(M+m+i-1)]t\} = 1.$$

$P_t(x, A)$ is a semigroup of Markov kernels on Q_p. By Kolmogorov theorem for any start measure ν on Q_p there is a Markov process x_t on Q_p associated with it, in the sense that $P_t(x, A)$ are the transition probabilities of x_t.
Hence we have proven the following

Theorem 8. Let a be as in Theorem 6. Then there is a continuous time Markov stochastic process $x_t, t \geq 0$ with state space Q_p given by the transition probabilities $P_t(x, A), t \geq 0, x \in Q_p, A$ a countable union of p-adic spheres, defined by (30) - (36). □

Remark In the theory of Markov chains usually only the nearest neighbors transitions are considered i.e. one assumes that for some $m_0 \in \mathbb{N}$ one has

$$u(M, m) = 0, \quad m > m_0. \tag{37}$$

As a consequence the summations in (8) and (27) are finite.
By properties of the p-adic metric such an assumption results in a "confinement" of the process. Indeed if (37) holds then by (20), (21) and property b) one gets $a(M + m - 1) = 0$ for all $m > m_0$. This according to formula (29) yields $P_{M,m}(t) = P_{M,m_0}(t)$. Hence by (35) $P_t(x, K(a, p^M)) \equiv 0$ whenever $\text{dist}_p(x, K(a, p^M)) > p^{M+m}$.

Remark Let us sketch shortly how the diffusion on p-adics we have constructed can be looked upon as a diffusion on a regular, infinite tree, following a procedure of [BrO]. Consider a regular tree, in which at each "level" every branch splits into p other branches, numbered 0 to $p-1$. After n levels, the endpoints $z_j, j = 1, ..., p^n$ can be mapped naturally into the field Q_p, in the following way. Label each level of the tree by an integer i, with $i = m$ for the apex of the tree and $i = m+n-1$ for the last branch. To the branch number $a \in \{0, ..., p-1\}$ associate the number ap^i. The map τ from endpoints to p-adics is defined by $\tau(z_i) \equiv \sum_{j=0}^{n-1} a_{i,m+j} p^{m+j}$. Between the endpoints we can define an ultrametric distance $|||z_i - z_j|||$ by setting it equal to the number of levels one must move up the tree before z_i and z_j are connected to a common branch. We then have, for $|||z_i - z_j||| \neq 0$, $\|\tau(z_i) - (z_j)\|_p = p^{|||z_i - z_j||| - m - n}$, with $\| \quad \|_p$ the p-adic norm. For $|||z_i - z_j||| = 0$ we have $\|\tau(z_i) - \tau(z_j)\|_p = \infty$. In the limit $n \to \infty$ the $\tau(z_i)$ become the subset $K(0, p^m)$ of Q_p, hence by letting m run over $\mathbb{Z}$, we recover Q_p.

Remark It is possible to study the properties of the transition semigroup of the process and its generator in more details. Also it is possible to handle the limit $t \to \infty$ rigorously, by a p-adic Laplace method (confirming heuristic results in [BrO]). Furthermore it is possible to study path properties of the process. For these topics see [AK].

Acknowledgements

We gratefully acknowledge stimulating discussions with Philippe Blanchard, Philippe Combe, Tyll Krüger, John Lafferty, Andrew T. Ogielski, Hernando Sierra, B.L. Spokoiny, V.I. Volovich. It is a pleasure to thank Prof. Ana Bela Cruzeiro and Prof. Takeyuchi Hida for kind invitations to give lectures at Lisbon resp. Nagoya. The second named author would like to thank SFB 237 (Essen - Bochum - Düsseldorf) and the German - Polish project (Karlsruhe) for financial support.

References

[AK] S. Albeverio, W. Karwowski, in preparation.

[BeC] E.G. Beltrametti, G. Cassinelli, Quantum mechanics and p-adic numbers, Found. Phys. 2, 1-7 (1972).

[Bha-Re] A.T.Bharucha-Reid. Elements of the Theory of Markov Processes and Their Applications, McGraw-Hill, New York, Toronto, London 1960.

[BrO] L. Brekke, M. Olson, p-adic diffusion and relaxation in glasses, Preprint EFI (1989).

[Ca] P. Cartier, Harmonic analysis on trees, Proc. Symp. Pure Math. 26, 419-424 (1972). P. Cartier, Géométrie et analyse sur les arbres, pp 123-140 in Sem. Bourbaki 1971/72, Lect. Notes Maths. 317, Springer (1973).

[DoS] C. De Dominicis, M. Schreckenberg; Diffusion in ultrametric spaces, pp. 255-274 in Lect. Notes Phys. 275 , Springer, Berlin (1986).

[Dos] P.G. Doyle, T.L. Snell, Random walks and electrical networks, AAM, 1984.

[EU] C.J. Everett, S. Ulam, On some possibilities of generalizing the Lorentz group in the special relativity theory, J. Comb. Th.1, 248-270 (1966).

[FiT] A. Figá - Talamanca, Analisi armonica su strutture discrete, Boll UMI 3A, 313-334 (1984).

[Gr] B. Grossman, Membranes in string theory, trees, the Weil conjectures and the Ramanujan numbers, Rockefeller Prepr. (1989).

[GrWH] S. Grossmann, F. Wegner, K.H. Hoffman; J. Phys. Lett. 46, L575 (1985).

[Ko] N. Koblitz, p-adic numbers, p-adic analysis and zeta-functions, 2nd Ed., Springer, New York (1984).

[Ly1] T. Lyons, Transience of reversible Markov chains, Ann. Prob. 11, 393-402 (1983).

[Ly2] R. Lyons, Random walks and percolation on trees, Ann. Prob. 18, 931-958 (1990).

[Me] Y. Meurice, Quantum mechanics with p-adic numbers, Int. J. Mod. Phys. A 4, 5133-5147 (1989).

[OgS] J.A.T. Ogielski, D.L.Stein, Dynamics on ultrametric spaces, Phys. Rev. Lett., 55, 1634-1637 (1985).

[Pa] G. Parisi, On p-adic functional integrals. Mod. Phys. Letts., A, 639-643 (1988).

[Pr] Ch. Preston, Gibbs states on countable sets, Cambridge Univ. Press (1974).

[RaT] R. Rammal, G. Toulouse, Ultrametricity for physicists, Rev. Mod. Phys. 58, 765-788 (1986).

[Sa] S. Sawyer, Isotropic random walks in a tree, Z. Wahrsch. Verw. Geb. 42, 279-292 (1987).

[Sch] W.H. Schikhof, Ultrametric calculus. An introduction to p- adic analysis, Cambridge University Press, Cambridge (1984).

[Schr] M. Schreckenberg, Long range diffusion in ultrametric spaces, Z. Phys. B 60, 483-488 (1985).

[Se] J.P. Serre, Arbres amalgames et SL_2, Astérisque 46 (1977).

[VlV1] V.S. Vladimirov, I.V. Volovich, p-adic Schrödinger type equation, Letts. Math. Phys. 18, 43-53 (1989).

[VlV2] V.S. Vladimirov, V.I. Volovich, p-adic quantum mechanics, Commun. Math. Phys. 123, 659-678 (1989).

THE HIDA CALCULUS APPROACH TO STOCHASTIC INTEGRATION

A.N. AL-HUSSAINI
Department of Statistics and Applied Probability, University of Alberta
Edmonton, Alberta, Canada T6G 2G1

ABSTRACT

Through homogeneous chaos, Hida's basic operators ∂_t, ∂_t^* of white noise calculus, $\dot{W}_t$ derivative of Wiener process W_t at time t and multiplication by $\dot{W}_t$ are introduced. ∂_t is related to the corresponding operator D_t of Malliavin calculus.

A unified account of these ideas shows at least as far as the problems considered here are concerned there is no real difference between the two theories. Employing the operators ∂_t, ∂_t^* in some instances is very effective and may be more natural, e.g. the result of Gaveau and Trauber concerning the Skorohod integral becomes more transparent and easy to prove.

Also, we prove that $\dot{W}_t \cdot \varphi = \partial_t \varphi + \partial_t^* \varphi$ which further clarifies white noise integration.

1. Preliminaries

One of the superficial differences between the Malliavin calculus and the calculus of Hida lies in the use made in the latter theory of the Schwartz space of distributions with Gaussian white noise measure as its fundamental probability space. It seems to us that Hida's ideas can just as well be formulated starting with homogeneous chaos expansions on classical Wiener spaces. Indeed, by keeping the basic probability space as simple and small as possible, it helps one to focus attention on the generalized Brownian functionals that form an indispensable part of Hida's calculus. Another advantage of this procedure, in our opinion, is that it facilitates comparison between the theories of Malliavin and Hida. The present article is written by way of comment along these lines.

Since our main interest is in the Skorohod integral and the ideas connected with stochastic integration, we shall only consider $H = L^2[0,1]$ the usual Hilbert space of square integrable functions relative to the Lebesgue measure over the interval $[0,1]$.

Let $\widehat{L}^2([0,1]^n)$ be the space of symmetric functions in $L^2([0,1]^n)$; $H^k([0,1]^n)$ be the Sobolev space of order k over $[0,1]^n$. Put $\widehat{H}^k([0,1]^n) =$

$H^k([0,1]^n) \cap \widehat{L}^2([0,1]^n)$.

In the sequel W will stand for the standard Wiener space; $\mathbb{L}^2(W)$ is the space of square integrable functions over W. P will be the corresponding probability measure. Duality will be denoted by $\langle\ ,\ \rangle$, sometimes without subscripts. Context should eliminate any confusion.

Every $\varphi \in \mathbb{L}^2(W)$ can be written (using Ito-Wiener chaos representation) as:

$$\varphi = \sum_{n=0}^{+\infty} \int_0^1 \cdots \int_0^1 f_n(t_1,\ldots,t_n)dW_{t_1}\ldots dW_{t_n}$$

where $f_n \in \widehat{L}^2([0,1]^n)$. Following Hida ([4],[5]), let:

$$K_n = \{\int_0^1 \cdots \int_0^1 f_n(t_1,\ldots,t_n)dW_{t_1}\ldots dW_{t_n} : f \in \widehat{L}^2([0,1]^n)\}$$

$$K_n^{(n)} = \{\int_0^1 \cdots \int_0^1 f_n(t_1,\ldots,t_n)dW_{t_1}\ldots dW_{t_n} : f \in \widehat{H}^{\frac{n+1}{2}}([0,1]^n)\}.$$

Let $K_n^{(-n)}$ be the dual of $K_n^{(n)}$ ([5], [8]). For $\phi \in K_n^{(n)}$, given by $\int_0^1 \cdots \int_0^1 f_n(t_1,\ldots,t_n)dW_{t_1}\ldots dW_{t_n}$, where $f_n \in \widehat{H}^{\frac{n+1}{2}}([0,1]^n)$, define:

$$\|\phi\|_+ = \sqrt{n!}\ \|f_n\|^{\widehat{H}^{\frac{n+1}{2}}([0,1]^n)}.$$

Similarly for corresponding functionals $\psi \in K_n^{(-n)}$ and $g_n \in \widehat{H}^{-\frac{n+1}{n}}([0,1]^n)$, define: $\|\psi\|_- = \sqrt{n!}\ \|g_n\|_{\widehat{H}^{-\frac{n+1}{2}}([0,1]^n)}$.

Thus

$$\begin{array}{ccccc} K_n^{(n)} & \subset & K_n & \subset & K_n^{(-n)} \\ \downarrow & & \downarrow & & \downarrow \\ \sqrt{n!}\ \widehat{H}^{\frac{n+1}{2}}([0,1]^n) & \subset & \sqrt{n!}\ \widehat{L}^2([0,1]^n) & \subset & \sqrt{n!}\ \widehat{H}^{-\frac{n+1}{2}}([0,1]^n). \end{array}$$

The inclusions are continuous and the arrows are unitary maps; details are given in ([8]). We define:

$$(L^2)^+ = \sum_n \oplus K_n^{(n)}, \quad (L^2)^- = \sum_n \oplus K_n^{(-n)}.$$

We have ([5],[8]) the following Gelfand triple $(L^2)^+ \subset \sum \oplus K_n \subset (L^2)^-$. The latter is the space of generalized Wiener functions. $(L^2)^+$ is the space of test Wiener functionals. Noting that $H^1 \subset H \subset H^{-1}$, let $\mathcal{H}^1 \subset \mathcal{H} \subset \mathcal{H}^{-1}$ be the corresponding exponential symmetric Hilbert spaces.

Now for $F \in \mathbb{L}^2(W)$, $\xi \in H^1([0,1])$, define $(UF)(\xi) = E(F \cdot \rho(\xi))$, where:

$$\rho(\xi) = e^{\int_0^1 \xi(t)dW_t - \frac{1}{2}\int_0^1 \xi^2(t)dt}.$$

If $F = \int_0^1 \cdots \int_0^1 f_n(t_1,\ldots,t_n)dW_{t_1}\ldots dW_{t_n} \in K_n$ then it can be shown that:

$$(UF)(\xi) = \int_0^1 \cdots \int_0^1 f_n(t_1,\ldots,t_n)\xi(t_1)\ldots\xi(t_n)dt_1\ldots dt_n.$$

Thus $K^1 = \{F \in \mathbb{L}^2 : (UF)(\cdot) \in \mathcal{H}^1\}$ and K^{-1} is the dual of K^1, and we have Gelfand triple $K^1 \subset \mathbb{L}^2 \subset K^{-1}$. For $F \in K^i$, define $\|F\|_i = \|UF(\cdot)\|_{\mathcal{H}_i}$, $i = 1, -1$.

An important property is that the set $\{\rho(\xi) : \xi \in H\}$ forms a total set in $\mathbb{L}^2(W)$, ([10], pp. 144-146).

2. Operators ∂_t, ∂_t^*;

Main aim in this section, besides introducing Hida's operators ∂_t, ∂_t^* and Malliavin-Sobolev operator D_t is to show that $\partial_t = D_t$; (Theorem 1).

In what follows we write $U_\phi(\xi)$ in place of $U\phi(\xi)$. Let $U_\phi'(\xi,t) = \frac{d}{d\epsilon} U_\phi(\xi + \epsilon\delta_t)|_{\epsilon=0}$ whenever the right hand side exists. δ_t is the Dirac measure at t.

DEFINITION 1: Define ([4],[5],[7]), $\partial_t\phi = U^{-1}U_\phi'(\cdot,t)$. Then ∂_t is Hida's annihilation operator at t. ∂_t^* the adjoint of ∂_t is called the creation operator at t.

Consider $\phi = \int_0^1 \cdots \int_0^1 f_n(t_1,\ldots,t_n)dW_{t_1}\ldots dW_{t_n}$ where $f_n \in \widehat{H}^{\frac{n+1}{2}}([0,1]^n)$. By the Sobolev imbedding theorem ([1]), we may and do assume that f_n is a continuous function. It is easy to see that

$$\frac{U_\phi(\xi+\varepsilon h)-U_\phi(\xi)}{\varepsilon}$$

$$\longrightarrow n\int_0^1 \cdots \int_0^1 f_n(t_1,\ldots,t_{n-1},t)\xi(t_1)\ldots\xi(t_{n-1})h(t)dt_1\ldots dt_{n-1}dt$$

$$= U'_\phi(\xi)[h] = \big(U'_\phi(\xi,\cdot),h(\cdot)\big)_{L^2}$$

where $U'_\phi(\xi,t) = n\int_0^1 \cdots \int_0^1 f_n(t_1,\ldots,t_{n-1},t)\xi(t_1)\ldots\xi(t_{n-1})dt_1,\ldots,dt_{n-1}$. Note that

$$|U'_\phi(\xi)[h]| \le C_1\|h\|; \quad \text{and}$$
$$|U'_\phi(\xi;t)| \le C_1\|f_n(\ldots,t)\|.$$

It holds that:

$$U'_\phi(\xi)[h] = \langle J_t(\xi),h\rangle_{H^{-1},H^1}$$

where

$$J_t(\xi) = \int_0^1 U'_\phi(\xi;t)\delta_t dt \in H^{-1},$$

an integral in the sense of Pettes. Note that $t \to \delta_t$ is continuous. Based on ([15], Corollary 3.2) or by a similar proof $\|\delta_t - \delta_s\|_{-1} \le C|t-s|^{1/2}$. Summarizing:

$$U'_\phi(\xi,t) = \frac{d}{d\varepsilon}\,U(\xi+\varepsilon\delta_t)|_{\varepsilon=0}, \quad \phi \in K_n^{(n)},$$

and $\partial_t\phi = U^{-1}U'_\phi(\cdot,t)$.

$\partial^*_\psi = U^{-1}\xi(t)U_\psi(\xi)$ by similar calculation i.e. using Wiener-Ito chaos expansion.

To define Malliavin-Sobolev operator D_t we follow ([16]). Let $e_1,e_2,\ldots$ be a complete orthonormal set in $L^2([0,1])$. Define:

$D_{e_i}\phi = \sum_1^{+\infty} n \int_0^1 \cdots \int_0^1 f_n(t_1,\ldots,t_{n-1},t)e_i(t)dtdW_{t_1}\ldots dW_{t_{n-1}}$, where $\phi = \sum_1^{+\infty} \int_0^1 \cdots \int_0^1 f_n(t_1,\ldots,t_n)dW_{t_1}\ldots dW_{t_n}$.

DEFINITION 2.: $D\phi = \sum_1^{+\infty}(D_{e_i}\phi)e_i$, provided that,

$$\sum_1^{+\infty} nn! \, \|f_n\|^2 < +\infty.$$

This D is called Malliavin-Sobolev derivative.

THEOREM 1. $\partial = D$.

PROOF: We will show only that $\partial = D$ on $K_n^{(n)}$. Now $D_{e_i}\phi = n\int_0^1 \cdots \int_0^1 f_n(t_1,\ldots,t_{n-1}t)e_i(t)dtdW_{t_1}\ldots dW_{t_n}$, therefore

$$\begin{aligned} U_{D_{e_i}\phi}(\xi) &= \left(U'_\phi(\xi;\cdot), e_i\right)_{L^2} \\ &= U'_\phi(\xi)[e_i]. \end{aligned}$$

So for $h \in H$,

$$\begin{aligned} U_{D_h F}(\xi) &= U_{\sum(h,e_i)D_{e_i}\phi}(\xi) \\ &= U_{\sum_i (h, D_{e_i}\phi e_i)}(\xi) \\ &= U_{(D\phi,h)}(\xi) \\ &= \sum_i (h,e_i)U_{D_{e_i}\phi}(\xi) \\ &= \sum_i (h,e_i)(U'_\phi(\xi),e_i) \\ &= U'_\phi(\xi)[h]; \quad \text{that is} \\ \left(U_{D\phi}(\xi),h\right) &= U_{(D\phi,h)}(\xi) = \left(U'_\phi(\xi;\cdot),h\right). \end{aligned}$$

Thus $U_{D\phi}(\xi) = U'_\phi(\xi;t)$ implying

$$D_t\phi = U^{-1}U'_\phi(\xi;\cdot) = \partial_t\phi \qquad \text{Q.E.D.}$$

3. $\dot{W}_t$ and Its Relation to ∂_t, ∂_t^*

In defining $\dot{W}$, the arguments here are close to ([7]).
Let

$$\begin{aligned} F_\varepsilon &= \frac{1}{\varepsilon}\,[W_{t+\varepsilon} - W_t] \in K_1^{(-1)} \\ &= \frac{1}{\varepsilon}\int_0^1 1_{[t,t+\varepsilon]}(s)dW_s \\ &= \int_0^1 f(t)dW_t, \quad f \in H \subset H^{-1} \end{aligned}$$

$$\begin{aligned} \therefore \quad U_{F_\varepsilon}(\xi) &= \int_0^1 f(t)\xi(t)dt, \quad \xi \in H^1 \\ &= \frac{1}{\varepsilon}\int_t^{t+\varepsilon} \xi(s)ds \\ &\to \langle \delta_t, \xi\rangle \quad \text{by continuity of } \delta \quad \text{(Section 1).} \end{aligned}$$

Thus according to the topology at the end of Section 1, $F_\varepsilon \to \dot{W}_t$ say, in $K_1^{(-1)} \subset (L^2)^-$. Also note that

$$U_{\dot{W}_t}(\xi) = \langle \delta_t, \xi\rangle.$$

PROPOSITION (1). $\partial_t^* 1 = \dot{W}_t$.

PROOF: Apply U to both sides.

In the white noise literature multiplication by $\dot{W}_t$ is defined ([7],[8]) by .

$$\dot{W} \cdot \varphi = \partial_t \varphi + \partial_t^* \varphi. \qquad *$$

In this paper we adopt a definition, then prove $*$ immediately. At the end of this section we will justify $*$ using Ito's decomposition formula ([6]).

But first let $\widehat{C}_n = C^\infty([0,1]^n)$ be the bounded, with bounded derivatives, and symmetric functions on $[0,1]^n$, and
$S_n = \{\int_0^1 \cdots \int_0^1 f_n(t_1, \ldots, t_n)dW_{t_1}\ldots dW_{t_n} : f_n \in \widehat{C}_n\}$.

DEFINITION: $\langle (\dot{W}_t \cdot \varphi), F\rangle \stackrel{\text{def}}{=} \langle \dot{W}_t, \varphi F\rangle$. $\phi \in K_n^{(n)}$, $F \in S_m$, for all n, m.

This definition is obviously motivated by a similar definition in Schwartz distribution theory. But there is a difference.

THEOREM (2). $\dot{W}_t \cdot \varphi = \partial_t \varphi + \partial_t^* \varphi.$

PROOF: For $F \in S_m$, and $\phi \in K_n^{(n)}$, for all n, m.

$$\begin{aligned} \langle \dot{W}_t \cdot \varphi, F \rangle &= \langle \dot{W}_t, \varphi F \rangle, \quad \text{by definition above} \\ &= \langle \partial_t^* 1, \varphi F \rangle, \quad \text{using the above proposition} \\ &= \langle 1, \partial_t(\varphi F) \rangle \\ &= \langle 1, \varphi \partial_t F + F \partial_t \varphi \rangle \\ &= \langle 1, \varphi \partial_t F \rangle + \langle 1, F \partial_t \varphi \rangle \\ &= \langle \partial_t^* \varphi, F \rangle + \langle \partial_t \varphi, F \rangle \\ &= \langle \partial_t \varphi + \partial_t^* \varphi, F \rangle \\ &\Longrightarrow \dot{W}_t \cdot \varphi = \partial_t \varphi + \partial_t^* \varphi. \end{aligned}$$

∂_t obeys Leibniz rule ([7]).

We give another proof of Theorem 2, which uses Ito's decomposition formula:

$$I_1(\psi) I_p(\varphi) = I_{p+1}(\varphi \psi) + \sum_{k=1}^{p} I_{p-1}(\varphi \underset{(k)}{\times} \psi)$$

([6]) where,

$$\begin{aligned} \varphi \underset{(k)}{\times} \psi(t_1, \ldots, t_{k-1}, t_{k+1}, \ldots, t_p) &= \int_0^1 \varphi(t_1, \ldots, t_p) \psi(t_k) dt_k \\ &= \frac{1}{\varepsilon} \int_t^{t+\varepsilon} \varphi(t_1, \ldots, s, \ldots, t_p) ds = \varphi^\varepsilon \quad \text{say.} \end{aligned}$$

Here $\psi_\varepsilon = \frac{1}{\varepsilon} 1_{[t,t+\varepsilon]}$ and $\frac{1}{\varepsilon} [W_{t+\varepsilon} - W_t] = I(\psi_\varepsilon)$.

$$E I_{p-1}(\varphi^\varepsilon) \rho(\xi) = \int_0^1 \cdots \int_0^1 \frac{1}{\varepsilon} \int_t^{t+\varepsilon}$$

$$\times \varphi(t_1, \ldots, s, \ldots, t_p) ds \xi(t_1) \ldots \xi(t_{k-1}) \xi(t_{k+1}) \ldots \xi(t_p) dt_1 \ldots dt_p;$$

φ is taken to be smooth or at least continuous and symmetric.

$$= \frac{1}{\varepsilon} \int_t^{t+\varepsilon} \int_0^1 \cdots \int_0^1$$

$$\times \varphi(t_1, \ldots, s, \ldots, t_p)\xi(t_1)\ldots\xi(t_{k-1})\xi(t_{k+1})\ldots\xi(t_p)dt_1\ldots dt_p ds$$

$$\longrightarrow \int_0^1 \cdots \int_0^1 \varphi(t_1, \ldots t_{p-1}, t)\xi(t_1)\ldots\xi(t_{p-1})dt_1\ldots dt_{p-1}.$$

$$\mathbb{E}I_{p+1}(\varphi\psi_\varepsilon)\rho(\xi) = \int_0^1 \cdots \int_0^1 \varphi(t_1\ldots t_p)\psi_\varepsilon(t_{p+1})\xi(t_1)\ldots\xi(t_{p+1})dt_1\ldots dt_{p+1}$$

$$= \int_0^1 \cdots \int_0^1 \varphi(t_1\ldots t_p)\xi(t_1)\ldots\xi(t_p) \int_0^1 \psi_\varepsilon(t_{p+1})\xi(t_{p+1})dt_{p+1}dt_1\ldots dt_p$$

$$= \int_0^1 \cdots \int_0^1 \varphi(t_1\ldots t_p)\xi(t_1)\ldots\xi(t_p)dt_p\ldots dt_p \cdot \frac{1}{\varepsilon}\int_t^{t+\varepsilon} \xi(t_{p+1})dt_{p+1}$$

$$\longrightarrow \xi(t) \int_0^1 \cdots \int_0^1 \varphi(t_1, \ldots, t_p)\beta(t_1)\ldots\xi(t_p)dt_1\ldots dt_p \quad \text{if} \quad \xi \in H^1.$$

Hence by the preceeding Ito's decomposition formula, we have:

$$\lim_{\varepsilon\to 0} E\{I_1(\psi_\varepsilon)I_p(\varphi)\rho(\xi)\} = \xi(t)\int_0^1 \cdots \int_0^1 \varphi(t_1, \ldots, t_p)\xi(t_1)\ldots\xi(t_p)dt_1\ldots dt_p$$

$$+ p\int_0^1 \cdots \int_0^1 \varphi(t_1, \ldots, t_{p-1}, t)\xi(t_1)\ldots\xi(t_{p-1})dt_1\ldots dt_{p-1}$$

i.e.

$$\lim_{\varepsilon\to 0} U_{\frac{1}{\varepsilon}[W_{t+\varepsilon}-W_t]\cdot\Phi}(\xi) = \xi(t)U_\Phi(\xi) + U_{\partial_t\Phi}(\xi) \quad \text{which implies} \quad *;$$

where $\xi \in H^1$ and $\Phi = I_p(\varphi)$.

4. Skorohod Integral

This integral was introduced by A.V. Skorohod in ([13]). Several works have appeared since then ([13],[8],[12]), and references therein. The theorem below relates this integral to ∂_t^* in a straightforward way.

THEOREM (2). *If* $u : [0,1] \times W \to R$ *is square integrable, then:*

$$\int_0^1 \partial_t^* u_t dt = \sum \int_0^1 \cdots \int_0^1 \widetilde{f}(t_1, \ldots, t_n, t) dW_{t_1} \ldots dW_{t_n} dW_t$$

i.e. if one side is finite, then so is the other side, and the the equality holds.

Here $u_t = \sum \int_0^1 \cdots \int_0^1 f(t_1, \ldots, t_n, t) dW_{t_1} \ldots dW_{t_n}$ *is the usual Wiener-Ito decomposition, and* $\widetilde{f}$ *is the symmetrization of* f *which is symmetric in* $t_1, \ldots, t_n$.

PROOF: Suppose $u_t = \int_0^1 \cdots \int_0^1 f(t_1, \ldots, t_n, t) dW_{t_1} \ldots dW_{t_n}$. Thus

$$\partial_t^* u_t = U^{-1} \xi(t) U_{u_t}(\xi)$$

$$= U^{-1} \xi(t) \int_0^1 \cdots \int_0^1 f(t_1, \ldots, t_n, t) \xi(t_1) \ldots \xi(t_n) dt_1 \ldots dt_n,$$

$$\int_0^1 \partial_t^* u_t dt = U^{-1} \int_0^1 \cdots \int_0^1 f(t_1, \ldots, t_n, t) \xi(t_1) \ldots \xi(t_n) \xi(t) dt_1 \ldots dt_n dt$$

$$= U^{-1} \int_0^1 \cdots \int_0^1 \widetilde{f}(t_1, \ldots, t_n, t) \xi(t_1) \ldots \xi(t_n) \xi(t) dt_1 \ldots dt_n dt$$

$$= \int_0^1 \cdots \int_0^1 \widetilde{f}(t_1, \ldots, t_n, t) dW_{t_1} \ldots dW_{t_n} dW_t.$$

The result follows.

REMARK 1: In ([8],[9]), the Skorohod integral is expressed terms of ∂_t^*, by utilizing the following formula: $\partial_t^* \phi = \int \delta_t \widehat{\otimes} f(t_1, \ldots, t_{n+1}) dW_{t_1} \ldots, dW_{t_{n+1}}$, where $\phi \in K_n$ with kernel f. $\widehat{\otimes}$ stands for the symmetric tensor product.

In view of Theorem 2, we can use the conventional notation

$$\delta u = \int_0^1 \partial_t^* u_t dt.$$

The well known result of Gaveau and Trauber [3] can be recast using ∂ as follows. First write $\langle u, v\rangle = E\int_0^1 u_t v_t dt$. Then $\langle \partial F, u\rangle = E\int_0^1 \partial_t F u_t dt$.

Write $\mathcal{D}^{2,1} = \{F \in \mathbb{L}^2 : E(\|\partial F\|_H^2) < +\infty\}$.

REMARK 2: If $F \in \mathcal{D}^{2,1}$, and u is (t, w) measurable, and $\int_0^1 Eu_t^2 dt < +\infty$. Then $\langle \partial F, u\rangle = \langle F, \delta u\rangle$.

PROOF:

$$\begin{aligned}
\langle \partial_t F, u_t\rangle &= \langle F, \partial_t^* u_t\rangle \\
\int_0^1 E(\partial_t F \cdot u_t)dt &= \int_0^1 \langle F, \partial_t^* u_t\rangle dt \\
\langle \partial F, u\rangle &= \langle F, \int_0^1 \partial_t^* u_t dt\rangle \\
\langle \partial F, u\rangle &= \langle F, \delta u\rangle.
\end{aligned}$$

Integrated objects are measurable; and the integral is Pettis'.

5. Absolute Continuity

The theorem below is in ([11]), see also [1]. The proof is perhaps more lucid! However the topic is not central to the article. It is within the circle of the ideas presented here.

Write

$$\partial_u F = \int_0^1 \partial_t F \cdot u_t dt,$$

THEOREM (3). *Let* $\varphi \in \mathbb{L}^2(W), u$ *be integrable. If* $\varphi, \partial_u\varphi \in \mathcal{D}^{2,1}$, $\partial_u \varphi \neq 0$ *a.e. Then the probability law of* φ *has a density.*

PROOF: For $f \in C_b^\infty(R)$, let $F = f \circ \varphi$ using "integration by parts formula" proved above,

$$\begin{aligned}
E\int_0^1 u_t\partial_t(F\cdot\partial_u\varphi)dt &= E[\delta u\cdot F\partial_u\varphi] \\
E[\partial_u(F\cdot\partial_u\varphi)] &= E[\delta u\cdot F\partial_u\varphi] \\
E[F\cdot\partial_u^{(2)}\varphi + f'\circ\varphi\cdot\partial_u^2\varphi] &= E[\delta u\cdot F\partial_u\varphi] \\
E[f'\circ\varphi\cdot\partial_u^2\varphi] &= E[\delta u\cdot F\cdot\partial_u\varphi] - E[F\cdot\partial_u^{(2)}\varphi] \\
|E[f'\circ\varphi\cdot\partial_u^2\varphi]| &\le |E[\delta u\cdot F\cdot\partial_u\varphi]| + |E[F\cdot\partial_u^{(2)}\varphi]|
\end{aligned}$$

so by Holders inequality

$$\begin{aligned}
&|E[f'\circ\varphi\cdot\partial_u^2\varphi]| \le C_1\|f\|_\infty + C_2\|f\|_\infty \le C\|f\|_\infty \\
&\qquad\Longrightarrow \quad \text{absolute continuity relative to} \quad \partial_u^2\varphi dP \\
&\qquad\Longrightarrow \quad \text{absolute continuity relative to} \quad dP \\
&\qquad\qquad \text{since} \quad \partial_u^2\varphi > 0 \quad \text{a.e. by hypothesis.}
\end{aligned}$$

6. White Noise Integral

The following brief outline is in the spirit of Section 2.

Theorem 2 suggests "white noise" approach to stochastic integration, i.e.

$$\int \dot{W}_t\cdot u_t dt = \int \partial_t u_t dt + \int \partial_t^* u_t dt.$$

To enlarge and include predictable integrands Kuo and Russek ([9]) have exploited trace maps from right and left (using ([1]) to define $\int_0^1 \partial_{t_+} u_t dt$ right integral and $\int_0^1 \partial_{t_-} u_t dt$ left integral. We use their set up and the theorem (2) of the Section 2. For simplicity let $u_t = \int_0^1 \cdots \int_0^1 f(t; u_1, \ldots, u_n) dW_{u_1} \ldots dW_{u_n} = I(f(t;u))$.

Define:

$$\begin{aligned}
D_+^n &= \{(t,u) \in (0,1)\times(0,1)^n;\ u = (u_1,\ldots,u_n), u_1 > t\}, \quad \text{and} \\
D_-^n &= \{(t,u) \in (0,1)\times(0,1)^n:\ u = (u_1,\ldots,u_n), u_1 < t\}.
\end{aligned}$$

Let $H^\alpha(D^n_+)$ and $H^\alpha(D^n_-)$ be the Sobolev spaces of order $\alpha \geq 0$ on D^n_+ and D^n_- respectively. If $f \in H^\alpha(D^n_\pm)$, $\alpha = 1$, then f has a version ([1]) which is denoted by f again such that

$$t \to f(t; t_+, \cdot) \in L^2([0,1] \times [0,1]^{n-1}) \quad \text{and}$$
$$t \to f(t; t_-, \cdot) \in L^2([0,1] \times [0,1]^{n-1})$$

Furthermore $u_1 \to f(t,u)$ is continuous with values in $H^{\alpha-\frac{1}{2}}([0,1] \times [0,1]^n)$. We shall use $\partial_t^+, \partial_t^-$ instead of ∂_{t+} and ∂_{t-} respectively.

Let $\partial_t^+ I(f(t,u)) = nI(f(t;t_+,\cdot)$ and $\dot{W}_t^+ \cdot I(f(t,u)) = \lim_{u_1 \searrow t} \dot{W}_{u_1} \cdot I\big(f(t,u)\big)$, using the forgoing continuity, and replacing ϕ in Theorem 2 by $I\big(f(t,u)\big)$. ∂_t^- and $\dot{W}_t^-$ are defined similarly.

By Theorem (1) of Section 2, and the trace theorem mentioned above,

$$\dot{W}_t^+ \cdot u_t = \partial_t^+ u_t + \partial_t^* u_t,$$
$$\dot{W}_t^- \cdot u_t = \partial_t^- u_t + \partial_t^* u_t,$$

observe that $(\partial_t^+)^* = (\partial_t^-)^* = \partial_t^*$. We obtain two integrals:

$$\int_0^1 \dot{W}_t^+ \cdot u_t dt = \int_0^1 \partial_t^+ u_t dt + \int_0^1 \partial_t^* u_t dt, \tag{1}$$

$$\int_0^1 \dot{W}_t^- \cdot u_t dt = \int_0^1 \partial_t^- u_t dt + \int_0^1 \partial_t^* u_t dt. \tag{2}$$

We interpret integrals (1), (2), using the Theorem (2) of Section 2 as "Pettis" integrals. More precisely, equalities (1), (2) are to hold when duality operation is restricted to S_m, for every m, (Section 3).

REMARK 3: By modifying Sections 1, 2 and 3, $\alpha > \frac{1}{2}$ will suffice. Details of this will be given elsewhere.

7. Acknowledgement

I would like to thank Prof. G. Kallianpur for the many discussion I have had with him. The idea of using Ito's decomposition to prove $\dot{W}_t \cdot \varphi = \partial_t \varphi + \partial_t^* \varphi$ is due to him.

I would like to thank Professor T. Hida for the invitation to participate in the Nagoya Conference.

Research supported by NSERC grant #55-46375 and the Air Force Office of Scientific Research Contract No. F49620 85C0144.

8. References

1. R.A. Adams, *Sobolev Spaces*, (Academic Press, 1975).
2. J.M. Bismut, Martingales, the Malliavin Calculus and hypoellipticity under general Hormonder's conditions, *Z. Wahrs.*, **56** (1981) 469-505.
3. H. Gaveau and P. Trauber, L'integrale stoch. comme opérateur de divergence dans l'espace fonctionnel, *J. Funct. Anal.*, **46** (1982) 230-238.
4. T. Hida, Analysis of Brownian functionals, in *Carleton Math. Lecture Notes*, No. **13** (Carleton University, 1975).
5. T. Hida, *Brownian Motion*, (Springer-Verlag, 1980).
6. K. Ito, Multiple Wiener integral, *J. Math. Soc. Japan*, (1951) 157-169.
7. I. Kubo and S. Takenaka, Calculus on Gaussian White noise, I and II, *Proc. Japan Acad.*, (1980).
8. H.H. Kuo, Brownian motion, diffusions and infinite dimensional calculus, *Springer-Verlag Lec. Notes*, **1316** (1988) 130-169.
9. H.H. Kuo and A. Russek, White noise approach to stochastic integration, *J. Mult. Anal.*, **24** (1988) 218-236.
10. J. Neveu, Processus Aléatoires Gaussiens, Les Press. De L'Université De Montréal (1968).
11. D. Nualart, Noncausal stochastic integrals, *Springer-Verlag Lec. Notes*, **1316** (1988) 80-129.
12. P. Malliavin, Calculus of variations and hypoelliptic operators, *Proc. Inter. Symp. on Stochastic Diff. Eqs. Kyoto 1976*, (1978) 195-263.
13. D. Nualart and M. Zakai, Generalized stochastic integrals and Malliavin calculus, *Prob. Theory and Related Fields*, **73** (1986) 255-280.

14. A.V. Skorohod, On a generalization of stochastic integral, *Theory of Prob. and Appl.*, **XX** (1975) 219-233.
15. M.E. Taylor, Pseudodifferential Operators, Princeton Univ. Press (1981).
16. M. Zakai, The Malliavin calculus, *Acta App. Math.* (1985), 175-207.

Projection Spectral Theorem and Its Applications to the Infinite-dimensional Harmonic Analysis

YU. M. BEREZANSKY
Institute of Mathematics, Ukrainian Academy of Sciences
Repin str. 3, Kiev, Ukrainian SSR
252601 USSR

ABSTRACT. For every selfadjoint operator A with discrete spectrum $(\lambda_j)_{j=1}^{\infty}$ it is possible to write the representation $A = \sum_{j=1}^{\infty} \lambda_j P(\lambda_j)$ where $P(\lambda_j)$ is the projector to the eigenspace with eigenvalue λ_j. The projection spectral theorem gives generalization of this type of formulae for general selfadjoint operators and for commutative families of such operators.

By means of this spectral theorem it is possible to obtain wide generalization of Stone's theorem concerning a group of unitary operators and a family of normal operators connected by means of some relations. The second application of the projection spectral theorem to harmonic analysis consists of representations of positive definite kernels where we have the solutions of some equations instead of the exponents (theorems of Bochner's type). The third application is connected with construction of quantum stochastic integrals. The proofs of these results require the theory of generalized functions of infinitely many variables which is developed in the talk.

§1. Projection spectral theorem. Introduction. Let A be a selfadjoint operator defined on a Hilbert space H_0. To it there corresponds the resolution of identity (RI) E, i.e., a projection-valued measure on the σ-algebra $\mathcal{B}(\mathbf{R}^1)$ of Borel subsets of $\mathbf{R}^1$ for which we have the spectral integral representations

$$1 = \int_{\mathbf{R}^1} dE(\lambda), \quad A = \int_{\mathbf{R}^1} \lambda \, dE(\lambda). \tag{1}$$

In the case of an operator A with a discrete spectrum $(\lambda_j)_{j=1}^{\infty}$ we have $E(\alpha) = \sum_{\lambda_j \in \alpha} P(\lambda_j)$ $(\alpha \in \mathcal{B}(\mathbf{R}^1))$, where $P(\lambda_j)$ is the projection on the eigenspace corresponding to λ_j, and the relations (1) become equalities

$$1 = \sum_{j=1}^{\infty} P(\lambda_j), \quad A = \sum_{j=1}^{\infty} \lambda_j P(\lambda_j). \tag{2}$$

In the case of a continuous spectrum such equalities can no longer be written down, but it is possible to generalize them by means of generalized eigenfunction expansions in the case of a *separable* H_0.

Let

$$H_- \supseteq H_0 \supseteq H_+ \supseteq D \tag{3}$$

be a rigging of H_0 by Hilbert spaces H_+ and H_- of "test" and "generalized" vectors with a quasinuclear (i.e. Hilbert-Schmidt) imbedding $H_+ \subseteq H_0$, and let D be a linear topological space topologically (i.e. densely and continuously) imbedded in H_0. Then in (2) the "positive" space H_+ is dense in H_0 and the imbedding operator $H_+ \to H_0$ is quasinuclear; the "negative" space H_- consists of all continuous antilinear functionals ξ on H_+, i.e. H_- is the dual space of H_+ with respect to H_0: $\xi(\phi) = (\xi, \phi)_{H_0}$ $(\phi \in H_+)$.

Assume that the operator A is connected with the chain (3) in the standard way: the domain $\mathfrak{D}(A)$ contains D and the restriction $A \upharpoonright D$ acts continuously from D to H_+. Then the RI, as an operator-valued measure whose values are operators from H_+ to H_-, is differentiable with respect to a certain scalar measure $\mathcal{B}(\mathbf{R}^1) \ni \alpha \mapsto \rho(\alpha) \in [0, \infty)$ (the so-called spectral measure): $(dE/d\rho)(\lambda) = P(\lambda) : H_+ \to H_-$ exists, and the equalities (1) are rewritten in the form

$$1 = \int_{\mathbf{R}^1} P(\lambda) d\rho(\lambda), \quad A = \int_{\mathbf{R}^1} \lambda P(\lambda) d\rho(\lambda). \tag{4}$$

It can be shown that for ρ-almost all $\lambda \in \mathbf{R}^1$ the range $\mathfrak{R}(P(\lambda))$ consists of generalized eigenvectors $\xi \in H_-$ corresponding to λ, i.e. $A\xi = \lambda\xi$ in a natural generalized sense. More precisely, the following equality is true: for every $\phi \in D$

$$(\xi, A\phi)_{H_0} = \lambda(\xi, \phi)_{H_0}. \tag{5}$$

Thus, $P(\lambda)$ becomes a "projector" onto the corresponding generalized eigenspace and the equalities (4), being a generalization of the equalities (2) in the case of a discrete spectrum, constitute the spectral projection theorem for a single operator.

It is often convenient to consider a somewhat more special situation that the "quasinuclear" chain (3) is replaced by a "nuclear" chain

$$\Phi' \supseteq H_0 \supseteq \Phi, \tag{6}$$

where Φ is a nuclear space topologically imbedded in H_0 and Φ' is its dual space consisting of continuous antilinear functionals on Φ. A picture analogous to (3) holds for a selfadjoint operator A defined on H_0 and satisfies more stringent restriction: $\Phi \subseteq \mathfrak{D}(A)$ and $A \upharpoonright \Phi$ acts continuously on Φ. More precisely, the required chain (3) can be chosen for the chain (6) and the results (4) described above are preserved for A.

§2. The formulation of projection spectral theorem. Our aim is to establish the results of type (4) for an arbitrary family of commuting selfadjoint (or more generally normal) operators.

Let $A = (A_x)_{x \in X}$ be a family of selfadjoint operators A_x acting on a Hilbert space H_0, and let E_x be the RI of A_x. The cardinality of X is arbitrary. We suppose that any pair of operators A_x, A_y $(x, y \in X)$ commute in the strong sense, i.e., for any $\alpha, \beta \in \mathcal{B}(\mathbf{R}^1)$ the operators $E_x(\alpha)$ and $E_y(\beta)$ commute.

It can be shown that on the space $\mathbf{R}^X$ of all functions $X \ni x \mapsto \lambda(x) \in \mathbf{R}^1$ there exists an RI given as the direct product of the RI E_x : $E = \prod_{x \in X} E_x$. More precisely, let $\mathcal{C}_\sigma(\mathbf{R}^X)$ be the σ-algebra spanned by all cylindrical subsets $C(x_1, \cdots, x_n; \delta) = \{\lambda(\cdot) \in \mathbf{R}^X | (\lambda(x_1), \cdots, \lambda(x_n)) \in \delta\}$ where $x_1, \cdots, x_n$ are distinct points in X and $\delta \in \mathcal{B}(\mathbf{R}^n)$ $(n = 1, 2, \cdots)$. Then the RI on $\mathcal{C}_\sigma(\mathbf{R}^X)$ is defined by

$$E\left(C(x_1, \cdots, x_n; \delta_1 \times \cdots \times \delta_n)\right) = E_{x_1}(\delta_1) \cdots E_{x_n}(\delta_n), \tag{7}$$

where $x_1, \cdots, x_n \in X$ and $\delta_1, \cdots, \delta_n \in \mathcal{B}(\mathbf{R}^1)$ $(n = 1, 2, \cdots)$.

It is easy to conclude from (7) that for any $x \in X$

$$A_x = \int_{\mathbf{R}^X} \lambda(x) dE(\lambda(\cdot)), \quad 1 = \int_{\mathbf{R}^X} dE(\lambda(\cdot)) \tag{8}$$

(where we integrate the function $\mathbf{R}^X \ni \lambda(\cdot) \mapsto \lambda(x) \in \mathbf{R}^1$.) This formula, of course, is a generalization of (1). But in (1) (as in (4)) we can replace the domain of integration $\mathbf{R}^1$ with the spectrum $s(A) = \operatorname{supp} E$ of A. Although we can introduce the notion of a joint spectrum $s(A)$ of the family $A = (A_x)_{x \in X}$ by setting $s(A) = \operatorname{supp} E$ ($\mathbf{R}^X$ is endowed with the Tychonoff topology), the replacement of $\mathbf{R}^X$ in (8) with $s(A)$ is incorrect in general: if X is more than countable and the operators A_x are unbounded, it can happen that the joint spectrum $s(A)$ is empty for $A = (A_x)_{x \in X}$.

It is essential for us to make a similar replacement for a general family $A = (A_x)_{x \in X}$ of commuting selfadjoint operators A_x standardly connected with the chain (3). For this purpose we introduce the notion of generalized joint spectrum $g(A) \subseteq \mathbf{R}^X$ of the family A: the collection of all functions $X \ni x \mapsto \lambda(x) \in \mathbf{R}^1$ (eigenvalues) satisfying $A_x \xi = \lambda(x) \xi$ for any $x \in X$ in the generalized sense, where ξ is a generalized (joint) eigenvector of A_x (see (5)). Of course the set $g(A)$ depends on the rigging (3): if we change the rigging, the $g(A)$ changes too.

Roughly speaking, the projection spectral theorem asserts that in formulae (8) and in generalization of (4) for a family $A = (A_x)_{x \in X}$ we can replace $\mathbf{R}^X$ with $g(A)$. We shall give now the exact formulation.

Let D in (3) be a projective limit of Hilbert spaces, $O : H_+ \to H_0$, $O^+ : H_0 \to H_-$ be the imbedding operators and $\rho(\alpha) = \operatorname{Tr}(O^+ E(\alpha) O)$ $(\alpha \in \mathcal{C}_\sigma(\mathbf{R}^X))$ be the spectral measure of the family A in the space $\mathbf{R}^X$.

THEOREM 1. *In the generalized joint spectrum $g(A)$ there exists a subset π of full outer ρ-measure such that for each $\lambda(\cdot) \in \pi$ the operator $P(\lambda(\cdot)) : H_+ \to H_-$ exists with range $\mathfrak{R}(P(\lambda(\cdot)))$ consisting of generalized joint eigenvectors of A corresponding to the* *eigenvalue* $\lambda(\cdot)$ *("generalized projector onto the corresponding generalized eigenspace"). The following representations are valid as functional spectral integrals:*

$$
(9)\qquad \begin{aligned} O^+E(\alpha)O &= \int_\alpha dE(\lambda(\cdot)) = \int_\alpha P(\lambda(\cdot))d\rho(\lambda(\cdot)) \qquad (\alpha \subseteq \pi), \\ A_x &= \int_\pi \lambda(x)dE(\lambda(\cdot)) = \int_\pi \lambda(x)P(\lambda(\cdot))d\rho(\lambda(\cdot)) \qquad (x \in X). \end{aligned}
$$

Here the operator A_x *is understood to be an operator with domain* $\mathfrak{D}(A_x) \cap H_+$*, the measures* E *and* ρ *are the restrictions of the previous measures on* π *(such restrictions exist and are well-defined because* π *has the full outer measure with respect to* ρ *and therefore* E*).*

REMARK 1. Fix an arbitrary subset $\tau \subseteq \mathbf{R}^X$ containing π (for instance $\tau \supseteq g(A)$). The representations in (9) are valid if we replace π with τ and set $P(\lambda(\cdot)) = 0$ for $\lambda(\cdot) \in \tau \setminus \pi$.

REMARK 2. Assume that the index set X is a topological space and that the operators A_x depend continuously on $x \in X$ in the following sense: for each $\phi \in D$ the vector-valued function

$$
(10)\qquad X \ni x \mapsto A_x\phi \in H_+
$$

is weakly continuous. Then each eigenvalue $\lambda(\cdot) \in g(A)$ is a continuous function and the integrals in (9) are continual. If X is a differentiable manifold and (10) is k-times weakly continuously differentiable, then $\lambda(\cdot) \in g(A)$ is k-times continuously differentiable.

REMARK 3. In the above results the selfadjoint operators A_x can be replaced with commuting normal operators. The assertions remain valid with the following changes. The space $\mathbf{R}^X$ is replaced clearly by $\mathbf{C}^X$. The notion of standard connection is understood as follows: for any $x \in X$ the restrictions $A_x \restriction D$ and $A_x^* \restriction D$ act continuously from D to H_+. Instead of (5) we now have

$$
(11)\qquad (\xi, A_x^*\phi)_{H_0} = \lambda(x)(\xi,\phi)_{H_0}, \qquad (\xi, A_x\phi)_{H_0} = \overline{\lambda(x)}(\xi,\phi)_{H_0}
$$

$$
(\xi \in H_-, \phi \in D; x \in X).
$$

REMARK 4. An analogue of Theorem 1 is valid if we change the quasinuclear rigging (3) for a nuclear one (6). The case of nuclear rigging can be reduced to the quasinuclear one.

REMARK 5. In the case of X being countable the proof of Theorem 1 is as simple as in the case of a single operator. Essential difficulties arise when X is more than countable.

§3. The applications to the theorems of Stone type. The projection spectral theorem is a convenient tool of proving spectral representations of families of operators which are connected by means of some relations. Suppose that the operators A_x of our family A of commuting selfadjoint or normal operators are bounded and are connected in a definite way for different $x \in X$. For example, X is an abelian group and $A_{x+y} = A_x A_y (x, y \in X)$, or X is a linear space and $A_{\alpha x+\beta y} = \alpha A_x + \beta A_y$ $(x, y \in X; \alpha, \beta \in \mathbf{C}^1)$; i.e., the function $X \ni x \mapsto A_x$ effects a representation of such algebraic objects — a group, a linear space, a semigroup, an algebra, etc. Moreover, the relations can be less traditional, for instance, $X = \mathbf{R}^1$ and $A_x A_y = (A_{x+y} + A_{x-y})/2$ $(x, y \in \mathbf{R}^1)$, or A_x satisfying a differential equation with respect to x, and so on.

First I want to explain how to get from Theorem 1 the classical Stone theorem on a group A_x, $x \in \mathbf{R}^1$, of unitary operators in a separable Hilbert space H_0 (A_x depending continuously on x). Suppose we are given a family $A = (A_x)_{X \in \mathbf{R}^1}$ of commuting unitary operators. In addition, for the present, we suppose that the operators A_x are connected in a standard manner with some quasinuclear chain (3).

Let $\mathbf{R}^1 \ni x \mapsto \lambda(x) \in \mathbf{C}^1$ be the eigenvalue of A and ξ the corresponding generalized joint eigenvector. Then for any $x, y \in \mathbf{R}^1$

$$\lambda(x+y)\xi = A_{x+y}\xi = A_x A_y \xi = A_x(\lambda(y)\xi) = \lambda(x)\lambda(y)\xi,$$

and therefore $\lambda(x+y) = \lambda(x)\lambda(y)$, i.e., $\lambda(\cdot)$ is a character of the group $\mathbf{R}^1$ ($|\lambda(x)| = 1$ because $\lambda(x)$ belongs to the spectrum of A_x). From continuity of $\mathbf{R}^1 \ni x \mapsto A_x$ and Remark 2 we conclude that this character is continuous and therefore of the form $\lambda(x) = e^{i\lambda x}$, where λ is the corresponding point in $\mathbf{R}^1$. Hence, in our case the generalized spectrum $g(A) \subseteq \mathfrak{X}$, where $\mathfrak{X}$ is the set of all functions $\mathbf{R}^1 \ni x \mapsto e^{i\lambda x}$, $\lambda \in \mathbf{R}^1$. Setting $\tau = \mathfrak{X}$, we apply Theorem 1 and Remark 1. The second formula in (9), where π is replaced with $\tau = \mathfrak{X}$, becomes:

$$A_x = \int_{\mathfrak{X}} \lambda(x) dE(\lambda(\cdot)) = \int_{\mathfrak{X}} e^{i\lambda x} dE(\lambda(\cdot)) = \int_{\mathbf{R}^1} e^{i\lambda x} dF(\lambda) \quad (x \in \mathbf{R}^1). \tag{12}$$

Here F is the RI on $\mathbf{R}^1$ transported from E by the mapping: $\mathfrak{X} \ni \lambda(\cdot) = e^{i\lambda \cdot} \mapsto \lambda \in \mathbf{R}^1$. Thus the representation (12) is a usual representation of Stone's theorem. Our assumption on existence of a rigging is not essential: some general facts ensure the existence of such a rigging for our family A (see Remark 9 below).

Of course, such a proof of Stone's theorem is not simple, but it is very general: if we have some relation among operators A_x then after the application of this relation to a generalized joint eigenvector ξ we shall get an equation for the corresponding eigenvalue $\lambda(\cdot)$. Therefore $g(A)$ consists of solutions of this equation. We can take the set of all such solutions as a set τ and then (9) gives the required representation. This idea is illustrated by some examples.

THEOREM 2 (THE STONE THEOREM FOR HILBERT SPACE). *Let X be a real separable Hilbert space regarded as an additive group. Then a unitary representation $X \ni x \mapsto A_x$*

on H_0 has the form

$$A_x = \int_X e^{i(\lambda,x)_X}\, dE(\lambda) \qquad (x \in X), \tag{13}$$

where E is an RI on the σ-algebra $\mathcal{B}(X)$ of Borel subsets of X, if and only if there exists a rigging (3) such that (i) the restriction $A_x \restriction D$ acts continuously from D to H_+ ($x \in X$); and (ii) for each $\phi \in D$ the vector-valued function $X \ni x \mapsto A_x\phi \in H_+$ is weakly continuous. The RI in (13) is uniquely determined by $(A_x)_{x\in X}$.

THEOREM 3 (THE SZ.-NAGY-HILLE THEOREM FOR HILBERT SPACE). *Let $X \ni x \mapsto A_x$ be a representation of X as in Theorem 2, but assume that the operators A_x are normal and unbounded (in general). Then there exists a decomposition $H_0 = H_1 \oplus H_2$ where the subspaces H_1, H_2 are invariant under A_x ($x \in X$) and*

$$A_x \restriction H_1 = 0, \quad A_x \restriction H_2 = \int_{X_c} e^{(\lambda,x)_X}\, dE(\lambda) \quad (x \in X), \tag{14}$$

if and only if there exists a rigging (3) such that (i) both $A_x \restriction D$ and $A_x^* \restriction D$ act continuously from D to H_+ ($x \in X$); and (ii) for each $\phi \in D$ the vector-valued function $X \ni x \mapsto A_x\phi \in H_+$ is weakly continuous (in (14) E is an RI on the σ-algebra $\mathcal{B}(X_c)$ of Borel subsets of the complexification X_c of X). The RI in (14) is uniquely determined by $(A_x)_{x\in X}$.

Suppose that X is a real separable Hilbert space as above and that a family of commuting selfadjoint operators $X \ni x \mapsto A_x$ in the space H_0 (from the rigging (3)) is a selfadjoint representation of X, i.e., $\mathfrak{D}(A_x) \supseteq D$ for any $x \in X$ and $A_{\alpha x+\beta y}\phi = \alpha A_x\phi + \beta A_y\phi$ for any $\phi \in D$ ($x, y \in X$; $\alpha, \beta \in \mathbf{R}^1$).

THEOREM 4. *A selfadjoint representation of a real separable Hilbert space X has the form*

$$A_x = \int_X (\lambda, x)_X\, dE(\lambda) \qquad (x \in X), \tag{15}$$

where E is an RI on $\mathcal{B}(X)$ if the conditions (i) and (ii) in Theorem 2 hold for A_x. Conversely, the existence of a representation (15) for some family $(A_x)_{x\in X}$ implies the existence of a rigging (3) and the remaining requirements of the theorem. The RI is uniquely determined by $(A_x)_{x\in X}$.

Let $(e_j)_{j=1}^{\infty}$ be a fixed orthonormal basis of X. Then the set $X_+ = \{x = \sum_{j=1}^{\infty} x_j e_j \in X \,|\, x_j > 0\}$ is a topological semigroup with respect to addition. Let $X_+ \ni x \mapsto A_x$ be a representation of X_+ by means of normal operators (i.e., $A_{x+y} = A_x A_y$ for any $x, y \in X_+$).

THEOREM 5. *Theorem 3 is true for a representation of the semigroup X_+.*

By means of the projection spectral theorem we can investigate the representations of general abelian groups, algebras, etc. For example, let X be an abelian group, $\mathfrak{X}_g$ the set of all generalized characters of this group, i.e., all functions $X \ni x \mapsto \chi(x) \in \mathbf{C}^1$ such that $\chi(x+y) = \chi(x)\chi(y)$ $(x, y \in X)$. Let $X \ni x \mapsto A_x$ be a representation of X by means of normal operators A_x which act in the space H_0 of the nuclear rigging (6). Such a representation is called nuclear if $\Phi \subseteq \mathfrak{D}(A_x)$ and $A_x \upharpoonright \Phi$ acts continuously on Φ for any $x \in X$.

THEOREM 6. *The operators A_x of a nuclear representation of X has the form*

$$A_x = \int_{\mathfrak{X}_g} \chi(x) dE(\chi) \qquad (x \in X), \tag{16}$$

where E is an RI on the σ-algebra $C_\sigma(X) = \{\alpha \cap \mathfrak{X}_g | \alpha \in C_\sigma(\mathbf{C}^X)\}$.

REMARK 6. If the group X is topological and the representation is continuous (for any $\phi \in \Phi$ vector-valued function $X \ni x \mapsto A_x\phi \in \Phi$ is weakly continuous) then the domain of integration in (16) is replaced with the set $\mathfrak{X}_{g,c}$ of all continuous generalized characters. If the operators A_x are unitary then the domain of integration consists of all characters or of all continuous characters.

Similar results to Theorem 6 are true for representations of commutative algebras. In this case the domain of integration consists of multiplicative functionals.

If we are given a family $A = (A_x)_{x \in \mathbf{R}^1}$ of commuting selfadjoint operators connected by means of the relation $A_x A_y = (A_{x+y} + A_{x-y})/2$ $(x, y \in \mathbf{R}^1)$, then the corresponding equation for the eigenvalues $\lambda(\cdot)$ has the form

$$\lambda(x)\lambda(y) = \frac{1}{2}(\lambda(x+y) + \lambda(x-y)) \qquad (x, y \in \mathbf{R}^1). \tag{17}$$

If we suppose some continuity of the mapping $\mathbf{R}^1 \ni x \mapsto A_x$, then we must use only continuous real solutions of (17), i.e., the functions $\lambda(x) = \cos\sqrt{\lambda}x$ $(\lambda \in \mathbf{R}^1)$. For our operators A_x the representation of type (13) includes these functions. The example (13) is a particular case of more general situation where instead of $\cos\sqrt{\lambda}x$ we have characters of a commutative hypercomplex system with continuous basis. The results of Theorems 2–6 are true for representations of such systems and hypergroups.

REMARK 7. It is noticeable that the second part of the assertion is absent in theorems of type Theorem 6 (which requires a nuclear rigging): we can not assert the existence of a rigging from the representation (16). This situation is due to impossibility of constructing a space of test functions on $\mathfrak{X}_g$ with necessary properties (see below).

I want to explain now the idea of proof and the necessity of a rigging in Theorems 2–6 (ascending to construction of the Gårding domain in the theory of group representations).

We shall deal with Theorem 2. Being given operators (13) on H_0, we want to construct a rigging (3) which is standardly connected with A_x for any $x \in X$. Suppose that we can construct a quasinuclear rigging of type (3)

$$\mathcal{H}_-(X) \supseteq L_2(X, d\sigma(\lambda)) \supseteq \mathcal{H}_+(X) \supseteq \mathcal{D}(X) \tag{18}$$

with spaces consisting of complex-valued functions on X. Here $d\sigma(\lambda) = d(E(\lambda)e, e)_{H_0}$ where $e \in H_0$ with $\|e\|_{H_0} = 1$ is fixed; the operator of multiplication $\mathcal{D}(X) \ni \phi(\cdot) \mapsto e^{i(\cdot,x)} \times \phi(\cdot) \in \mathcal{H}_+(X)$ must be continuous for any $x \in X$. In addition, we suppose that the linear set of vectors

$$f_\phi = \int_X \phi(\lambda)\, dE(\lambda)e \in H_0 \qquad (\phi \in \mathcal{H}_+(X)) \tag{19}$$

is dense in H_0. Let $H_0 \supseteq H_+ \supseteq D$ be the image of the chain $L_2(X, d\sigma(\lambda)) \supseteq \mathcal{H}_+(X) \supseteq \mathcal{D}(X)$ under the mapping $L_2(X, d\sigma(\lambda)) \ni \phi \mapsto f_\phi \in H_0$ (the image of $L_2(X, d\sigma(\lambda))$ is equal to H_0). We endow H_+ and D with the topologies of the preimages; thus H_+ becomes a Hilbert space with inner product $(f_\phi, f_\psi)_{H_+} = (\phi, \psi)_{\mathcal{H}_+(X)}$ $(\phi, \psi \in \mathcal{H}_+(X))$. Then $H_- \supseteq H_0 \supseteq H_+ \supseteq D$ is the required quasinuclear rigging because the application of operator A_x to f_ϕ is equivalent to that of multiplication of $e^{i(\cdot,x)x}$ to $\phi(\cdot)$.

Our additional assumption on the density of vectors (19) in H_0 is not essential. Taking a new vector e in the orthogonal complement to these vectors, we can repeat our procedure.

In reality this construction is more complicated: it can be changed in such a manner that the measure $d\sigma(\lambda)$ in (18) is fixed and independent of the family $(A_x)_{x \in X}$. For example, we can take the Gaussian measure $d\gamma(\lambda)$ on the real separable Hilbert space X.

Thus, we have reduced our problem to construction of a quasinuclear rigging (18) with some properties, where the test function space consists of functions of infinitely many variables. We have used two connected methods of constructing such a rigging: 1) by means of infinite tensor product; 2) by means of construction of a rigging of Fock spaces and application of the Segal isomorphism. Let us explain these two ways briefly.

1) Let $H_{n,-} \supseteq H_{n,0} \supseteq H_{n,+} \supseteq D_n$ $(n = 1, 2, \cdots)$ be a sequence of quasinuclear riggings of type (3). It is possible to introduce the notion of an infinite tensor product $\otimes$ of Hilbert spaces and their projective limits. Accordingly, we obtain a chain

$$H_- = \bigotimes_{n=1}^{\infty} H_{n,-} \supseteq H_0 = \bigotimes_{n=1}^{\infty} H_{n,0} \supseteq H_+ = \bigotimes_{n=1}^{\infty} H_{n,+} \supseteq \bigotimes_{n=1}^{\infty} D_n = D, \tag{20}$$

which is again a quasinuclear rigging of type (3).

We shall realize the space $L_2(X, d\gamma(x))$ as $L_2(\ell_2, d\gamma(x))$ where $d\gamma(x)$ is the Gaussian product measure: $d\gamma(x) = d\gamma_1(x_1) \times d\gamma_2(x_2) \times \cdots$, $x = (x_n)_{n=1}^{\infty} \in \ell_2$. Then

$$\begin{aligned} L_2(X, d\gamma(x)) &= L_2(\ell_2, d\gamma(x)) = L_2(\mathbf{R}^\infty, d\gamma(x)) \\ &= L_2(\mathbf{R}^1, d\gamma_1(x_1)) \otimes L_2(\mathbf{R}^1, d\gamma_2(x_2)) \otimes \cdots \end{aligned}$$

and according to (20) we can construct a quasinuclear rigging of type (18) for $L_2(X, d\gamma(x))$.

2) Let $\mathcal{F}(Y)$ be the symmetric Fock space over a real space Y (which is a Hilbert space or a projective limit of Hilbert spaces). Given a nuclear chain of type (6) of real spaces $\Phi' \supseteq X \supseteq \Phi$, we apply the functor $\mathcal{F}$ to get the rigging $(\mathcal{F}(\Phi))' \supseteq \mathcal{F}(X) \supseteq \mathcal{F}(\Phi)$. Let $d\gamma(\xi)$ be a Gaussian measure on Φ' which is concentrated on X, and let $\mathcal{F}(X) \ni f \mapsto If \in L_2(\Phi', d\gamma(\xi))$ be the corresponding Segal isomorphism. The application of I gives the rigging $(I\mathcal{F}(\Phi))' \supseteq L_2(\Phi', d\gamma(\xi)) = I\mathcal{F}(X) \supseteq I\mathcal{F}(\Phi)$ which is again a chain of type (18).

REMARK 8. During the above discussion we can prove a converse result of Theorem 1: if the representation

$$A_x = \int_\tau \lambda(x)\, dE(\lambda(\cdot)) \qquad (x \in X)$$

exists with some "good" $\tau \subseteq \mathbf{C}^X$, then there exists a rigging (3) which is connected in a standard manner with A_x $(x \in X)$ (see also Remark 3). In particular, since $\tau = \mathbf{C}^n, \mathbf{C}^\infty$ are good, there exists such a rigging for any at most countable family of commuting normal operators.

REMARK 9. Modelled after the proof of existence of a rigging for operators A_x as in (13), we can give analogous results for a family of normal operators $A = (A_x)_{x \in X}$ where X is a locally compact group and A realizes the representation of this group: $A_x A_y = A_{xy}$ for any $x, y \in X$. Now instead of formula (19) we can write

$$f_\phi = \int_X \phi(y) A_y\, dg(y) e \in H_0, \tag{21}$$

where $dg(y)$ is the invariant measure on the group X and ϕ belongs to some space $\mathcal{H}_+(X)$ of test functions on X. The application of operator A_x to vector (21) is reduced to the transformation $\phi(\cdot) \mapsto \phi(x^{-1}\cdot)$. Its continuity depends on properties of the space $\mathcal{H}_+(X)$. In case $X = \mathbf{R}^n$ we can take, for example, the Sobolev space with some weight $p(x) : \mathcal{H}_+(\mathbf{R}^n) = W_2^\ell(\mathbf{R}^n, p(x)dx)$, $\ell > n/2$; as $\mathcal{D}(\mathbf{R}^n)$ we can choose the classical space of test functions $\mathcal{D}(\mathbf{R}^n)$ or $\mathcal{S}(\mathbf{R}^n)$.

This method is also applicable to some nonlocally compact groups, hypercomplex systems and hypergroups. In the case of a nonlocally compact group the procedure can be carried out if it admits a quasi-invariant measure. Then in (21) we take such a measure as $dg(y)$ and the application of A_x to f_ϕ under the integral yields some factor– the Radon-Nikodym derivative $dg(x^{-1}\cdot)/dg(\cdot)$. The space $\mathcal{H}_+(X)$, as a rule, consists of test functions of infinitely many variables.

§4. The applications to commutation relations (commutative models). I only mention application of Theorem 1 to commutation relations. Let $B = (B_y)_{y \in Y}$ be a family of operators connected by means of some permutation relations. It is often possible to construct another family $A = (A_x)_{x\in X}$ of commuting normal operators with the following property. Let

$$\phi = \int_\tau P(\lambda(\cdot))d\rho(\lambda(\cdot))\phi \qquad (\phi \in H_+) \tag{22}$$

be the representation of type (9) for this family. Then for any $y \in Y$ the action of B_y on H_+ is reduced to the action $B_yP(\lambda(\cdot))$ and the last operator can have the form $m(\lambda(\cdot))P(F_y(\lambda(\cdot)))(\cdot)$. Here $m(\lambda(\cdot))$ is a scalar factor and $\tau \ni \lambda(\cdot) \mapsto (F_y(\lambda(\cdot)))(\cdot)$ is a transformation of the space τ. Hence the vector $B_y\phi$ has the form

$$B_y\phi = \int_\tau m(\lambda(\cdot))P\left((F_y(\lambda(\cdot)))(\cdot)\right)d\rho(\lambda(\cdot))\phi = \int_\tau P(\mu(\cdot))d\rho_y(\mu(\cdot))\phi. \tag{23}$$

Here we have used the change of variables $\left(F_y(\lambda(\cdot))\right)(\cdot) = \mu(\cdot)$; $d\rho_y(\mu(\cdot))$ is the corresponding new measure obtained from the Radon-Nikodym derivative.

According to the formulae (22), (23) the action of B_y is equivalent to the transformation $\rho \mapsto \rho_y$ of the spectral measure and the permutation relations among operators B_y can be simpler in the language of this measure. In a series of important cases we can describe the relations among transformations of the spectral measure, and therefore, get representations of the initial permutation relations.

§5. The representation of positive-definite kernels. Now I should like to describe applications of the projection spectral theorem to integral representations of positive-definite kernels, in particular, to the theory of positive-definite functions of infinitely many variables and to the infinite-dimensional moment problem. Let us explain the idea of such application (ascending to M.Krein) in the simplest case of positive-definite functions of one variable.

Remind that a continuous function $(-2\ell, 2\ell) \ni x \mapsto k(x) \in \mathbb{C}^1$ $(0 < \ell \le \infty$ is fixed) is called positive-definite if the kernel $K(x,y) = k(y-x)$ $(x, y \in (-\ell, \ell))$ is positive-definite. On the set $C_0^\infty((-\ell,\ell))$ of finite and infinite differentiable functions f, g on $(-\ell, \ell)$ we introduce a quasi-inner product

$$(f,g)_{H_0} = \int_{-\ell}^{\ell}\int_{-\ell}^{\ell} k(y-x)f(y)\overline{g(x)}\,dxdy. \tag{24}$$

Let H_0 be the Hilbert space corresponding to (24) and A be the closure of the operator $C_0^\infty((-\ell,\ell)) \ni f \mapsto -idf/dt \in C_0^\infty((-\ell,\ell))$. This operator is Hermitian and admits a selfadjoint extension A_1 in H_0. It is easy to construct a rigging of type (3) or (6) for A_1 (for example, $\Phi = C_0^\infty((-\ell,\ell)))$ and hence it is possible to apply Theorem 1 with $X = \{1\}$. According to the first equality in (9) (for $\alpha = \mathbf{R}^1$) and (24), we have

$$\int_{-\ell}^{\ell}\int_{-\ell}^{\ell} k(y-x)f(y)\overline{g(x)}\,dxdy = (f,g)_{H_0} = \int_{\mathbf{R}^1}(P(\lambda)f,g)_{H_0}d\rho(\lambda) \tag{25}$$

$$(f, g \in C_0^\infty((-\ell, \ell)))$$

where $P(\lambda)$ is the corresponding generalized projector. By a simple calculation we find:

$$(P(\lambda)f, g)_{H_0} = c(\lambda) \int_{-\ell}^{\ell} \int_{-\ell}^{\ell} e^{i\lambda(y-x)} f(y)\overline{g(x)}\, dx dy \qquad (c(\lambda) \geq 0). \tag{26}$$

Inserting (26) into (25) and taking it into account that f, g are arbitrary, we get

$$k(t) = \int_{\mathbf{R}^1} e^{i\lambda t} d\sigma(\lambda) \qquad (t \in (-2\ell, 2\ell)) \tag{27}$$

(here $y - x = t$, $c(\lambda)d\rho(\lambda) = d\sigma(\lambda)$). The formula (27) is the classical result of Bochner (if $\ell = \infty$) and M.Krein (if $\ell < \infty$; in this case (27) asserts that every positive-definite function on $(-2\ell, 2\ell)$ can be extended to such a function on $\mathbf{R}^1$.)

The above mentioned method of proof is applicable also to the discrete situation where the difference is taken instead of d/dx. In particular, such approach gives a solution of the classical moment problem which consists in the following. A sequence $s = (s_n)_{n=0}^\infty$ ($s_n \in \mathbf{R}^1$) is called a moment sequence if the matrix $(a_{jk})_{j,k=0}^\infty$, $a_{jk} = s_{j+k}$, is positive-definite. Then it admits a representation

$$s_n = \int_{\mathbf{R}^1} \lambda^n \, d\sigma(\lambda) \qquad (n = 0, 1, \cdots) \tag{28}$$

with some measure $d\sigma(\lambda)$ (of course, the moment sequence $s = (s_n)_{n=0}^\infty$ follows from (28)).

Below I shall speak about infinite-dimensional generalization of (27) and (28). The method of proof is previous but now we apply Theorem 1 in the case of a countable family of operators A_x. The spaces in the rigging (3) (or (6)) consist of functions of infinitely many variables and are constructed by means of the procedure mentioned in Section 3.

Let X be a real separable Hilbert space and $e \in X$ a fixed normal vector. The strip X_ℓ in X is defined as $X_\ell = \{x \in X | (x, e)_X \in (-\ell, \ell)\}$ $(0 < \ell \leq \infty$ is fixed). A function $X_{2\ell} \ni x \mapsto k(x) \in \mathbf{C}^1$ is called positive-definite if for any $x^{(1)}, \cdots, x^{(n)} \in X_\ell$ and $\xi_1, \cdots, \xi_n \in \mathbf{C}^1$ $(n = 1, 2, \cdots)$ the inequality holds:

$$\sum_{j,k=1}^{n} k(x^{(j)} - x^{(k)}) \xi_j \overline{\xi_k} \geq 0.$$

The J-topology of X is by definition created by neighbourhoods of zero of the form $\{x \in X | (Ax, x)_X < \epsilon\}$ where A runs over the positive nuclear operators and $\epsilon > 0$.

THEOREM 7. *Every positive-definite function $k(x)$ ($x \in X_{2\ell}, \ell \leq \infty$) which is continuous at 0 with respect to the J-topology has a representation*

$$(29) \qquad k(x) = \int_X e^{i(x,\lambda)_X} d\sigma(\lambda) \qquad (x \in X_{2\ell}),$$

where $d\sigma(\lambda)$ is a positive finite measure on $\mathcal{B}(X)$. Conversely, every integral (29) is such a function. In case of $\ell = \infty$ the measure $d\sigma(\lambda)$ is uniquely determined.

In case of $\ell = \infty$ this result is the theorem of Minlos-Sazonov-Gross. In case of $\ell < \infty$ the result is a generalization of M.Krein's theorem for a Hilbert space.

REMARK 10. For the proof of Theorem 7 we first got a more general result on representation of a positive-definite function $k(x)$ on the space $\mathbf{R}^\infty$ with a product measure $p_1(x_1)dx_1 \times p_2(x_2)dx_2 \times \cdots$. The arguments are similar to the case (27) but we use a family of commuting operators $(\partial/\partial x_n)_{n=1}^\infty$ instead of a single operator d/dx. By this result and the J-continuity of k we can go on to a function on $X = \ell_2 \subset \mathbf{R}^\infty$.

For the formulation of the infinite-dimensional generalization of the moment problem (28) we first introduce a family of riggings of type (3):

$$(30) \qquad \bigcup_{\tau \in T} X_{-\tau} = \Phi' \supseteq X_{-\tau} \supseteq X_0 \supseteq X_\tau \supseteq \Phi = \operatorname{pr}\lim_{\tau \in T} X_\tau = \bigcap_{\tau \in T} X_\tau \quad (\tau \in T).$$

Here all the spaces are real, $X_{-\tau} = (X_\tau)_-$ and the cardinality of the index set T is arbitrary. We do not suppose the imbeddings $X_\tau \subseteq X_0$ to be quasinuclear. For $\tau \in T$ and $n = 0, 1, 2, \cdots$ let $X_\tau^{\otimes n}$ be the tensor product $X_\tau \otimes \cdots \otimes X_\tau$ (n times). We define $\Phi^{\otimes n}$ as $\operatorname{pr}\lim_{\tau \in T} X_\tau^{\otimes n}$, then the conjugate space $(\Phi^{\otimes n})' = \cup_{\tau \in T}(X_\tau^{\otimes n})' = \cup_{\tau \in T} X_{-\tau}^{\otimes n}$. In all these formulae we can go on to the complexification. The complexification is indicated by index c.

A sequence $s = (s_n)_{n=0}^\infty$, $s_n \in (\Phi^{\otimes n})'$, is called a moment sequence if it is symmetric and for every finite sequence $(\phi_n)_{n=0}^\infty$, $\phi_n \in \Phi_c^{\otimes n}$, the following inequality holds:

$$\sum_{j,k=0}^{\infty} s_{j+k}(\phi_j \otimes \overline{\phi_k}) \geq 0.$$

For s we want to get an integral representation of type (28) but, in general, such a representation does not exist because the corresponding operators are not commuting selfadjoint. Therefore we must demand the fulfillness of some estimates for s_n which guarantee such selfadjointness. One of the simplest estimates has the following form: there exists some $\tau \in T$ such that

$$(31) \qquad |s_{2n}(\phi^{(1)} \otimes \cdots \otimes \phi^{(2n)})| \leq m_n^2 \prod_{j=1}^{2n} \|\phi^{(j)}\|_{H_\tau}$$

$$(\phi^{(1)}, \cdots, \phi^{(2n)} \in \Phi; n = 0, 1, \cdots)$$

and the class $C\{m_n\}$ is quasi-analytic (for example, $m_n = n!$).

THEOREM 8. *Let $s = (s_n)_{n=0}^{\infty}$ be a moment sequence satisfying the estimate (31). Then the following representation holds:*

$$s_n = \int_{\Phi'} \lambda \otimes \cdots \otimes \lambda \, d\sigma(\lambda) \qquad (n = 0, 1, \cdots), \tag{32}$$

where $d\sigma(\lambda)$ is a positive finite measure on $\mathcal{B}(\Phi')$ and the function under integral has the form: $\Phi' \ni \lambda \mapsto \lambda \otimes \cdots \otimes \lambda \in (\Phi^{\otimes n})'$ (n times). Every sequence (32) is, of course, a moment sequence. The measure $d\sigma(\lambda)$ is uniquely determined.

REMARK 11. For some choice of spaces of type of (30), consisting of functions on $\mathbf{R}^d$ ($d = 1, 2, \cdots$), the representation (32) has the form

$$s_n(x_1, \cdots, x_n) = \int_{C(\mathbf{R}^d)} \lambda(x_1) \cdots \lambda(x_n) \, d\sigma(\lambda(\cdot)) \quad (n = 0, 1, \cdots). \tag{33}$$

We have so far given only two applications of Theorem 1 to the proofs of integral representations of positive-definite kernels. It is possible to get analogous representations in which under integrals the functions stand which are the generalized joint eigenfunctions of some family of operators $A = (A_x)_{x \in X}$ (in (29), (32) and (33) we have eigenfunctions of the operators of derivatives and shifts). Our kernel K must be closely connected with this family A. For example, if $X = \{1, 2, \cdots\}$ and $A = (\mathcal{L}_{x_n})_{n=1}^{\infty}$ with $\mathcal{L}_{x_n}$ being a linear differential operator with respect to the variable $x_n \in \mathbf{R}^1$, then the positive-definite kernel $K(x, y)$ $(x, y \in \mathbf{R}^{\infty})$ must satisfy the following equalities: $\mathcal{L}_{x_n} K = \overline{\mathcal{L}_{y_n}} K$ $(n = 1, 2, \cdots)$. If K is smooth and its growth at ∞ is controlled for every x_n, then we get a representation of K in terms of the solutions of the equations $\mathcal{L}_{x_n} \phi = \lambda_n \phi$ $(n = 1, 2, \cdots)$.

§6. Quantum stochastic integrals. We give now a short exposition of application of Theorem 1 to construction and investigation of the operator (quantum) stochastic integral. Let $A = (A(t))_{t \in (0,\infty)}$ be a family of commuting normal operators in a separable Hilbert space H_0 and let $\mathcal{B}((0, \infty)) \ni \alpha \mapsto E(\alpha)$ be an RI. We suppose that A and E partially commute in the sense that $A(t)$ and $E(\alpha)$ commute for any $t > 0$ and $\alpha \in \mathcal{B}((t, \infty))$. Then under some simple technical assumptions we can define the integral

$$B(\tau) = \int_0^{\tau} A(t) \, dE(t) \qquad (\tau \in (0, \infty)) \tag{34}$$

for any $\tau \in (0, \infty)$ in the following manner. Let K be the RI of the family A defined on the σ-algebra $\mathcal{C}_\sigma(\mathbf{C}^{(0,\infty)})$. According to (8) we have

$$A(t) = \int_{\mathbf{C}^{(0,\infty)}} \lambda(t) \, dK(\lambda(\cdot)) \qquad (t \in (0, \infty)).$$

Inserting the last formula into (34) and changing the order of integration formally, we obtain

$$B(\tau) = \int_{(0,\tau)\times\mathbf{C}^{(0,\infty)}} \lambda(t)\, d(E \times K)((t, \lambda(\cdot))) \qquad (\tau \in (0,\infty)). \tag{35}$$

It follows from the partial commutativity of A and E that the direct product $E \times K$ exists and is the RI. Therefore (35) is an ordinary spectral integral and we can define the integral (34) by means of (35).

Thus, (34)–(35) give a family $B = (B(\tau))_{\tau\in(0,\infty)}$ of commuting normal operators. This family has some interesting properties, for example: 1) $B(\tau)B(\sigma) = B^2(\min\{\tau,\sigma\})$ for $\tau,\sigma \in (0,\infty)$; 2) $A(s)$ and $B(\tau)$ commute for $s,\tau \in (0,\infty)$ with $s \le \tau$; 3) for a function $\mathbf{C}^1 \ni z \mapsto F(z) \in \mathbf{C}^1$

$$F\left(\int_0^\tau A(t)dE(t)\right) = \int_0^\tau F(A(t))dE(t) \qquad (\tau \in (0,\infty)).$$

Let A be a family which is connected in a standard manner with the quasinuclear rigging (3). Then the family B is also connected in a standard manner with this rigging and we can describe the generalized spectrum $g(B)$ and properties of the generalized joint eigenvectors of B.

I should like to explain how an ordinary stochastic integral $\int_0^\tau \xi_t d\mu_t = \eta_\tau$ $(\tau \in (0,\infty))$, where $\mu = (\mu_t)_{t\in(0,\infty)}$ is a square integrable martingale, is obtained within the conception of the integral (34). Let $(\Omega, \mathcal{F}, P)$ be a probability space and put $H_0 = L_2(\Omega, \mathcal{F}, dP(\omega))$. Let $A(t)$ be the multiplication operator by a random variable $\xi_t(\omega)$, $\omega \in \Omega$, and let E be an RI generated by the flow of σ-subalgebras $(\mathcal{F}_t)_{t\in(0,\infty)}$ of the σ-algebra $\mathcal{F}$, i.e. $E(t)$ is equal to the projector from H_0 onto the subspace $L_2(\Omega, \mathcal{F}_t, dP(\omega))$. For each $\mu_\infty \in H_0$, $\mu_t = E(t)\mu_\infty$ is a square integrable martingale; conversely, under some minimal conditions on the original martingale, it can be represented in the indicated form. It is easy to see that the condition of partial commutativity is here equivalent to $\mathcal{F}_t$-consistency of the process $\xi = (\xi_t)_{t\in(0,\infty)}$, i.e. to the measurability of the function $\xi_t(\omega)$ with respect to $\mathcal{F}_t$ for any $t > 0$. Then the following equality holds:

$$\int_0^\tau \xi_t d\mu_t = \left(\int_0^\tau A(t)dE(t)\right)\mu_\infty \qquad (\tau \in (0,\infty)).$$

We have thus generalized the notion of stochastic integral to the "quantum" case where we take general commuting normal operators instead of multiplication operators.

§7. **Comments.** The results of this lecture are contained in the books [1], [2] and in the articles [3], [4] (Section 6); these papers contain a complete bibliography. Theorem 1 is published in [5]-[7] and extends more early results of L.Gårding, I.M.Gelfand,

A.G.Kostyuchenko, Yu.M.Berezansky, K.Maurin, G.I.Kats. Theorems 2–6 also extend the results of R.S.Phillips, A.Devinatz, A.E.Nussbaum, R.K.Getoor, C. Ionescu Tulcea, S.Kurepa, G.Maltese. The theory of generalized functions of infinitely many variables as in Section 3 was developed in the papers **[8]**-**[14]**. The works **[15]**-**[18]** are connected with Theorem 7. The results of Section 6 are connected with **[19]**-**[21]**.

REFERENCES

1. Yu.M.Berezansky, "Selfadjoint operators in spaces of functions of infinitely many variables," Amer. Math. Soc., Providence, 1986.
2. Yu.M.Berezansky, Yu.G.Kondrat'ev, "Spectral methods in infinite-dimensional analysis," (in Russian), Naukova Dumka, Kiev, 1988. English transl., Kluver Academic Publishers Group, Holland, in print)
3. Yu.M.Berezansky, N.V.Zhernakov, G.F.Us, *Stochastic operator integrals*, Ukrainian Math. J. **39** (1987), 120–124.
4. Yu.M.Berezansky, N.W.Zhernakov, G.F.Us, *A spectral approach to quantum stochastic integrals*, Rep. Math. Phys. **27** (1989).
5. Yu.M.Berezansky, *An expansion in terms of generalized eigenvectors and an integral representation for positive definite kernels in the form of a continual integral*, Siberian Math. J. **9** (1968), 741–751.
6. Yu.M.Berezansky, *On the expansion in simultaneous generalized eigenvectors of an arbitrary family of commuting normal operators*, Soviet Math. Dokl. **17** (1976), 1042–1045.
7. Yu.M.Berezansky, *The projection spectral theorem*, Russian Math. Surveys **39-4** (1984), 1–62.
8. Yu.M.Berezansky, Yu.S.Samojlenko, *Nuclear spaces of functions of infinitely many variables*, Ukrainian Math. J. **25** (1973), 599–609.
9. T. Hida, "Analysis of Brownian functionals," Carleton Mathematical Lecture Notes Vol. 13, Carleton University, Ottawa, 1975.
10. Yu.G.Kondrat'ev, Yu.S.Samojlenko, *The spaces of trial and generalized functions of infinite number of variables*, Rep. Math. Phys. **14-3** (1978), 325–350.
11. Yu.G.Kondrat'ev, *Nuclear spaces of entire functions in problems of infinite-dimensional analysis*, Soviet Math. Dokl. **22** (1980), 588–592.
12. Yu.G.Kondrat'ev, *Spaces of entire functions of infinitely many variables associated with a rigging of Fock space*,, (in Russian), in "Spectral Analysis of Differential Operators," Inst. Mat. Akad. Nauk Ukrain., Kiev, 1980, pp. 18–37.
13. T.Hida, J.Potthoff, L.Streit, *Dirichlet forms and white noise analysis*, Commun. Math. Phys. **116** (1988), 235–245.
14. S.Albeverio, T.Hida, J.Potthoff, M.Röckner, L.Streit, *Dirichlet forms in terms of white noise analysis I: Construction and QFT examples*, Rev. Math. Phys. **1** (1990), 291–312.
15. L. Gross, *Harmonic analysis on Hilbert space*, Mem. Amer. Math. Soc. **46** (1963), 1–62.
16. Hui-Hsiung Kuo, "Gaussian measures in Banach spaces," Lect. Notes in Math. Vol. 463, Springer-Verlag, 1975.
17. J.Friedrich, L.Klotz, *On expansions of positive definite operator-valued functions*, Rep. Math. Phys. **26** (1988), 45–65.
18. J.Friedrich, *Integral representations of positive definite matrix-valued distributions on cylinders*, Trans. Amer. Math. Soc. **313** (1989), 275–299.
19. I.Cuculescu, *Spectral families and stochastic integrals*, Rev. Roum. Math. Pures. Appl. **15** (1970), 201–223.
20. C.Barnett, R.F.Streater, I.F.Wilde, *The Itô-Clifford integral*, J. Funct. Anal. **48** (1982), 172–212.
21. R.F.Streater, *Quantum stochastic integrals*, Acta Phys. Austr. **26** (1984), 53–74.

ON THE CONVERGENCE OF FUNCTIONALS OF RANDOM WALKS TO THE LOCAL TIMES OF BESSEL PROCESSES

A.N.BORODIN
Leningrad Branch Steklov Institute of Mathematics
Fontanka 27, Leningrad 191011 USSR

ABSTRACT

The paper deals with the asymptotic behaviour of some functionals of multi-dimensional random walk. The well-known invariance principle for random walks imply the convergence of the distributions of a wide class of functionals of random walks to the distributions corresponding functionals of Wiener processes. But the invariance principle is not applicable directly for the investigation of limit behaviour of some interesting functions of random walks. Two such functionals are considered in this paper.

1. The Results

Let $\vec{\nu}_k$ be d-dimensional random walk with finite variance and with independent coordinates, i.e. $\vec{\nu}_k=(\nu_{1k},\dots,\nu_{dk})$, $\nu_{jk}=\sum_{l=1}^{k}\xi_{jl}$, where $\{\xi_{jl}\}_{j=1}^{d}{}_{l=1}^{\infty}$ - are independent random variables, identically distributed for each fixed j, with $E\xi_{jl}=0$, $E\xi_{jl}^2=D_j<\infty$. Let $\vec{W}_n(s)=n^{-1/2}\vec{\nu}_{[ns]}$, $s\in[0,\infty)$, where $[a]$ denotes the largest integer not exceeding a. It is well-known that the process $\vec{W}_n(s)$, $s\in[0,\infty)$, converges weakly as $n\to\infty$ to Brownian motion $\vec{W}(s)$ ("weak invariance principles"), where $\vec{W}(s)=(W_1(s),\dots,W_d(s))$, $\{W_j(s)\}_{j=1}^{d}$ - are independent Brownian motions with mean zero and variance $EW_j^2(s)=D_j s$.

For a wide class of functionals of random walk $\vec{\nu}_k$ the weak convergence of $\vec{W}_n(S)$ to $\vec{W}(S)$ imply the convergence of the distributions of the functionals of random walk to the distributions of the corresponding functionals of Brownian motion. For example, if $f(\vec{y})$, $y \in R^d$, - is continuous function, then the distributions of functional $n^{-1}\sum_{k=1}^{n} f(n^{-1/2}\vec{\nu}_k)$ converge to those of $\int_0^1 f(\vec{W}(s))ds$.
This statement holds true not only for continuous functions, but for the indicator functions, and of cause for its linear combinations. So as $n \to \infty$

$$n^{-1}\sum_{k=1}^{n} \mathbb{1}_A(n^{-1/2}\vec{\nu}_k) \longrightarrow \int_0^1 \mathbb{1}_A(\vec{W}(s))ds \tag{1.1}$$

in the sence of the convergence of the distributions, where $\mathbb{1}_A(\cdot)$ is the indicator function of set A.

The "weak invariance principle" can not be applicable directly for all interesting functionals of random walk. There exists a functionals of random walk $\vec{\nu}_k$ which have a limiting distribution, but this fact is not an immediate consequence of the weak convergence of the processes $\vec{W}_n(S)$ to $\vec{W}(S)$. In this paper we consider in detail only two such functionals. But even these examples can illustrate some difficulties and unexpected effects which appear in the problems of a such type. Briefly we also describe some othere functionals of $\vec{\nu}_k$, which can not be investigated immediately with the help of "weak invariance principle". But here there are many unsolved problems. In many cases we can't even describe the limiting functional.

The first functional considered in the paper is the functional

$$\ell_n(t,x) = n^{-1/2}\sum_{k=0}^{[nt]} \mathbb{1}_{[0,\delta]}(\|\vec{\nu}_k\| - x\sqrt{n}), \quad (t,x)\in(0,\infty)^2,$$

where $\|\cdot\|$ is Euclidean norm in R^d , $\delta>0$ is some constant. This functional depends on two parameters, and we shall consider it as a two parameter process. The variable $n^{1/2}\ell_n(t,x)$ is the number of visits of $\vec{\nu}_k$ at the domain $\{\vec{y}: x\sqrt{n}\le\|\vec{y}\|\le x\sqrt{n}+\delta\}$ in $[nt]$ steps. We can represent the functional $\ell_n(1,x)$ in the following form

$$\ell_n(1,x)=n^{-1/2}\sum_{k=1}^{n}1\!\!1_{A_n}(n^{-1/2}\vec{\nu}_k),$$

where $A_n=\{\vec{a}: x\le\|\vec{a}\|\le x+\delta n^{-1/2}\}$. The assential difference of this functional from that in (1.1) consist in the fact that the set A_n is shrink with the speed $n^{-1/2}$. As a consequence of this fact we have another normalizing multiple $n^{-1/2}$ and another limit.

The limiting process for the processes $\ell_n(t,x)$ will be the so called local time of the Bessel process of the rank d multiplied by the constant δ. The Bessel process $b(s)$ of the rank d can be defined as follows $b(s)=\|\vec{W}(s)\|$, i.e. it is a distance of d-dimensional Wiener process $\vec{W}(s)$ from zero. The local time $\ell(t,x)$ of the Bessel process $b(s)$ can be defined as a limit

$$\ell(t,x)=\lim_{\varepsilon\downarrow 0}\varepsilon^{-1}\int_0^t 1\!\!1_{[x,x+\varepsilon)}(b(s))ds,\quad (t,x)\in(0,\infty)^2. \qquad (1.2)$$

This limit exists with probability one ([1], [2]) and $\ell(t,x)$ is the continuous process in $(t,x)\in(0,\infty)^2$.

For one-dimensional random walk ν_k the asymptotic behaviour of the process $\ell_n(t,x)$ was investigated in [3]. Some previous results were presented in monograph [4]. The fact that for d-dimensional random walk $\vec{\nu}_k$ the limit for the process $\ell_n(t,x)$ will be the process $\delta\ell(t,x)$ is not unexpectable, because there is a strong analogy with one-dimensional case. To understand it one should consider the one dimensional discreat time process $\|\vec{\nu}_k\|$, $k=0,1,2,\dots$. Then $\sqrt{n}\,\ell_n(t,x)$ be the number of visits of $\|\vec{\nu}_k\|$ at the interval $[x\sqrt{n}, x\sqrt{n}+\delta]$ in $[nt]$ steps. According to the "weak invariance principle" the process $b_n(s)=n^{-1/2}\|\vec{\nu}_{[ns]}\|$ converge weakly to the Bessel process $b(s)$. So we have the situation which is analogous to one dimensional case. The main difference here is that the process $\|\vec{\nu}_k\|$ is not a random walk.

Such a strong analogy in asymptotic behaviour is not true for some other functionals of d-dimensional random walk. We illustrate it now. The second functional which will be considered in the paper is the functional

$$\rho_n(t,x)=n^{-1/2}\sum_{k=0}^{[nt]} \mathbb{1}_{(-\infty,0)}\big((\|\vec{\nu}_k\|-x\sqrt{n})(\|\vec{\nu}_{k+1}\|-x\sqrt{n})\big).$$

The variable $n^{1/2}\rho_n(t,x)$ is the number of crossings of $\vec{\nu}_k$ over the sphere of radious $x\sqrt{n}$ for $[nt]$ steps. For one-dimensional random walk the limiting process for $\rho_n(t,x)$ is the local time of $|W(s)|$ multiplied by the constant equal to the first absolute moment of the first step of the random walk (see, for example, [5]). What will be the limit for $\rho_n(t,x)$ in the case of d-dimensional random walk. From the first point of view the limit will be again the local time of the Bessel process $b(s)$ multiplied by some constant. But it is not so. The limiting process $\rho(t,x)$ can be described as follows. Let $f(\vec{y}) = E|\sum_{l=1}^{d} y_l \xi_{1l}| = E|(\vec{y},\vec{\nu}_1)|$. Then

$$\rho(t,x)=x^{-1}\int_0^t f(\vec{W}(s))\,\ell(ds,x), \tag{1.3}$$

where $\ell(t,x)$ is the local time of the Bessel process $b(s)=\|\vec{W}(s)\|$. Note that according to the definition (1.2) $\ell(s,x)$ is increasing process in s, so the integral (1.3) is a usual integral with respect to increasing function. From (1.2) one can conclude that the points of growth of function $\ell(s,x)$ are the points s for which $b(s)=x$.

So the meaning of the integral is accomulated only in that moments s when $\vec{W}(s)$ crosses the sphere of radious x. In one-dimensional case this fact imply that

$$\rho(t,x) = x^{-1}\int_0^t |W(s)| E|\xi_1|\, \ell(ds,x) = E|\xi_1|\, \ell(t,x),$$

where $\ell(t,x)$ is the local time of $|W(s)|$.

Let us describe briefly some other functionals of d-dimensional random walk. Instead of sphere of radious x we can take some surface in R^d. We can consider the number of visits of the normalized random walk $n^{-1/2}\vec{\nu}_k$ at the $\delta n^{-1/2}$ -neighbourhood of this surface in $[nt]$ steps. We believe that this number of visits normalized by $\sqrt{n}$ will have the limit, which is equal the local time of Brownian motion $\vec{W}(s)$ on the surface multiplied by 2δ. The local time of Brownian motion $\vec{W}(s)$ on the surface can be determing in a usual way. It is a limit as $\varepsilon \downarrow 0$ the normalized by ε the total time up to time t spend by $\vec{W}(s)$ in ε -neighbourhood of the surface. More intricate will be the asymptotic behaviour of the number of crossings of $n^{-1/2}\vec{\nu}_k$ over the surface for $[nt]$ steps. How to express the limiting process in this case is not quite clear.

Before formulating the main results of the paper we introduce some additional assumptions. It seems to us that these assumptions are technical ones. We shall consider only the random walks with continuous values. Let $\varphi_j(t) = E\exp(it\xi_{j1})$. It is assumed that

$$\int_{-\infty}^{\infty} |\varphi_j(t)|^2 dt < \infty. \tag{1.4}$$

Moreover we suppose that $E|\xi_{j1}|^3 < \infty$, $j = 1, 2, \ldots, d$.

Theorem 1. The finite-dimensional distributions of the process $\ell_n(t,x)$, $(t,x) \in (0,\infty)^2$, converge to those of the process $\delta\ell(t,x)$.

Theorem 2. The finite-dimensional distributions of the process $\rho_n(t,x)$, $(t,x)\in(0,\infty)^2$, converge to those of the process $\rho(t,x)$.

2. Proof of Theorem 1.

The "weak invariance principle" stated that the process $\vec{W}_n(s)=n^{-1/2}\vec{\nu}_{[ns]}$, $s\in[0,\infty)$, converges weakly as $n\to\infty$ to Brownian motion $\vec{W}(s)$. Using the Skorokhod embedding scheme ([6], ch.7, § 2) this fact can be expressed in a following way. By Brownian motion $\vec{W}(s)$ one can construct such random walks $\vec{\nu}_k^n$ for each n having the same distributions as the random walk $\vec{\nu}_k$ that the processes $\vec{W}_n(s)=n^{-1/2}\vec{\nu}^n_{[ns]}$ satisfy the relation

$$\sup_{0\le s\le t}\|\vec{W}_n(s)-\vec{W}(s)\|\to 0 \tag{2.1}$$

in probability for any $t>0$. Furthermore we deal only with the random walks $\vec{\nu}_k^n$, therefore we omit the index n . Also the variables $\vec{\nu}_k^n-\vec{\nu}_{k-1}^n$ will be denoted by $\vec{\xi}_k=(\xi_{1k},\dots,\xi_{dk})$.

To prove the theorem we shall derive the special representations for the processes $\ell(t,x)$ and $\ell_n(t,x)$. For this purpose we choose one of the coordinate of the Wiener process $\vec{W}(s)$ and of the random walk $\vec{\nu}_k$, for example the last one, and consider it separately. Denote $\vec{W}^c(s)=(W_1(s),\dots,W_{d-1}(s))$, $\alpha(s)=\|\vec{W}^c(s)\|$, $\vec{\nu}_k^c=(\nu_{1k},\dots,\nu_{d-1k})$, $\alpha_n(s)=n^{-1/2}\|\vec{\nu}^c_{[ns]}\|$.

Since we must prove only the convergence of finite-dimensional distributions some trajectories of the Wiener process $\vec{W}(s)$ and the random walk $\vec{\nu}_k$ can be excluded from the consideration. This is motivated by technical reasons. The probability of the set of excluded trajectories must be small. Let $0<\Delta<1$ and $Y_\Delta=\{\vec{y}: x\le\|\vec{y}\|\le x+1-\Delta,\ y_1^2+y_2^2+\dots+y_{d-1}^2\ge x^2\Delta^2\}$. Let V_Δ be the "set of those trajectories $\vec{W}(s)$ which for $0 \leq s \leq t$ doesn't visit Y_Δ. Let also V_Δ^n be the set of those trajectories $\vec{W}_n(s)$ which for $0 \leq s \leq t$ doesn't visit Y_Δ. One can choose Δ so close to one and n_0 so large that in view of (2.1) the probabilities of the sets V_Δ and V_Δ^n, $n \geq n_0$, will be so close to one as necessary. Furthermore we can consider the trajectories only from the sets V_Δ and V_Δ^n since there probabilities are close to one.

The following lemma supply us with the suitable representation of the local time of Bessel process.

Lemma 2.1. For $d \geq 2$ and for $\vec{W} \in V_\Delta$

$$\ell(t,x) =$$
$$= \frac{x}{\pi} \int_{-\infty}^{\infty} \int_0^t e^{ivW(s)} \frac{\cos(v(x^2 - a^2(s))^{1/2})}{(x^2 - a^2(s))^{1/2}} \mathbb{1}_{(0,x\Delta)}(a(s))\, ds\, dv. \qquad (2.2)$$

Proof. Using the obvious relation

$$\mathbb{1}_{[a,b)}(z) = \lim_{m \to \infty} \pi^{-1} \int_a^b \frac{\sin(m(x-z))}{(x-z)} dx =$$

$$= \lim_{m \to \infty} (2\pi)^{-1} \int_{-m}^{m} e^{ivz} \int_{-\infty}^{\infty} e^{-ivy} \mathbb{1}_{[a,b)}(y)\, dy\, dv. \qquad (2.3)$$

one can wright

$$\mathbb{1}_{[x,x+\varepsilon)}(\sqrt{a^2 + z^2}) = \lim_{m \to \infty} \frac{1}{2\pi} \int_{-m}^{m} e^{ivz} \int_{-\infty}^{\infty} e^{-ivy} \mathbb{1}_{[x,x+\varepsilon)}(\sqrt{a^2 + y^2})\, dy\, dv.$$

Let

$$q_\varepsilon(m,M) =$$
$$= \frac{1}{2\pi\varepsilon} \int_0^t ds \int_m^M dv\, e^{ivW(s)} \int_{-\infty}^{\infty} dy\, e^{-ivy} \mathbb{1}_{[x,x+\varepsilon)}(\sqrt{a^2(s) + y^2})\, \mathbb{1}_{(0,x\Delta)}(a(s)).$$

Then for the trajectories from V_Δ and $\varepsilon < \Delta$

$$\frac{1}{\varepsilon}\int_0^t \mathbb{1}_{[x,x+\varepsilon)}(\ell(s))ds =$$

$$= \frac{1}{\varepsilon}\int_0^t \mathbb{1}_{[x,x+\varepsilon)}\left(\sqrt{a^2(s)+W^2(s)}\right)ds = \lim_{m\to\infty} q_\varepsilon(-m,m). \tag{2.4}$$

By the definition of local time the left hand side of this equation tends to $\ell(t,x)$ when $\varepsilon \downarrow 0$, so it is necessary to prove that the right hand side tends to that of (2.2).

Let us estimate $Eq_\varepsilon^2(m,M)$, $m<M$. Denote

$$D_\varepsilon(s) = \mathbb{1}_{(0,x\Delta)}(a(s))\,\varepsilon^{-1}\int_{-\infty}^{\infty} e^{-ivy}\,\mathbb{1}_{[x,x+\varepsilon)}\left(\sqrt{a^2(s)+y^2}\right)dy.$$

Using the fact that the Bessel process $a(s)$ at each point s has the density function $a^{d-1}e^{-a^2/2s}/$ $/s^{d/2}2^{d/2-1}\Gamma(d/2)$ one can fined

$$ED_\varepsilon^2(s) \leq 4C(1-\Delta^2)^{-1},$$

where the constant C independent of ε. Then taking into account the assumption that $\vec{W}(s)$ is the process with independent coordinates we have

$$Eq_\varepsilon^2(m,M) = \frac{1}{(2\pi)^2}\int_m^M dv_1\int_m^M dv_2\int_0^t ds_1\int_0^t ds_2\, Ee^{iv_1W(s_1)+iv_2W(s_2)}ED_\varepsilon(s_1)D_\varepsilon(s_2) \leq$$

$$\leq \frac{2C}{D_d^2(1-\Delta^2)\pi^2}\int_m^M dv_1\int_m^M dv_2\; v_1^{-2}\left(1-e^{-tD_d v_1^2/2}\right)(v_1+v_2)^{-2}\left(1-e^{-tD_d(v_1+v_2)^2/2}\right) \leq$$

$$\leq \frac{2C}{(1-\Delta^2)\pi^2}\int_{-\infty}^{\infty}\frac{dv}{1+D_d v^2}\int_m^M\frac{dv}{1+D_d v^2} \leq \frac{2C}{(1-\Delta^2)\pi D_d^{3/2}\,m} \tag{2.5}$$

It is important that this estimate independent of ε , M and tends to zero when $m \to \infty$. Let

$$q_0(m,M)=\frac{x}{\pi}\int_m^M\int_0^t e^{ivw(s)}\frac{\cos(v\sqrt{x^2-a^2(s)})}{\sqrt{x^2-a^2(s)}}\,1\!\!1_{(0,x\Delta)}(a(s))\,ds\,dv.$$

Since for $0\leq a<x\Delta$

$$\frac{1}{\varepsilon}\int_{-\infty}^{\infty}e^{-ivy}1\!\!1_{[x,x+\varepsilon)}(\sqrt{a^2+y^2})\,dy \underset{\varepsilon\to 0}{\longrightarrow} \frac{2x\cos(v\sqrt{x^2-a^2})}{\sqrt{x^2-a^2}}$$

then for any fixed m

$$\lim_{\varepsilon\downarrow 0} q_\varepsilon(-m,m) = q_0(-m,m). \tag{2.6}$$

Analogously to (2.5) one can obtain the estimate

$$E q_0^2(m,M) \leq \frac{2C}{(1-\Delta^2)\pi D_d^{3/2} m} \tag{2.7}$$

It is clear that the same estimates holds true for the values $Eq_\varepsilon^2(-M,-m)$ and $Eq_0^2(-M,-m)$, $0<m<M$. These estimates together with (2.6) imply that

$$\lim_{\varepsilon\downarrow 0}\lim_{m\to\infty} q_\varepsilon(-m,m) = \lim_{m\to\infty} q_0(-m,m)$$

and consequently (2.2).

Continue the proof of the theorem. We will consider only that trajectories of random walk $\vec{\nu}_k$ for which $\vec{W}_n \in V_\Delta^n$. Let $\mu_k = \left(\sum_{j=1}^{d-1} \nu_{jk}^2\right)^{1/2}$. Then for $\delta n^{-1}<\Delta$

$$\ell_n(t,x)=n^{-1/2}\sum_{k=1}^{[nt]} 1\!\!1_{[x,x+\delta/n)}\left(\sqrt{n^{-1}\mu_k^2+n^{-1}\nu_k^2}\right)1\!\!1_{(0,x\Delta)}(n^{-1/2}\mu_k).$$

Using the notations $a_n(s)=n^{-1/2}\|\vec{\nu}_{[ns]}^{\,c}\|=n^{-1/2}\mu_{[ns]}$, $W_n(s) = n^{-1/2}\nu_{[ns]}$, $t_n=[nt]/n$, one can write

$$\ell_n(t,x)=\sqrt{n}\int_0^{t_n}1\!\!1_{[x,x+\delta/n)}\left(\sqrt{a_n^2(s)+W_n^2(s)}\right)1\!\!1_{(0,x\Delta)}(a_n(s))\,ds.$$

Let

$$q_n(m,M)=\frac{\sqrt{n}}{2\pi}\int_0^{t_n} ds \int_m^M dv\, e^{ivW_n(s)} \times$$

$$\times \int_{-\infty}^{\infty} dy\, e^{-ivy} \mathbb{1}_{[x,x+\delta/n)}\left(\sqrt{a_n^2(s)+y^2}\right) \mathbb{1}_{(0,x\Delta)}(a_n(s))$$

Analogously to (2.4) we have

$$l_n(t,x) = \lim_{m\to\infty} q_n(-m,m). \tag{2.8}$$

According to the "weak invariance principle" as $n\to\infty$ the process $a_n(s)$, $s\in[0,\infty)$, converges weakly to the process $a(s)$ and $W_n(s)$, $s\in[0,\infty)$, converges weakly to $W(s)$. Then for those s , for which $0\leq a_n(s)\leq x\Delta$ we have

$$\sqrt{n}\int_{-\infty}^{\infty} e^{-ivy}\mathbb{1}_{[x,x+\delta/n)}\left(\sqrt{a_n^2(s)+y^2}\right)dy \longrightarrow \frac{2x\delta\cos\left(v\sqrt{x^2-a^2(s)}\right)}{\sqrt{x^2-a^2(s)}},$$

and, consequently, for any fixed m

$$\lim_{n\to\infty} q_n(-m,m) = \delta q_0(-m,m). \tag{2.9}$$

Now to prove the theorem it is sufficient to obtain the estimates for $Eq_n^2(m,M)$, $Eq_n^2(-M,-m)$, $m<M$, which de - crease when n and m growth. Denote

$$D_n(s) = \mathbb{1}_{(0,x\Delta)}(a_n(s))\sqrt{n}\int_{-\infty}^{\infty} e^{-ivy}\mathbb{1}_{[x,x+\delta/n)}\left(\sqrt{a_n^2(s)+y^2}\right)dy.$$

Then one can write

$$Eq_n^2(m,M) = \frac{2}{(2\pi)^2}\int_m^M dv_1\int_m^M dv_2\int_0^{t_n} ds_1\int_{s_1}^{t_n} ds_2\, E e^{iv_1W_n(s_1)+iv_2W_n(s_2)} ED_n(s_1)D_n(s_2) \leq$$

$$\leq \frac{1}{2\pi^2}\int_m^M dv_1\int_m^M dv_2\left(\int_0^{2/n} ds_1\int_{s_1}^{t_n} ds_2 + \int_{t_n-2/n}^{t_n} ds_1\int_{s_1}^{t_n} ds_2 + \int_{2/n}^{t_n-2/n} ds_1\int_{s_1}^{s_1+2/n} ds_2 + \right.$$

$$+\int_{2/n}^{t_n-2/n} ds_1 \int_{s_1+2/n}^{t_n} ds_2\, E e^{iv_1 W_n(s_1)+iv_2 W_n(s_2)} ED_n(s_1)D_n(s_2) = \sum_{l=1}^{4} L_l .$$

The essential term of the sum is the last one. The previous terms can be estimated with the help of the formula (2.3). It is possible to obtain the estimates with necessary properties. We consider in detail only L_4. Using the fact that for independent random vectors $\vec{\xi}_k$ with independent coordinates, satisfying $E\|\vec{\xi}_k\|^2 < \infty$, the random variable $n^{-1/2}\sum_{k=0}^{[ns]}\vec{\xi}_k$ has a bounded density function, it is not hard to show that

$$ED_n^2(s) \leqslant C_\Delta \delta^2. \tag{2.10}$$

Then we have

$$|L_4| \leqslant \frac{1}{2\pi^2 n}\int_{m/\sqrt{n}}^{M/\sqrt{n}} dv_1 \int_{m/\sqrt{n}}^{M/\sqrt{n}} dv_2 \sum_{k=2}^{[nt]-2}\sum_{l=k+2}^{[nt]} |\varphi^k(v_1+v_2)\varphi^{k-l}(v_1)| \leqslant$$

$$\leqslant \frac{1}{2\pi^2 n}\int_{m/\sqrt{n}}^{M/\sqrt{n}} dv_1 \int_{m/\sqrt{n}}^{M/\sqrt{n}} dv_2\, |\varphi^2(v_1+v_2)||\varphi^2(v_1)| \frac{(1-|\varphi^n(v_1+v_2)|)(1-|\varphi^n(v_1)|)}{(1-|\varphi(v_1+v_2)|)(1-|\varphi(v_1)|)} \leqslant$$

$$\leqslant \frac{1}{2\pi^2\sqrt{n}}\int_{m/\sqrt{n}}^{\infty} dv|\varphi^2(v)|([nt]\wedge(1+v^{-2}))\frac{1}{\sqrt{n}}\int_0^{\infty} dv|\varphi^2(v)|([nt]\wedge(1+v^{-2})) \leqslant C\left(\frac{1}{\sqrt{n}}+\frac{1}{m}\right).$$

The complete estimate will have a form

$$E q_n^2(m,M) = o(1/n) + o(1/m),$$

where $o(1/n)$ tends to zero as $n\to\infty$ uniformly in m and $o(1/m)$ tends to zero as $m\to\infty$ uniformly in n. The same estimate holds true for $Eq_n^2(-M,-m)$. These estimates, the estimate (2.7) together with (2.9) imply that

$$\lim_{n\to\infty}\lim_{m\to\infty} q_n(-m,m) = \delta \lim_{m\to\infty} q_c(-m,m).$$

In view of (2.8), (2.2) and the definition of $q_0(m, M)$ this relation is just the statement of theorem 1.

3. Proof of Theorem 2

Since the proof of theorem 2 is very similar to that of theorem 1 we point out only some essential aspects. We retain all notations. As it was explained in the proof of theorem 1 it is sufficient to consider only that trajectories of random walk $\vec{\nu}_k$ for which $\vec{W}_n \in V^n_\Delta$. Since the last coordinate of $\vec{\nu}_k$ is considered separately we set for simplicity $\xi_k = \xi_{dk}$. Let $\vec{\xi}^c_k = (\xi_{1k}, \dots, \xi_{d-1k})$, $\vec{W}^c_n(s) = n^{-1/2}\vec{\nu}^c_{[ns]}$. Using (2.3) one can obtain

$$\rho_n(t,x) =$$

$$= \frac{1}{\sqrt{n}} \sum_{k=0}^{[nt]} \mathbb{1}_{(-\infty,0)}\left(\left(\sqrt{\|\vec{\nu}^c_k\|^2 + \nu_k^2} - x\sqrt{n}\right)\left(\sqrt{\|\vec{\nu}^c_k + \vec{\xi}^c_{k+1}\|^2 + (\nu_k + \xi_{k+1})^2} - x\sqrt{n}\right)\right) =$$

$$= \frac{1}{2\pi} \lim_{m\to\infty} \int_{-m}^{m} dv \int_0^{t_n} ds\, e^{ivW_n(s)} \sqrt{n} \int_{-\infty}^{\infty} e^{-ivy} \mathbb{1}_{(-\infty,0)}\left(\left(\sqrt{\|\vec{W}^c_n(s)\|^2 + y^2} - x\right) \times\right.$$

$$\left. \times \left(\sqrt{\|\vec{W}^c_n(s) + \vec{\xi}_{[ns]+1}/\sqrt{n}\|^2 + (y + \xi_{[ns]+1}/\sqrt{n})^2} - x\right)\right) \mathbb{1}_{(0,x\Delta)}(\|\vec{W}^c_n(s)\|)\, dy .$$

For the fixed s let us consider the asymptotic behaviour of the integral

$$I_n(v,s) = \sqrt{n} \int_{-\infty}^{\infty} e^{-ivy} \mathbb{1}_{(-\infty,0)}\left(\left(\sqrt{\|\vec{W}^c_n(s)\|^2 + y^2} - x\right) \times\right.$$

$$\left. \times \left(\sqrt{\|\vec{W}^c_n(s) + \vec{\xi}_{[ns]+1}/\sqrt{n}\|^2 + (y + \xi_{[ns]+1}/\sqrt{n})^2} - x\right)\right) dy .$$

Transforming the indicator function, it is not hard to show that

$I_n(v,s) \approx$

$$\approx \frac{\exp(-iv\sqrt{x^2-\|\vec{W}^c(s)\|^2})}{\sqrt{x^2-\|\vec{W}^c(s)\|^2}} \left|(\vec{W}^c(s), \vec{\xi}^{\,c}_{[ns]+1}) + \xi_{[ns]+1}\sqrt{x^2-\|\vec{W}^c(s)\|^2}\right| +$$

$$+ \frac{\exp(iv\sqrt{x^2-\|\vec{W}^c(s)\|^2})}{\sqrt{x^2-\|\vec{W}^c(s)\|^2}} \left|(\vec{W}^c(s), \vec{\xi}^{\,c}_{[ns]+1}) - \xi_{[ns]+1}\sqrt{x^2-\|\vec{W}^c(s)\|^2}\right|.$$

It is importent that here the vector $(\vec{\xi}^{\,c}_{[ns]+1}, \xi_{[ns]+1})$ is independent of Brownian motion $\vec{W}^c(u)$, when $u \leq s$. Using this fact, one can let $f(\vec{y}) = E|(\vec{y}^c, \vec{\xi}^{\,c}_{[ns]+1}) + y_d\,\xi_{[ns]+1}|$ and applying the special version of the law of large numbers, to obtain

$$\int_0^{t_n} e^{ivW_n(s)} \mathbb{1}_{(0,x\Delta)}(\|\vec{W}^c_n(s)\|) I_n(v,s)\,ds \approx$$

$$\approx \int_0^t e^{ivW(s)} \left[\frac{\exp(-iv\sqrt{x^2-\|\vec{W}^c(s)\|^2})}{\sqrt{x^2-\|\vec{W}^c(s)\|^2}} f(\vec{W}^c(s), \sqrt{x^2-\|\vec{W}^c(s)\|^2}) + \right.$$

$$\left. + \frac{\exp(iv\sqrt{x^2-\|\vec{W}^c(s)\|^2})}{\sqrt{x^2-\|\vec{W}^c(s)\|^2}} f(\vec{W}^c(s), -\sqrt{x^2-\|\vec{W}^c(s)\|^2}) \right] ds.$$

Now repeating the main arguments which lead to the proof of theorem 1 and taking into account the notation $\|\vec{W}^c(s)\| = a(s)$ it is possible to establish that

$\lim_{n\to\infty} \rho_n(t,x) =$

$$= \frac{1}{2\pi} \lim_{m\to\infty} \int_{-m}^{m} dv \int_0^t e^{ivW(s)} \left[\frac{\exp(-iv\sqrt{x^2-a^2(s)})}{\sqrt{x^2-a^2(s)}} f(\vec{W}^c(s), \sqrt{x^2-a^2(s)}) + \right.$$

$$+\frac{\exp(iv\sqrt{x^2-a^2(s)})}{\sqrt{x^2-a^2(s)}} f(\vec{W}^c(s), -\sqrt{x^2-a^2(s)})\Big] \mathbb{1}_{(0,x\Delta)}(a(s))ds. \quad (3.1)$$

The problem is to prove that this limiting process coinside with the process represented by (1.3). In the right hand side of (3.1) there are two terms. In the first term in view of the structure of the integral the argument $\sqrt{x^2-a^2(s)}$ in function f can be replaced by $W(s)$, since the multiple before iv in the power of exponent is equal $W(s)-(x^2-a^2(s))^{1/2}$. In the second term by the same reason the argument $-\sqrt{x^2-a^2(s)}$ in function f can also be replased by $W(s)$. Finally we have

$$\lim_{n\to\infty} \rho_n(t,x) =$$

$$=\frac{1}{\pi}\int_{-\infty}^{\infty}\int_0^t e^{ivW(s)} \frac{\cos(v\sqrt{x^2-a^2(s)})}{\sqrt{x^2-a^2(s)}} f(\vec{W}(s)) \mathbb{1}_{(0,x\Delta)}(a(s))ds\,dv.$$

Now the desired representation (1.3) is the consequence of formula (2.2).

4. References

1. E.S.Boylan, Local times for a class of Markov processes. Illinois J.Math. 8 (1964), 19-39.
2. R.M.Blumenthal and R.K.Getoor, Local times for Markov processes, Z.Wahrsch. Verw. Gebiete 3 (1964), 50-74.
3. A.N.Borodin, On the asymptotic behaviour of local times of recurrent random walks with finite variance, Teor.Veroyatnost. i Primenen. 26 (1981), 769-783.
4. A.V.Skorokhod and N.P.Slobodenyuk, Limit theorems for random walks, Naukova Dumka, Kiev 1970, p.304.
5. A.N.Borodin, On the character of convergence to Browni-

an local time. II, Probab. Theory Relat.Fields 72 (1986), 251-277.

6. A.V.Skorokhod, Investigations on Theory of Random Processes. Kiev University, Kiev 1961, p.216.

FEYNMAN INTEGRAL OF VARIATIONS OF FUNCTIONALS

R.H. Cameron* and D.A. Storvick
Department of Mathematics, University of Minnesota
Minneapolis, Minnesota 55455

Abstract

In this paper we establish a basic relationship for the analytic Feynman integral of the first variation of a functional. The functional is defined on the space $C[a,b]$ of continuous functions $x(t)$ defined on $a \leq t \leq b$ and vanishing at $t=a$.

We apply our basic theorem to establish the analytic Feynman integrability of new classes of functionals and give formulas for their Feynman integrals. In particular, these new classes contain functionals which are neither bounded nor analytic.

Definition: A functional F defined on Wiener space, $C[a,b]$, is said to be an analytic functional at a point $x_o \in C[a,b]$ if $F(\rho x_o)$ has an analytic extension in ρ to a complex neighborhood of $\rho = 1$.

In papers [1,3,4,8,9] the functionals considered are all bounded. In papers [5] & [6] we considered products of bounded functionals with functions of $x(b)$ that are integrable on $\mathbf{R}^1$. In [7] we considered analytic functionals which are unbounded and whose rate of growth can be faster than exponential. The classes of functionals considered in the present paper contain non-analytic functionals that are unbounded with a rate of growth equivalent to the rate of growth of finite products of linear functionals.

The basic theorem of this paper expresses the analytic Feynman integral of the first variation of a functional F in terms of the analytic Feynman integral of the product of F with a linear functional. The proof is based on a theorem of

---

* Robert Horton Cameron died on 17 June 1989. The second author expresses his appreciation for the many years of close association and joint research with Professor Cameron

one of the authors [2;p. 919] giving the Wiener integral of the first variation of a functional F in terms of the Wiener integral of the product of F by a linear functional.

We begin by presenting that theorem rephrased so that the functional is defined on $C[a,b]$ rather than on $C[0,1]$ and with the Wiener integral normalized on the basis of "unit variance per unit time" instead of "half unit variance per unit time".

<u>Theorem A</u>: *Let $y(t)$ be a real absolutely continuous function on $[a,b]$ which vanishes at $t=a$, and let $y'(t)$ be essentially of bounded variation on $[a,b]$. Let $F(x)$ be a Wiener integrable functional on $C[a,b]$. If $F(x)$ has a first variation $\delta F = \delta F(x|y)$ for all $x \in C[a,b]$, such that for some $\eta > 0$, $\sup_{|h|\leq\eta} |\delta F(x+hy|y)|$ is Wiener integrable on $C[a,b]$, then both members of equation (1) exist and they are equal:*

$$\int_{C[a,b]} \delta F(x|y)dx = \int_{C[a,b]} F(x)\{\int_a^b y'(t)dx(t)\}dx. \tag{1}$$

We shall use the definition of the analtyic Wiener integral of a functional F defined on $C[a,b]$ given in [3]:

<u>Definition</u>: Let F be a functional such that the Wiener integral

$$J(\lambda) = \int_{C[a,b]} F(\lambda^{-1/2}x)dx$$

exists for all real $\lambda > 0$. If there exists a function $J^*(\lambda)$ analytic in the half-plane $Re\lambda > 0$ and such that $J^*(\lambda) = J(\lambda)$ for all real $\lambda > 0$, then we define J^* to be the <u>analytic Wiener integral</u> of F over $C[a,b]$ with parameter λ, and for $Re\lambda > 0$ we write

$$\int_{C[a,b]}^{anw_\lambda} F(x)dx = J^*(\lambda).$$

<u>Definition</u>. Let q be a real parameter ($q \neq 0$) and let F be a functional whose analytic Wiener integral exists for $Re\lambda > 0$. Then if the following limit exists, we call it the <u>analytic Feynman integral of</u> F <u>over</u> $C[a,b]$ <u>with parameter</u> q, and we write

$$\int_{C[a,b]}^{anf_q} F(x)dx = \lim_{\lambda \to -iq_{Re\lambda>0}} \int_{C[a,b]}^{anw_\lambda} F(x)dx.$$

§1. In order to prove our basic theorem which is the theorem for Feynman integrals corresponding to Theorem A for Wiener integrals, we first present a lemma which allows us to change scale in Theorem A.

__Lemma 1__: *Let $y(t)$ be a real absolutely continuous function on $[a,b]$ which vanishes at $t=a$, and let $y'(t)$ be essentially of bounded variation on $[a,b]$.*

For some real $\rho > 0$, let $F(\rho x)$ be Wiener integrable on $C[a,b]$. If $F(\rho x)$ has a first variation $\delta F(\rho x|\rho y)$ for all $x \in C$ such that for some $\eta > 0$

$$\sup_{|h|\leq\eta} |\delta F(\rho x + \rho h y|\rho y)|$$

is Wiener integrable in x, then

$$\int_{C[a,b]} \delta F(\rho x|\rho y)dx = \int_{C[a,b]} F(\rho x)[\int_a^b y'(t)dx(t)]dx. \tag{2}$$

Proof. We apply Theorem A to the functional after a change of scale. To do this we set

$$G(x) = F(\rho x) \tag{3}$$

and note that

$$G(x+hy) = F(\rho x + h\rho y)$$

and

$$\frac{\partial}{\partial h} G(x+hy)|_{h=0} = \frac{\partial}{\partial h} F(\rho x + h\rho y)|_{h=0}$$

or

$$\delta G(x|y) = \delta F(\rho x|\rho y), \tag{4}$$

and the existence of either member implies that of the other.

The lemma follows by replacing F by G in equation (1) and then applying (3) and (4).

We now state our basic theorem on the Feynman integral of the variation of a functional

__Theorem 1__: *Let $y \in C[a,b]$ be absolutely continuous and let y' be essentially of bounded variation on $[a,b]$. For every $\rho > 0$, let $F(\rho x)$ be Wiener integrable on $C[a,b]$, and let $F(\rho x)$ have a first variation $\delta F(\rho x|\rho y)$ for all $x \in C$ such that for some positive function $\eta(\rho) > 0$,*

$$\sup_{|h|\leq\eta(\rho)} |\delta F(\rho x + \rho h y|\rho y)|$$

is Wiener integrable in x for each $\rho > 0$. Then if either member of the following equation exists, both analytic Feynman integrals below exist and we have for each real $q \neq 0$,

$$\int_{C[a,b]}^{anf_q} \delta F(x|y)dx = iq \int_{C[a,b]}^{anf_q} F(x)[\int_a^b y'(t)dx(t)]dx. \tag{5}$$

<u>*Proof*</u>: To establish this theorem for the analytic Feynman integral, we consider the following change of variable in equation (2) of Lemma 1. We let ρ be positive and set $z(t) \equiv \rho y(t)$. Then (2) becomes

$$\int_{C[a,b]} \delta F(\rho x|z)dx = \rho^{-2} \int_{C[a,b]} F(\rho x)[\int_a^b z'(t)d(\rho x(t))]dx. \tag{6}$$

If we let $\rho = \lambda^{-1/2}$, (6) becomes

$$\int_{C[a,b]} \delta F(\lambda^{-1/2}x|z)dx = \lambda \int_{C[a,b]} F(\lambda^{-1/2}x)[\int_a^b z'(t)d(\lambda^{-1/2}x(t))]dx. \tag{7}$$

Thus by the definition of the analytic Wiener integral, if we now let λ be any complex number such that $Re\lambda > 0$ and if either side of the following equation exists, both exist and we have

$$\int_{C[a,b]}^{anw_\lambda} \delta F(x|z)dx = \lambda \int_{C[a,b]}^{anw_\lambda} F(x)[\int_a^b z'(t)dx(t)]dx. \tag{8}$$

Thus from the definition of the analytic Feynman integral, Theorem 1 follows by letting $\lambda \to -iq$, $Re\lambda > 0$.

§**2:** To apply these results on the analytic Feynman integral, to specific classes of functionals, we consider the class S' of functionals defined in [3]:

<u>*Definition*</u>: Let $\mathcal{B} = \mathcal{B}[a,b]$ be the space of real right continuous functions of bounded variation on $[a,b]$ that vanish at b. Let $\mathcal{A}'$ be the σ-algebra of subsets of $\mathcal{B}$ generated by the class of sets of the form

$$\{v|v \in \mathcal{B}, \quad \int_a^b v(t)\phi(t)dt < \lambda\}$$

where $\phi \in L_2([a,b])$ and $-\infty < \lambda < \infty$. Let $\mathcal{M}' \equiv \mathcal{M}'(\mathcal{B}[a,b])$ be the class of complex measures of finite variation defined on $\mathcal{B}[a,b]$ with $\mathcal{A}'$ as its class of measurable sets. If $\nu \in \mathcal{M}'$ we set $\|\nu\| = var\nu$ over $\mathcal{B}$. We shall also use another norm on $\mathcal{B}$, namely

$$|||v||| = \sup_{t\in[a,b]} |v(t)|.$$

Let $S' \equiv S'(\mathcal{B})$ be the space of functionals of the form

$$F(x) \equiv \int_{\mathcal{B}} exp\{i \int_a^b v(t)dx(t)\}d\mu(v) \tag{9}$$

for $x \in C[a,b]$ and $\mu \in \mathcal{M}'$. In [3] we proved that for $F \in S'$, there is a unique measure $\mu \in \mathcal{M}'$ related to F by equation (9). We shall call μ the measure associated with F.

In [3] a number of properties of S' were established. We showed that S' is a Banach algebra with norm $\|F\| = \|\mu\|$, and we proved that if $F \in S'$, F is analytic Feynman integrable.

For the convenience of the reader, we present a lemma which will be used to establish the existence of the first variation of a functional.

Lemma 2: *Let $X \equiv (X, \Sigma, \mu)$ be a general measure space with a σ-algebra of subsets Σ, and a measure μ. Let $f(x,a)$ be defined for $(x,a) \in X \times [a_1, a_2]$. For each a, $a_1 < a < a_2$, let $f(x,a)$ be integrable with respect to μ over X. For each $(x,a) \in X \times (a_1 a_2)$, let $\frac{\partial f}{\partial a}(x,a)$ exist. Let $a_o \in (a_1, a_2)$. For μ-almost every $x \in X$, let $\frac{\partial f}{\partial a}(x,a)$ be continuous with respect to a at a_o. Let there exist a real positive ν-integrable function $g(x)$ such that for $(x,a) \in X \times (a_1, a_2)$,*

$$|\frac{\partial f}{\partial a}(x,a)| \leqq g(x).$$

Then if

$$F(a) \equiv \int_X f(x,a)d\mu(x),$$

$F(a)$ is differentiable at $a = a_o$, and

$$F'(a_o) = \int_X \frac{\partial f}{\partial a}(x, a_o)d\mu(x) \tag{10}$$

Proof: Let α_o be fixed, $\alpha_1 < \alpha_o < \alpha_2$. Then if $\alpha_1 < \alpha_o + \Delta\alpha < \alpha_2$, for each $x \in X$, there exists $\theta \equiv \theta(x)$, $0 \leq \theta \leq 1$, so that

$$f(x, \alpha_o + \Delta\alpha) - f(x, \alpha_o) = \frac{\partial f}{\partial \alpha}(x, \alpha_o + \theta\Delta\alpha)\Delta\alpha.$$

Then

$$\int_X [f(x,\alpha_o+\Delta\alpha)-f(x,\alpha_o)]d\nu = \Delta\alpha\int_X \frac{\partial f}{\partial\alpha}(x,\alpha_o+\theta\Delta\alpha)d\mu$$

or

$$\frac{1}{\Delta\alpha}[F(\alpha_o+\Delta\alpha)-F(\alpha_o)] = \int_X \frac{\partial f}{\partial\alpha}(x,\alpha_o+\theta(x)\Delta\alpha)d\mu.$$

Thus for ν-almost every $x \in X$, as $\Delta\alpha \to 0$,

$$\frac{\partial f}{\partial\alpha}(x,\alpha_o+\theta(x)\Delta\alpha) \to \frac{\partial f}{\partial\alpha}(x,\alpha_o)$$

and by dominated convergence,

$$\int_X \frac{\partial f}{\partial\alpha}(x,\alpha_o+\theta(x)\Delta\alpha)d\mu \to \int_X \frac{\partial f}{\partial\alpha}(x,\alpha_o)d\mu$$

as $\Delta\alpha \to 0$ and consequently the left member below exists and we have

$$\frac{dF}{d\alpha}|_{\alpha=\alpha_o} = \int_X \frac{\partial f}{\partial\alpha}(x,\alpha_o)d\mu$$

and the Lemma is proved.

We now present our first application of our basic theorem, Theorem 1.

<u>Theorem 2</u>. *Let $F \in S'$ so that*

$$F(x) = \int_{\mathcal{B}[a,b]} exp\{i\int_a^b v(t)dx(t)\}d\mu(v) \tag{11}$$

where μ is a complex measure of bounded variation on $\mathcal{B}[a,b]$. Assume further that

$$\int_{\mathcal{B}[a,b]} |||v||||d\mu(v)| < \infty. \tag{12}$$

Let $z(t)$ be a function of bounded variation on $[a,b]$. Then the following integrals exist and we have for real $q \neq 0$,

$$\int_{C[a,b]}^{anf_q} F(x)[\int_a^b z(t)dx(t)]dx = \tag{13}$$

$$= -\frac{1}{q}\int_{C[a,b]}^{anf_q}\int_{B[a,b]} exp\{i\int_a^b v(t)dx(t)\}\int_a^b v(t)z(t)d + d\mu(v)dx$$

$$= \frac{i}{q}\int_{B[a,b]} exp\{\frac{1}{2qi}\int_a^b v^2(t)dt\}\int_a^b v(t)z(t)dtd\mu(v).$$

Proof: We first apply our lemma 2 on differentiating under the integral sign to the case $X \equiv \mathcal{B}[a,b]$, and begin with $y(t) = \int_a^t z(s)ds$ with z of bounded variation so that y is absolutely continuous and $y(a) = 0$ and thus

$$\delta F(x|y) = \frac{\partial}{\partial h}F(x+hy)|_{h=0}$$

$$= \frac{\partial}{\partial h}\int_{B[a,b]} exp\{i\int_a^b v(t)dx(t) + ih\int_a^b v(t)dy(t)\}d\mu(v)|_{h=0}$$

$$= \int_{B[a,b]} \frac{\partial}{\partial h} exp\{i\int_a^b v(t)dx(t) + ih\int_a^b v(t)dy(t)\}|_{h=0}d\mu(v)$$

$$= i\int_{B[a,b]} exp\{i\int_a^b v(t)dx(t)\}\int_a^b v(t)dy(t)d\mu(v).$$

Let ρ be real and positive, then

$$\delta F(\rho x + \rho hy|\rho y) = i\int_{B[a,b]} exp\{i\int_a^b v(t)d(\rho(t) + \rho hy(t)\}\int_a^b v(t)d(\rho y(t))d\mu(v)$$

So

$$|\delta F(\rho x + \rho hy|\rho y)| \leqq |\rho|\int_{B[a,b]}^b v(t)||dy(t)||d\mu(v)$$

$$\leqq |\rho|\int_{B[a,b]} |||v||| \cdot V_a^b y|d\mu(v)|$$

$$= |\rho|V_a^b y\int_{B[a,b]} |||v||||d\mu(v)| < \infty.$$

Thus

$$\sup_{|h|\leqq h(\rho)} |\delta F(\rho x + \rho hy|\rho y)|$$

is finite and independent of x and hence is Wiener integrable.

We now define a new measure $\nu \in \mathcal{M}$ by means of

$$\nu_y(E) \equiv \int_E [\int_a^b v(t)dy(t)]d\mu(v)$$

for $E \in \mathcal{A}'$ the σ-algebra of subsets of $\mathcal{B}$ described at the beginning of Section 2 so we have

$$\delta F(x|y) = i\int_{\mathcal{B}[a,b]} exp\{i\int_a^b v(t)dx(t)\}d\mu(v) \in S'.$$

Thus $\delta F(x|y)$ is an element of S' and by Theorem 5.2 of [3], we see that $\delta F(x|y)$ is Feynman integrable. Therefore by Theorem 1 above,

$$\begin{aligned} J &\equiv \int_{C[a,b]}^{anf_q} i\int_{\mathcal{B}[a,b]} exp\{i\int_a^b v(t)dx(t)\}\int_a^b v(t)dy(t)d\mu(v)dx \\ &= -iq\int_{C[a,b]}^{anf_q}\int_{\mathcal{B}[a,b]} exp\{i\int_a^b v(t)dx(t)\}d\mu(v)[\int_a^b y'(t)dx(t)]dx. \end{aligned} \tag{14}$$

By Theorem 5.2 of [3], if we utilize the measure ν_y on $\mathcal{B}[a,b]$ defined above, we see that

$$\begin{aligned} J &= \int_{C[a,b]}^{anf_q} i\int_{\mathcal{B}[a,b]} exp\{i\int_a^b v(t)dx(t)\}d\nu_y(v)dx \\ &= i\int_{\mathcal{B}[a,b]} exp\{\frac{1}{2qi}\int_a^b v^2(t)dt\}d\nu_y(v) \\ &= i\int_{\mathcal{B}[a,b]} exp\{\frac{1}{2qi}\int_a^b v^2(t)dt\}\int_a^b v(t)dy(t)d\mu(v) \end{aligned} \tag{15}$$

The left member of equation (14) exists because the integrand is in Class S' and hence the right member exists. The integrand of the right member of (14) is the product of an element of S' by a linear and hence an unbounded function of x . In equation (15) we evaluate the left member of (14). Replacing $y'(t)$ by $z(t)$ in (15) we obtain equation (13) and Theorem 2 is proved.

Thus we observe that Theorems 1 and 2 establish the analytic Feynman integrability of a large class of unbounded functionals.

§3. _Example_: We now construct a functional $F \in S'$ which satisifies the hypotheses of Theorem 2 and these three additional conditions:

i) its associated measure μ satisfies condition (12): $\int_{\mathcal{B}}(1+|||v|||)|d\mu(v)| < \infty$.

ii) F fails to be an analytic functional for almost every $x \in C[a,b]$.

iii) $G(x) \equiv F(x)\{\int_a^b p(s)dx(s)\}$ is unbounded in x for p of bounded variation on $[a,b]$.

Thus $G(x)$ is not a member of the class we discussed previously in [7] nor is $G \in S'$. However we will establish that $G(x)$ is analytic Feynman integrable.

Let $\mu \in \mathcal{M}'$ be defined thus: for $n = 1, 2, \ldots$ let the set consisting of the one function

$$E_n = \{v | v(t) \equiv 2^n(b-t) \text{for } a \leq t \leq b\} \tag{16}$$

have measure

$$\mu(E_n) = \frac{1}{n^2 2^n} , \tag{17}$$

and let

$$\mu[\mathcal{B} - \cup_{n=1}^{\infty} E_n] = 0.$$

Thus

$$\mu[\mathcal{B}] = \sum_{n=1}^{\infty} \frac{1}{n^2 2^n} < \infty.$$

Then

$$\begin{aligned} F(x) &\equiv \int_{\mathcal{B}[a,b]} exp\{i \int_a^b v(t)dx(t)\}d\mu(v) \\ &= \sum_{n=1}^{\infty} \frac{1}{n^2 2^n} exp\{i \int_a^b 2^n(b-t)dx(t)\} \\ &= \sum_{n=1}^{\infty} \frac{1}{n^2 2^n} exp\{i 2^n \int_a^b x(t)dt\} , \end{aligned} \tag{18}$$

and

$$F \in S' .$$

To prove that the measure μ satisfies

$$\int_{\mathcal{B}[a,b]} (1 + |||v|||)d\mu(v) < \infty$$

we note that

$$\int_{\mathcal{B}[a,b]}(1+|||v|||)d\mu(v) = \int_{\cup_{n=1}^{\infty}E_n}(1+|||v|||)d\mu(v)$$
$$= \sum_{n=1}^{\infty}\int_{E_n}(1+|||v|||)d\mu(v)\ .$$

If $v \in E_n$ by (16)

$$|||v||| = \max_{t\in[a,b]}|v(t)| = \max_{t\in[a,b]}|2^n(b-t)| = 2^n(b-a).$$

Thus

$$\sum_{n=1}^{\infty}\int_{E_n}(1+|||v|||)d\mu(v) = \sum_{n=1}^{\infty}\int_{E_n}\int_{E_n}[1+2^n(b-a)]d\mu(v)$$
$$= \sum_{n=1}^{\infty}[1+2^n(b-a)]\mu(E_n)$$
$$= \sum_{n=1}^{\infty}[1+2^n(b-a)]n^{-2}2^{-n}$$
$$= \sum_{n=1}^{\infty}n^{-2}2^{-n} + (b-a)\sum_{n=1}^{\infty}n^{-2} < \infty.$$

Hence

$$\int_{\mathcal{B}[a,b]}(1+|||v|||)|d\mu(v)| < \infty$$

and μ satisfies condition (i) above. Since μ is a positive measure it satisfies (12). Thus

$$F(\rho x) = \sum_{n=1}^{\infty}n^{-2}2^{-n}exp\{i2^n\rho\int_a^b x(t)dt\}, \tag{19}$$

and if we let $exp\{i\rho\int_a^b x(t)dt\} = z$, we have

$$F(\rho x) = \sum_{n=1}^{\infty}n^{-2}2^{-n}z^{2^n} \equiv H(z). \tag{20}$$

By the Hadamard gap theorem see Titchmarsh [10; p. 223], $H(z)$ is analytic for $|z| < 1$ but NOT analytic on $|z| = 1$.

Replacing z by $exp\{i\rho\int_a^b x(t)dt\}$ we see that $F(\rho x)$ fails to be analytic in ρ on the real axis (unless $\int_a^b x(t)dt = 0$). Thus for almost every x , $F(\rho x)$ is not analytic in ρ , in particular the functional appearing in the 2nd member of (14) above,

$$\int_{B[a,b]} exp\{i\int_a^b v(t)dx(t)\}d\mu(v)[\int_a^b p(t)dx(t)], \tag{21}$$

is an unbounded non-analytic functional of x which by Theorem 2 is analytic Feynman integrable. To establish that (21) is unbounded, we note that its first factor (18) is bounded away from zero because $F(x) = \sum_{n=1}^{\infty} n^{-2}2^{-n}exp\{i2^n \int_a^b x(t)dt\}$ is in absolute value greater than 1/4 since its first term has absolute value 1/2 and the sum of the absolute values of the remaining terms is less than 1/4 . The second factor is unbounded in x since it is linear in x (assuming $p(t)$ is not almost everywhere zero).

We have thus shown that for fixed x , $F(\rho x)$ fails to be analytic in ρ in the neighborhood of any point ρ on the real axis (unless $\int_a^b x(t)dt = 0$). Thus we have constructed a non-analytic unbounded functional $G(x)$ which is analytic Feynman integrable.

§4: We next evaluate the analytic Feynman integral of functionals which are products of elements of S' by <u>two</u> linear factors.

<u>*Theorem 3*</u>: *If $G \in S'$, so for $x \in C[a,b]$*

$$G(x) \equiv \int_{B[a,b]} exp\{i\int_a^b v(t)dx(t)\}d\mu(v) \tag{22}$$

where μ is a complex measure of finite variation with

$$\int_{B[a,b]} |||v|||\,|d\mu(v)| < \infty \tag{23}$$

and let $z_1(t)$ and $z_2(t)$ be essentially of bounded variation. Then the following integrals exist and we have

$$\begin{aligned} -iq &\int_{C[a,b]}^{anf_q} G(x)\int_a^b z_1(t)dx(t)\int_a^b z_2(t)dx(t)dx \\ &= \int_a^b z_1(t)z_2(t)dt\int_{B[a,b]} exp\{\frac{1}{2qi}\int_a^b v^2(t)dt\}d\mu(v) \\ &-\frac{1}{q}\int_{B[a,b]} exp\{\frac{1}{2qi}\int_a^b v^2(t)dt\}\int_a^b v(t)z_1(t)dt\int_a^b v(t)z_2(t)dtd\mu(v). \end{aligned} \tag{24}$$

Proof of Theorem 3. Let $y_1(t)$ and $y_2(t)$ be real absolutely continuous functions on $[a,b]$ which vanish at $t=a$ and such that y_1', y_2' are esentially of bounded variation. We let

$$F(x) \equiv G(x) \int_a^b y_1'(t)dx(t) \tag{25}$$

for $x \in C[a,b]$. Then

$$F(x+hy_2) = G(x+hy_2) \int_a^b y_1'(t)d[x(t)+hy_2(t)]$$

and applying Lemma 2 we see

$$\begin{aligned} \frac{\partial}{\partial h} F(x+hy_2) &= G(x+hy_2) \int_a^b y_1'(t)d[y_2(t)] \\ &+ \int_a^b y_1'(t)d[x+hy_2] \frac{\partial}{\partial h} G(x+hy_2) \\ \delta F(x|y_2) &= G(x) \int_a^b y_1'(t)y_2'(t)dt + \delta G(x|y_2) \int_a^b y_1'(t)dx(t) \end{aligned}$$

and from (22)

$$\delta G(x|y_2) = \int_{B[a,b]} exp\{iv(t)dx(t)\} i \int_a^b v(t)dy_2(t)d\mu(v),$$

so

$$\delta F(x|y_2) = \int_a^b y_1'(t)y_2'(t)dt \int_B exp\{i \int_a^b v(t)dx(t)\}d\mu(v) \tag{26}$$

$$+i \int_a^b y_1(t)dx(t) \int_B exp\{i \int_a^b v(t)dx(t)\} \int_a^b v(t)dy_2(t)d\mu(v).$$

We now substitute this expression into equation (5) of Theorem 1 to obtain

(27)
$$\int_{C[a,b]}^{anf_q} \int_a^b y_a'(t)y_2'(t)dt \int_{B[a,b]} exp\{i \int_a^b v(t)dx(t)\}d\mu(v)dx$$

$$+i \int_{C[a,b]}^{anf_q} \int_a^b y_1'(t)dx(t) \int_B exp\{i \int_a^b v(t)dx(t)\} \int_a^b v(t)dy_2(t)d\mu(v)dx$$

$$= -iq \int_{C[a,b]}^{anf_q} \int_{B[a,b]} exp\{i \int_a^b v(t)dx(t)\}d\mu(v) \int_a^b y_1'(t)dx(t) \int_a^b y_2' dx(t)dx.$$

From the right hand side of this equation we have

$$
\begin{aligned}
Q \equiv & -iq \int_{C[a,b]}^{anf_q} G(x) \int_a^b y_1'(t)dx(t) \int_a^b y_2'(t)dx(t)dx \\
= & \int_a^b y_1'(t)y_2'(t)dt \int_{C[a,b]}^{anf_q} G(x)dx \\
& + i \int_{C[a,b]}^{anf_q} [\int_a^b y_1'(t)dx(t)] \int_{B[a,b]} exp\{i \int_a^b v(t)dx(t)\}d\mu(v)dx
\end{aligned} \tag{28}
$$

where ν is the measure such that for every Borel measurable set $E \subset L_2[a,b]$,

$$
\nu_{y_2}(E) = \int_E (\int_a^b v(t)dy_2(t))d\mu(v). \tag{29}
$$

We now evaluate the integrals in the right hand side of equation (28) by using Theorem 5.2 of [3] and equation (14) and (16) above. Thus

$$
\begin{aligned}
Q = & \int_a^b y_1'(t)y_2'(t)dt \int_{B[a,b]} exp\{\frac{1}{2qi} \int_a^b v^2(t)dt\}d\mu(v) \\
& - \frac{1}{q} \int_{B[a,b]} exp\{\frac{1}{2qi} \int_a^b v_2'(t)dt\} \int_a^b v(t)dy_1(t)d\nu(v) \\
= & \int_a^b y_1'(t)y_2'(t)dt \int_{B[a,b]} exp\{\frac{1}{2qi} \int_a^b v^2(t)dt\}d\mu(v) \\
& - \frac{1}{q} \int_{B[a,b]} exp\{\frac{1}{2qi} \int_a^b v^2(t)dt\} \int_a^b v(t)dy_1(t) \int_a^b v(t)dy_2(t)d\mu(v)
\end{aligned} \tag{30}
$$

and setting $y_1'(t) = z_1(t)$ and $y_2'(t) = z_2(t)$, the theorem is proved.

Bibliography

[1] S. Albeverio and R. Hoegh-Krohn, *Mathematical theory of Feynman path integrals*, Lecture Notes in Math., 523, Springer-Verlag, Berlin, 1976.

[2] R.H. Cameron, *The first variation of an indefinite Wiener integral*, Proc. Amer. Math. Soc. 2 (1951), 914-924.

[3] R.H. Cameron and D.A. Storvick, *Some Banach algebras of analytic Feynman integrable functionals*, Analytic functions, (Kozubnik, 1979) Lecture Notes in Math., 798, Springer- Verlag, Berlin, 1980, 18-67.

[4] R.H. Cameron and D.A. Storvick, *Analytic Feynman integral solutions of an integral equation related to the Schroedinger equation*, J. Analyse Math. 38 (1980), 34-66.

[5] R.H. Cameron and D.A. Storvick, *New existence theorems and evaluation formulas for sequential Feynman integrals*, Proc. London Math. Soc. (3) 52 (1986), 557-581.

[6] R.H. Cameron and D.A. Storvick, *New existence theorems and evaluation formulas for analytic Feynman integrals*, Deformations of mathematical structures, Complex analysis with physical applications, Kluwer Acad. Publ., Dordrecht (1989), 297- 308.

[7] R.H. Cameron and D.A. Storvick, *Unbounded Feynman integrable functionals defined in terms of analytic functions*, Complex analysis (Joensuu 1987) Lecture Notes in Math., 1351, Springer-Verlag, Berlin, (1987), 78-92.

[8] R.P. Feynman, *Space-time approach to non- relativistic quantum mechanics*, REv. Modern Phys. 20 (1948), 115- 142.

[9] G.W. Johnson and D.L. Skoug, *Notes on the Feynman integral, III: the Schroedinger equation*, Pacific J. Math. 105 (1983), 321-358.

[10] E.C. Titchmarsh, *The theory of functions*, 2nd Ed. Oxford Univ., Press (London), 1939.

Department of Matheamtics
University of Minnesota
Minneapolis, MN 55455
U.S.A.

STABILITY THEOREMS FOR THE OPERATOR-VALUED FUNCTION SPACE INTEGRAL*

KUN SOO CHANG

Department of Mathematics, Yonsei University
Seoul 120-749, Korea

and

KUN SIK RYU

Department of Mathematics, Hannam University
Daejon 300-791, Korea

ABSTRACT

Stability theorems for the operator-valued function space integral as an operator on $L_p(1 \leq p \leq 2)$ has been studied for certain functionals involving the Lebesgue measure. Johnson and Lapidus established stability theorems for the integral as an operator on L_2 for certain functionals involving any Borel measure. We give theorems insuring stability with respect to potentials and wave functions for the function space integral as an operator on $L_p(1 < p < 2)$ for certain functionals involving some Borel measures.

1. Introduction

In 1984, Johnson proved a bounded convergence theorem for the operator-valued function space integral [4]. This is the first stability

---

* Research partially supported by Yonsei University, the Korea Science and Engineering Foundation, and the Ministry of Education.

theorem for the integral as a bounded linear operator on $L_2(\mathbb{R}^n)$ where n is any positive integer. In [6], Johnson and Skoug introduced stability theorems for the integral as an $\mathcal{L}(L_p(\mathbb{R}^N), L_{p'}(\mathbb{R}^N))$ theory, $1 < p \leq 2$, where N is a positive integer such that $N < \frac{2p}{2-p}$ and $\frac{1}{p} + \frac{1}{p'} = 1$. Let $C_0(\mathbb{R})$ be the complex-valued continuous functions on $\mathbb{R}$ which vanish at ∞. Chang studied stability theorems for the integral as a bounded linear operator from $L_1(\mathbb{R})$ to $C_0(\mathbb{R})$ [1]. In those papers mentioned above, they treat certain functionals which involve only the Lebesgue measure on the interval $(0,t)$.

In [5]. Johnson and Lapidus established stability theorems for the integral as an $\mathcal{L}(L_2(\mathbb{R}^N), L_2(\mathbb{R}^N))$ theory for certain functionals involving any Borel measure on $(0,t)$. In this paper, we give theorems insuring stability with respect to potentials and wave functions for the operator-valued function space integral for certain functionals involving some Borel measures as an $\mathcal{L}(L_p(\mathbb{R}^N), L_{p'}(\mathbb{R}^N))$ theory, $1 < p < 2$.

2. Preliminaries and Notations

In this section we present some necessary notations, definitions, and lemmas which are needed in the next section.

Let $\mathbb{N}$ be the set of all natural numbers and let $\mathbb{R}$ be the set of all real numbers. Let $\mathbb{C}$, $\mathbb{C}_+$, and $\tilde{\mathbb{C}}_+$ be the set of all complex numbers, all complex numbers with positive real part and all nonzero complex numbers with nonnegative real part, respectively.

Let $1 < p < 2$ be given and let p' be such that $\frac{1}{p} + \frac{1}{p'} = 1$. Let α

in $(1, \infty)$ be such that

$$\alpha = \frac{p}{2-p}$$

In our theorems, N will be a positive integer restricted so that

$$N < 2\alpha$$

Let γ be a real number such that

$$\frac{2\alpha}{2\alpha-N} < \gamma < \infty.$$

For $1 \leq s < \infty$, $L_s(\mathbb{R}^N)$ is the space of the Borel measurable, $\mathbb{C}$-valued functions ψ on $\mathbb{R}^N$ such that $|\psi|^s$ is integrable with respect to the Lebesgue measure m_L on $\mathbb{R}^N$. More formally, the elements of $L_s(\mathbb{R}^N)$ are equivalent classes of functions, with ψ_1 and ψ_2 said to be equivalent if they are equal almost everywhere with respect to m_L.

For $1 \leq s,t < \infty$, $L(L_s(\mathbb{R}^N), L_t(\mathbb{R}^N))$ denote the space of bounded linear operators from $L_s(\mathbb{R}^N)$ into $L_t(\mathbb{R}^N)$.

For λ in $\tilde{\mathbb{C}}_+$, ψ in $L_p(\mathbb{R}^N)$, ξ in $\mathbb{R}^N$ and a positive real number s, let

$$(C_{\lambda/s}\,\psi)(\xi) = \left(\frac{\lambda}{2\pi s}\right)^{N/2} \int_{\mathbb{R}^N} \psi(u)\exp\left(-\frac{\lambda\|u-\xi\|^2}{2s}\right) dm_L(u)$$

where if N is odd we always cloose $\lambda^{\frac{1}{2}}$ with nonnegative real part and if Re $\lambda = 0$ the integral in the above should be interpreted in the mean just as in the theory of the L_p Fourier transform.

Let $t > 0$ be given. $M(0,t)$ will denote the space of complex Borel measures on the interval $(0,t)$. A measure μ in $M(0,t)$ is said to be continuous if $\mu(\{\tau\}) = 0$ for every τ in $(0,t)$ and a measure ν in $M(0,t)$ is said to be discrete if there is an at most countable subset $\{\tau_i | i=1,2,\ldots\}$ of $(0,t)$ and a summable sequence $\langle\omega_i\rangle$ from $\mathbb{C}$ such that $\nu = \sum_{i=1}^{\infty} \omega_i \delta_{\tau_i}$ where δ_{τ_i} is the Dirac measure with total mass one concentrated at τ_i. Then every measure η in $M(0,t)$ has a unique decomposition, $\eta = \mu + \nu$ into a continuous part μ and a discrete part ν.

$M(0,t)^*$ will denote the subset of $M(0,t)$ which satisfies the following conditions:

(a) If μ is the continuous part of η in $M(0,t)$, then the Radon-Nikodym derivative $d|\mu|/dm$ exists and is essentially bounded where m is the Lebesgue measure on $(0,t)$.

(b) If $\nu = \sum_{i=1}^{\infty} \omega_i \delta_{\tau_i}$ is the discrete part of η in $M(0,t)$, then $\sum_{i=1}^{\infty} |\omega_i| \tau_i^{-\gamma'\delta}$ converges where $\delta = N/2\alpha$, $\frac{1}{\gamma} + \frac{1}{\gamma'} = 1$.

For η in $M(0,t)^*$, let $L_{\alpha\gamma:\eta}([0,t] \times \mathbb{R}^N) \equiv L_{\alpha\gamma:\eta}$ be the space of $\mathbb{C}$-valued Borel measurable functionals θ on $[0,t] \times \mathbb{R}^N$ such that

$$\|\theta\|_{\alpha\gamma:\eta} \equiv \{\int_{(0,t)} \|\theta(s,\cdot)\|_\alpha^\gamma \, d|\eta|(s)\}^{1/\gamma} < \infty.$$

Note that $L_{\alpha\gamma:\eta} \subset L_{\alpha s:\eta}$ if $1 \le s \le \gamma \le \infty$. If θ is in $L_{\alpha\gamma:\eta}$ and if $\eta = \mu + \nu$ is the Lebesgue decomposition, it is not difficult to show that θ is in $L_{\alpha\gamma:\mu} \cap L_{\alpha\gamma:\nu}$ and $\|\theta\|_{\alpha\gamma:\eta} = \|\theta\|_{\alpha\gamma:\mu} + \|\theta\|_{\alpha\gamma:\nu}$.

Let θ be in $L_\alpha(\mathbb{R}^N)$. From [8], the function $M_\theta : L_{p'}(\mathbb{R}^N) \to L_p(\mathbb{R}^N)$ defined by $M_\theta(\psi) = \psi\theta$ is in $L(L_{p'}(\mathbb{R}^N), L_p(\mathbb{R}^N))$ and $\|M_\theta\| \le \|\theta\|_\alpha$. It will be convenient to let $\theta(s)$ denote $M_{\theta(s,\cdot)}$ for θ in $L_{\alpha\gamma:\eta}$.

Let $C_0[0,t] \equiv C_0$ be the space of R^N-valued continuous functions x on $[0,t]$ such that $x(o) = 0$. We consider C_0 as equipped with N-dimensional Wiener measure m_ω which is just the product of N one-dimensional Wiener measures. Let $C[0,t]$ be the space of R^N-valued continuous functions on $[0,t]$.

Let F be a functional on $C[0,t]$. Given $\lambda > 0$, ψ in $L_p(R^N)$, and ξ in R^N, let

$$[I_\lambda(F)\psi](\xi) = \int_{C_0} F(\lambda^{-\frac{1}{2}} x + \xi)\psi(\lambda^{-\frac{1}{2}} x(t) + \xi)dm_\omega(x).$$

If for m_L-a.e. ξ in R^N, $[I_\lambda(F)\psi](\xi)$ exists in $L_{p'}(R^N)$ and if the map $\psi \to I_\lambda(F)\psi$ gives an element of $L(L_p(R^N), L_{p'}(R^N))$, we say that the operator-valued function space integral $I_\lambda(F)$ exists for λ. Suppose there exists λ_0 $(0 < \lambda_0 \leq \infty)$ such that $I_\lambda(F)$ exists for all $0 < \lambda < \lambda_0$ and there exists an $L(L_p(R^N), L_{p'}(R^N))$-valued function which is analytic in $C_{+,\lambda_0} \equiv C_+ \cap \{z \in C : |z| < \lambda_0\}$ and agrees with $I_\lambda(F)$ on $(0,\lambda_0)$, then this $L(L_p,L_{p'})$-valued function is called the operator-valued function space integral of F associated with λ and in this case, we say that $I_\lambda(F)$ exists for λ in C_{+,λ_0}. If $I_\lambda(F)$ exists for λ in C_{+,λ_0} and $I_\lambda(F)$ is strongly continuous in $\tilde{C}_{+,\lambda_0} \equiv \tilde{C}_+ \cap \{z \in C : |z| < \lambda_0\}$, we say that $I_\lambda(F)$ exists for λ in $\tilde{C}_{+,\lambda_0}$. When λ is purely imaginary, $I_\lambda(F)$ is called the analytic operator-valued Feynman integral of F.

Let (Ω, μ) be a measure space and let $L(X,Y)$ be the space of bounded linear operators from X into Y, where X and Y are Banach spaces.

Let $G : \Omega \to L(X,Y)$ be a function such that for each x in X, $\{G(s)\}(x)$ is Bochner integrable with respect to μ. Then there exists a linear operator J from X into Y such that

$$J(x) = (B) \int_{\Omega} \{G(s)\}(x)\ d\mu(s) \quad \text{for } x \text{ in } X,$$

where $(B) \int_{\Omega} \{G(s)\}(x)\ d\mu(s)$ refers to the Bochner integral. This linear operator J is denoted by

$$(BS) \int_{\Omega} G(s)\ d\mu(s)$$

and it is called the Bochner integral in the strong operator sense. When $X = Y$, J is called the strong integral of G.

We finish this section with three lemmas. From [6], we obtain the following lemma.

Lemma 1. Let h be in $L_\alpha(R^N)$ and let $\langle g_n \rangle$ be a sequence from $L_\alpha(R^N)$ such that g_n converges to g m_L-a.e. as $n \to \infty$ and such that $|g_n(u)| \le h(u)$ m_L-a.e. u for all n. Then g is in $L_\alpha(R^N)$ and M_{g_n} converges to M_g in the strong operator topology on $L(L_{p'}(R^N), L_p(R^N))$.

Let η be in $M(0,t)^*$ and let θ be in $L_{\alpha\gamma:\eta}$. Let $\eta = \mu + \nu$ be the decomposition of η into its continuous part μ and discrete part ν with $\nu = \sum_{i=1}^{\infty} \omega_i \delta_{\tau_i}$. Let

$$F_n(y) = \left(\int_{(0,t)} \theta(s,y(s))\ d\eta(s)\right)^n \text{ for } y \text{ in } C[0,t].$$

And for each h in N, put σ be a permutation on $\{1,2,\ldots,h\}$ such that

$$\tau_{\sigma(1)} < \tau_{\sigma(2)} < \cdots < \tau_{\alpha(h)}.$$

From [2], we have the following lemma.

Lemma 2. Suppose that $\theta(\tau_i,\cdot)$, $i = 1,2,\ldots$, are essentially bounded. Then the operator $I_\lambda(F_n)$ exists for all λ in $\tilde{C}_+$ and for all λ in $\tilde{C}_+$,

$$I_\lambda(F_n) = n! \sum_{h=0}^{\infty} \sum_{\substack{q_0+\ldots+q_h=n \\ q_0\neq 0}} \frac{\omega_1^{q_1}\ldots\omega_h^{q_h}}{q_1!\ldots q_h!} \Big[\sum_{j_1+\ldots+j_{h+1}=q_0}$$

$$(BS) \int_{\Delta_{q_0;j_1,\ldots,j_{h+1}}} L_0 \circ L_1 \circ \ldots \circ L_h \, d \prod_{u=1}^{q_0} \mu(s_u)\Big]$$

where for each h in $\mathbb{N}$, $\Delta_{q_0;j_1,\ldots j_{h+1}} = \{(s_1,\ldots,s_{q_0}) \in (0,t)^{q_0} |$

$0 < s_1 < \ldots < s_{j_1} < \tau_{\sigma(1)} < s_{j_1+1} < \ldots < s_{j_1+j_2} < \tau_{\sigma(2)} < \ldots < s_{q_0} < t\}$

and where for $(s_1,\ldots,s_{q_0}) \in \Delta_{q_0;j_1,\ldots,j_{h+1}}$ and $m \in \{0,1,2,\ldots,h\}$,

$$L_m = [\theta(\tau_{\alpha(m)})]^{q_{\sigma(m)}} \circ C_{\lambda/(s_{j_1+\ldots+j_m+1}-\tau_{\sigma(m)})} \circ \theta(s_{j_1+\ldots+j_m+1}) \circ \ldots$$

$$\circ\, \theta(s_{j_1+\ldots+j_{m+1}}) \circ C_{\lambda/(\tau_{\sigma(m+1)}-s_{j_1+\ldots+j_{m+1}})}.$$

(We use the convention $\tau_{\sigma(0)} = 0$, $\tau_{\sigma(h+1)} = t$ and $[\theta(\tau_{\sigma(0)})]^{q_{\sigma(0)}} = 1$, an identity map on $L_{p'}(R^N))$. Moreover.

$$\| I_\lambda(F_n)\| \leq n! \sum_{h=0}^{\infty} \sum_{\substack{q_0+\ldots+q_h=n \\ q_n\neq 0}} \frac{|\omega_1|^{q_1}\ldots|\omega_h|^{q_h}}{q_1!\ldots q_h!}(q_0!)^{-\frac{1}{r}} \left(\frac{(q_0+h)!}{q_0!\,h!}\right)^{\frac{1}{2r}}$$

$$\times\left(\frac{|\lambda|}{2\pi}\right)^{(q_0+h+1)\delta}\left[\prod_{n=1}^{h}\left(\|\theta(\tau_{\sigma(n)},\cdot)\|_{\infty}^{q_n-1}\|\theta(\tau_{\sigma(n)},\cdot)\|_{\alpha}\right)^{\rho(q_n)}\right](\text{ess sup } d|\mu|/dm)^{\frac{q_0}{r'}}$$

$$\times(\|\theta\|_{\alpha r:\mu})^{q_0}\left[\sum_{j_1+\dots+j_{h+1}=q_0}\left\{\prod_{n=0}^{h} E(\tau_{\sigma(n)},\tau_{\sigma(n+1)};j_{n+1};r'\delta)^{\frac{2}{r'}}\right]^{\frac{1}{2}}.$$

Denote this norm estimate $B_n(|\lambda|)$.

Let $\lambda_0 > 0$ be given and let $f(z) = \sum_{n=1}^{\infty} a_n z^n$ be an analytic function in $C^{\sim}_{+,\lambda_0}$ such that $\sum_{n=0}^{\infty} |a_n|\, B_n(|\lambda|)$ is finite for all λ in $C^{\sim}_{+,\lambda 0}$. Let η be in $M(0,t)^*$, θ be in $L_{\alpha r:\eta}$, and let

$$F(y) = f\left(\int_{(0,t)} \theta(s,y(s))\, d\eta(s)\right) \text{ for } y \text{ in } C[0,t].$$

From [2], we have the following lemma.

Lemma 3. Suppose $\theta(\tau_i,\cdot)$, $i=1,2,\dots,$ are essentially bounded. Then $I_\lambda(F)$ exists for all λ in $C^{\sim}_{+,\lambda_0}$ and is given by

$$I_\lambda(F) = \sum_{n=0}^{\infty} a_n\, I_\lambda(F_n),$$

where F_n is a functional defined in Lemma 2. Moreover, for λ in $C^{\sim}_{+,\lambda_0}$, the series $\sum_{n=0}^{\infty} a_n\, I_\lambda(F_n)$ converges in operator norm and

$$\|I_\lambda(F)\| \le \sum_{n=0}^{\infty} |a_n|\, B_n(|\lambda|).$$

3. Stability Theorems

Some functionals considered in the theory of operator-valued function space integrals are defined in term of potentials, wave functions and measures. It is natural to ask if the corresponding operators are stable under permutations of these objects.

First, we consider stability with respect to the potentials.

Let η be in $M(0,t)^*$ and let H belong to $\mathcal{L}_{\alpha r:\eta}$. Let $\eta = \mu + \sum_{i=1}^{\infty} \omega_i \delta_{\tau_i}$ be the decomposition of η into its continuous part and discrete part and let $<\theta_n>$ be a sequence in $\mathcal{L}_{\alpha r:\eta}$.

For nonnegative integer n, we let

$$F_n(y) = (\int_{(o,t)} \theta(s,y(s))\, d\eta(s))^n \quad \text{for } y \text{ in } C[0,t]$$

and

$$F_n^{(m)}(y) = (\int_{(0,t)} \theta_m(s,y(s))\, d\eta(s))^n \quad \text{for } y \text{ in } C[0,t].$$

Theorem 1. Suppose $\theta_m(s,u) \to \theta(s,u)$ as $m \to \infty$ for $\eta \times m_L$ - a.e. (s,u) and $|\theta_m(s,u)| \leq H(s,u)$ for all positive integer m and for $\eta \times m_L$ - a.e. (s,u). Then θ is also in $\mathcal{L}_{\alpha r:\eta}$, for all positive integers m,n, $I_\lambda(F_n)$ and $I_\lambda(F_n^{(m)})$ exist for all λ in $\tilde{C}_+$ and

$$I_\lambda(F_n^{(m)}) \to I_\lambda(F_n) \quad \text{strongly as } m \to \infty$$

Proof. It is easy to see that θ is in $\mathcal{L}_{\alpha r:\eta}$. By Lemma 2, for all positive integers m,n, $I_\lambda(F_n)$ and $I_\lambda(F_n^{(m)})$ exist for all λ in $\tilde{C}_+$. For $(s_1,s_2,\ldots,s_{q_0}) \in \Delta_{q_{j_0,j_1,\ldots,j_{h+1}}}$ and $\ell \in \{0,1,2,\ldots,h\}$, L_ℓ is defined in Lemma 2. Let $L_\ell^{(m)}$ be defined as in Lemma 2 except with θ replaced by θ_m. Fix ψ in $L_p(R^N)$. For $\ell = 1,2,\ldots,h$, let

$$A_\ell = (L_0^{(m)} \circ L_1^{(m)} \circ \ldots \circ L_{\ell-1}^{(m)} \circ L_\ell \circ \ldots \circ L_h)\psi \text{ and}$$

$$B_\ell = (L_\ell \circ L_{\ell+1} \circ \dots \circ L_h)\psi.$$

And for $\ell = 1,2,\dots,h$ and $v = 1,2,\dots, j_{\ell+1}$, let

$$C_{\ell+1,v} = \{\theta_m(\tau_{\sigma(\ell)})\}^{q_{\sigma(\ell)}} \circ C_{\lambda/(S_{j_1+j_2+\dots+j_\ell+1}-\tau_{\sigma(\ell)})} \circ$$

$$\theta_m(S_{j_1+j_2+\dots+j_\ell+1}) \circ C_{\lambda/(S_{j_1+j_2+\dots+j_\ell+2} - S_{j_1+\dots+j_\ell+1})} \circ \dots\dots \circ$$

$$\theta_m(S_{j_1+\dots+j_\ell+v-1}) \circ C_{\lambda/(S_{j_1+\dots+j_\ell+v} - S_{j_1+\dots+j_\ell+v-1})} \circ$$

$$\theta(s_{j_1+\dots+j_\ell+v}) \circ \dots \circ \theta(S_{j_1+j_2+\dots+j_{\ell+1}}) \circ C_{\lambda/(\tau_{\sigma(\ell+1)}-S_{j_1+\dots+j_{\ell+1}})} \circ B_{\ell+1}.$$

Then $\| (L_0 \circ L_1 \circ \dots \circ L_h)\psi - (L_0^{(m)} \circ L_1^{(m)} \circ \dots \circ L_h^{(m)})\psi\|_p$,

$$\leqq \sum_{\ell=1}^{h} \|A_\ell - A_{\ell-1}\|_p,$$

$$\leqq \sum_{\ell=1}^{h} (\prod_{k=1}^{\ell-1} \|L_k^{(m)}\|)\| L_\ell \circ B_{\ell+1} - L_\ell^{(m)} \circ B_{\ell+1}\|_p,$$

$$\leqq \sum_{\ell=1}^{h} \{(\prod_{k=1}^{\ell-1} \|L_k^{(m)}\|)(\sum_{v=1}^{j_{\ell+1}} \|C_{\ell+1,v} - C_{\ell+1,v-1}\|_p,)$$

By Lemma 1, the right-hand side of the last inequality above converges to zero as $m \to \infty$. Thus, $L_0^{(m)} \circ L_1^{(m)} \circ \dots \circ L_h^{(m)} \to L_0 \circ L_1 \circ \dots \circ L_h$ strongly as $m \to \infty$. Now we claim that

$$(*) \int_{\Delta_{q_0,j_1,\dots,j_{h+1}}} (L_0^{(m)} \circ L_1^{(m)} \circ \dots \circ L_h^{(m)})\psi \mathop{X}_{u=1}^{q_0} d_\mu(s_u) \to$$

$$\to \int_{\Delta_{q_0,j_1,\ldots,j_{h+1}}} (L_0 \circ L_1 \circ \ldots \circ L_h)\psi \mathop{X}_{u=1}^{q_0} d\mu(s_u)$$

as $m \to \infty$.

Since for every positive integer m, $|\theta_m(s,u)| \leq H(s,u)$ for $\eta \times m_L$ - a.e. (s,u), for a.e. $(s_1,s_2,\ldots,s_{q_0}) \in \Delta_{q_0,j_1,\ldots,j_{h+1}}$

$$\|L_0^{(m)} \circ L_1^{(m)} \circ \ldots \circ L_h^{(m)} \psi\|_p,$$

$$\leq \|\psi\|_p \left(\frac{|\lambda|}{2\pi}\right)^{(q_0+h+1)\delta} \left\{\left(\prod_{\ell=1}^{h} \|\theta_m(\tau_{\sigma(\ell)},\cdot)\|_\infty^{q_{\sigma(\ell)}-1}\right)\|\theta_m(\tau_{\sigma(\ell)},\cdot)\|_\alpha\right\}^{\rho(q_0)}$$

$$\times \left(\prod_{u=1}^{q_0} \|\theta_m(s_u,\cdot)\|_\alpha\right) \{s_1(s_2-s_1)\ldots(\tau_1-s_{j_1})(s_{j_1+1}-\tau_1)\ldots(t-s_{q_0})\}^{-\delta}$$

$$\leq \|\psi\|_p \left(\frac{|\lambda|}{2\pi}\right)^{(q_0+h+1)\delta} \left\{\left(\prod_{\ell=1}^{h} \|H(\tau_{\sigma(\ell)},\cdot)\|_\infty^{q_{\sigma(\ell)}-1}\right) \|H(\tau_{\sigma(\ell)},\cdot)\|_\alpha\right\}^{\rho(q_{\sigma(\ell)})}$$

$$\times \left(\prod_{u=1}^{q_0} \|H(s_u,\cdot)\|_\alpha\right) \{s_1(s_2-s_1)\ldots(\tau_1-s_{j_1})(s_{j_1+1}-\tau_1)\ldots(t-s_{q_0})\}^{-\delta}.$$

By the same method as in the proof of Lemma 2 in [2], the right-hand side in the last inequality above is $\mathop{X}_{u=1}^{q_0} \mu$- integrable. Hence, by the dominated convergence theorem for the Bochner integral, (*) is established. Hence $I_\lambda(F_n^{(m)}) \to I_\lambda(F_n)$ strongly as $m \to \infty$. Thus, the proof of theorem is complete.

Let $f(z) = \sum_{n=0}^{\infty} a_n z^n$ be an analytic function in $C_+^\sim \cap \{z \in C \mid |z| < \lambda_0\}$ for some positive real number λ_0. For positive integer m and y in $C[0,t]$, let

$$F(y) = f\left(\int_{(0,t)} \theta(s,y(s))\, d\eta(s)\right) \quad \text{and}$$

$$F^{(m)}(y) = f\left(\int_{(0,t)} \theta_m(s,y(s))\, d\eta(s)\right).$$

Theorem 2. Suppose for all λ in $\tilde{C}_{+,\lambda_0}$ and for all positive integer m, $\sum_{n=1}^{\infty} |a_n| \, \| I_\lambda(F_n^{(m)}) \|$ and $\sum_{n=1}^{\infty} \| I_\lambda(F_n) \|$ are convergent. Then $I_\lambda(F)$ and $I_\lambda(F^{(m)})$, m = 1,2,..., exist for all λ in $\tilde{C}_{+,\lambda_0}$ and

$$I_\lambda(F^{(m)}) \to I_\lambda(F) \text{ strongly as } m \to \infty \text{ in } \tilde{C}_{+,\lambda_0}.$$

Proof. By Lemma 3, $I_\lambda(F^{(m)})$ and $I_\lambda(F)$ exist for all λ in $\tilde{C}_{+,\lambda_0}$ and $I_\lambda(F) = \sum_{n=0}^{\infty} a_n I_\lambda(F_n)$ and $I_\lambda(F^{(m)}) = \sum_{n=0}^{\infty} a_n I_\lambda(F_n^{(m)})$ for all positive integer m. And for ψ in $L_p(R^N)$,

$$\begin{aligned}
&\lim_{m\to\infty} I_\lambda(F^{(m)})\, \psi \\
&= \lim_{m\to\infty} \lim_{k\to\infty} \sum_{n=0}^{k} a_n I_\lambda(F_n^{(m)})\, \psi \\
&= \lim_{k\to\infty} \lim_{m\to\infty} \sum_{n=0}^{k} a_n I_\lambda(F_n^{(m)})\, \psi \\
&= \sum_{n=0}^{\infty} a_n I_\lambda(F_n)\, \psi \\
&= I_\lambda(F)\, \psi
\end{aligned}$$

for all λ in $\tilde{C}_{+,\lambda_0}$. Thus, the proof of theorem is complete.

Now, we consider stability with respect to the wave functions.

By the triangle inequality of the norm and Theorem 2, directly we obtain the following theorem.

Theorem 3. Suppose $\psi^{(m)} \to \psi$ in $L_p(R^N)$ as $m \to \infty$. Then for positive integer m, $I_\lambda(F)\,\psi$ and $I_\lambda(F^{(m)})\,\psi^{(m)}$ exist in $L_{p'}(R^N)$ for λ in $C^{\sim}_{+,\lambda_0}$ and $I_\lambda(F^{(m)})\,\psi^{(m)} \to I_\lambda(F)\,\psi$ in $L_{p'}(R^N)$ as $m \to \infty$.

Lastly, we treat the stability theorem with respect to the measures. Let η and η_m, $m = 1,2,\ldots,$ be in $M(0,t)^*$ such that η_m converges to η in the total variation norm and let θ be bounded which is in

$$L_{\alpha r:\eta} \cap \left(\bigcap_{m=1}^{\infty} L_{\alpha r:\eta_m}\right). \quad \text{Let}$$

$$F_n(y) = \left(\int_{(0,t)} \theta(s,y(s))\, d\eta(s)\right)^n \text{ and}$$

$$F_n^m(y) = \left(\int_{(0,t)} \theta(s,y(s)\, d\eta_m(s)\right)^n \text{ for } y \text{ in } C[0,t].$$

And let $f(z) = \sum_{n=0}^{\infty} a_n z^n$, let

$$F(y) = f\left(\int_{(0,t)} \theta(s,y(s))\, d\eta(s)\right) \text{ and let}$$

$$F^m(y) = f\left(\int_{(0,t)} \theta(s,y(s))\, d\eta_m(s)\right) \text{ for } y \text{ in } C[0,t].$$

By the same method as in the proof of Theorem 4.4 in [5], we obtain,

Theorem 4. Suppose that there are positive real number λ_0 and K such that for all λ in $C^{\sim}_{+,\lambda_0}$ and all positive real number m,

$$\sum_{n=0}^{\infty} |a_n|\, \| I_\lambda(F_n^m) \| < K \text{ and } \sum_{n=0}^{\infty} |a_n|\, \| I_\lambda(F_n) \| < K$$

Then for λ in $\tilde{C}_{+,\lambda_0}$ and for positive integer m, $I_\lambda(F)$ and $I_\lambda(F^m)$ exist and $I_\lambda(F^m) \to I_\lambda(F)$ uniformly in λ on all compact subsets of C_{+,λ_0} in the operator norm topology. Moreover for all $0 < \lambda < \lambda_0$,

$$\| I_\lambda(F^m) - I_\lambda(F) \| \leqq \frac{\lambda_0}{2\pi t} T_m \text{ for } m = 1,2,\ldots$$

where $T_m = \sup\{|f(z_1) - f(z_2)| \,|\, |z_1 - z_2| \leqq M_m\}$ with $M_m = \|\eta - \eta_m\| \, \|\theta\|_\infty$. Here, $\|\theta\|_\infty$ denotes the supremum norm of θ on $(0,t) \times \mathbb{R}^N$.

Reference

1. J.S. Chang, Stability theorems for the Feynman integral; the $\mathcal{L}(L_1(\mathbb{R}), C_0(\mathbb{R})$ theory, *Supplemento ai Rendiconti del Circolo Mathematico di Palermo, Serie II, 17(1987), 135-151.*

2. K.S. Chang and K.S. Ryu, Analytic operator-valued function space integral as an $\mathcal{L}(L_p, L_{p'})$ theory, to appear in Trans. *Amer. Math. Soc.*

3. E. Hille and R.S. Phillips, *Functional Analysis and Semi-group,* A.M.S. Colloq, Pub. 31 (1957).

4. G.W. Johnson, A bounded convergence theorem for the Feynman integral, *J. Math. Phy., 25, (1984), 1323-1326.*

5. G.W. Johnson and M.L. Lapidus, Generalized Dyson series, Generalized Feynman Diagram, the Feynman integral and Feynman's operational Calculus, Mem. *Amer. Math. Soc., 62, No. 351(1986).*

6. G.W. Johnson and D.L. Skoug, Stability theorem for the Feynman integral, *Supplemento ai Rondiconti, del Circolo Mathematico di Palermo, Serie II, 8(1985), 361-377.*

CONDITIONAL FEYNMAN INTEGRALS FOR THE FRESNEL CLASS OF FUNCTIONS ON ABSTRACT WIENER SPACES

DONG MYUNG CHUNG
Department of Mathematics
Sogang University
Seoul 121-742, Korea

ABSTRACT

In this paper we establish formulas for the conditional Feynman integral of functions in the Fresnel class of functions on an abstract Wiener space and then use them to provide the fundamental solution to the Schroedinger equation for a class of potentials. Some illustrative examples for the fundamental solution are given in the last section.

1. Introduction

We consider the Schrödinger equation of quantum mechanics (for a single particle of mass m)

$$(1.1)\qquad i\hbar\frac{\partial}{\partial t}\Gamma(t,\vec{\eta}) = -\frac{\hbar^2}{2m}\Delta\Gamma(t,\vec{\eta}) + V(\vec{\eta})\Gamma(t,\vec{\eta})$$

$$\Gamma(0,\vec{\eta}) = \psi(\vec{\eta}),\quad \vec{\eta}\in\mathbb{R}^n$$

where Δ is the Laplacian on $\mathbb{R}^n$, $\hbar$ is Planck's constant and V is a suitable potential. Let $K(t,\vec{\eta},0,\vec{\xi})$ denote the fundamental solution to Schrödinger equation (1.1), i.e.,

$$\Gamma(t,\vec{\eta}) = \int_{\mathbb{R}^n} K(t,\vec{\eta},0,\vec{\xi})\psi(\vec{\xi})d\vec{\xi}$$

According to Feynman [10], $K(t,\vec{\eta},0,\vec{\xi})$ is given by the formal path integral:

$$(1.2) \qquad K(t,\vec{\eta},0,\vec{\xi}) = \int_{C_{\vec{\xi},\vec{\eta}}[0,t]} \exp\left\{\frac{i}{\hbar} S(x)\right\} \mathfrak{D}(x)$$

where $C_{\vec{\xi},\vec{\eta}}[0,t]$ is the space of paths x such that $x(0)=\vec{\xi}$ and $x(t)=\vec{\eta}$, $\mathfrak{D}(x)$ is a uniform "measure" which does not exist, and S(x) is the action integral associated with the path x; i.e.,

$$S(x) = \int_0^t \left[\frac{m}{2}\left(\frac{dx}{ds}\right)^2 - V(x(s))\right] ds$$

The basic problem of quantum mechanics is to find the solution $\Gamma(t,\vec{\eta})$ or the fundamental solution $K(t,\vec{\eta},0,\vec{\xi})$ to Eq. (1.1).

In [11] Gelfand and Yaglom made an attempt to give sense to the formal integral in Eq. (1,2) by introducing a Wiener measure with complex variance parameter. Unfortunately their attempt was failed as pointed out by Cameron in [3, p. 126].

There has been several rigorous approaches to Eq. (1.2) to provide the fundamental solution to Eq. (1.1) (see for examples, [13], [17], [19]).

In [7] and [9] the concept of the conditional Feynman integral has been introduced and is used to provide a method of getting the fundamental solution to the Schrödinger equation and to obtain the kernel of operator valued Feynman integrals of various functions.

This paper is in line with the work by the author [9]. We establish formulas for the conditional Feynman integral of functions involving (time independent or dependent) potentials and then use them to obtain the fundamental solution to the Schrödinger equation for a class of potentials.

2. Definitions and Preliminaries

Let H be a real separable infinite dimensional Hilbert space with inner product $\langle\cdot,\cdot\rangle$ and norm $|\cdot|$. Let $\|\cdot\|$ be a measurable norm on H with respect to the Gaussian cylinder set measure m on H (see [12, 18]). Let B denote the completion of H with respect to $\|\cdot\|$. Let i denote the natural injection from H into B. The adjoint operator i^*

of i is one-to-one and maps B^* continuously onto a dense subset of H^*. By indentifying H with H^* and B^* with i^*B^*, we have a triple $B^* \subset H = H^* \subset B$ and $\langle x,y\rangle = (x,y)$ for all x in H and y in B^*, where $(\cdot,\cdot)$ denotes the B-B^* pairing. By a well known result of Gross, $m\circ i^{-1}$ has a unique countably additive extension ν to the Borel σ-algebra $\mathcal{B}(B)$ of B. The triple (H,B,ν) is called an abstract Wiener space.

Let $\{e_j; j\ge 1\}$ be a complete orthonormal set in H such that the e_j's are in B^*. For each $h \in H$ and $x \in B$, we define

$$(h,x)^{\sim} = \lim_{n\to\infty} \sum_{j=1}^{n} \langle h,e_j\rangle (e_j,x)$$

if the limit exists and = 0 otherwise. It is well known that for each $(h\neq 0)$ in H, $(h,\cdot)^{\sim}$ is a Gaussian random variable on B with mean zero, variance $|h|^2$, and that if $\{h_1,h_2,\cdots,h_n\}$ is an orthonormal set in H, then $(h_j,x)^{\sim}$'s are independent. Furthermore, it is easy to see that $(h,\lambda x)^{\sim} = (\lambda h,x)^{\sim} = \lambda(h,x)^{\sim}$ for all $\lambda > 0$.

Let M(H) be the class of all complex valued Borel measures on H with bounded variation. Then M(H) is a Banach algebra under the total variation norm and with convolution as multiplication.

Given two complex valued measurable functions F and G on B, F is said to be equal to G s-almost surely (s-a.s.) if for each $\alpha > 0$, $\nu\{x\in B: F(\alpha x) \neq G(\alpha x)\} = 0$ (see [4]). For a measurable function F on B, let [F] denote the equivalence class of functions which are equal to F s-a.s.. The class of equivalence classes given by

$$\mathcal{F}(B) = \{[F]: F(x) = \int_H e^{i(h,x)^{\sim}} d\sigma(h),\ \sigma \in M(H)\}$$

is called the Fresnel class of functions on B. As is customary we will identify a function with its s-equivalence class and think of $\mathcal{F}(B)$ as a class of functions on B rather than as a class of equivalence classes. It is known [16] that $\mathcal{F}(B)$ is a Banach algebra over the complex field with the norm $\|F\| = \|\sigma\|$ and the mapping $\sigma \longrightarrow F$ is a Banach algebra isomorphism, where σ and F are related by

$$F(x) = \int_H e^{i(h,x)^{\sim}} d\sigma(h). \tag{2.1}$$

It is known [16] that for each F in $\mathcal{F}(B)$ given by (2.1), the analytic Feynman integral $E^{\mathrm{anf}_q}(F)$ of F with parameter $q(\neq 0)$ exists and is given by the formula $E^{\mathrm{anf}_q}(F) = \int_H \exp\{-\frac{i}{2q}|h|^2\}d\sigma(h)$.

Let $\mathbb{R}^n$, $\mathbb{C}$, $\mathbb{C}^+$ denote respectively the n-dimensional Euclidean space, the complex numbers, and the complex numbers with positive real part. Let X be an $\mathbb{R}^n$-valued function and Y a $\mathbb{C}$-valued integrable function on $(B,\mathcal{B}(B),\nu)$. Let P_X be the probability distribution of X. Then by the definition of conditional expectation, the conditional expectation of X, is any $\mathbb{R}^n$-valued Borel measurable and P_X-integrable function ϕ on $\mathbb{R}^n$ such that for all $E \in \mathcal{B}(\mathbb{R}^n)$,

$$\int_{X^{-1}(E)} Y(x)d\nu(x) = \int_E \phi(\vec{\eta})dP_X(\vec{\eta}).$$

By the Radon-Nikodym theorem such a function ϕ exists and is unique up to Borel null sets in $\mathbb{R}^n$. Following Yeh [20] the function $\phi(\vec{\eta})$, written $E[Y|X=\vec{\eta}]$, is called the <u>conditional abstract Wiener integral</u> of Y given X (see [6]).

We are now ready to give the definition of the conditional analytic Feynman integral of a function F given X.

Let $X:B \longrightarrow \mathbb{R}^n$ be measurable and let $F:B \longrightarrow \mathbb{C}$ be such that for each $\lambda > 0$, $F(\lambda^{-1/2}\cdot)$ is integrable on B. For $\lambda > 0$, let

$$J_\lambda(\vec{\eta}) = E[F(\lambda^{-1/2}\cdot)|X(\lambda^{-1/2}\cdot) = \vec{\eta}]$$

denote the conditional abstract Wiener integral of $F(\lambda^{-1/2}\cdot)$ given $X(\lambda^{-1/2}\cdot)$. If for a.e. $\vec{\eta} \in \mathbb{R}^n$, there exists a function $J_\lambda^*(\vec{\eta})$, analytic in λ on $\mathbb{C}^+$ such that $J_\lambda^*(\vec{\eta}) = J_\lambda(\vec{\eta})$ for all $\lambda > 0$, then J_λ^* is defined to be the conditional analytic Wiener integral of F over B given X with parameter λ and for $\lambda \in \mathbb{C}^+$, we write

$$E^{\mathrm{anw}_\lambda}[F|X = \vec{\eta}] = J_\lambda^*(\vec{\eta}).$$

If for fixed real $q \neq o$, the limit

$$\lim_{\lambda \to -iq} E^{\mathrm{anw}_\lambda}[F|X = \vec{\eta}]$$

exists for a.e. $\vec{\eta} \in \mathbb{R}^n$ where λ approaches $-iq$ through $\mathbb{C}^+$, we will denote this limit by $E^{anf_q}[F|X=(\cdot)]$ and call it the conditional analytic Feynman integral of F over B given X with parameter q.

3. Conditional Feynman Integral of functions in $\mathcal{F}(B)$

In this section we consider $\mathbb{R}^n$-valued conditioning functions on B of the form

$$X(x) = ((g_1,x)^\sim, (g_2,x)^\sim, \cdots, (g_n,x)^\sim) \tag{3.1}$$

where $\{g_1, g_2, \cdots, g_n\}$ is an orthonormal subset of H, and then give the evaluation formula for the conditional Feynman integral of F in $\mathcal{F}(B)$ given X of the form (3.1). We note that if h is in H, then

$$X(h) = (\langle g_1,h\rangle, \langle g_2,h\rangle, \cdots, \langle g_n,h\rangle) \tag{3.2}$$

Lemma 3.1. Let $h \in H$ and assume that $Y(x) = f((h,x)^\sim)$ is integrable on B. Then

$$E[Y|X = \vec{\eta}] = \frac{1}{\sqrt{2\pi|p|^2}} \int_{-\infty}^{\infty} f\left(u + \sum_{j=1}^{n} \langle h,g_j\rangle\eta_j\right) \exp\left\{-\frac{u^2}{2|p|^2}\right\} du$$

where $\vec{\eta} = (\eta_1, \eta_2, \cdots, \eta_n)$ and $|p|^2 = |h|^2 - \sum_{j=1}^{n} \langle h,g_j\rangle^2$.

Proof. Let $p = h - \sum_{j=1}^{n} \langle h,g_j\rangle g_j$. Then $\langle p, \sum_{j=1}^{n} \langle h,g_j\rangle g_j\rangle = 0$ and so $(p,x)^\sim$ and $(\sum_{j=1}^{n} \langle h,g_j\rangle g_j, x)^\sim$ are independent. So we obtain

$$\begin{aligned}
&E[f((h,x)^\sim)|\ X(x) = \vec{\eta}] \\
&= E[f((p,x)^\sim + (\sum_{j=1}^{n} \langle h,g_j\rangle g_j, x)^\sim) | (\sum_{j=1}^{n} \langle h,g_j\rangle g_j, x_j)^\sim = \sum_{j=1}^{n} \langle h,g_j\rangle \eta_j] \\
&= E[f((p,x)^\sim) + \sum_{j=1}^{n} \langle h,g_j\rangle \eta_j)] \\
&= \int_{-\infty}^{\infty} f(u + \sum_{j=1}^{n} \langle h,g_j\rangle \eta_j) \frac{1}{\sqrt{2\pi|p|^2}} \exp\left\{-\frac{u^2}{2|p|^2}\right\} du.
\end{aligned}$$

Hence we complete the proof.

Theorem 3.2. [9] Let $F \in \mathcal{F}(B)$ be given by (2.1), and let $X(x)$ be as in (3.1). Then for all $q \neq 0$ and $\vec{\eta} \in \mathbb{R}^n$,

$$(3.3)\qquad E^{anf_q}[F|X=\vec{\eta}] = \int_H \exp\left\{-\frac{i}{2q}\,(|h|^2 - |X(h)|^2) + i\langle X(h),\vec{\eta}\rangle\right\} d\sigma(h).$$

Proof. We first note that

$$E[F(\lambda^{-1/2}\cdot)\,|\,X(\lambda^{-1/2}\cdot) = \vec{\eta}]$$
$$= \int_H E[\exp\{i(h,\lambda^{-1/2}x)^{\sim}\}\,|\,X(\lambda^{-1/2}\cdot) = \vec{\eta}]\,d\sigma(h).$$

Thus we have, using Lemma 3.1

$$E[\exp\{i(h,\lambda^{-1/2}x)^{\sim}\,|\,X(\lambda^{-1/2}\cdot) = \vec{\eta}\ ]$$
$$= \int_{-\infty}^{\infty} \exp\left\{i\sum_{j=1}^{n}\langle h,g_j\rangle\eta_j\right\} \exp\{i\lambda^{-1/2}u\}\frac{1}{\sqrt{2\pi|p|^2}}\exp\left\{-\frac{u^2}{2|p|^2}\right\}du$$
$$= \exp\left\{i\sum_{j=1}^{n}\langle h,g_j\rangle\eta_j\right\} \exp\left\{-\frac{1}{2\lambda}\,|p|^2\right\}$$

where $|p|^2 = |h|^2 - \sum_{j=1}^{n}\langle h,g_j\rangle^2 = |h|^2 - |X(h)|^2$. Hence we obtain

$$(3.4)\qquad E[F(\lambda^{-1/2}\cdot)\,|\,X(\lambda^{-1/2}\cdot) = \vec{\eta}]$$
$$= \int_H \exp\{i\langle X(h),\vec{\eta}\rangle\}\cdot\exp\left\{-\frac{1}{2\lambda}\,(|h|^2 - |X(h)|^2)\right\}d\sigma(h)$$

Hence since $\sigma \in M(H)$, the right hand side of (3.4) is an analytic function of λ throughout $\mathbb{C}^+$ and is continuous of λ for $\mathrm{Re}\lambda \geq 0$, $\lambda \neq 0$. Hence we establish Eq. (3.3) as desired.

Corollary 3.3. Let $h \in H$ and let S be a bounded linear operator on H. Then for all $q \neq 0$, $\vec{\eta} \in \mathbb{R}^n$ and $a \in \mathbb{R}$, we have

$$E^{anf_q}[\exp\{ia(Sh,x)^{\sim}\}\,|\,X(x) = \vec{\eta}]$$
$$= \exp\left\{-\frac{ia^2}{2q}\,(|Sh|^2 - \sum_{j=1}^{n}\langle g_j,Sh\rangle^2) + ia\sum_{j=1}^{n}\langle g_j,Sh\rangle\eta_j\right\}$$

Proof. Apply Theorem 3.2 with $F \in \mathcal{F}(B)$ associated with the unit mass at $Sh \in H$.

Let $B = C_0[0,t]$, the Banach space with sup norm of continuous functions x on $[0,t]$ with $x(0) = 0$, and $H = C_0'[0,t]$, the Hilbert space $\{x \in C_0[0,t] \mid x(s) = \int_0^s f(u)du,\ f \in L^2[0,t]\}$ with $|x| = |f|_{L^2}$. It is well known [12] that $(C_0'[0,t], C_0[0,t], m_w)$ is an abstract Wiener space where m_w is the classical Wiener measure and that if $h \in C_0'[0,t]$, and $x \in C_0[0,t]$, then

$$(h,x)^{\sim} = \int_0^t \frac{dh}{ds}(s)\, \tilde{d}x(s)$$

where the integral is the Paley-Wiener-Zygmund integral.

We now specialize our results to the space $(C_0'[0,t], C_0[0,t], m_w)$ to obtain conditional Feynman integrals over the Wiener space.

Let's fix a partition $\{0 = t_0 < t_1 < \cdots < t_n = t\}$ of $[0,t]$, and let $g_j \in C_0'[0,t]$ be defined by

$$(3.5) \qquad g_j(s) = (t_j - t_{j-1})^{-1/2} \int_0^s 1_{[t_{j-1}, t_j)}(u)du, \quad j = 1, 2, \cdots, n.$$

Then $\{g_1, g_2, \cdots, g_n\}$ is an orthonormal set in $C_0'[0,t]$ and $|X(h)|^2 = \sum_{j=1}^n \langle g_j, h\rangle^2 = \sum_{j=1}^n \frac{1}{t_j - t_{j-1}} \left(\int_{t_{j-1}}^{t_j} \frac{dh}{ds}(s)ds\right)^2$ for all $h \in C_0'[0,t]$. We note that for $x \in C_0[0,t]$, $(x(t_1), \cdots, x(t_n)) = (\eta_1, \cdots, \eta_n)$ if and only if $(g_j, x)^{\sim} = (t_j - t_{j-1})^{-1/2}(\eta_j - \eta_{j-1})$ for all $j = 1, \cdots, n$, where $\eta_0 = 0$.

Let $T : C_0'[0,t] \longrightarrow C_0'[0,t]$ be the linear operator defined by

$$(3.6) \qquad (Tg)(\tau) = \int_0^\tau g(u)du.$$

Then we see that the adjoint operator T^* of T is given by

$$(T^* g)(\tau) = g(t)\tau - \int_0^\tau g(u)du.$$

Example 3.1. [9] Let σ be a complex valued Borel measure on $L^2[0,t]$ with bounded variation. If $F : C_0[0,t] \longrightarrow \mathbb{C}$ is given by

$$(3.7) \qquad F(x) = \int_{L^2[0,t]} \exp\{i \int_0^t v(s)\tilde{d}x(s)\}\, d\sigma(v) \text{ s-a.s. } x$$

(see [2]), then using Theorem 3.2, we obtain

$$(3.8)\quad E^{\mathrm{anf}_q}[F|x(t_1)=\eta_1,\cdots, x(t_n)=\eta_n]$$

$$= \int_{L^2[0,t]} \exp\left\{-\frac{i}{2q}\left[\int_0^t |v(s)|^2 ds - \sum_{j=1}^n \frac{1}{t_j-t_{j-1}}\left(\int_{t_{j-1}}^{t_j} v(s)ds\right)^2\right] + i\sum_{j=1}^n \frac{\eta_j-\eta_{j-1}}{t_j-t_{j-1}}\int_{t_{j-1}}^{t_j} v(s)ds\right\} d\sigma(v).$$

In particular when n = 1, we have

$$E^{\mathrm{anf}_q}[F|x(t)=\eta]$$

$$= \int_{L^2[0,t]} \exp\left\{\frac{-i}{2q}\int_0^t v(s)^2 ds - \frac{1}{t}\left(\int_0^t v(s)ds\right)^2 + i\frac{\eta}{t}\int_0^t v(s)ds\right\} d\sigma(v).$$

Example 3.2. Let T be as in (3.6). Let $h \in C_0'[0,t]$ be defined by $h(\tau) = \tau$. Then $(T^*h,x)^\sim = \int_0^t x(s)ds$, $|T^*h|^2 = \frac{1}{3}t^3$ and $\langle g_j, T^*h\rangle = \int_0^T g_j(s)ds$ where g_j is as in (3.5). Hence by Corollary 3.3 with $S = T^*$,

$$E^{\mathrm{anf}_q}[\exp\{ia\int_0^t x(s)ds\}|x(t_1)=\eta_1,\cdots, x(t_n)=\eta_n]$$

$$= \exp\left\{-\frac{ia^2}{2q}\left(|T^*h|^2 - \sum_{j=1}^n \langle h, Tg_j\rangle^2\right) + ia\sum_{j=1}^n \langle h, Tg_j\rangle \frac{\eta_j - \eta_{j-1}}{\sqrt{t_j-t_{j-1}}}\right\}$$

$$= \exp\left\{-\frac{ia^2}{2q}\left[\frac{1}{3}t^3 - \frac{1}{4}\sum_{j=1}^n (t_j - t_{j-1})(2t - t_j - t_{j-1})^2\right] + \frac{ia}{2}\sum_{j=1}^n (\eta_j + \eta_{j-1})(t_j - t_{j-1})\right\}$$

where $a \in \mathbb{R}$ and $\eta_0=0$. In particular when n = 1, we have

$$E^{\mathrm{anf}_q}[\exp\{ia\int_0^t x(s)ds\}|x(t)=\eta\}] = \exp\left\{-\frac{iat^2}{24q} + \frac{ia\eta t}{2}\right\}.$$

Example 3.3. Let T be as in (3.6), $y \in L^2[0,t]$ and $h \in C_0'[0,t]$ be defined by $h(\tau) = \int_0^\tau y(s)ds$. Then $(T^*h,x)^\sim = \int_0^t y(s)x(s)ds$. Hence by Corollary 3.3 with $S = T^*$, we have

$$E^{anf_q}[\exp\{ia\int_0^t y(s)x(s)ds\}|x(t_1)=\eta_1,\cdots,x(t_n)=\eta_n]$$

$$= \exp\left\{-\frac{ia^2}{2q}\left[|T^*h|^2 - \sum_{j=1}^{n}\left(\int_0^t y(s)g_j(s)ds\right)^2\right]\right.$$

$$\left. + ia\sum_{j=1}^{n}\frac{\eta_j-\eta_{j-1}}{\sqrt{t_j-t_{j-1}}}\int_0^t y(s)g_j(s)ds\right\}$$

where $a \in \mathbb{R}$ and the g_j's are as in (3.5). In particular when $n = 1$, $y(s) = s$, or $= s^2$, we have, respectively

$$E^{anf_q}[\exp\{ia\int_0^t sx(s)ds\}|x(t)=\eta] = \exp\left\{-\frac{ia^2t^5}{90q} + \frac{ia\eta t^2}{3}\right\}$$

$$E^{anf_q}[\exp\{ia\int_0^t s^2x(s)ds\}|x(t)=\eta] = \exp\left\{-\frac{19a^2t^7}{224q} + \frac{ia\eta t^3}{4}\right\}$$

Furthermore, if $n = 1$, and $y(s) = \sum_{j=1}^{m} y_j\delta(s-s_j)$ where $0 < s_1 < \cdots < s_m < t$, then we have

$$E^{anf_q}[\exp\{ia\sum_{j=1}^{m} y_jx(s_j)\}|x(t)=\eta]$$

$$= \exp\left\{-\frac{ia^2}{2q}\sum_{p=1}^{m}\sum_{j=1}^{p}(2-\delta_{jp})y_jy_ps_j + \frac{ia^2}{2qt}\left(\sum_{j=1}^{m}s_jy_j\right)^2 + \frac{ia\eta}{t}\sum_{j=1}^{m}s_jy_j\right\}$$

wehre δ_{jp} is the Kronecker delta.

Example 3.4. [5,8] Let $V(u) = \int_{\mathbb{R}} e^{iuv}d\mu(v)$ where μ is a complex Borel measure on $\mathbb{R}$ with bounded variation. Then we have

$$E^{anf_q}[\exp\{ia\int_0^t V(x(s))ds\}|x(t_1)=\eta_1,\cdots,x(t_n)=\eta_n]$$

$$= \prod_{k=1}^{n} E^{anf_q}[\exp\{ia\int_{t_{k-1}}^{t_k} V(x(s))ds\}|x(t_{k-1})=\eta_{k-1},x(t_k)=\eta_k]$$

$$= \prod_{k=1}^{n} E^{anf_q}[\exp\{ia\int_0^{t_k-t_{k-1}} V(x(s)+\eta_{k-1})ds\}|x(t_k-t_{k-1})=\eta_k-\eta_{k-1}]$$

$$= \prod_{k=1}^{n} \sum_{m=0}^{\infty} (ia)^m \int_{\Delta_m(t_k)} \int_{\mathbb{R}^m} \exp\left\{ \frac{-i}{2q} \sum_{p=1}^{m} \sum_{j=1}^{p} (2-\delta_{jp}) y_j y_p s_j + \frac{i}{2q(t_k - t_{k-1})} \right.$$

$$\left. \cdot \left(\sum_{j=1}^{m} y_j s_j \right)^2 + \frac{i(\eta_k - \eta_{k-1})}{t_k - t_{k-1}} \sum_{j=1}^{m} y_j s_j + i\eta_{k-1} \sum_{j=1}^{m} y_j \right\} \prod_{j=1}^{m} (d\mu(y_j) ds_j).$$

where $\Delta_m(t_k) = \{(s_1, \cdots, s_m) \mid 0 = s_0 < s_1 < \cdots < s_m < t_k - t_{k-1}\}$ and $\eta_0 = 0$.

4. Fundamental solutions to the Schrödinger equation

In this section we use the conditional analytic Feynman integral obtained in Section 3 to provide the fundamental solutions to the Schrödinger equation for a class of potentials.

Theorem 4.1. Let F and X be as in Theorem 3.2, and let $\psi \in L^1(\mathbb{R}^n)$. For $\vec{\eta} \in \mathbb{R}^n$, let $G(x) \equiv G_{\vec{\eta}}(x) = F(x)\psi(X(x) + \vec{\eta})$. Then

$$(4.1) \qquad E^{\text{anf}_q}(G) = \int_{\mathbb{R}^n} E^{\text{anf}_q}[F \mid X = \vec{\xi} - \vec{\eta}] \left(\frac{q}{2\pi i}\right)^{n/2} \exp\left\{\frac{iq|\vec{\xi} - \vec{\eta}|^2}{2}\right\} \psi(\vec{\xi}) d\vec{\xi}$$

where $E^{\text{anf}_q}[F \mid X = (\cdot)]$ is given in Eq. (3.3)

Proof We observe that

$$\int_B F(\lambda^{-1/2}x)\psi(X(\lambda^{-1/2}x) + \vec{\eta}) d\nu(x)$$

$$= \int_{\mathbb{R}^n} E[F(\lambda^{-1/2}x) \mid X(\lambda^{-1/2}x) + \vec{\eta} = \vec{\xi}] \psi(\vec{\xi}) \left(\frac{\lambda}{2\pi}\right)^{n/2} \exp\left\{-\frac{\lambda|\vec{\xi} - \vec{\eta}|^2}{2}\right\} d\vec{\xi}$$

$$= \int_{\mathbb{R}^n} E[F(\lambda^{-1/2}x) \mid X(\lambda^{-1/2}x) = \vec{\xi} - \vec{\eta}] \psi(\vec{\xi}) \left(\frac{\lambda}{2\pi}\right)^{n/2} \exp\left\{-\frac{\lambda|\vec{\xi} - \vec{\eta}|^2}{2}\right\} d\vec{\xi}$$

for all $\lambda > 0$. Thus using Theorem 3.2 and Morera's Theorem, we obtain

$$(4.2) \qquad E^{\text{anw}_\lambda}(G) = \int_{\mathbb{R}^n} E^{\text{anw}_\lambda}[F \mid X = \vec{\xi} - \vec{\eta}] \left(\frac{\lambda}{2\pi}\right)^{n/2} \exp\left\{-\frac{\lambda|\vec{\xi} - \vec{\eta}|^2}{2}\right\} \psi(\vec{\xi}) d\vec{\xi}$$

for all $\lambda \in \mathbb{C}^+$. Since $E^{\text{anw}_\lambda}[F \mid X = \vec{\xi} - \vec{\eta}]$ is bounded and $\psi(\vec{\xi}) \in L^1(\mathbb{R}^n)$, the right hand side of Eq. (4.2) is continuous in λ for $\text{Re}\lambda \geq 0$, $\lambda \neq 0$

and hence $E^{\text{anf}_q}(G)$ exists and is given by Eq. (4.1).

In our next theorem, we need the following summation procedure (see [15], p. 340):

$$\bar{\int}_{\mathbb{R}^n} f(\vec{\eta})d\vec{\eta} = \lim_{A\to\infty} \int_{\mathbb{R}^n} f(\vec{\eta}) \exp\{-\frac{|\vec{\eta}|^2}{2A}\} d\vec{\eta} \tag{4.3}$$

whenever the expression on the right exists. Of course if $f \in L^1(\mathbb{R}^n)$, it is clear using the dominated convergence theorem that

$$\bar{\int}_{\mathbb{R}^n} f(\vec{\eta})d\vec{\eta} = \int_{\mathbb{R}^n} f(\vec{\eta})d\vec{\eta} .$$

Theorem 4.2. [9] Let F and X be as in Theorem 3.2. Assume that $\psi \in L^2(\mathbb{R})$ and is given by

$$\psi(\vec{\eta}) = \int_{\mathbb{R}^n} \exp\{i\langle\vec{y},\vec{\eta}\rangle\} \, d\phi(\vec{y}) \tag{4.4}$$

where ϕ is a complex Borel measure on $\mathbb{R}^n$ with bounded variation. For $\vec{\eta} \in \mathbb{R}^n$, let $G(x) \equiv G_{\vec{\eta}}(x) = F(x)\psi(X(x)+\vec{\eta})$ Then for all $q \neq 0$,

$$E^{\text{anf}_q}(G) = \bar{\int}_{\mathbb{R}^n} E^{\text{anf}_q}[F|X=\vec{\xi}-\vec{\eta}]\,(\frac{q}{2\pi i})^{n/2}\exp\left\{\frac{iq|\vec{\xi}-\vec{\eta}|^2}{2}\right\} \psi(\vec{\xi})d\vec{\xi} \tag{4.5}$$

where $E^{\text{anf}_q}[F|X=(\cdot)]$ is given by formula (3.3).

Remark We note that Eq. (4.1) and (4.5) may be written as

$$E^{\text{anf}_q}(G_{\vec{\eta}}) = E^{\text{anf}_q}[F|X= -\vec{\eta}]\,(\frac{q}{2\pi i})^{n/2} \exp\left\{\frac{iq\|\vec{\eta}\|^2}{2}\right\} * \psi(\vec{\eta})$$

where * denotes convolution of functions.

We now consider the time independent potential of the form

$$V(u) = cu + b\int_{\mathbb{R}} e^{iuv}d\mu(v) \quad (b, c \in \mathbb{R})$$

where μ is as in Example 3.4. For $\eta \in \mathbb{R}$, $\vec{\xi}_0 = (\xi_0, \xi_1, \cdots, \xi_{n-1})$ and $\vec{t}_0 = (0, t_1, \cdots, t_{n-1})(0<t_1<\cdots<t_n<t)$, let

$$K(t,\eta,\vec{t}_0,\vec{\xi}_0;q) = E^{\mathrm{anf}_q}\left[\exp\{-\frac{i}{\hbar}\int_0^t V(x(t-s)+\eta)ds\}\,\middle|\,x(t_k)=\xi_k,\ k=0,\cdots,n-1\right]$$

Then by Theorem 4.2 with the results of [1, Theorem 3.1] and [14], $K(t,\eta,\vec{t}_0,\vec{\xi}_0;\frac{m}{\hbar})$ is the fundamental solution to Eq.(1.1), i.e., the solution of Eq.(1.1) with the initial condition $\psi(\vec{\eta})=\delta(\vec{\eta}-\vec{\xi}_0)$.

Example 4.1. Let V = 0 (case of no force).

$$K_0(t,\eta,\vec{t}_0,\vec{\xi}_0;\frac{m}{\hbar}) \equiv \prod_{k=1}^{n}\left(\frac{m}{2\pi i\hbar(t_k-t_{k-1})}\right)^{1/2}\exp\left\{\frac{im}{2\hbar}\sum_{j=1}^{n}\frac{(\xi_k-\xi_{k-1})^2}{t_k-t_{k-1}}\right\},$$

where $\xi_n = \eta$.

Example 4.2. Let V(u) = cu (case of constant force).

$$K(t,\eta,\vec{t}_0,\vec{\xi}_0;\frac{m}{\hbar}) = K_0(t,\eta,\vec{t}_0,\vec{\xi}_0;\frac{m}{\hbar})\cdot\exp\left\{-\frac{ic^2}{2m\hbar}\left[\frac{1}{3}t^3-\frac{1}{4}\sum_{k=1}^{n}(t_k-t_{k-1})\cdot(2t-t_k-t_{k-1})^2\right]+\frac{ic}{2\hbar}\sum_{k=1}^{n}(\xi_k+\xi_{k-1})(t_k-t_{k-1})\right\},$$

where $\xi_n = \eta$. In particular when n = 1, we have

$$K(t,\eta,0,\xi_0;\frac{m}{\hbar}) = \left(\frac{m}{2\pi i\hbar t}\right)^{1/2}\exp\left\{\frac{i}{\hbar}\left[\frac{m(\eta-\xi_0)^2}{2t}-\frac{c^2t^3}{24m}+\frac{ct(\eta+\xi_0)}{2}\right]\right\}.$$

Example 4.3. Let $V(u) = \int_{\mathbb{R}} e^{iuv}d\mu(v)$; μ is as in Example 3.4

$$K(t,\eta,\vec{t}_0,\vec{\xi}_0;\frac{m}{\hbar}) = K_0(t,\eta,\vec{t}_0,\vec{\xi}_0;\frac{m}{\hbar})\cdot\prod_{k=1}^{n}\sum_{m=0}^{\infty}\left(-\frac{i}{\hbar}\right)^m\int_{\Delta_m(t_k)}\int_{\mathbb{R}^m}$$
$$\cdot\exp\left\{-\frac{i\hbar}{2m}\left[\sum_{p=1}^{m}\sum_{j=1}^{p}(2-\delta_{jp})y_jy_ps_j-\frac{1}{t_k-t_{k-1}}\left(\sum_{j=1}^{m}y_js_j\right)^2\right]+\frac{i(\xi_k-\xi_{k-1})}{t_k-t_{k-1}}\sum_{j=1}^{m}y_js_j+i\xi_{k-1}\sum_{j=1}^{m}y_j\right\}\prod_{j=1}^{m}(d\mu(y_j)ds_j),$$

where $\Delta_m(t_k)$ is as in Example 3.4 and $\xi_n = \eta$.

We next consider the time dependent potential of the form

$$\theta(s,u) = \int_{\mathbb{R}} e^{iuv} d\mu_s(v) \tag{4.6}$$

where $\{\mu_s | 0 \le s \le t\}$ is a family from $M(\mathbb{R})$ satisfying (i) $\mu_s(E)$ is a Borel measurable function of s for every $E \in \mathcal{B}(\mathbb{R})$, and (ii) $\|\mu_s\| \in L^1[o,t]$ for all $t>0$. For η, $\xi \in \mathbb{R}$, let

$$K(t,\eta,0,\xi;q) = \left(\frac{q}{2\pi it}\right)^{1/2} \cdot \exp\left\{\frac{iq}{2t}(\eta-\xi)^2\right\}$$
$$\cdot E^{anf_q}[\exp\{-\frac{i}{\hbar}\int_0^t \theta(s,x(s)+\xi)ds\} | x(t)+\xi=\eta]$$

Then by the result of [8,Theorem 6], $K(t,\eta,0,\xi;q)$ is the solution of the Schrödinger integral equation:

$$K(t,\eta,0,\xi;q) = \left(\frac{q}{2\pi it}\right)^{1/2} \exp\left\{\frac{iq}{2t}(\eta-\xi)^2\right\} + \int_0^t \left(\frac{q}{2\pi i(t-s)}\right)^{1/2}$$
$$\cdot \overline{\int_{\mathbb{R}}} \theta(s,\zeta)K(t,\eta,0,\zeta;q) \cdot \exp\left\{\frac{iq}{2(t-s)}(\eta-\zeta)^2\right\} d\zeta ds.$$

<u>**Example 4.4.**</u> Let $\theta(s,u) = \int_{\mathbb{R}} e^{iuv} d\mu_s(v)$; $\{\mu_s\}$ is as in Eq.(4.6).

$$K(t,\eta,0,\xi;\tfrac{m}{\hbar}) = K_0(t,\eta,0,\xi;\tfrac{m}{\hbar}) \sum_{m=0}^{\infty} (ia)^m \int_{\Delta_n(t)} \int_{\mathbb{R}^m} \exp\left\{\frac{-i}{2q} \sum_{p=1}^{m} \sum_{j=1}^{p} (2-\delta_{jk}) y_j y_p s_j\right.$$
$$\left. + \frac{i}{2qt}\left(\sum_{j=1}^{m} y_j s_j\right)^2 + \frac{i(\eta-\xi)}{t} \sum_{j=1}^{m} y_j s_j + i\xi \sum_{j=1}^{m} y_j\right\} \prod_{j=1}^{m} (d\mu_{s_j}(y_j) ds_j) .$$

where $\Delta_m(t) = \{(s_1, \cdots, s_m) : 0=s_0<s_1<\cdots<s_m<t\}$.

5. Final Remarks

We have studied here only the conditional Feynman integrals for the Fresnel class of functions on B. The conditional Feynman integral of functions involving the harmonic potential and the forced harmonic potential will be presented later.

As in the case of Feynman integrals (see [17]), the concept of conditonal Feynman integral can be transferred to functions defined on a white noise space $(\mathcal{S}^*, \mu)$ by replacing (H, B, ν) by $(L^2(\mathbb{R}), \mathcal{S}^*, \mu)$. Here $\mathcal{S}^*$ is the dual space of Schwartz space of rapidly decreasing smooth real valued functions on $\mathbb{R}$, and μ is a white noise measure on $(\mathcal{S}^*, \mathcal{B}(\mathcal{S}^*))$.

REFERENCES

1. A. Albeverio and R Høegh-Krohn, *Mathematical theory of Feynman path integrals*, Lecture Notes in Mathematics, No. **523** (Springer, Berlin, 1976).

2. R.H. Cameron and D.A. Storvick, *Some Banach algebras of analytic Feynman integrable functionals*. Analytic Functions, Kozubnik 1979, Lecture notes in Mathematics, No. **798** (Springer, Berlin, 1980), 18-67.

3. R.H. Cameron, *A family of integrals serving to connect the Wiener and Feynman integrals*, J. Math. and Physics **39** (1960) 126-141.

4. D.M. Chung, *Scale-invariant measurability in abstract Wiener spaces*, Pacific J. Math., **130** (1987), 27-40.

5. D.M. Chung and S.J. Kang, *Conditional Wiener integrals and an integral equation*, J. of Korean Math. Soc., **25** (1988), 37-52.

6. D.M. Chung and S.J. Kang, *Evaluation formulas for conditional abstract Wiener integrals*, Stochastic Analysis and Applications, **7** (1989), 125-144.

7. D.M. Chung and D. Skoug, *Conditional analytic Feynman integrals and a related Schroedinger integral equation*, SIAM on Mathematical Analysis, **20** (1989), 950-965.

8. D.M. Chung, C. Park and D. Skoug, *Operator-valued Feynman integral via conditional Feynman integrals*, Pacific J. Math. **146** (1990), 21-42.

9. D.M. Chung, *Conditional analytic Feynman integrals on abstract Wiener spaces*. to appear in Proc. Amer. Math. Soc.

10. R.P. Feynman, *Space-time approach to non-relavistic quantum mechanics*, Rieview of Modern Physics **20** (1948) 367-387.

11. I.M. Gelfand and A.M. Yaglom, *Integration in functional spaces*, J. Math. Physics **1** (1960) 48-69.

12. L. Gross, *Abstract Wiener space*, Proc. 5th, Berkeley Sym. Math. Stat. Prob. **2** (1965), 31-42.

13. K. Ito, *Generalized uniform complex measures in the Hilbertian metric space with their applications to Feynman path integrals*, Proc. Fifth Berkely Symposium on Math. Stat. and Prob. (1967) II, part **4**, 145-161

14. G.W. Johnson, *The equivalence of two approaches to the Feynman integral*, J. Math. Phys., **23** (1982), 2090-2096.

15. G.W. Johnson and D.L. Skoug, *Notes on the Feynman integral, III: The Schroedinger equation*, Pacific J. of Math., **105** (1983), 321-358.

16. G. Kallianpur and C. Bromley, *Generalized Feynman integrals using analytic continuation in several complex variables*, Stochastic Analysis, M. Pinsky, ed., (Marcel-Dekker, 1984).

17. G. Kallianpur, D. Kannan and R.L. Karandikar, *Analytic and sequential Feynman integrals on abstract Wiener and Hilbert spaces and a Cameron-Martin formula*, Ann. Inst. H. Poincaré, **21** (1985), 323-361.

18. H.H. Kuo, *Gaussian Measures in Banach Spaces*, Lecture Notes in Mathematics, No. **463** (Springer, Berlin, 1975.)

19. L. Streit and T. Hida, *Generalized Brownian functionals and the Feynman integral*, Stochastic Process and their Application, **16** (1983), 55-69.

20. J. Yeh, *Inversion of conditional Wiener integrals*, Pacific J. Math., **59** (1975), 623-638.

Ricatti and soliton equations[1]

Michiel Hazewinkel
CWI
POBox 40799, 1009AB Amsterdam
The Netherlands

Abstract. In this lecture I note that a number of integrable systems can be viewed as rational quotients of linear dynamical systems. This holds for the KdV equation, the Toda lattices and the KP equations. In particular I point out that the KP equation is in fact an infinite dimensional Riccati equation. Conjecturally all integrable systems are rational quotients of linear ones.

AMS classification: 58F07

1. Introduction.

The main purpose of this lecture is to point out that many integrable systems can be viewed as rational quotients of linear dynamical systems, to argue that this is a good way to look at them, and to suggest that this rational quotient property might be a characteristic (and defining) property for integrable systems.

2. A micro course on the formal structure of the inverse scattering transform method for the KdV equation.

Consider the Korteweg–de Vries equation for shallow water waves in the form

$$u_t = 6uu_x - u_{xxx} \tag{2.1}$$

where $u(t, x)$ is a real valued function of x (place) and t (time), and subscripts denote partial derivatives. It is desired to solve the Cauchy problem (initial value problem) for (2.1) with initial data

$$u(0, x) = a(x) \tag{2.2}$$

where $a(x)$ is a given smooth function that decays sufficiently fast to zero as $|x| \to \infty$. For this formal outline of ISTM (inverse scattering transform method) the analytical details are of lesser importance. It suffices to know that the scheme works (and where to find the proofs, cf e.g. [1,2]).

[1] I dedicate this paper to Prof. Takeyuki Hida, at the occasion of his retirement in 1991 from Nagoya University, in admiration and gratitude for his creation of white noise analysis (infinite dimensional stochastic calculus). Without him the world of stochastics would not have been the same.

One considers an associated linear eigenvalue problem

$$(2.3)\qquad (-\frac{d^2}{dx^2}+u)\phi=\lambda\phi$$

where u is a given function of x (often called a potential) also decaying fast enough as $|x|\to 0$. Then it is known that the spectrum of the Hill operator (some say Schrödinger operator), $-\frac{d^2}{dx^2}+u$, consists of two parts:

(i) a discrete part consisting of a finite number of negative eigenvalues

$$(2.4)\qquad \lambda=-\kappa_n^2$$

(ii) a continuous part

$$\lambda=k^2.$$

The corresponding eigenfunctions are denoted $\phi_n(x)$ and $\psi_k(x)$ and these can be chosen such that

$$(2.5)\qquad \begin{aligned}&\phi_n(x)\sim c_n e^{\kappa_n x}\ as\ x\to\infty,\\ &\int \phi_n(x)^2dx=1,\ \phi_n(x)>0\,\text{as}\ x\to\infty\end{aligned}$$

$$(2.6)\qquad \psi_k(x)\sim e^{-ikx}+b(k)e^{ikx}\ as\ x\to\infty$$

$$(2.7)\qquad \psi_k(x)\sim a(k)e^{-ikx}\ as\ x\to-\infty$$

Thus one associates to a potential u a set of *spectral data* $(\kappa_n, c_n, b(k))$. The $a(k)$ and $b(k)$ are called transmission and reflection coefficients, respectively, and (2.6), (2.7) are viewed as presenting the scattering of an incident wave from ∞ by the potential u. The ISTM derives its name from this way of looking at things.

The crucial observation is now the following. If $u(x)$ evolves in time according to the KdV equation (2.1) then the associated spectral data to u evolve as follows

$$(2.8)\qquad \dot{\kappa}_n=0$$

$$(2.9)\qquad \dot{c}_n=4\kappa_n^3c_n$$

$$(2.10)\qquad \dot{b}(k)=8ik^2b(k)$$

Thus, given the spectral data at time zero $\kappa_n(0)$, $c_n(0)$, $b(k,0)$ it is a trivial matter to write down the spectral data at any time $t>0$. It remains to recover the potential u from its spectral data. This is done by means of the Gelfand–Levitan–Marchenko equation. The procedure is as follows. Given κ_n, c_n, $b(k)$ (as a function of t), form the kernel

$$(2.11)\qquad B(\xi)=\sum_{n=1}^{N}c_n^2e^{-\kappa_n\xi}+\frac{1}{2\pi}\int_{-\infty}^{+\infty}b(k)e^{ik\xi}dk$$

Let

$$\bar{B}(x,y) = B(x+y) \tag{2.12}$$

Now consider the GLM equation for $K(x,y)$

$$K(x,y) + \bar{B}(z,y) + \int_x^\infty \bar{B}(z,y)K(x,z)dz = 0 \tag{2.13}$$

Then the potential u with the given scattering data κ_n, c_n, $b(k)$ is given by

$$u(x;t) = -2\frac{\partial}{\partial x}K(x,x;t) \tag{2.14}$$

Thus the procedure to solve the initial value problem for the KdV is as follows: find the initial scattering data (the direct scattering transform) from the initial potential $a(x) = u(x,0)$; calculate the scattering data at time t by equations $(2.8) - (2.10)$ (a triviality); calculate the potential at time t from the scattering data at time t (the inverse scattering transform). It is worth pointing out that the first step, finding the spectrum of $u(x,0)$, is the bottleneck for the calculation of actual concrete solutions.

Now let us take a very formal algebraic look at the inverse scattering calculation. To this end view the integral in (2.13) as a (convolution) product of $\bar{B}$ and K, so that (2.3) looks like

$$K + \bar{B} + K * \bar{B} = 0, \quad K = \bar{B} * (I + \bar{B})^{-1} \tag{2.15}$$

Thus the total picture of the transition $(\kappa_n, c_n^2, b(k) \mapsto u$ looks as follows $(\kappa_n, c_n^2, b(k) \overset{(1)}{\mapsto} \bar{B}(x,y) \overset{(2)}{\mapsto} K(x,y) \overset{(3)}{\mapsto} u(x))$ with (1) linear, (2) looks like taking a rational fraction, and (3) is linear again. So the KdV equation appears to be a rational quotient of the linear dynamical system

$$\dot{\kappa}_n = 0, \quad (c_n^2)^{\cdot} = 8\kappa_n^3 c_n^2, \quad \dot{b}(k) = 8ik^2 b(k).$$

I like to call this a rational covering linearization.

3. The matrix Riccati equation.

The presentation of the KdV equation as a rational quotient of a linear dynamical system in the preceding section is highly formal and from that it is far from clear that this can be made to work analytically. I have no doubt however, that the results of Segal, Wilson [5] can be seen to provide just such a picture (among other things).

Let me now turn to a covering linearization situation where the analytical details are trivial. Consider the matrix Riccati equation

$$\dot{P} = AP + PD + PBD + C \tag{3.1}$$

Here P is an unknown matrix of size $n \times m$ depending on time t and the A, B, C, D are known matrices of constants of sizes $n \times n$, $m \times n$, $n \times m$, $m \times m$, respectively.

Consider the linear dynamical system

$$\frac{d}{dt}\begin{pmatrix} X \\ Y \end{pmatrix} = \begin{pmatrix} A & C \\ -B & -D \end{pmatrix}\begin{pmatrix} X \\ Y \end{pmatrix}$$

where X is an $n \times m$ matrix and Y an $m \times m$ matrix. Assume for the moment that $Y(t)^{-1}$ exists and set $P = XY^{-1}$. Then an easy calculation shows that P satisfies (3.2) if $(X, Y)^T$ satisfies (3.2). Thus (3.2) is a rational covering linearization of (3.1).

There is a small snag in the picture as presented just above in that in order to form the rational quotient XY^{-1}, the inverse, Y^{-1}, must be assumed to exist. This arises because the space of $n \times m$ matrices is not really the right space on which the Riccati equation (3.1) should be studied. It should be considered instead on the Grassmann manifold of m dimensional subspaces of $n+m$ space.

For definiteness sake let's work over the reals (the complex numbers would work just as well). Let $\mathbf{R}^{(n+m)\times m}$ be the space of all real matrices of size $(n+m) \times m$ and let $\mathbf{R}^{(n+m)\times m}_{reg}$ be the open subspace of all $(n+m) \times m$ matrices of maximal rank m. Then (3.2) is a dynamical system on this open subspace of $\mathbf{R}^{(n+m)\times m}$.

Let $Gr_m(\mathbf{R}^{n+m})$ be the Grassmann manifold of m dimensional subspaces of $\mathbf{R}^{n+m}$. To each $M \in \mathbf{R}^{(n+m)\times m}_{reg}$ associate the subspace of $\mathbf{R}^{n+m}$ spanned by the columns of M. This defines a rational quotient map

$$\pi : \mathbf{R}^{(n+m)\times m}_{reg} \to Gr_m(\mathbf{R}^{n+m}) \tag{3.3}$$

Observe that $\pi(MS) = \pi(M)$ for all $S \in GL(m; \mathbf{R})$, and inversely if $\pi(M) = \pi(M')$ then $M' = MS$ for some $S \in GL(m; \mathbf{R})$. It follows from this that π is compatible with the dynamical system (3.2); i.e. if $M(t)$, $M'(t)$ are solutions of (3.2) and $\pi(M(0)) = \pi(M'(0))$, then $\pi(M(t)) = \pi(M'(t))$ for all t. Thus (3.2) induces a dynamical system on $Gr_m(\mathbf{R}^{n+m})$, the *Riccati flow*. The space of $n \times m$ matrices, $\mathbf{R}^{n\times m}$, imbeds as an open dense subspace in $Gr_m(\mathbf{R}^{n+m})$ by

$$P \mapsto \pi\begin{pmatrix} P \\ I_m \end{pmatrix} \tag{3.4}$$

where I_m is the $m \times m$ identity matrix. The restriction of the vector field on $Gr_m(\mathbf{R}^{n+m})$ that describes the Riccati flow to the open dense subset $\mathbf{R}^{n+m}$ is the right hand side of (3.1).

The equation (3.1) has finite escape time phenomena and these are perfectly described by the Riccati flow on the compactification $Gr_m(\mathbf{R}^{n+m})$ of $\mathbf{R}^{n\times m}$.

4. The Toda lattices.

The N–particle (non–periodic) Toda lattice is given by the equations

$$\begin{aligned} \ddot{q}_n &= -e^{q_{n+1}-q_n} + e^{q_n - q_{n-1}} \\ q_0 &= \infty, \quad q_{N+1} = -\infty, \; n = 1, \ldots, N \end{aligned} \tag{4.1}$$

That is, it is a Hamiltonian system with Hamiltonian

$$H = \frac{1}{2}\sum p_n^2 + \sum e^{q_{n+1}-q_n} \tag{4.2}$$

Let $a_k = -\frac{1}{2}p_k$, $b_k = \frac{1}{2}e^{\frac{1}{2}(q_k - q_{k+1})}$. Then, as is well known, the equations (4.1) transform into

$$\dot{A} = [A, B] \tag{4.3}$$

where A is the tridiagonal matrix with diagonal elements, $a_1, a_2, \dots, a_N$; supradiagonal elements $b_1, \dots, b_{N-1}$; and infradiagonal elements $b_1, \dots, b_{N-1}$ and B is the tridiagonal matrix with zero diagonal elements, the same supradiagonal elements as A, and $-b_1, \dots, -b_{N-1}$ as infradiagonal elements. I.e.

$$A = \begin{pmatrix} a_1 & b_1 & & 0 \\ b_1 & a_2 & \ddots & \\ & \ddots & \ddots & b_{N-1} \\ 0 & & b_{N-1} & a_N \end{pmatrix},$$
$$B = \begin{pmatrix} 0 & b_1 & & 0 \\ -b_1 & 0 & \ddots & \\ & \ddots & \ddots & b_{N-1} \\ 0 & & -b_{N-1} & 0 \end{pmatrix} \tag{4.4}$$

It is in this form (4.3) that I want to consider the Toda Lattice. In this setting there is the following theorem, [3,8]:

Consider the linear dynamical system

$$\dot{Y}(t) = A(0)Y, \quad Y(0) = I_N \tag{4.5}$$

so that $Y(t) = \exp(tA(0))$. Take a *QR decomposition* of $Y(t)$, i.e. write $Y(t)$ as a product

$$Y(t) = Z(t)X(t) \tag{4.6}$$

where $Z(t)$ is orthogonal and $X(t)$ is lower triangular with positive diagonal entries. Then

$$A(t) = Z(t)^{-1}A(0)Z(t) \tag{4.7}$$

solves (4.3) (with initial conditions $A(t) = A(0)$ at time zero). Note that $Y(t) \mapsto A(t)$ is again a fractional linear transformation.

The Toda lattices are one class of integrable systems that are obtained from semi-simple Lie algebras by what is known as the AKSRS (Adler Kostant Symes Reyman Semenov–Tian–Shansky) construction. In the general case much the same picture holds. The space on which the rational covering linearization lives is the socalled double of the underlying Lie algebra, [6].

5. Factorizing differential operators.

Consider a differential operator of order $n+m$ in x

$$Q = D^{n+m} + q_1 D^{n+m-1} + \cdots + q_{n+m-1} D + q_{n+m} \tag{5.1}$$

where D is short for $\frac{d}{dx}$ and the p_i are given functions of x. Let us consider the problem of factorizing Q into product

$$\begin{aligned} Q = R^*R, \quad R &= D^m + r_1 D^{m-1} + \cdots + r_m, \\ R^* &= D^n + r_1^* D^{n-1} + \cdots + r_n^* \end{aligned} \tag{5.2}$$

Here the $r_1, \ldots, r_m; r_1^*, \ldots, r_n^*$ are the unknowns. Writing this out one finds a set of differential equations for the r's and r^*'s involving the known functions $q_1, \ldots, q_{n+m}$. In case $n = m = 1$ one finds the scalar Riccati equation, and, though, as we shall see, the problem has much to do with matrix Riccati equations, this little fact is a total red herring.

Let

$$V = \{u : Qu = 0\} \tag{5.3}$$

be the solution space of Q. This is a space of dimension $n+m$. The crucial observation of Mikio Sato is that factorizations (5.2) of Q correspond bijectively with m–dimensional subspaces of V. The correspondence is simple. Let $Q = R^*R$ be a factorization; then the solution space of R

$$W_R = \{u : Ru = 0\} \tag{5.4}$$

is a subspace of dimension m of V. Inversily, let W be an m–dimensional subspace of V. Choose a basis $u_1, \ldots, u_m$ of W and define a differential operator R_W by the Wronskian formula

$$R_W u = \det \begin{pmatrix} u_1 & \ldots & u_m \\ Du_1 & \ldots & Du_m \\ \vdots & & \vdots \\ D^{m-1}u_1 & \ldots & D^{m-1}u_m \end{pmatrix}^{-1} \begin{pmatrix} u_1 & \ldots & u_m & u \\ Du_1 & \ldots & Du_m & Du \\ \vdots & & \vdots & \vdots \\ D^m u_1 & \ldots & D^m u_m & D^m u \end{pmatrix} \tag{5.5}$$

which defines a differential operator of the form $D^m + \ldots$, and which clearly has W as its space of solutions.

6. Finite chunks of the KP equation.

The next step is to consider an operator A which takes the solution space V of Q into itself and to consider the linear flow induced by e^{tA} on $Gr_m(V)$. The easiest and most natural case is the one where Q has constant coefficients. Then the operators D^k all take V into itself. So we are interested in the flows

$$W \mapsto e^{t_k D^k} W \tag{6.1}$$

on $Gr_m(V)$. The question is: how will the operators R corresponding to W evolve if W evolves according to (6.1). The answer is the following. The factors R and R^* of Q will evolve according to the operator equations

$$\frac{\partial R}{\partial t_k} = P_k R - R D^k \tag{6.2}$$

$$\frac{\partial R^*}{\partial t_k} = - R^* P_k + D^k R^* \tag{6.3}$$

Here P_k is a differential operator of order k of the form D^k+ lower degree, and it is uniquely determined (by R) by the requirenment that (6.2) makes sense, i.e. that the right hand side of (6.2), which is a priori of degree $m+k-1$, is in fact of degree $m-1$. This determines the coefficients of $P_k = D^k + p_1 D^{k-1} + \cdots + p_k$ uniquely as differential expressions in the coefficients $r_1, \ldots, r_m$ of R, so that (6.2) becomes a set of partial differential equations (nonlinear) for $r_1, \ldots, r_m$.

Once one has guessed at (6.2) it is a triviality to prove that this equation does the job, i.e. that if u is in W_R then $e^{t_k D^k} u$ is in $W_{R(t_k)}$; simply calculate $\frac{d}{dt_k}(R(t_k)\exp(t_k D^k)(u))$.

All the above is due to M. Sato, [4]; I learned about it through some notes of K. Takasaki for which I have no proper reference.

7. The KP Hierarchy.

The next step is to observe what happens if the degree of R in (6.2) is shifted. So let R satisfy (6.2) and let $\bar{R} = RD$. Then one easily checks that

$$\frac{\partial \bar{R}}{\partial t_k} = P_k \bar{R} - \bar{R} D^k \tag{7.1}$$

(with exactly the same P_k). So there is independence (of a kind) of the degree m of R. The only natural thing to do is to multiply R with D^{-m} to get a pseudodifferential operator and this gives then the ∞ hierarchy of equations

$$R = 1 + r_1 D^{-1} + r_2 D^{-2} + \cdots \tag{7.2}$$

$$\frac{\partial R}{\partial t_k} = P_k R - R D^k \tag{7.3}$$

where calculating with the D^{-i} is done by the rule

$$D^{-1} u = u D^{-1} - u^{(1)} D^{-2} + u^{(2)} D^{-3} - u^{(3)} D^{-4} + \cdots \tag{7.4}$$

with $u^{(1)} = \frac{du}{dx}$, etc.

The equations (7.3) in fact are (more precisely, cover) the Kadomptsev–Petviashvili hierarchy of equations as shall now be shown.

The, by now, standard way of writing down the KP hierarchy is as follows. Consider a pseudo-differential operator S of the form

$$S = D + \sum_{i=1}^{\infty} s_i D^{-i} \tag{7.5}$$

For every pseudodifferential operator $U = \sum_{i=-n}^{\infty} u_i D^{-i}$ define $U_+ = \sum_{i=-n}^{0} u_i D^{-i}$ (its differential operator part). Then the KP hierarchy is

$$\frac{\partial S}{\partial t_k} = [(S^k)_+, S] \tag{7.6}$$

which is a set of partial differential equations for the coefficients $s_1, s_2, \ldots$ occurring in (7.5).

The relation between (7.3) and (7.6) is as follows. Let R satisfy (7.3); define

$$S = RDR^{-1} \tag{7.7}$$

Note that S is of the form (7.5). An easy calculation shows that

$$\frac{\partial S}{\partial t_k} = [P_k, S] \tag{7.8}$$

The final step is the simple lemma that if P_k is a differential operator of the form $P_k = D^k + p_1 D^{k-1} + \cdots + p_k$ such that $[P_k, S]_+ = 0$ (which is what is necessary for (7.8) to make sense) then P_k is of the form

$$P_k = (S^k)_+ + c_1(S^{k-1})_+ + \cdots + c_k \tag{7.9}$$

for certain constants $c_1, \ldots, c_k$. Thus up to a triangular constant coefficient transformation in the times $t_1, t_2, \ldots$ equations (7.8) are the same as (7.6). And the equations (7.8) are rationally covered by the equations (7.3) which in turn are infinite dimensional Riccati flows and hence rationally covered by a linear flow.

So the KP hierarchy is also a rational quotient of a linear flow.

Very many integrable systems arise as specializations of the KP hierarchy. However, there is nothing, a priori, that guarantees that such a specialization is compatible with the rational quotient structure. This would be the case if the specialization relations are given by linear relations at the level of the rational covering linearization. This is the case in the case of the KdV hierarchy as a specialization of the KP hierarchy. This specialization, though quite well known, is, for completeness, briefly recalled below in section 8.

This specialization problem is akin to (but different, though undoubtedly related) the specialization problem in the Zaharov–Shabat dressing method, where the question is which specializations are compatible with the dressing transformations, [9].

8. Obtaining the KdV hierarchy from the KP hierarchy.

In the case of the KdV the basic operator is the Hill operator

$$L = D^2 + u \tag{8.1}$$

Let S be the unique operator of the form (7.5) such that $S^2 = L$.

$$S = D + s_1 D^{-1} + s_2 D^{-2} \tag{8.2}$$

One readily finds $s_1 = \frac{1}{2}u$, $s_2 = -\frac{1}{4}u_x$, $s_3 = -\frac{1}{8}(u^2 + u_{xx})$, ... so that in this case S is a very special operator of this form.

The KdV hierarchy is now

$$\frac{\partial L}{\partial t_k} = [(S^k)_+, L] \tag{8.3}$$

which is a consequence of $\frac{\partial S}{\partial t_k} = [(S^k)_+, L]$. The KdV equation itself arises for the case $k = 3$. Indeed

$$(S^3)_+ = D^3 + 3s_1 D + 3s_{1x} + 3s_2 = D^3 + \frac{3}{2}uD + \frac{3}{4}u_x \tag{8.4}$$

so that we find (writing $t = t_3$)

$$u_t = \frac{\partial L}{\partial t} = [D^3 + \frac{3}{2}uD + \frac{3}{4}u_x, D^2 + u] = \frac{3}{2}uu_x + \frac{1}{4}u_{xxx} \tag{8.5}$$

one of the well known forms of the KdV equation.

9. Conjecture and conclusion.

We have seen that several (classes of) integrable dynamical systems are in fact rational quotients of linear systems. I would like to suggest here that this could always be the case and that this might be a good defining property. The word rational here is important. For of course integrable dynamical system, in any case in the finite degree of freedom case, can be canonically transformed into a linear one (action – angle coordinates). However this transformation is not (as a rule) rational.

The remarks made above raise a rather large number of questions which would be interesting (and probably very rewarding) to sort out. Some of these are the following. Is the very formal rational covering linearization of the KdV of section 1 compatible with the KdV as a specialization of the KP and the rational covering linearization of the KP? To what extent are rational covering linearizations unique? Can canonical transformations be lifted in some sense to a rational covering linearization? How should superposition principles, Bäcklund transformations, and dressing transformations be interpreted at the rational covering linearization level?

Finally, to touch on another aspect: the matrix Riccati equation has a well understood phase portrait, [7]. Can something similar be done for the KP equations as projective limits of Riccati equations?

References

1. P.G.Drazin, *Solitons*, Cambridge Univ. Pr., 1983.
2. W.Eckhaus, A. van Harten, *The inverse scattering transformation and the theory of solitons*, North–Holland, 1981.
3. A.G.Reyman, M.A.Semenov–Tian–Shansky, *Reduction of Hamiltonian systems, affine Lie algebras, and Lax equations* I, Inv. Math. **54** (1980), 81–100.
4. M.Sato, *Soliton equations as dynamical systems on infinite dimensional Grassmann manifolds*, RIMS Kokyuroku **439** (1981), 30–40.
5. G.B.Segal, G.Wilson, *Loop group and equations of KdV type*, Publ. Math. IHES **61** (1985), 5–65.
6. M.A.Semenov–Tian–Shansky, *What is a classical R–matrix*, Funk. Anal. i Pril. **17**: 4 (1983), 17–33. (English translation: Functional Analysis and Applications **17** (1984), 259–272).
7. M.A.Shayman, *Phase portrait of the matrix Riccati equation*, SIAM J. Control and Opt. **24** (1986), 1–65.
8. W.W.Symes, *Hamiltonian group actions and integrable systems*, Physica **1D** (1980), 339–374.
9. V.E.Zakharov, S.V.Manakov, *Soliton theory*. In: Physics Reviews Vol. 1, Harwood Acad. Publ., 1979, 133–190.

STATIONARY RANDOM FIELDS OVER HYPERGROUPS

HERBERT HEYER
Mathematisches Institut, Universität Tübingen
Auf der Morgenstelle 10, 74 Tübingen, Germany

ABSTRACT

Results on random fields over homogeneous spaces are reinterpreted within the framework of commutative hypergroups. In particular, group invariance of the covariance kernel is replaced by stationarity with respect to the hypergroup structure of the corresponding double coset space. This report on the first steps towards a theory of random fields over hypergroups is oriented at R. Gangolli's program of studying Lévy-Schoenberg kernels on homogeneous spaces and the Gaussian random fields attached to them.

1. Introduction

There are various approaches leading to significant results on stationary random fields over double coset spaces $K = G//H$ arising from Gelfand pairs (G,H). In the works of R. Gangolli[5] (1967), J. Faraut[4] (1973), R. Askey and N.H. Bingham[2] (1976) or more recently in the paper of J.P. Arnaud and G. Letac[1] (1985) the notion of stationarity is given in terms of a distance function on the corresponding homogeneous space G/H. These approaches do not fully employ the commutativity of the convolution algebra of biinvariant measures on K. Since the beginning of a systematic study of K as a commutative hypergroup it appears reasonable to define stationarity directly on K in terms of the convolution axiomatically introduced in K. An initiation in this direction was given by R. Lasser and M. Leitner[9] (1989) and taken up also by M.M. Rao[12] (1990).

In this note both approaches will be discussed within the general framework of commutative hypergroups. At first we shall motivate hypergroup stationarity by illustrating how it comes about in the classical theory of weakly stationary random fields over $\mathbb{R}^d$ and $\mathbb{S}^d$. The general theory of commutative hypergroups then yields immediately analogues of the well-known theorems of Bochner, Cramér and Kolmogorov. Shift families attached to weakly stationary random fields over K can be applied to establish orthogonal decomposition results. More refined studies have been carried out for polynomial hypergroups.

2. R. Gangolli's program

Our reference to the origin of the Lévy Seminar organized over the years by T. Hida, is Lévy's Brownian motion of several variables. In his contribution of 1967 R. Gangolli[5] studied centered Gaussian random fields $\{X(a): a \in T\}$ over homogeneous spaces $T = G/H$ in terms of their covariance kernels f given by

$$f(a,b) = E(X(a)X(b))$$

for all $a,b \in T$. Within this set-up two classical situations appear as special cases:

(a) The Euclidean case $T = \mathbb{R}^d = M(d)/SO(d)$, with the motion group $M(d)$ and the special orthogonal group $SO(d)$, where

$$f(a,b) := \frac{1}{2}(||a|| + ||b|| - ||a-b||)$$

for all $a,b \in T$, $||.||$ denoting the Euclidean distance,
and

(b) the spherical case $T = S^d = SO(d+1)/SO(d)$, where

$$f(a,b) := \frac{1}{2}(d(a,o)+d(b,o)-d(a,b))$$

for all $a,b \in T$, d denoting the geodesic distance, o being an arbitrary point on S^d.

In both cases f is a *Lévy-Schoenberg (LS) kernel*, i.e. a mapping $f: T\times T \to \mathbb{R}$ satisfying the conditions

(LS1) (Symmetry) $f(a,b) = f(b,a)$.

(LS2) (Norming) There is an $o \in T$ such that $f(a,o) = o$ for all $a \in T$.

(LS3) The polarized kernel
$(a,b) \to r(a,b) := f(a,a)+f(b,b)-2f(a,b)$
is G-*invariant*, i.e.
$r(xa,xb) = r(a,b)$
for all $x \in G$, $a,b \in T$.

(LS4) f is *positive definite*, i.e.

$$\sum_{i,j=1}^{n} \alpha_i \bar{\alpha}_j f(a_i,a_j) \geq o$$

for all $\alpha_1,\ldots,\alpha_n \in \mathbb{C}$, $a_1,\ldots,a_n \in T$.

Gangolli characterizes LS kernels for homogeneous spaces arising from Abelian groups, motion groups, compact and noncompact Lie groups and discusses properties of sample functions, orthogonal decompositions, and white noise representations of the corresponding random fields. In particular he constructs LS kernels which provide Lévy's Brownian motion (BM) in several variables on S^d as centered Gaussian processes which restricted to a geodesic through o are one-dimensional BMs parametrized by the geodesic distance.

Extending Gangolli's work R. Askey and N.H. Bingham considered general compact Riemannian symmetric spaces of rank 1 which as compact two-point homogeneous space have been classified as the spheres $\mathbb{S}^d (d\geq 1)$ and the projective spaces $P^d(\mathbb{R}), P^d(\mathbb{C}), P^d(\mathbb{H}) (d\geq 2), P^d(\mathbb{O})$. For such spaces T the Laplace-Beltrami operator

$$\Delta := \frac{1}{A(\Theta)} \frac{d}{d\Theta} (A(\Theta)\frac{d}{d\Theta})$$

on [o,1] with

$$A(\Theta) := \sin^p(\frac{\pi\Theta}{2}) \sin^q(\pi\Theta)$$

(equal to the area of the sphere of radius Θ in T), where $p,q\geq o$ are determined by the structure of T, has eigenfunctions which on [-1,1] induce normalized Jacobi polynomials $Q_n^{\alpha,\beta}$ with $\alpha := \frac{1}{2}(p+q-1), \beta := \frac{1}{2}(q-1)$. We note that in the classical case

(b) $T = \mathbb{S}^d$ one has $p=o, q=d-1$, hence $\alpha=\beta= \frac{1}{2}(d-2)$, and the $Q_n^{\alpha,\beta}$ coincide with the ultraspherical (Gegenbauer) polynomials.

In extending Gangolli's program beyond the scope of Gelfand pairs we shall recognize this approach once we deal with Sturm-Liouville hypergroups of compact type.

3. Hypergroups and generalized translations

The double coset spaces G//H corresponding to the Gelfand pairs (G,H) of the motivating examples (a) and (b) are $\mathbb{R}_+$ and [o,1] respectively. In these spaces the group structure of the defining groups M(d) and SO(d+1) has vanished; what remains is the structure of a *hypergroup*. Here, a hypergroup is introduced as a locally compact space K together with a convolution * in the Banach space $M^b(K)$ of bounded (Radon) measures on K such that $(M^b(K),*)$ becomes a (commutative) Banach algebra (with neutral element and involution) and such that among others the following axioms hold

(HG1) For $x,y\in K$ the product $\varepsilon_x*\varepsilon_y$ (of Dirac measures)

belongs to the set $M_c^1(K)$ of probability measures on K with compact support.

(HG2) The mapping $(x,y) \to \mathrm{supp}(\varepsilon_x * \varepsilon_y)$ from $K \times K$ into the system of compact subsets of K, furnished with the Hausdorff-Michael topology, is continuous.

For every f in the space $C^b(K)$ of bounded continuous functions on K and $x \in K$ the *translate of* f *by* x is defined by

$$T^x f(y) := \int_K f(z)\, \varepsilon_x * \varepsilon_y(dz)$$

whenever $y \in K$. The *generalized translation operator* T^x can be applied to rewrite the convolution of arbitrary measures $\mu, \nu \in M^b(K)$ as

$$\mu * \nu(f) = \int\int T^x f(y)\, \mu(dx)\, \nu(dy)$$

valid for all $f \in C^b(K)$.

For more details on the axiomatic of a hypergroup and on basic properties of the generalized translation operator the reader is referred to R.I. Jewett[7].

The following selection of *standard examples* is taken from the list presented by the author in [6].

I Double coset hypergroups $K = G//H$ for pairs (G,H) consisting of a locally compact group G and a compact subgroup H of G, where the convolution is defined by

$$\varepsilon_{HgH} * \varepsilon_{Hg'H} := \int \varepsilon_{Hghg'H}\, \omega_H(dh)$$

for all $g, g' \in G$.

I.1 (corresponding to (a) above) $K = M(d)//SO(d) \cong \mathbb{R}_+$ where $\varepsilon_x * \varepsilon_y$ for $x, y \in \mathbb{R}_+$ is defined via Bessel functions B_λ $(\lambda \in \mathbb{R}_+)$ as a measure in $M_c^1(\mathbb{R}_+)$ such that $\mathrm{supp}(\varepsilon_x * \varepsilon_y) = [|x-y|, x+y]$ and

$$B_\lambda(x) B_\lambda(y) = T^x B_\lambda(y)$$

I.2 (corresponding to (b) above) $K = SO(d+1)//SO(d) \cong [-1,1]$ where $\varepsilon_x * \varepsilon_y$ for $x,y \in [-1,1]$ is defined via ultraspherical polynomials $Q_n (n \in \mathbb{Z}_+)$ as a measure in $M^1([-1,1])$ such that $\mathrm{supp}(\varepsilon_x * \varepsilon_y) = [-1,1]$ and

$$Q_n(x)Q_n(y) = T^x Q_n(y).$$

II <u>Sturm-Liouville hypergroups</u> $K = \mathbb{R}_+$ or $K = [o,1]$ defined by eigenvalue problems $L\phi_\lambda = \lambda\phi_\lambda (\lambda \in \mathbb{R}_+)$ involving the operator

$$L := -\frac{1}{A(x)} \frac{d}{dx} (A(x) \frac{d}{dx})$$

on $\mathbb{R}_+$ and $[o,1]$ respectively. Here the conditions to be put on the Sturm-Liouville functions A, and the boundary conditions have to be specified according to the non-compact and compact cases $\mathbb{R}_+$ and $[o,1]$ respectively. In both cases there exists for given $x,y \in K$ a measure $\varepsilon_x * \varepsilon_y \in M_c^1(K)$ satisfying

$$\phi_\lambda(x)\phi_\lambda(y) = T^x \phi_\lambda(y),$$

where $\mathrm{supp}(\varepsilon_x * \varepsilon_y)$ equals to the intervals $[|x-y|, x+y]$ and $[|x-y|, 1-|1-x-y|]$ respectively.

The following classifications are contained in [14].

II.1 If $G//H$ is isomorphic to a Sturm-Liouville hypergroup $\mathbb{R}_+$, then G/H is of the form $\mathbb{R}^d (d \geq 1)$, $H^d(\mathbb{R}), H^d(\mathbb{C}), H^d(\mathbb{H})$ $(d \geq 1)$, and $H^2(\mathbb{O})$.

II.2 If $G//H$ is isomorphic to a Sturm-Liouville hypergroup $[o,1]$, then G/H is of the form $S^d (d \geq 1)$, $P^d(\mathbb{R}), P^d(\mathbb{C}), P^d(\mathbb{H})$ $(d \geq 2)$, and $P^2(\mathbb{O})$.

The Sturm-Liouville hypergroups of Euclidean and spherical type with G/H equal to $\mathbb{R}^d$ and S^d correspond to the examples (a) and (b) with Sturm-Liouville functions A defined by $A(x) := x^{d-1}$ and $A(x) := \sin^{d-1}(\pi x)$ for all x in $\mathbb{R}_+$ and $[o,1]$ respectively.

III Polynomial hypergroups $K = \mathbb{Z}_+$ defined by polynomials Q_n on $\mathbb{R}(n\in\mathbb{Z}_+)$ via the linearization

$$Q_m(x)Q_n(x) = \sum_{r=|m-n|}^{m+n} Q_r(x)g(m,n,r)$$

with coefficients $g(m,n,r)\geq o$.

III.1 If the double coset hypergroup G//H is isomorphic to $\mathbb{Z}_+$, then it arises from a sequence of Cartier polynomials, and the corresponding homogeneous space G/H is an infinite distance-transitive graph. See [13].

The necessary *preparations from the harmonic analysis of commutative hypergroups* can be taken from the expository literature, f.e. from [6]. Very briefly we quote a few notions and facts.

On any commutative hypergroup K there exists a Haar measure ω_K. Characters of K are defined as bounded continuous mapping $\chi : K \to \mathbb{C}$ satisfying $\chi(x^-) = \overline{\chi(x)}$ and

$$\chi(x)\chi(y) = T^x\chi(y)$$

for all $x,y\in K$. The characters of K form the dual set $K^\wedge$ of K which furnished with the compact open topology is a locally compact space. Fourier-Stieltjes transforms and cotransforms are introduced as mappings $\mu \to \hat{\mu}$ and $\mu \to \check{\mu}$ from $M^b(K)$ to $C^b(K^\wedge)$ and from $M^b(K^\wedge)$ to $C^b(K)$ respectively, and they have the usual properties. In particular, there exists a Plancherel-Levitan measure $\pi\in M_+(K^\wedge)$ such that $L^2(K,\omega_K)$ can be embedded (isometrically) into $L^2(K^\wedge,\pi)$. However, $\text{supp}\pi$ does not necessarily equal $K^\wedge$. Also, positive definite functions can be defined on K in the usual way (and by a different method also on the space $K^\wedge$). However, positive definite functions are not necessarily bounded on a hypergroup. Still Bochner's theorem holds in the sense that a complex-valued function f on K is bounded, continuous and positive definite iff there exists a measure $\mu \in M^b(K^\wedge)$ satisfying $f = \check{\mu}$. The representing measure μ is uniquely determined by f.

4. Stationarity of random fields

Let $(\Omega, \mathfrak{A}, P)$ be a probability space and K an arbitrary commutative hypergroup. Any mapping $X: K \to L^2_{\mathbb{C}}(\Omega, \mathfrak{A}, P)$ is said to be a *random field over* K. For a random field X over K we assume that the first and second moments

$$M(a) := E(X(a))$$

and

$$f(a,b) := E[(X(a)-M(a))\overline{(X(b)-M(b))}]$$

exist for all $a, b \in K$. The random field X over K is called *centered* if $M(a) = M$ for all $a \in K$. If there is no special mention we assume that $M = o$.

Definition. A centered random field X over K is said to be *stationary* if the *covariance kernel* f of X belongs to $C^b(K \times K)$ and satisfies the formula

$$f(a,b) = \int_K f(x,e)\, \varepsilon_a * \varepsilon_{b^-}(dx)$$

valid for all $a, g \in K$.

In the case of stationarity this formula reads as

$$f(a,b) = T^a F(b^-)$$

where $F := f(.,e)$ denotes the *covariance function* of X.

The definition of stationary random fields over hypergroups together with an initiation of the theory is due to R. Lasser [8] and M. Leitner [9,10]. The subsequent basic theorems can be obtained directly from their analogues for groups.

4.1 Theorem (Bochner). For any stationary random field X over K with covariance kernel f there exists a *spectral measure* $\mu \in M_+^b(K^\wedge)$ such that

$$f(a,b) = \int_{K^\wedge} \chi(a)\overline{\chi(b)}\mu(d\chi)$$

whenever $a,b \in K$.

4.2 Theorem (Cramér). For every random field X over K the following statements are equivalent:

(i) X is stationary (over K).

(ii) There exists a unique orthogonal *stochastic measure* $Z : \mathcal{B}(K^\wedge) \to L^2(\Omega,\mathcal{A},P)$ such that

$$X(a) = \int_{K^\wedge} \chi(a)Z(d\chi)$$

for all $a \in K$.

Clearly, the relationship between the spectral measure μ and the stochastic measure Z of X is given by

$$\mu(A) = ||Z(A)||_2^2$$

for all $A \in \mathcal{B}(K^\wedge)$.

4.3 Theorem (Kolmogorov). For any function $F \in C^b(K)$ the following statements are equivalent.

(i) F is positive definite.

(ii) There exists a stationary random field X over K whose covariance function equals F.

From M.M. Rao [12] we quote the subsequent generalization of stationarity.

Definition. A centered random field X over K is called (weakly) *harmonizable* if the covariance kernel f of X admits the representation

$$f(a,b) = \int_{K^\wedge}\int_{K^\wedge} \chi(a)\overline{\rho(b)}\beta(d\chi,d\rho)$$

for all $a,b \in K$, where β is a positive definite bimeasure on $\mathfrak{B}(K^\wedge) \times \mathfrak{B}(K^\wedge)$ and the integral is to be understood in the sense of Morse and Transue.

For the necessary technical details see [3]. Evidently harmonizability reduces to stationarity over K if β is concentrated on the diagonal of $K^\wedge \times K^\wedge$.

4.4 <u>Theorem</u>. A centered random field X over K is harmonizable iff it is V-*bounded* in the sense that the set

$$\{\int_K \phi(a)X(a)\omega_K(da) \; : \; \phi \in L^1 \cap L^2(K,\omega_K) \; : \; \|\hat{\phi}\|_\infty \leq 1\}$$

is norm bounded in $L^2(\Omega,\mathfrak{A},P)$.

In order to show the significance of hypergroup stationarity we consider classical centered weakly stationary time series $(X(a))_{a\in T}$ with $T = \mathbb{R}$ or $\mathbb{Z}$ and the corresponding mean estimating processes $(Y(a))_{a\in K}$ with $K := \mathbb{R}_+$ or $\mathbb{Z}_+$ respectively. It turns out that theses processes lose the (classical) weak stationarity; what they retain is stationarity with respect to the Bessel and ultraspherical hypergroup structures on K respectively. In the case (a) one has

$$Y(a) := \begin{cases} \frac{1}{2a}\int_{-a}^{a} X(r)dr & \text{if } a \in \mathbb{R}_+^\times \\ X(o) & \text{if } a=o. \end{cases}$$

Let μ be the classical spectral measure of X and $B_\lambda := B^\alpha$ the modified Bessel function of order $\alpha = \frac{1}{2}$. Then for all $a,b \in \mathbb{R}_+$ we have

$$f_Y(a,b) = \frac{1}{2a}\frac{1}{2b}\int_{-a}^{a}\int_{-b}^{b} E[(X(r)-M)\overline{(X(s)-M)}]drds$$

$$= \int_{\mathbb{R}} [\frac{1}{2a}\int_{-a}^{a} e^{iru}dr \, \frac{1}{2b}\int_{-b}^{b} e^{isu}ds] \, \mu(du)$$

$$= \int_{\mathbb{R}} B_a(u) B_b(u) \mu(du)$$

$$= \int_{\mathbb{R}} [\int_{R_+} B_c(u) \varepsilon_a * \varepsilon_b(dc)] \mu(du)$$

$$= \int_{\mathbb{R}_+} f_Y(c,o) \varepsilon_a * \varepsilon_b(dc).$$

In the case

(b) we define

$$Y(n) := \frac{1}{2n+1} \sum_{k=-n}^{n} X(k)$$

for all $n \in \mathbb{Z}_+$. Denoting again the classical spectral measure of X by μ and the ultraspherical polynomial of order (α,β) with $\alpha = \frac{1}{2}$ and $\beta = -\frac{1}{2}$ by Q_n we obtain for all $m,n \in \mathbb{Z}_+$ that

$$f_Y(m,n) = \frac{1}{2m+1} \frac{1}{2n+1} \sum_{k=-m}^{m} \sum_{k=-n}^{n} E[(X(m)-M)\overline{(X(n)-M)}]$$

$$\int_{-\pi}^{\pi} [\frac{1}{2m+1} \sum_{k=-m}^{m} e^{ikt} \frac{1}{2n+1} \sum_{k=-n}^{n} e^{ikt}]\mu(dt)$$

$$= \int_{-\pi}^{\pi} Q_m \circ \cos t \, Q_n \circ \cos t \, \mu(dt)$$

$$= \sum_{k=|m-n|}^{m+n} g(m,n,k) \int_{-\pi}^{\pi} Q_k \circ \cos t \, \mu(dt)$$

$$= \sum_{k=|m-n|}^{m+n} g(m,n,k) f_Y(k,o)$$

$$= \sum_{k \in \mathbb{Z}_+} f_Y(k,o) \varepsilon_m * \varepsilon_n(\{k\}).$$

5. Shift families and orthogonal decompositions

Let $X : K \to L^2_{\mathbb{C}}(\Omega,A,P)$ be a stationary random field over K with spectral measure μ and stochastic measure Z. We define $H_o := \langle\{X(a) : a \in K\}\rangle$ and $H := \overline{H_o}$. For $a,b \in K$ we introduce the $\mathbb{C}$-valued random variable

$$\varepsilon_a * X(b) := \int_{K^\wedge} \chi(a)\chi(b)Z(d\chi)$$

Definition. For $a \in K$ the *shift by* a is given as the mapping $T^a : H_o \to H$ defined for $a_o, \ldots, a_n \in \mathbb{C}$ and $b_o, \ldots, b_n \in K$ by

$$T^a(\sum_{k=o}^{n} a_k X(b_k)) := \sum_{k=o}^{n} a_k \varepsilon_y * X(b_k).$$

By Theorem 4.2 T^a is well-defined.
The *shift family* $\{T^a : a \in K\}$ corresponding to X has the following properties:

(SF1) T^a is (the extension of) a normal contraction on H with $(T^a)^* = T^{a^-}$ for all $a \in K$ and $T^e = Id$.

(SF2) The mapping $a \to T^a$ is continuous.

(SF3) For all $a, b \in K$

$$T^a o T^b = \int_K T^r \varepsilon_a * \varepsilon_b(dr)$$

The proofs of these properties follow from the fact that the mapping $f \to \int_{K^\wedge} f dZ$ is an isometric isomorphism from $L^2(K^\wedge, \mu)$ onto H and hence T^a corresponds to the multiplication operator $f \to \hat{\varepsilon}_{a^-} f$ from $<\{\hat{\varepsilon}_{b^-} : b \in K\}>$ into $L^2(K^\wedge, \mu)$.

We also note that given a Hilbert subspace H of $L^2(\Omega, A, P)$ such that $E\xi = o$ for all $\xi \in H$ and given $\eta \in H$, any family $\{T^a : a \in K\}$ of operators on H satisfying (SF1) to (SF3) defines a stationary random field X over K through $X(a) := T^a \eta$ for all $a \in K$.

5.1 Theorem (Lamperti). Let X be a stationary random field over K with spectral measure μ and stochastic measure Z, and let $\{A, B\}$ be a measurable partition of $K^\wedge$. Then there are stationary random fields X^A and X^B over K of the form

$$X^C(a) := \int_C \chi(a) Z(d\chi)$$

for all $a \in K, C \in \{A, B\}$ such that

(i) $X = X^A + X^B$ and
(ii) $X^A \perp X^B$.

Conversely, if $X^{(1)}$ and $X^{(2)}$ are stationary random fields over K with $X^{(i)}(a) \in \langle\{X(a) : a \in K\}\rangle^-$ for all $a \in K$, $i=1,2$ satisfying
(i) $X = X^{(1)} + X^{(2)}$ and
(ii) $X^{(1)} \perp X^{(2)}$,
then there exists a measurable partition $\{A,B\}$ of $K^\wedge$ such that $X^{(1)} = X^A$ and $X^{(2)} = X^B$.

In a recent paper [11] M. Leitner applied this theorem in order to obtain an orthogonal decomposition of a stationary random field X over K, in the sense of (i) and (ii) of Theorem 5.1, where A equals the annihilator $G(K)^\perp$ of the maximum subgroup G(K) of K, B equals its complement in $K^\wedge$, and the random fields X^A and X^B are stationary random fields over the double coset hypergroup $K//G(K)$ and K respectively. For the definition of the maximum subgroup of a hypergroup and of the double coset hypergroup of a hypergroup see [7].

Applying the ideas of translation and decomposition some extrapolation problems can be discussed along the lines of M. Leitner's thesis [10]. Let $\underline{A}$ be a subfamily of $\underline{P}(K)$. For a given stationary random field X over K we introduce the space

$$H_X := \bigcap_{A \in \underline{A}} H_X(A),$$

where $H_X(A) := \langle\{X(a) : a \in A\}\rangle^-$ for every $A \in \underline{A}$. X is said to be $\underline{A}$- *adapted* if $T^a H_X \subset H_X$ for all $a \in K$, $\underline{A}$- *regular* if $H_X = \{o\}$, and $\underline{A}$-*singular* if $H_X = H_X(K)$.

5.2 Theorem (Wold). Let X be an $\underline{A}$-adapted random field over K. There exist uniquely stationary random fields R and S over K with the following properties
(i) R is $\underline{A}$-regular, and S is $\underline{A}$-singular.

(ii) $X = R + S$.

(iii) $R \perp S$.

(iv) $H_R(A) \subset H_X(A)$, and $H_S(A) \subset H_X(A)$ for all $A \in \underline{A}$.

6. Further developments for polynomial hypergroups

Let $K = \mathbb{Z}_+$ be a polynomial hypergroup (of type III) defined by a sequence $(Q_n)_{n \geq 0}$ of polynomials Q_n on $\mathbb{R}$ which are orthogonal with respect to an orthogonality measure π. In fact, the polynomials Q_n are defined recursively in terms of sequences $(a_n)_{n\geq 0}$, $(b_n)_{n \geq 0}$ and $(c_n)_{n \geq 1}$ in $\mathbb{R}_+^\times$, $\mathbb{R}$ and $\mathbb{R}_+^\times$ respectively such that $a_n + b_n + c_n = 1$, $b_n \geq 0$ for all $n \geq 1$, and $a_0 + b_0 = 1$. The set

$$D_S := \{x \in \mathbb{R} : (Q_n(x))_{n \geq 0} \text{ is bounded}\}$$

is homeomorphic to $\mathbb{Z}_+^\wedge$ under the mapping $x \to (n \to (Q_n(x))$, and D_S is a compact subset of $[1-2a_0, 1]$.

Let now X be a stationary random field over $\mathbb{Z}_+$ with shift family $\underline{T} := \{T^n : n \in \mathbb{Z}_+\}$. Then $W := \langle \underline{T} \rangle^-$ is a commutative subalgebra of bounded linear operators on the Hilbert space H which is isometrically isomorphic to the space $C(\Delta)$ of continuous functions on the structure space Δ of W. Since T^1 is symmetric and contractive, we have $h(T^1) \in [-1,1]$ for all $h \in \Delta$. Choosing $t_h \in [1-2a_{o,1}]$ such that $Q_1(t_h) = h(T^1)$ we obtain $Q_n(t_h) = h(T^n)$ for all $n \geq 1$, $\sup_{n \geq 0} |Q_n(t_h)| \leq 1$ and hence $t_h \in D_S$. The mapping $h \to t_h$ from Δ into D_S is one-to-one. Thus we have established a homeomorphism from Δ onto the compact subset $D_S(\underline{T}) := \{t_h : h \in \Delta\}$ of D_S and an isometric isomorphism $\rho \to \hat{\rho}$ from W onto $C(D_S(\underline{T}))$, where $\hat{\rho}(t_h) := h(\rho)$ for all $t_h \in D_S(\underline{T})$. In particular $\widehat{T^n}$ and Q_n coincide on $D_S(\underline{T})$. Applying the spectral theorem for C^*-algebras we see that $D_S(\underline{T}) = \operatorname{supp}\mu$, where μ is the spectral measure of X. As a consequence there exists an operator A $(= a_0 T^1 + b_0 \mathrm{Id}) \in W$ such that $\underline{T}^n = Q_n(A)$ for all $n \in \mathbb{Z}_+$.

6.1 *White noise* (over $\mathbb{Z}_+$) can be introduced as a centered random field Z over $\mathbb{Z}_+$ with the property that

$$E(Z(m)\overline{Z(n)}) = \delta_{mn} g(m,n,o)$$

for all $m,n \in \mathbb{Z}_+$. Z turns out to be a stationary random field over $\mathbb{Z}_+$ with spectral measure μ equal to the orthogonality (Plancherel) measure π on $D_S \cong \mathbb{Z}_+^{\wedge}$.

6.2 Let Z be white noise with shift family $\{T^n : n \in \mathbb{Z}_+\}$. A random field X over $\mathbb{Z}_+$ is said to be a *moving average* with respect to Z if there exists a sequence $(\alpha_n)_{n \geq o}$ in $L_C^2(\mathbb{Z}_+, \omega_{\mathbb{Z}_+})$ such that

$$X(n) = \sum_{k \geq o} \alpha_k T^n Z(k) \omega_{\mathbb{Z}_+}(\{k\})$$

whenever $n \in \mathbb{Z}_+$. Moving averages X with respect to white noise Z are stationary random fields over $\mathbb{Z}_+$ with spectral measure $|\hat{\alpha}|^2 . \pi$, where

$$\hat{\alpha} := \sum_{k \geq o} \alpha_k Q_k \omega_{\mathbb{Z}_+}(\{k\}).$$

They admit representations of the form

$$X(n) = \sum_{k \geq o} \alpha_{k*n} Z(k) \omega_{\mathbb{Z}_+}(\{k\})$$

for all $n \in \mathbb{Z}_+$. Regularity of random fields over $\mathbb{Z}_+$ will be considered with respect to the family $\underline{A}$ of sets $\{n, n+1, \ldots\}$ with $n \in \mathbb{Z}_+$. On the occasion of a meeting in Oberwolfach M. Leitner communicated the following result. If the polynomial hypergroup $\mathbb{Z}_+$ admits a diffuse orthogonality measure π on $\mathbb{Z}_+^{\wedge}$ then any regular random field X over $\mathbb{Z}_+$ whose spectral measure μ admits a density in $L^1(\mathbb{Z}_+^{\wedge}, \pi)$ is in fact a moving average over $\mathbb{Z}_+$.

References

1. J.-P. Arnaud and G. Letac, *La formule de représentation spectrale d'un processus gaussien stationnaire sur un arbre homogène*, in: Probabilités sur les structures géométriques, pp. 1-11, Publications du Laboratoire de Statistique et Probabilités, Université Paul Sabatier, Toulouse 1985

2. R. Askey and N.H. Bingham, *Gaussian processes on compact symmetric spaces*, Z. Wahrscheinlichkeitstheorie verw. Gebiete 37 (1976), 127-143

3. D.K. Chang and M.M. Rao, *Bimeasures an nonstationary processes*, in: Real and Stochastic Analysis, ed. M.M. Rao, pp. 7-118, John Wiley & Sons, New York 1986

4. J. Faraut, *Fonction Brownienne sur une variété Riemannienne*, in: Séminaire de Probabilité VII, pp. 61-76, Lecture Notes in Mathematics Vol. 321, Springer 1973

5. R. Gangolli, *Positive definite kernels on homogeneous spaces and certain stochastic processes related to Lévy's Brownian motion of several variables*, Ann.Inst. Henri Poincaré, Section B, Vol. III, n$^{\underline{o}}$2 (1967), 121-225

6. H. Heyer, *Convolution semigroups and potential kernels on a commutative hypergroup*, in: The Analytical and Topological Theory of Semigroups, ed. K.H. Hofmann, J.D. Lawson, J.S. Pym, pp. 279-312, Walter de Gruyter, Berlin, New York 1990

7. R.I. Jewett, *Spaces with an abstract convolution of measures*, Advances in Math. 18 (1975), 1-101

8. R. Lasser, *Applications of the theory of hypergroups*, Math.Comput.Modelling 11 (1988), 210-211

9. R. Lasser and M. Leitner, *Stochastic processes indexed by hypergroups* I, J. Theoret.Probab. 2 (1989), 301-311

10. M. Leitner, K-*schwach stationäre Prozesse;* K *eine Hypergruppe,* Dissertation TU München 1989

11. M. Leitner, *Stochastic processes indexed by hypergroups* II, J. Theoret.Prob. (to appear)

12. M.M. Rao, *Bimeasures and harmonizable processes,* in: Probability Measures on Groups IX, ed. H. Heyer, pp. 254-298, Lecture Notes in Mathematics Vol. 1379, Springer 1989

13. M. Voit, *Central limit theorems for random walks on* $\mathbb{N}_0$ *that are associated with orthogonal polynomials,* J. Multivariate Anal. 34 (1990), 290-322

14. Hm. Zeuner, *One-dimensional hypergroups,* Advances in Math. 76 (1989), 1-18

CANONICAL REPRESENTATIONS OF GAUSSIAN PROCESSES AND INTEGRAL OPERATORS

MASUYUKI HITSUDA
Department of Mathematics,
Kumamoto University,
Kurokami, 2-39-1
Kumamoto 860, Japan

Dedicated to Professor T. Hida

Abstract. A genararization of the innovation theorem is presented in the first part of the paper. Applying the result, the canonical representation is systematically obtained for some class of Gaussian processes.

1. INTRODUCTION

We will consider a Gaussian process $X = \{X(t); \quad t \in I\}$ defined on a probability space $(\Omega, \mathbf{B}, P)$ with parameter t of an interval I . In the present paper, it is assumed that the process X is centered and (A) the filtering $\mathfrak{B}(X) = \{\mathbf{B}_t(X); t \in I\}$ for X is a continuously increasing system of sub-σ -fields $\mathbf{B}_t(X)$ generated by $\{X(s); s \leq t\}$. It is known by Hida[3] that a Gaussian process $X = \{X(t); t \in I\}$ has a canonical representation of the form

$$X(t) = \sum_{i=1}^{N} \int^{t} F_i(t,u)dB_i(u), \tag{1}$$

under the condition (A), where $B = \{B(t) = (B_1(t), B_2(t), \cdots, B_N(t));$ $t \in I\}$ is an N-vector valued Gaussian martingale with independent components and $F_i(t,u)$ is a function satisfying

$$\sum_{i=1}^{N} \int^{t} F_i(t,u)^2 m_i(du) < \infty,$$

with $m_i(dt) = E(|dB_i(t)|^2)$, $i = 1, 2, \cdots, N$. The term of canonical representation means that the coincidence of the filterings $\mathfrak{B}(X)$ for X and $\mathfrak{B}(B)$ for B, in other words $\mathbf{B}_t(X) = \mathbf{B}_t(B)$, $t \in I$. The number N is determined uniquely when the spectral measures $m_i(dt)$, $i = 1, 2, \cdots, N$, are arranged in the order of absolute continuity : $m_i(dt) \gg m_{i+1}(dt)$, $i = 1, 2, \cdot\cdot, N-1$. Such an operation is always permitted and the number N in (1) is possible to be ∞. Such an example is constructed in [6]. The aim of the paper is to construct the innovation B for a special class of Gaussian processes and discuss whether the multiplicity N is one or not.

In the beginning, we try to generalize the so-called innovation theorem which is proved by Shiryaev and Kailath (see[11]) independently. The theorem is well applied to many kinds of Gaussian semimartingales, which appear in the typical noise plus message problems. In the last two sections, some aspects of the applications are presented.

2. A GENERALIZATION OF INNOVATION THEOREM.

Let $X = \{X(t); t \in I\}$, $I = [0,1]$ or $[0,\infty)$, be a stochastic process described in the form:

$$X(t) = B(t) + \int_0^t \varphi(s,\omega)dA(s), t \in I, \tag{2}$$

where (i) $B = \{B(t); t \in I\}$ is a Brownian motion with respect to a filteration $\mathcal{F} = \{\mathbf{F}_t; t \in I\}$, (ii) A is an increasing and continuous function defined on I and (iii) the random variable $\varphi(t, \omega)$ is $\mathbf{F}_t$-measurable and (t, ω)-measurable for each $t \in I$. In this section, we do not assume that $\varphi(t, \omega)$ is Gaussian.

THEOREM 1. *Let X be a stochastic process given by (2) and let $\mathcal{G} = \{\mathbf{G}_t; t \in I\}$ be an increasing system of σ-fields satisfying $\mathbf{B}_t(X) \subset \mathbf{G}_t \subset \mathbf{F}_t$, $t \in I$. If an integrability condition*

$$E\left(\int_0^t |\varphi(s,\omega)| dA(s)\right) < \infty, t \in I, \tag{3}$$

is satisfied, then the process $\bar{B} = \{\bar{B}(t); t \in I\}$ defined by

$$\bar{B}(t) = X(t) - \int_0^t \hat{\varphi}(s,\omega) dA(s), t \in I, \tag{4}$$

is a Brownian motion with respect to the filteration $\mathcal{G} = \{\mathbf{G}_t; t \in I\}$, where $\hat{\varphi}(t,\omega) = E(\varphi(t,\omega)|\mathbf{G}_t)$.

[Remark] In case of $\mathcal{G} = \mathcal{F}$ (or $\mathbf{G}_t = \mathbf{F}_t, t \in I$) the result is trivial because $\hat{\varphi}(t,\omega) = \varphi(t,\omega), t \in I$. In case of $\mathcal{G} = \mathfrak{B}(X)$ (or $\mathbf{G}_t = \mathbf{B}_t(X), t \in I$) and $A(t) = t$, Theorem 1 is a usual innovation theorem proved by Shiryaev and Kailath (see[11]).

PROOF: The idea of the proof is almost analogous to the one of the usual innovation theorem, so it is only shown in the outline. Let us apply the generalized Itô formula (see Kunita and Watanabe[10]) to the function $\exp\{i\lambda(\bar{B}(t)-\bar{B}(s))\}$. As

$$\bar{B}(t)=B(t)+\int_0^t(\varphi(s,\omega)-\hat{\varphi}(s,\omega))dA(s),$$

we get

$$\begin{aligned}\exp\{i\lambda(\bar{B}(t)-\bar{B}(s))\} &= 1+i\lambda\int_s^t \exp\{i\lambda(\bar{B}(u)-\bar{B}(s))\}dB(u)\\ &+i\lambda\int_s^t \exp\{i\lambda(\bar{B}(u)-\bar{B}(s))\}(\varphi(u,\omega)-\hat{\varphi}(u,\omega))dA(u)\\ &-\frac{\lambda^2}{2}\int_s^t \exp\{i\lambda(\bar{B}(u)-\bar{B}(s))\}du. \qquad (5)\end{aligned}$$

Taking the conditional expectation on both side of (5), a simple equality

(6)

$$E[\exp\{i\lambda(\bar{B}(t)-\bar{B}(s))\}|\mathbf{G}_s]=1-\frac{\lambda^2}{2}\int_s^t E[\exp\{i\lambda(\bar{B}(u)-\bar{B}(s))\}|\mathbf{G}_s]du,$$

holds with probability one for each $s\le t, s,t\in I$. In fact, the conditional expectation of the right hand side of (5) with respect to $\mathbf{F}_s$ is equal to zero with probability one and so is with respect to $\mathbf{G}_s$ $(\subset \mathbf{F}_s)$, and the third term is also zero because of

$$E[\varphi(u,\omega)-\hat{\varphi}(u,\omega)|\mathbf{G}_s]=E[E[\varphi(u,\omega)|\mathbf{G}_u]-\hat{\varphi}(u,\omega)|\mathbf{G}_s]=0.$$

Then we get from (6)

$$E[\exp\{i\lambda(\bar{B}(t)-\bar{B}(s))\}|\mathbf{G}_s]=\exp\{-\frac{\lambda^2}{2}(t-s)\},$$

with probability one for each $s \le t$, $s,t \in I$. Thus we know that $\bar{B}$ is a Brownian motion with respect to $\mathcal{G}$. ∎

3. Separation problem for a Gaussian channel.

In the present section, we consider a type of Gaussian channel $X = \{X(t); t \in I\}$, $I = [0,1]$ or $[0,\infty)$, which is described by (2) in Section 2. with properties (i), (ii) and (iii). In addition, we assume that (iv) $\varphi = \{\varphi(t,\omega); t \in I\}$ is a Gaussian system independent of B . The problem is to give a condition that

$$\mathfrak{B}(X) = \{\mathbf{B}_t(B) \vee \mathbf{B}_t(\Phi); t \in I\}, \tag{7}$$

where $\Phi = \{\Phi(t) = \int_0^t \varphi(s,\omega) dA(s); t \in I\}$. In other words, we receive the complete information about the message $\{\varphi(t,\omega); t \in I\}$ through the channel (1) at each moment , if the condition is satisfied. In order to describe the result, we refer the basic result from [7] which the author obtained, suggested by the principal paper[3] by Hida. The book[4] include general information about the fact. It is convenient to introduce two kind of Hilbert spaces, the one is the reproducing kernel Hilbert space $\mathcal{H}(X)$ which is introduced by [1] and the other is the linear span $H(X)$ of the process X . A typical correspondence between these spaces is stated in [3],[4] and [7].

Notation. 1. The space $\mathcal{H}_t(X), t \in I$, means the reproducing kernel

Hilbert space spanned by the covariance function $\{\Gamma_X(s,u); s,u \le t\}$ of X.

2. The space $\mathcal{H}_t(X)$, $t \in I$, means the Hilbert space spanned by $\{X(s); s \le t\}$, and $H(X)$ means the linear span by $X = \{X(t); t \in I\}$.

LEMMA 1. *Let X and Y be a pair of independent Gaussian processes. Then*

$$\mathbf{B}_t(X+Y) = \mathbf{B}_t(X) \vee \mathbf{B}_t(Y), t \in I, \tag{8}$$

if and only if $\mathcal{H}_t(X) \cap \mathcal{H}_t(Y) = \{0\}, t \in I$.

[Remark] Since (X, Y) constitutes a Gaussian system, the property (7) is equivalent to

$$H_t(X+Y) = H_t(X) \oplus H_t(Y) (\text{direct sum}), t \in I. \tag{9}$$

From Lemma 1, we get the following criterion on the separation of information.

LEMMA 2. *For a Gaussian channel (2) having properties (i) ~ (iv), the separation of filtering (7) holds if and only if*

(10)

$\mathcal{H}_t(\Phi) - \{0\} \subset \{a; a(s), s \le t$ *is a continuous funtion and* $\mathcal{Q}(a; [0,t]) = \infty\}$, *for any* $t \in I$, *where*

$$Q(a;[0,t]) = \sup\{\sum_{i=0}^{n-1} \frac{|a(t_{i+1}) - a(t_i)|^2}{t_{i+1} - t_i};$$

$\Delta : 0 = t_0 < t_1 < \cdots < t_n = t$ *is a finite decomposition of* $[0,t]\}$.

PROOF is easy based on a known fact that the reproducing kernel Hilbert space $\mathcal{H}_t(B)$ is the set $\{a; a(s) = \int_0^s \alpha(u)du, \alpha \in L^2([0,t])\}$. ∎

By the use of Lemma 2, the more concrete result is derived.

PROPOSITION 1. *If the message* $\varphi = \{\varphi(t,\omega); t \in I\}$ *in (2) satisfies a condition that a function*

(11) $$\beta : \beta(t) = E[\varphi(t,\omega)\xi], t \in I,$$

is continuous for any ξ *of* $H(\Phi)$ *and the increasing function* A *satisfies*

(12) $$Q(A;[s,t]) = \infty, t > s > 0,$$

then the property (7) holds.

PROOF: For a fixed moment $t_0 > 0$, a member of $\mathcal{H}_{t_0}(\Phi)$ is a function $b : b(t) = \int_0^t \beta(s)dA(s)\,(= E[\Phi(t)\xi])$, where β is of Eq. (11). In case of $b \neq 0$ (eqivalently $\beta \neq 0$) , there exists a small interval $I_1 = [s_1, t_1]$, $s_1 < t_1$, such that $|\beta(s)| > C > 0, s \in I_1$, since β is continuous from the assumption. Therefore, we get $Q(b;[0,t_0]) > CQ(A;[s_1,t_1]) = \infty$ from Eq.(12). Thus the result (7) is derived by virtue of Lemma 1. ∎

[Example] Let $B_1 = \{B_1(t); t \in I\}$ be a Brownian motion independent of B. In the Gaussian channel X :

$$(13) \qquad X(t) = B(t) + \int_0^t B_1(s)dA(s), t \in I,$$

the separation (7) is varid if A satisfies the condition in the Proposition 1 . The Gaussian process X of (13) is of mutiplicity two in the sence of Hida[3] and the spectral measures are both Lebesgue. This kind of example appeared in §8, Chap.4 of [12].

4. Stationary white Gaussian channels.

In Section 3, we discussed the case of the complete separation of the message and the noise. However, it is difficult to get innovations of a Gaussian channel (2) in general. Here we pick up the stationary white Gaussian channel from which a single innovation is well extracted. Let us consider a Gaussian channel $\dot{X}$ described by distribution-valued (or generalized) processes :

$$(13) \qquad \dot{X}(t) = \dot{B}(t) + \varphi(t,\omega), t \in (-\infty, \infty),$$

where (i) $\dot{B} = \{B(t); t \in (-\infty, \infty)\}$ is the Gaussian white noise with respect to a filteration $\mathcal{F} = \{\mathbf{F}_t; t \in I\}$, (ii) $\varphi = \{\varphi(t,\omega); t \in (-\infty, \infty)\}$ is a purely non-deterministic stationary Gaussian process correlated stationarily with $\dot{B}$ and (iii) the random variable $\varphi(t,\omega)$ is $\mathbf{F}_t$-measurable for each $t \in I$. The exact meaning of Eq.(13) is that the equality

$$(13') \qquad (\zeta, \dot{X}) = (\zeta, \dot{B}) + (\zeta, \varphi)$$

holds for any test function $\zeta \in (S)$, where (S) is the Schwartz space. It is better to note that Eq.(13') is equivalent to

$$(13'') \qquad \int \zeta(t)dX(t) = \int \zeta(t)dB(t) + \int \zeta(t)\varphi(t,\omega)dt.$$

[Remark 1] The integral of the second term of the right hand side of (13") is well defined for any stationary process φ. Moreover, under the assumption that φ is purely non-deterministic, the integral is well defined for ζ of $L^2(-\infty, \infty)$. In fact, the covariance $\Gamma_\varphi(t,s) = \Gamma_\varphi(t-s)$ of φ is subordinate to the one $\Gamma_{\dot{B}}(t,s) = \delta_0(t-s)$ of the white noise, in other words, there exists a constant $C > 0$ such that $C\delta_0(t-s) - \Gamma_\varphi(t-s)$ is positive definite. The detailed story is in [7].

[Remark 2] The property (i) for the channel (13) means that the process

$$B_\zeta = \{B_\zeta(t) = \int_{-\infty}^{t} \zeta(s)dB(s); t \in (-\infty, \infty)\}$$

is a Gaussian martingale with respect to the filteration $\mathcal{F}$, for each $\zeta \in L^2(-\infty, \infty)$. Let us take note that

$$E[B_\zeta(t)^2] = \int_{-\infty}^{t} \zeta^2(s)ds.$$

Notation. A notation $\mathbf{B}_t(\dot{X})$ means the sub-σ -field generated by $\{\dot{X}(s); s \leq t\}$. $\mathbf{B}_t(\dot{X})$ is equivalent to the sub-σ -field generated by $\{\int \zeta(s)dX(s); \zeta \in (S)$ and ζ is supported in $(-\infty, t]\}$.

THEOREM 2. *In the Gaussian channel (13), the generalized process* $\dot{\bar{B}} = \{\dot{\bar{B}}(t); t \in (-\infty, \infty)\}$:

$$(14) \qquad \dot{\bar{B}}(t) = \dot{X}(t) - \hat{\varphi}(t, \omega), \textit{where } \hat{\varphi}(t, \omega) = E[\varphi(t, \omega) | \mathbf{B}_t(\dot{X})]$$

is a Gaussian white noise with respect to the filteration $\mathfrak{B}(\dot{X}) = \{\mathbf{B}_t(\dot{X}); t \in (-\infty, \infty)\}$. *Furthermore, the equivalence of the filterings* $\mathfrak{B}(\dot{X}) = \mathfrak{B}(\dot{\bar{B}})$ *(or* $\mathbf{B}_t(\dot{X}) = B_t(\dot{\bar{B}}), t \in (-\infty, \infty)$ *) holds.*

PROOF: We can apply Theorem 1 to the process X_a for a fixed $a \in (-\infty, \infty)$:

$$X_a(t) = \int_a^t dX(s) = (1_{[a,t]}, \dot{X}), t \geq a,$$

for the filtering $\mathcal{G} = \{\mathbf{B}_t(\dot{X}); t \geq a\}$. Here, 1_A is the indicator function of the set $A : 1_A(s) = 1, s \in A; = 0$ otherwise. In fact,

$$X_a(t) = \int_a^t dB(t) + \int_a^t \varphi(s, \omega) ds$$

defines a Gaussian channel with respect to the filtering $\mathcal{F} = \{\mathbf{F}_t; t \geq a\}$, and moreover it is obvious that $\mathbf{B}_t(X_a) \subset \mathbf{B}_t(\dot{X}) \subset \mathbf{F}_t, t \geq a$. Thus the process $\bar{B}_a$ defined by

$$\bar{B}_a(t) = X_a(t) - \int_a^t \hat{\varphi}(s,\omega)ds$$

is a Brownian motion of the starting time a due to Theorem 1. On the other hand, it is easy to prove $\bar{B}_a(t) = \bar{B}_b(t) - \bar{B}_b(a)$, with probability one, for any $b \leq a$. Thus we can define $\dot{\bar{B}}$ independently of selection of the starting time. As the generalized process $\dot{\bar{B}}$ is a white noise with respect to $\mathfrak{B}(\dot{X})$, $\dot{\bar{B}}$ is one of the innovations of $\dot{X}$. But it is known that a Gaussian stationary purely non-deterministic process has a single innovation [9], and the circumstance is same for a stationary generalized process. Therefore $\dot{X}$ has a single innovation $\dot{\bar{B}}$. ∎

COROLLARY. *The generalized process $\dot{X}$ has the canonical representation*

$$\dot{X}(t) = \dot{\bar{B}}(t) - \int_{-\infty}^t k(t-s)d\bar{B}(s), t \in (-\infty, \infty), \tag{14}$$

where $k \in L^2([0,\infty))$. Conversely, $\dot{\bar{B}}$ is represented as

$$\dot{\bar{B}}(t) = \dot{X}(t) - \int_{-\infty}^t \ell(t-s)dX(s), t \in (-\infty, \infty), \tag{15}$$

where $\ell \in L^2([0,\infty))$.

PROOF: The random variable $\hat{\varphi}(t,\omega)$ belongs to $H_t(\dot{X})$, because the pair of generalized process is a Gaussian, so the conditional expectation is a linear function of $\dot{X}$. As $H_t(\bar{B}) = H_t(X), \hat{\varphi}(t,\omega)$ is expressed as $\int_{-\infty}^{t} -k(t,s)d\bar{B}(s)$, $\int_{-\infty}^{t} k(t,s)^2 ds < \infty$. Because of the stationarity of $\hat{\varphi}$, $k(t,s)$ is a fuction of $t-s$. The representation (15) is obtained in the same manner as in (14). ∎

[Concluding remark] In general, it is difficult to state the equivalence of filterings for $\dot{X}$ and $\dot{\bar{B}}$, which are connected by $\dot{\bar{B}} = (I-L)\dot{X}$, where L is an integral operator of Volterra type. Another case where the equivalence of filterings holds has been given in [5] , connecting with the equivalence of induced measures of two processes. Some related topix are in the papers [2],[7] and [8].

Acknowledgements : The author expresses his hearty thanks to Professor Hida for his constant advice on the subject as the pioneer of the representation theory of Gaussian processes.

References

1. N. Aronsjain, *Theory of reproducing kernels*, Trans. Amer. Math. Soc. **68** (1950), 337–404.
2. O. Enchev, "Linear estimation of L^2 -processes," Publication of Centre de Recherchs Mathématiques, Université de Montréal, 1989.
3. T. Hida, *Canonical representation of Gaussian processes and their applications*, Memoirs of College of Sci. Univ. of Kyoto **A. 33, Math.** (1960), 109–155.
4. T. Hida and M. Hitsuda, "Gaussian processes : Representation and Applications," (to appear from American Mathematical Society).
5. M. Hitsuda, *Representation of Gaussian processes equivalent to Wiener process*, Osaka J. Math. **5** (1968), 229–312.

6. M. Hitsuda, *Multiplicity of some class of Gaussian process*, Nagoya Math. J. **52** (1973), 39–46.
7. M. Hitsuda, *Wiener-like integrals for Gaussian processes and the linear estimation problems*, Stochastic Analysis and Applications (Ed. by M. Pinsky), Dekker Inc. (1984), 167–177.
8. M. Hitsuda, *Canonical representation of a Gaussian semimartingale and the innovation*, (to appear).
9. K. Karhunen, *Über die Struktur stationärer zufälliger Functionen*, Arkiv för Mat. 1, nr.3 (1950), 141–160.
10. H. Kunita and S. Watanabe, *On square integrable martingales*, Nagoya Math.J. **30** (1967), 209–245.
11. R. Sh. Liptzer and A. N. Shiryaev, "Statistics of Stochastic processes," 1974 Nauka. (English version 1977 ~ 1978 Springer).
12. R. Sh. Liptzer and A. N. Shiryaev, "Theory of martingales. (in Russian)," 1986 Nauka.

STOCHASTIC CALCULUS OF VARIATION ON GAUSSIAN SPACES AND WHITE NOISE ANALYSIS*

ZHIYUAN HUANG
Department of Mathematics, Wuhan University
Wuchang, Hubei 430072 P. R. China

ABSTRACT

Let (X, H, μ) be the Gaussian space with any locally convex topological vector space X. Most results of Malliavin calculus on abstract Wiener spaces are extended to this more general case. Especially, for the classical Wiener space $(W, \underset{\sim}{H}, \underset{\sim}{\mu})$ and Hida's white noise space $(\mathcal{S}', H, \mu)$, the differential operator $J: W \to \mathcal{S}'$ is a continuous injection and W is Borel isomorphic to its range $J(W)$ under J. In terms of Malliavin calculus, we extend the chaos decomposition to Hida's distributions over white noise spaces. By virtue of this intrinsic connection between Malliavin calculus and Hida's white noise calculus, we simply recapture some known results on Watanabe's distributions over Wiener spaces.

1. Introduction

Since N. Wiener introduced the Wiener space in 1923, the Wiener functional integration has played an important role in mathematical physics. However, a satisfactory differential calculus for these functionals had not been established until 1976 when P. Malliavin[11] launched out his famous work on stochastic calculus of variation. Almost at the same time, in 1975, T. Hida[4,5] initiated a new approach to the study of Brownian functionals by regarding them as functionals of Gaussian white noise. The connection between these two approaches has been investigated by Meyer and Yan[13,14] as well as Potthoff[15].

A mathematically reasonable framework for infinite dimensional differential calculus is the abstract Wiener space (B, H, μ) introduced by L. Gross[3], where H, the set of "directions", is only a dense subspace of B, the set of "points", while in finite dimensional case they are merged into one set. This is essentially due to the fact that the Gaussian measure μ is only quasi-invariant. However, as already pointed out by L. Gross[3], although the Cameron-Martin subspace H is of Wiener measure zero, its structure completely determines the nature of differential calculus for Wiener functionals while the space B itself is

* Work supported by the National Natural Science Foundation of China.

only auxiliary and in some instances even unnecessary. It is the relation between B and H which remains important. This suggested that one could replace the Banach space B by any locally convex topological vector space X in his framework provided that H is densely imbedded in X and that μ has a σ-additive extension to the Borel σ-algebra of X. This more general framework is closely related to the theory of quantum fields[1]. One of the typical cases is the rigged Hilbert space: $X' \hookrightarrow H \hookrightarrow X$, where X is the strong topological dual space of Fréchet nuclear space X' (hence X is reflexive). This framework is exactly what Hida[4,5] has adopted to develop his white noise analysis.

In the present paper, we extend the Malliavin calculus to general Gaussian spaces and investigate the intrinsic connection between Hida's white noise calculus and Malliavin calculus. This work enables us to combine these two kinds of calculus into a framework via a simple isomorphism. In terms of Malliavin calculus, we extend Stroock's representation of homogeneous chaos decomposition[18] for Wiener functionals to Hida's distributions over white noise spaces. By virtue of the nice properties of Hida's testing functionals and the above mentioned isomorphism, we simply recapture some known results on redefinitions of Watanabe's testing functionals obtained by Malliavin[12] and on positive distributions given by Sugita[20] over Wiener spaces.

2. Stochastic calculus of variation on Gaussian spaces

Let H be a real separable Hilbert space which is continuously and densely imbedded in a locally convex complete vector space X. Let X' be the strong topological dual space of X. By identifying H' with H, we have the rigged Hilbert space $X' \hookrightarrow H \hookrightarrow X$. Let μ be a Gaussian measure on X defined by

$$\int_X \exp\{i\langle \xi, x\rangle\} d\mu(x) = \exp\{-\ \|\xi\|_H^2/2\} \qquad \xi \in X'$$

where $\langle \cdot , \cdot \rangle$ stands for the natural pairing between X' and X (one sufficient condition for existence and σ-additivity of μ is that the nuclear topology of X' is stronger than the topology in H). The triplet (X, H, μ) is referred to as Gaussian space. In the special case when X is a Banach space equipped with measurable norm, it is known as the abstract Wiener space.

Example 1 (Classical Wiener space) Let H be the Hilbert space $L^2(R; R^d)$. For every h in H, denote by $\tilde{h}$ its indefinite integral. All the $\tilde{h}$'s constitute a Hilbert space $\tilde{H}$ equipped with the inner product

$$(\tilde{h}, \tilde{g})_{\tilde{H}} = (h, g)_H.$$

Let W be the space of all continuous functions $w \in C(R; R^d)$ with $w(0) = 0$ such that

$$\lim_{|t|\to\infty} (1+t^2)^{-1/2}|w(t)| = 0. \tag{2.1}$$

Obviously, W is a separable Banach space equipped with the norm

$$\| w \|_W = \sup_{t\in R}(1+t^2)^{-1/2}|w(t)|. \tag{2.2}$$

Denote by $\underset{\sim}{\mu}$ the Wiener measure on W, then $(W, H, \underset{\sim}{\mu})$ is a Gaussian space. Under $\underset{\sim}{\mu}$, the coordinate processes $\{w(t), t\geqq 0\}$ and $\{w(t), t\leqq 0\}$ are two independent Brownian motions.

Example 2 (Hida's white noise space) Let H be as in Example 1, $\mathcal{S}$ be the Schwartz space of rapidly decreasing functions in $C^\infty(R;R^d)$ and $\mathcal{S}'$ the space of tempered distributions. Then we have the rigged Hilbert space $\mathcal{S} \subsetneq H \subsetneq \mathcal{S}'$. Let μ be the standard Gaussian white noise measure on $\mathcal{S}'$ introduced by Hida[4]. Then, $(\mathcal{S}', H, \mu)$ is a Gaussian space. Under μ, the elements of $\mathcal{S}'$ can be thought of as "sample paths" of Gaussian white noise.

Most results of Malliavin calculus on abstract Wiener spaces can be extended to general Gaussian spaces. For simplicity of presentation, here we only give an outline of extensions. For example, to define operators on Gaussian functionals, firstly we consider polynomials:

$$\varphi(x) = p(\langle \xi_1, x\rangle, \cdots, \langle \xi_n, x\rangle) \tag{2.3}$$

where $\xi_1, \cdots, \xi_n \in X'$ and $p: R^n \to R$ is a real polynomial. Note that in the Eq. (2.3), we can always assume that $\{\xi_k\}$ forms an orthonormal system in H. Hence we can fix an arbitrarily chosen orthonormal basis (abbr. ONB) in all the expressions for polynomials over certain Gaussian space. Since $\mathcal{P}$, the set of polynomials, is dense in functional space $(L^p)(\equiv L^p(X,\mu))$ for $p\in[1,\infty)$, it follows that the functional space (L^p) over any two (infinite-dimensional) Gaussian spaces are isomorphic to each other. This is quite evident for $p=2$, because (L^2) is isomorphic to the symmetric Fock space $\Gamma(H)$ by a standard result and any two infinite dimensional separable Hilbert spaces are isomorphic to each other.

Let E be a separable Hilbert space. Denote by $\mathcal{P}(E)$ the totality of E-valued polynomials. As in Malliavin calculus over an abstract Wiener space [19,22], we define operators ∇, δ and N, L as follows:

For $\psi\in\mathcal{P}(E)$ and $h\in H$, we define $\nabla_h\psi\in\mathcal{P}(E)$ by

$$(\nabla_h\psi(x), e)_E = \frac{\partial}{\partial\varepsilon}(\psi(x+\varepsilon h), e)_E\Big|_{\varepsilon=0} \qquad \forall\, e\in E \tag{2.4}$$

and $\nabla \psi \in \mathscr{P}(H \otimes E)$ by

$$(\nabla \psi(x), h \otimes e)_{H \otimes E} = (\nabla_h \psi(x), e)_E \quad \forall\, e \in E,\ \forall\, h \in H. \quad (2.5)$$

Let $\delta_h = \nabla_h^*$, $\delta = \nabla^*$ be the dual operators of ∇_h and ∇ respectively, $N = \delta \nabla$ be the number operator and $L = -\frac{1}{2} N$ the Ornstein-Uhlenbeck operator. Using Cameron-Martin formula, for $\psi \in \mathscr{P}(E)$ and $h \in H$, we have

$$\delta_h \psi(x) = \langle h, x \rangle \psi(x) - \nabla_h \psi(x) \quad \mu - \text{a.e. } x \quad (2.6)$$

where $\langle h, x \rangle$ $(h \in H)$ is the unique linear and continuous extension of $\langle \xi, x \rangle$ $(\xi \in X')$ in (L^2). Moreover, the following canonical commutation relations (abbr. CCR) hold:

$$\begin{aligned}{}[\nabla_h, \nabla_g] &= 0 = [\delta_h, \delta_g] \\ [\nabla_h, \delta_g] &= (h, g)_H, \quad \forall\, h, g \in H \end{aligned} \quad (2.7)$$

Finally, for $\psi \in \mathscr{P}(H)$, we have

$$\delta \psi(x) = \langle \psi(x), x \rangle - \operatorname{tr} \nabla \psi(x) \quad \mu - \text{a.e. } x \quad (2.8)$$

where $\operatorname{tr} \nabla \psi$ is the trace of $\nabla \psi$ (as an element in $\mathscr{P}(H \otimes H)$). Note that the operators δ_h and ∇_h, δ and ∇ being densely defined, N and L being essentially selfadjoint, all of them are closable. since for every $n \in N$, $\varphi \in \mathscr{P}$ implies that $\nabla^n \varphi \in \mathscr{P}(H^{\hat{\otimes} n})$ where $\hat{\otimes}$ stands for symmetric tensor product, it follows that the dual operator $\delta^n = (\nabla^n)^*$ is actually defined on $\mathscr{P}(H^{\hat{\otimes} n})$.

Now let $\{\xi_k\} \subset X'$ be an ONB of H. For $\alpha = (\alpha_1, \cdots, \alpha_n) \in I_0^n$, the set of naturally ordered elements in N_0^n, denote by $n_j(\alpha)$ the number of entries of α equal to j $(j \in N_0)$. Put $n(\alpha)! = \prod_{j \in N_0} n_j(\alpha)!$ and

$$\hat{\xi}_\alpha = (n(\alpha)!)^{-1/2} \hat{\bigotimes}_{i=1}^n \xi_{\alpha_i}. \quad (2.9)$$

Then, $\{\hat{\xi}_\alpha, \alpha \in I_0^n\}$ forms an ONB in $H^{\hat{\otimes} n}$. Using Eq. (2.6) and (2.7), we can prove that

$$\delta^n \hat{\xi}_\alpha = (n(\alpha)!)^{-1/2} \prod_j H_{n_j(\alpha)}(\langle \xi_j, x \rangle) \quad (2.10)$$

where $\{H_n\}$ are Hermite polynomials and for $\beta = (\beta_1, \cdots, \beta_m) \in I_0^m$, we have

$$(\delta^n \hat{\xi}_\alpha, \delta^m \hat{\xi}_\beta)_{(L^2)} = \delta_{\alpha\beta}. \tag{2.11}$$

Since the linear span of polynomials $\{\delta^n \hat{\xi}_\alpha, \alpha \in I_0^0, n \in N_0\}$ is dense in (L^2), it follows that every φ in (L^2) has a unique decomposition:

$$\varphi = \sum_n (n!)^{-1} \delta^n \varphi_n \tag{2.12}$$

where $\varphi_n \equiv \nabla^n \varphi(1)$ is the unique element in $H^{\hat{\otimes} n}$ such that for $\forall\ \alpha \in I_0^0$

$$(\varphi_n, \hat{\xi}_\alpha)_{H^{\otimes n}} = (\varphi, \delta^n \hat{\xi}_\alpha)_{(L^2)} \tag{2.13}$$

This representation of Wiener-Itô chaos decomposition was obtained by Stroock[18]. It is remarkable that, unlike the one in terms of multiple Wiener-Itô integrals, this expression enables us to extend the homogeneous chaos to generalized functionals in a mathematically rigorous way (see Theorem 4. 3 below).

3. The intrinsic connection between Malliavin calculus and Hida's white noise calculus

Now we consider the classical Wiener space (W, H, μ) and the white noise space $(\mathscr{S}', \tilde{H}, \tilde{\mu})$ (Example 1 and 2 above). If we define a map

$$J: \quad W \to \mathscr{S}'$$

as differentiation in sense of distribution, that is, for $w \in W$ and $\xi \in \mathscr{S}$

$$\langle \xi, Jw \rangle = - \langle \dot{\xi}, w \rangle \tag{3.1}$$

where $\cdot$ stands for the usual derivative with respect to t, then we have the following basic result:

Theorem 3. 1 The map $J: W \to \mathscr{S}'$ defined by Eq. (3. 1) is a continuous injection, its image $J(W)$ is a Borel set in $\mathscr{S}'$ and W is Broel isomorphic to $J(W)$ under J. Moreover, $J(H) = \tilde{H}$, the restriction of J to H is an isomorphism between Hilbert spaces H and $\tilde{H}$.

Proof Note that the Eq. (2. 1) implies that every element w in W is a slowly increasing continuous function. Therefore, w together with all its derivatives are temperate distributions (e. g. cf. Schwartz[17], Theorem VII. 4. VI, p. 239). It is easy to see from the condition $w(0) = 0$ that J is an injection. To prove its continuity, we assume that $d = 1$ for notational simplicity. Since the Hermite functions

$$\xi_n(t) = (2^n n!\sqrt{\pi})^{-1/2}(-1)^n e^{t^2/2}\frac{d^n}{dt^n}e^{-t^2}, \quad n \in N_0 \tag{3.2}$$

constitute an ONB in $H=L^2(R)$ and

$$\Lambda\xi_n = 2(n+1)\xi_n, \qquad n \in N_0 \tag{3.3}$$

where $\Lambda=1+t^2-\frac{d^2}{dt^2}$ is the harmonic oscillator. Then for $\xi\in\mathscr{S}\subset H, \xi=\sum_n a_n\xi_n$, there exists $k\in N$ big enough such that

$$\|\xi\|_{2,-k}^2 \equiv \|\Lambda^{-k/2}\xi\|_H^2 = \sum_n (2(n+1))^{-k}a_n^2$$
$$\leq \int_{-\infty}^{\infty} |\xi(t)|^2(1+t^2)^{-2}dt \leq \sup_{t\in R}|\xi(t)|^2(1+t^2)^{-1}\int_{-\infty}^{\infty}(1+t^2)^{-1}dt$$
$$= \pi\|\xi\|_W^2 \tag{3.4}$$

which implies that J is a continuous map from W to $\mathscr{S}'$. Hence, it is also $\mathscr{B}(W)/\mathscr{B}(\mathscr{S}')$ measurable. Since both W and $\mathscr{S}'$ are standard measurable spaces, it follows from a well known theorem (e. g. cf. Itô[6], Theorem 2. 1. 1, P. 14) that $J(W)\in\mathscr{B}(\mathscr{S}')$ and $J(W)$ is Borel isomorphic to W under J. The remaining part of the theorem is obvious. Q. E. D.

Denote by $\langle \cdot , \cdot \rangle_\sim$ the natural pairing between W' and W, then, integration by part implies that

$$\langle\xi, Jw\rangle = -\langle\dot\xi, w\rangle = -\int_{-\infty}^{\infty}(w(s), \dot\xi(s))ds$$
$$= \int_{-\infty}^{\infty}(\xi(s), dw(s)) = \langle\underset{\sim}{\xi}, w\rangle_\sim \qquad w\in W, \qquad \xi\in\mathscr{S}. \tag{3.5}$$

Hence

$$\int_{\mathscr{S}'}\exp\{i\langle\xi,x\rangle\}d\underset{\sim}{\mu}\circ J^{-1}(x) = \int_{J^{-1}(\mathscr{S}')}\exp\{i\langle\xi,Jw\rangle\}d\underset{\sim}{\mu}(w)$$
$$= \int_W \exp\{i\langle\underset{\sim}{\xi},w\rangle_\sim\}d\underset{\sim}{\mu}(w) = \exp\{-\|\underset{\sim}{\xi}\|_{\underset{\sim}{H}}^2/2\}$$
$$= \exp\{-\|\xi\|_H^2/2\} \tag{3.6}$$

which implies that $\mu=\underset{\sim}{\mu}\circ J^{-1}$. Thus we have

Corollary 3. 2 $\mu=\underset{\sim}{\mu}\circ J^{-1}$ and $\mu(J(W))=1$.

Now for $p\in[1,\infty)$, define a map

$$J_\star : L^p(\mathscr{S}', \mu) \to L^p(W, \underset{\sim}{\mu})$$

by setting $(J_\star\varphi)(w) = \varphi(Jw)$. Then we have

Corollary 3. 3 For every $p \in [1, \infty)$, the Banach space $L^p(\mathscr{S}', \mu)$ is isomorphic to $L^p(W, \underset{\sim}{\mu})$ under $J_\star$.

Through this isomorphism, we can transform a white noise functional into a Wiener functional, and vice versa. By the Eq. (3. 5) we see that if φ is a polynomial of the form (2. 3), then

$$(J_\star\varphi)(w) = p(\langle \underset{\sim}{\xi_1}, w\rangle_\sim, \cdots, \langle \underset{\sim}{\xi_n}, w\rangle_\sim)$$

is a polynomial on Wiener space since $\underset{\sim}{\xi_1}, \cdots, \underset{\sim}{\xi_n} \in W'$. If we work on the Wiener space $(W, \underset{\sim}{H}, \underset{\sim}{\mu})$ and define operators $\underset{\sim}{\underset{\sim}{\nabla_h}}, \underset{\sim\sim}{\delta_h}$ etc. similarly (one exception is that for $\psi \in \mathscr{D}(E)$, instead of defining $\underset{\sim}{\nabla}\psi \in \mathscr{D}(\underset{\sim}{H} \otimes E)$, we define $\underset{\sim}{\nabla}\psi \in \mathscr{D}(H \otimes E)$ so that

$$(\underset{\sim}{\nabla}\psi(w), h \otimes e)_{H\otimes E} = (\underset{\sim\sim}{\nabla_h}\psi(w), e)_E$$

holds for every $h \in H$ and $e \in E$). Then, it is easy to prove the following

Corollary 3. 4 For $h \in H$, it holds that

$$\begin{aligned}
\underset{\sim\sim}{\nabla_h} &= J_\star \circ \nabla_h \circ J_\star^{-1} \\
\underset{\sim\sim}{\delta_h} &= J_\star \circ \delta_h \circ J_\star^{-1} \\
\underset{\sim}{\nabla} &= J_\star \circ \nabla \circ J_\star^{-1} \\
\underset{\sim}{\delta} &= J_\star \circ \delta \circ J_\star^{-1} \\
\underset{\sim}{L} &= J_\star \circ L \circ J_\star^{-1}.
\end{aligned} \tag{3. 7}$$

In next section, we shall mainly deal with white noise functionals and denote the E-valued functional space $L^p(\mathscr{S}', \mu; E)$ by $(L^p)(E)$ or simply by (L^p) in case $E = R$.

4. Representations for Hida's distributions over white noise spaces

Now we work on the white noise space $(\mathscr{S}', H, \mu)$. For the sake of notational convenience, we always assume that $d = 1$. According to S. Watanabe[21,22] (also cf. Sugita[19]), the Sobolev spaces over $(\mathscr{S}', H, \mu)$ are defined as follows:

For $p \in (1, \infty)$, $k \in N_0$ and $\varphi \in \mathscr{D}$, define

$$\| \varphi \|_{p,k} \equiv \| (I + N)^{k/2}\varphi \|_{(L^p)} \tag{4. 1}$$

where N is the number operator. Denote by D_p^k the Banach space obtained by completing $\mathscr{D}$ with respect to the norm (4. 1) and D_q^{-k} its dual space with $p^{-1} + q^{-1} = 1$. Then we have a conutably normed space $D^{\infty}=\bigcap_{k,p} D_p^k$ and its dual $D^{-\infty}=\bigcup_{k,p} D_p^{-k}$. The elements in D^{∞} and $D^{-\infty}$ are referred to as Watanabe's testing and generalized functionals respectively.

Hida, Kubo, Takenaka and Yokoi[4,9,10] (as well as Kondrat'ev and Samoylenko[7,8]) proposed another framework for testing and generalized functionals which was based on the second quantization of harmonic oscillator Λ. More specifically, for $k\in N_0$ and φ, $\psi \in \mathscr{P}$, define the inner product

$$(\psi, \varphi)_{(k)} \equiv (\psi, \Gamma(\Lambda)^k\varphi)_{(L^2)} \tag{4.2}$$

where

$$\Gamma(\Lambda)^k\varphi = \sum_n (n!)^{-1}\delta^n((\Lambda^{\otimes n})^k\varphi_n) \tag{4.3}$$

if φ has decomposition (2.12). Denote by $(\mathscr{S}_k)$ the Hilbert space obtained by completing $\mathscr{P}$ with respect to inner product (4.2) and $(\mathscr{S}_{-k})$ its dual. Thus we have a countably Hilbertian space $(\mathscr{S})=\bigcap_k(\mathscr{S}_k)$ and its dual $(\mathscr{S})'=\bigcup_k(\mathscr{S}_{-k})$. The elements in $(\mathscr{S})'$ are referred to as Hida's generalized functionals (or distributions).

Since the spectrum of Λ is $\sigma(\Lambda)=\{2(n+1), n\in N_0\}$, hence inf $\sigma(\Lambda)\geqq 2$, it is easy to prove that

$$(\mathscr{S}) \subsetneq D^{\infty} \subsetneq (L^2) \subsetneq D^{-\infty} \subsetneq (\mathscr{S})' \tag{4.4}$$

(e. g. cf. Potthoff & Yan[16]). Moreover, this important dual pair of spaces possesses some very nice properties:

1° $(\mathscr{S})$ is a Fréchet nuclear algebra (cf. Kubo & Takenaka[9]);

2° every φ in $(\mathscr{S})$ has a unique point wise defined, strongly continuous representative $\tilde{\varphi}$, hence the evaluation map $\varphi \to \varphi(x)$ $(x\in\mathscr{S}')$ belongs to $(\mathscr{S})'$ (cf. Kubo & Yokoi[10]);

3° every positive distribution in $(\mathscr{S})'$ is a finite Borel measure on $\mathscr{S}'$ (cf. Kondrat'ev[8] and Yokoi[23]).

We will establish representations for Hida's distributions which can be considered as a generalization of the result obtained by Stroock[18]. To this end we firstly extend domains of ∇, δ etc. to the space $(\mathscr{S})'$.

Theorem 4.1 For $m\in N_0$, the linear operator

$$\nabla^m: \mathscr{P} \to \mathscr{P}(H^{\hat{\otimes} m})$$

extends to $(\mathscr{S})' \to (\mathscr{S})'(H^{\hat{\otimes} m})$ and is continuous from $(\mathscr{S}_{k+1})$ to $(\mathscr{S}_k)(H^{\hat{\otimes} m})$ for every $k \in Z$.

Proof Suppose that $\varphi = \sum_n (n!)^{-1}\delta^n \varphi_n \in \mathscr{P}$. For $\alpha = (\alpha_1, \cdots, \alpha_n) \in I_0^n$, let $\hat{\xi}_\alpha$ be as in (2.9) and

$$\varphi_n = \sum_{\alpha \in I_0^n} a_\alpha \hat{\xi}_\alpha. \tag{4.5}$$

For $m \leq n$ put

$$\varphi_n^{(-m)}(t_1, \cdots, t_{n-m}) = \sum_{\alpha \in I_0^n} a_\alpha \hat{\xi}_\alpha(t_1, \cdots, t_{n-m}, \cdot)$$

where $\hat{\xi}_\alpha(t_1, \cdots, t_{n-m}, \cdot)$ is regarded as an element in $L^2(R^{n-m}; H^{\hat{\otimes} m})$. Then

$$\nabla^m \varphi = \sum_{n \geq m} ((n-m)!)^{-1} \delta^{n-m} \varphi_n^{(-m)}.$$

In view of Eq. (3.3), we have

$$\begin{aligned}\| \nabla^m \varphi \|_{(k)}^2 &= \sum_{n \geq m} ((n-m)!)^{-1} \| (\Lambda^{\otimes (n-m)})^{k/2} \varphi_n^{(-m)} \|_{L^2(R^{n-m}, H^{\hat{\otimes} m})}^2 \\ &\leq \sum_{n \geq m} ((n-m)!)^{-1} \sum_{\alpha \in I_0^n} a_\alpha^2 (n(\alpha)!)^{-1} (2^{n-m}(\alpha+1))^k m!\end{aligned}$$

where $\alpha + 1 = (\alpha_1 + 1) \cdots (\alpha_n + 1)$. On the other hand,

$$\begin{aligned}\| \varphi \|_{(k+1)}^2 &= \sum_n (n!)^{-1} \| (\Lambda^{\otimes n})^{(k+1)/2} \varphi_n \|_{L^2(R^n)}^2 \\ &= \sum_n (n!)^{-1} \sum_{\alpha \in I_0^n} a_\alpha^2 (n(\alpha)!)^{-1} (2^n(\alpha+1))^{k+1}.\end{aligned}$$

Choosing a constant $C = C(k, m)$ independent of φ such that $m!\, 2^{-mk} \leq C 2^n / n(n-1) \cdots (n-m+1)$ for all $n \geq m$, it is easy to see that

$$\| \nabla^m \varphi \|_{(k)}^2 \leq C \| \varphi \|_{(k+1)}^2 \tag{4.6}$$

which implies that $\nabla^m : (\mathscr{S}_{k+1}) \to (\mathscr{S}_k)(H^{\hat{\otimes} m})$ is continuous and the proof is complete.

Q. E. D.

By duality, we have

Corollary 4. 2 For $m \in N_0$, the linear operator δ^m has an extension to $(\mathscr{S})'(H^{\hat{\otimes} m}) \to (\mathscr{S})'$ and is continuous from $(\mathscr{S}_{k+1})(H^{\hat{\otimes} m})$ to $(\mathscr{S}_k)$ for every $k \in Z$.

Furthermore, we can extend the operator δ^n in following way. Since δ^n is an isometry from $H^{\hat{\otimes} n}$ into (L^2), $\Lambda^{k/2}$ is an isometry from H onto $\mathscr{S}_{-k}$, consequently, $(\Lambda^{k/2})^{\otimes n}$ is an isometry from $H^{\hat{\otimes} n}$ onto $\mathscr{S}_{-k}^{\hat{\otimes} n}$ and $\Gamma(\Lambda^{k/2})$ is an isometry from (L^2) onto $(\mathscr{S}_{-k})$. Look at the following diagram;

$$\begin{array}{ccc} H^{\hat{\otimes} n} & \xrightarrow{\delta^n} & (L^2) \\ (\Lambda^{k/2})^{\otimes n} \downarrow & & \downarrow \Gamma(\Lambda^{k/2}) \\ \mathscr{S}_{-k}^{\hat{\otimes} n} & \xrightarrow{\delta^n_{(k)}} & (\mathscr{S}_{-k}) \end{array} \tag{4. 7}$$

where $\delta^n_{(k)}$ is defined over the Gaussian space $(\mathscr{S}', \mathscr{S}_{-k}, \mu_{-k})$ and μ_{-k} is the Gaussian measure on $\mathscr{S}'$ defined by

$$\int_{\mathscr{S}'} \exp\{i\langle \xi, x\rangle\} d\mu_{-k}(x) = \exp\{-\ \|\xi\|^2_{,-k}/2\}$$
$$= \exp\{-\ (\xi, \Lambda^{-k}\xi)_H/2\}.$$

It is easy to see that $(\mathscr{S}_{-k}) = L^2(\mathscr{S}', \mu_{-k})$ and the diagram (4. 7) is commutative. Hence

$$\delta^n_{(k)} = \Gamma(\Lambda^{k/2}) \circ \delta^n \circ (\Lambda^{-k/2})^{\otimes n}, \quad k \in Z \tag{4. 8}$$

are consistent and we will denote the extension by the same symbol δ^n. Thus we have

Theorem 4. 3 Any $F \in (\mathscr{S})'$ has a unique decomposition;

$$F = \sum_n (n!)^{-1} \delta^n F_n. \tag{4. 9}$$

If $F \in (\mathscr{S}_{-k})$ for some $k \in N$, then for every $n, F_n = \nabla^n F(1)$ is the unique element in $\mathscr{S}_{-k}^{\hat{\otimes} n}$ such that

$$\langle F_n, \varphi_n \rangle = \langle F, \delta^n \varphi_n \rangle \quad \forall\ \varphi_n \in \mathscr{S}_k^{\hat{\otimes} n} \tag{4. 10}$$

where $\langle \cdot , \cdot \rangle$ on the left (resp. right) hand side is the natural pairing between $\mathscr{S}_{-k}^{\hat{\otimes} n}$ and $\mathscr{S}_k^{\hat{\otimes} n}$ (resp. $(\mathscr{S}_{-k})$ and $(\mathscr{S}_k)$).

Proof Eq. (4. 9) follows from above argument and the chaos decomposition for Gaussian functionals established in Eq. (2. 12). It remains to derive Eq. (4. 10) from

Eq. (2.13).

For $\alpha \in I_0^n$ and $k \in Z$, put

$$\hat{\xi}_\alpha^{(k)} = (2^n(\alpha+1))^{-k/2}\hat{\xi}_\alpha.$$

Then $\{\hat{\xi}_\alpha^{(k)},\ \alpha \in I_0^n\}$ constitute an ONB in $\mathscr{S}_k^{\hat{\otimes} n}$. Note that $\Gamma(\Lambda^k)$: $(\mathscr{S}_k) \to (\mathscr{S}_{-k})$ is an isometry such that

$$\langle F, G\rangle = (F, \Gamma(\Lambda^k)G)_{(-k)} \tag{4.11}$$

for $F \in (\mathscr{S}_{-k}), G \in (\mathscr{S}_k)$; and $(\Lambda^{-k})^{\otimes n}: \mathscr{S}_{-k}^{\hat{\otimes} n} \to \mathscr{S}_k^{\hat{\otimes} n}$ is an isometry such that

$$\langle F_n, (\Lambda^{-k})^{\otimes n} G_n\rangle = (F_n, G_n)_{\mathscr{S}_{-k}^{\hat{\otimes} n}} \tag{4.12}$$

for $F_n, G_n \in \mathscr{S}_{-k}^{\hat{\otimes} n}$. Let

$$\varphi_n = \sum_{\alpha \in I_0^n} a_\alpha \hat{\xi}_\alpha^{(k)} \in \mathscr{S}_k^{\hat{\otimes} n}.$$

Then we have

$$\begin{aligned}
\langle F, \delta^n\varphi_n\rangle &= (F, \Gamma(\Lambda^k)\delta^n\varphi_n)_{(-k)} && \text{(by 4.11)}\\
&= \sum_{\alpha \in I_0^n} a_\alpha (F, \delta^n \hat{\xi}_\alpha^{(-k)})_{(-k)} \\
&= \sum_{\alpha \in I_0^n} a_\alpha (F_n, \hat{\xi}_\alpha^{(-k)})_{\mathscr{S}_{-k}^{\hat{\otimes} n}} && \text{(by 2.13)}\\
&= \sum_{\alpha \in I_0^n} a_\alpha \langle F_n, \hat{\xi}_\alpha^{(k)}\rangle && \text{(by 4.12)}\\
&= \langle F_n, \varphi_n\rangle. && \text{Q.E.D.}
\end{aligned}$$

Corollary 4.4 If $F = \sum_n (n!)^{-1}\delta^n F_n \in (\mathscr{S})'$ and $\varphi = \sum_n (n!)^{-1}\delta^n\varphi_n \in (\mathscr{S})$, then

$$\langle F, \varphi\rangle = \sum_n (n!)^{-1}\langle F_n, \varphi_n\rangle. \tag{4.13}$$

5. Capacities, redefinitions of Watanabe's testing functionals and results on positive generalized functionals

Following Malliavin[12], for $p \in (1, \infty)$ and $k \in N$, we define the (p,k)-capacity of an open set O in $\mathscr{S}'$ as

$$C_p^k(O) = \inf\{ \|\varphi\|_{p,k};\quad \varphi \in D_p^k, \varphi \geqq 1 \quad \mu\text{-a. e. on } O\} \tag{5.1}$$

and that of an arbitrary subset A of $\mathscr{S}'$ as

$$C_p^k(A) = \inf\{C_p^k(O);\quad O \text{ is open and } O \supset A\}. \tag{5.2}$$

Denote by $\tilde{C}_p^k$ the corresponding capacities on Wiener space $(W, \underset{\sim}{H}, \underset{\sim}{\mu})$, it is easy to see that

$$\tilde{C}_p^k(J^{-1}A) \leqq C_p^k(A). \tag{5.3}$$

Hence, if A is a slim set (i. e. $C_p^k(A)=0$ for all p and k) in $\mathscr{S}'$, then $J^{-1}A$ is slim in W.

A functional φ on $\mathscr{S}'$ is said to be (p,k)-quasicontinuous (abbr. (p,k)-qc) if for any $\varepsilon>0$ there exists an open set O_ε with $C_p^k(O_\varepsilon)<\varepsilon$ such that φ is continuous on $\mathscr{S}'\backslash O_\varepsilon$. If φ is (p,k)-qc for all p and k, then it is said to be ∞-quasicontinuous (abbr. ∞-qc). Moreover, the term "(p,k)-q. e." (resp. "∞-q. e.") means "except on a set of (p,k)-capacity zero" (resp. "except on a slim set").

Lemma 5.1 If φ is (p,k)-qc (resp. ∞-qc) on $\mathscr{S}'$, then $J_*\varphi$ is (p,k)-qc (resp. ∞-qc) on W.

Proof Suppose that φ is (p,k)-qc on $\mathscr{S}'$, then by definition for any $\varepsilon>0$ there exists an open set O_ε with $C_p^k(O_\varepsilon)<\varepsilon$ such that φ is continuous on $\mathscr{S}'\backslash O_\varepsilon$. Since $(J_*\varphi)(w)=\varphi(Jw)$ and $J: W\to\mathscr{S}'$ is continuous, it follows that $J_*\varphi$ is continuous on $W\backslash J^{-1}O_\varepsilon$. Note that $J^{-1}O_\varepsilon$ is an open set in W and $\tilde{C}_p^k(J^{-1}O_\varepsilon)=C_p^k(O_\varepsilon)<\varepsilon$, this implies that $J_*\varphi$ is (p,k)-qc on W.

Lemma 5.2 If $\{\varphi_n\}$ converges in D_p^k and φ_n is continuous for every n, then we can take a suitable subsequence $\{\varphi_{n_j}\}$ of $\{\varphi_n\}$ so that for any $\varepsilon>0$ there exists an open set O_ε with $C_p^k(O_\varepsilon)<\varepsilon$ such that $\{\varphi_{n_j}\}$ converges uniformly on $\mathscr{S}'\backslash O_\varepsilon$.

Proof By Chebyshev's inequality[12] we can choose $\{\varphi_{n_j}\}$ so that

$$C_p^k(O_j) \leqq 2^{-j} \qquad \forall\, j \in N_0$$

where

$$O_j = \{x;\, |\varphi_{n_{j+1}}(x) - \varphi_{n_j}(x)| > 2^{-j}\} \tag{5.4}$$

is an open set in $\mathscr{S}'$ for every j.

Given $\varepsilon>0$, choose j_0 such that $2^{-j_0+1}<\varepsilon$ and set $O_\varepsilon=\bigcup_{j=j_0}^{\infty}O_j$, then

$$C_p^k(O_\varepsilon) \leqq \sum_{j=j_0}^{\infty} 2^{-j} < \varepsilon$$

and $\{\varphi_{n_j}\}$ converges uniformly on $\mathscr{S}'\backslash O_\varepsilon$ as desired. Q. E. D.

Remark Taking $A = \bigcap_n O_{1/n}$ in the lemma, we have $C_p^k(A) = 0$ and $\{\varphi_{n_j}\}$ converges on $\mathscr{S}'\backslash A$ which implies the (p,k)-q. e. convergence of this subsequence. Moreover, as we see in the proof, it suffices to assume that $\{\varphi_n\}$ converges in (p,k)-capacity, i. e. for any $\varepsilon>0$

$$C_p^k(|\varphi_n - \varphi_m| > \varepsilon) \to 0 \qquad \text{as } n,m \to \infty. \tag{5. 5}$$

Now let $\varphi \in D_p^k$. Since $(\mathscr{S}) \subsetneq D_p^k$, we can choose a sequence $\{\varphi_n\}$ in $(\mathscr{S})$ which converges to φ in D_p^k. Let $\{\tilde{\varphi}_n\}$ be the continuous versions of $\{\varphi_n\}$. By Chebyshev's inequality

$$C_p^k(|\tilde{\varphi}_n - \tilde{\varphi}_m| > \varepsilon) \leqq \varepsilon^{-1} \| \varphi_n - \varphi_m \|_{p,k} \to 0$$

as $n,m\to\infty$. Therefore, there exists a subsequence $\{\tilde{\varphi}_{n_j}\}$ which converges to some $\tilde{\varphi}$ (p,k)-q. e. and uniformly outside an open set of arbitrary small (p,k)-capacity. It is easy to see that $\tilde{\varphi}$ is a (p,k)-qc version of φ and uniquely defined (p,k)-q. e.. In view of Lemma 5. 1, coming to the Wiener space via the above mentioned isomorphism, we simply recapture Malliavin's theorem of redefinition of Watanabe's testing functionals[12,20].

Theorem 5. 3 (Malliavin) For $p \in (1,\infty)$ and $k \in N$, each $\varphi \in D_p^k$ has a version $\tilde{\varphi}$ which is (p,k)-qc and uniquely defined (p,k)-q. e.; Each $\varphi \in D^\infty$ has a version $\tilde{\varphi}$ which is ∞-qc and uniquely defined ∞-q. e..

Finally, we give a short remark on positive distributions. An element F in $(\mathscr{S})'$ is said to be positive if $\langle F, \varphi\rangle \geqq 0$ for every $\varphi \in (\mathscr{S})$ with continuous version $\tilde{\varphi} \geqq 0$. Yokoi[23] has proved that every positive element F in $(\mathscr{S})'$ corresponds to a finite Borel measure ν_F on $\mathscr{S}'$ so that for all $\varphi \in (\mathscr{S})$

$$\langle F, \varphi\rangle = \int_{\mathscr{S}'} \tilde{\varphi}(x) d\nu_F(x). \tag{5. 6}$$

Results for positive distributions in $D^{-\infty}$ over Wiener space have been obtained by Sugita[20]. Since the space D^∞ is not nuclear, the proof is much more complicate than Yokoi's proof. But in view of the above mentioned isomorphism, we can define J^*F in $(\mathscr{S})'$ for positive F in $D^{-\infty}$ over Wiener space such that

$$\langle J^*F,\quad \varphi\rangle = \langle F,\quad J_*\varphi\rangle \tag{5.7}$$

for every φ in $(\mathscr{S})$. According to (5.6), we have

$$\langle J^*F,\quad \varphi\rangle = \int_{\mathscr{S}'} \tilde{\varphi}(x)d\nu_{J^*F}(x) \tag{5.8}$$

Note that if $\tilde{\varphi}$ vanishes on $J(W)$, then

$$\langle J^*F,\quad \varphi\rangle = \langle F,\quad J_*\varphi\rangle = 0.$$

Consequently ν_{J^*F} is supported by $J(W)$ and $\tilde{\nu}_F = \nu_{J^*F} \circ J$ makes sense. Using an argument similar to that in the proof of Theorem 5.3, we will simply recapture Sugita's results;

Theorem 5.4 (Sugita) Let $F \in D^{-\infty}$ be a positive generalized Wiener functional. Then, there exists a unique finite Borel measure $\tilde{\nu}_F$ on W such that for all $\varphi \in D^{\infty}$,

$$\langle F,\ \varphi\rangle = \int_W \tilde{\varphi}(w)d\tilde{\nu}_F(w) \tag{5.9}$$

where $\tilde{\varphi}$ is the ∞-qc version of φ.

Acknowledgement. The author is very grateful to Professor T. Hida for his very kind invitation and wonderful hospitality. Also he gratefully acknowledges the helpful comments from Professor P. Krée and the information concerning references[7, 8] given by Professor Yu. M. Berezansky.

References

1. S. Albeverio & R. Høegh-Krohn, Dirichlet forms and diffusion processes on rigged Hilbert spaces, *Z. Wahrsch. verw. Gebiete*, 40, 1-57 (1977)

2. I. M. Gel'fand & N. Ya. Vilenkin, Generalized Functions, Vol. 4, Academic, NY (1968)

3. L. Gross, Abstract Wiener spaces, *Proc. 5th Berkeley Symp. on Math. Stat. and Prob.* II, Part I, 31-42, Univ. of Calif. (1967)

4. T. Hida, Analysis of Brownian functionals, *Carleton Math. Lect. Notes*, No. 13. Carleton Univ., Ottawa (1975)

5. T. Hida, Brownian Motion, Springer, NY (1980)

6. K. Itô, Foundations of Stochastic Differential Equations in Infinite Dimensional

Spaes, *CBMS-NSF Regional Conf. Series in Applied Math.*, Vol. 47, SIAM, Philadelphia (1984)

7. Yu. G. Kondrat'ev & Yu. S. Samoylenko, The space of trial and generalized functions of infinite number of variables, *Reports on Math. Phys.*, Vol. 14, No. 3, 325-350 (1978)

8. Yu. G. Kondrat'ev, Nuclear spaces of entire functions in problems of infinite-dimensional analysis, *Soviet Math. Dokl.*, Vol. 22, No. 2, 588-592(1980)

9. I. Kubo & S. Takenaka, Calculus on Gaussian white noise, I - VI, *Proc. Japan Acad.*, 56, 376-380 (1980); 56, 411-416 (1980); 57, 433-437 (1981); 58, 186-189 (1982)

10. I. Kubo & Y. Yokoi, A remark on the space of testing random variables in the white noise calculus, *Nagoya Math. J.*, 115, 139-149(1989)

11. P. Malliavin, Stochastic calculus of variation and hypoelliptic operators, *Proc. Intern. Symp. SDE Kyoto* 1976 (ed. by K. Itô), 195-263, Kinokuniya (1978)

12. P. Malliavin, Implicit functions in finite corank on the Wiener space, *Proc. Taniguchi Intern. Symp. on Stochastic Analysis*, Katata and Kyoto 1982, 353-370, Kinokuniya (1984)

13. P. A. Meyer & J. A. Yan, A propos des distributions sur l'espace de Wiener, *Lecture Notes in Math.*, Vol. 1247, 8-26, Springer(1987)

14. P. A. Meyer & J. A. Yan, Distributions sur l'espace de Wiener (suite), *lecture Notes in Math.*, Vol. 1372, Springer(1989)

15. J. Potthoff, White noise approach to Malliavin calculus, *J. Funct. Anal.*, 71, 207-217(1987)

16. J. Potthoff & J. A. Yan, Some results about test and generalized functionals of white noise, to appear in : *Proc. Singapore Probab. Conf.* (1989), L. Y. Chen ed.

17. L. Schwartz, Théorie des Distributions, Hermann, Paris(2ème éd.)(1966)

18. D. W. Stroock, Homogeneous chaos revisited, *Lecture Notes in Math.*, Vol. 1247, 1-7, Springer(1987)

19. H. Sugita, Sobolev spaces of Wiener functionals and Malliavin's calculus, *J. Math. Kyoto Univ.*, 25, 31-48(1985)

20. H. Sugita, Positive generalized Wiener functions and potential theory over abstract Wiener spaces, *Osaka J. Math.*, 25, 665-696(1988)

21. S. Watanabe, Malliavin's calculus in terms of generalized Wiener functionals, In: *Theory and Application of Random Fields* (ed. by G. Kallianpur), Springer, Berlin(1983)

22. S. Watanabe, Lectures on Stochastic Differential Equations and Malliavin Calculus, *Tata Inst. Fund. Research*, Springer(1984)

23. Y. Yokoi, Positive generalized functionals, Preprint(1987), to appear in *Hiroshima Math. J.* (1989)

Mutual Information and Capacity of the Continuous Time Gaussian Channel with Feedback

SHUNSUKE IHARA
Department of Mathematics
College of General Education
Nagoya University
Nagoya, JAPAN

1. Introduction

In this paper, we treat the Gaussian channel (GC) presented by

$$Y(t) = X(t) + Z(t), \qquad 0 \le t \le T, \tag{1}$$

where the channel input $X(t)$ is interfered by the Gaussian noise $Z(\cdot)$. The GC is with feedback if the channel input $X(t)$ is a causal functional of the message ξ and the channel output $Y(\cdot)$. On the other hand, the GC without feedback is represented by

$$Y^0(t) = X^0(t) + Z(t), \qquad 0 \le t \le T, \tag{2}$$

where the channel input $X^0(\cdot)$ is independent of $Z(\cdot)$. Denote by $I_T(\xi, Y)$ (resp. $I_T(X^0, Y^0)$) the mutual information between the message ξ and the output $Y = \{Y(t); 0 \le t \le T\}$ (resp. between X^0 and Y^0). Over the GC's (1) and (2) the mutual information $I_T(\xi, Y)$ and $I_T(X^0, Y^0)$, respectively, are transmitted. The aim of the paper is to show that feedback can increase the mutual information by at most twice. More precisely, we will prove that the inequality

$$I_T(\xi, Y) \le 2I_T(X^0, Y^0) \tag{3}$$

holds if the covariance functions of X and X^0 are same and $X^0(\cdot)$ is a Gaussian process (Theorem 1).

With this basic result, we can show that the feedback capacity of the GC is never more than twice of the capacity of the same GC without feedback. Let $\mathcal{R}$ be a class of covariance functions. We denote by $C_T^F(\mathcal{R})$ the feedback capacity of the GC (1) under the constraint that the input X is a process with a covariance function belonging to $\mathcal{R}$. Note that the so-called average power constraint is a kind of the constraint. We denote by $C_T^0(\mathcal{R})$ the capacity of the GC (2) without feedback under the same constraint. Then we will show the inequality

$$C_T^F(\mathcal{R}) \le 2C_T^0(\mathcal{R}) \tag{4}$$

(Theorem 2).

The inequality (4) is due to Pinsker and Ebert. However no proof by Pinsker has been published (see comments in [1, 2]). Although Ebert [2] gave a proof of (4) in a special case, his proof is somewhat complicated. In the discrete time case Cover and Pombra [1] have shown the inequalities (3) and (4) with the aid of certain matrix inequalities. Our derivation of (3) and (4) is quite different from the method by Ebert, and depends on the results by Cover and Pombra [1] and also depends on the results given in [6, 7].

Furthermore we show that the factor two in the inequalities (3) and (4) is best possible in the sense that the number 2 can not be replaced any other numbers less than 2 (Theorem 3).

2. Main Results

In this section, we introduce the basic notations and give the statements of our main results.

Let ξ and η be random variables defined on the basic probability space $(\Omega, \mathcal{B}, P)$ taking values on measurable space $(G, \mathcal{G})$ and $(H, \mathcal{H})$, respectively. Denote by μ_ξ and $\mu_{\xi\eta}$ the probability distribution of ξ and the joint probability distribution of ξ and η, respectively. The mutual information $I(\xi, \eta)$ between ξ and η is defiend by

$$I(\xi, \eta) = \int_{G\times H} \log \frac{d\mu_{\xi\eta}}{d\mu_\xi \times \mu_\eta} d\mu_{\xi\eta}$$

if $\mu_{\xi\eta}$ is absolutely continuous with respect to the product measure $\mu_\xi \times \mu_\eta$ $(\mu_{\xi\eta} \ll \mu_\xi \times \mu_\eta)$, where $d\mu_{\xi\eta}/d\mu_\xi \times \mu_\eta$ is the Radon-Nikodym derivative; otherwise $I(\xi, \eta)$ is infinite.

Let the noise $Z = \{Z(t);\ 0 \le t \le T\}$ of the GC (1) be a separable Gaussian process defined on $(\Omega, \mathcal{B}, P)$. We assume that a message ξ is a random variable defined on $(\Omega, \mathcal{B})$, taking values in an arbitrary measurable space. So ξ may be a stochatic process as well as a finite dimensional random variable. We say that the GC is with feedback if the channel input $X(\cdot)$ is a causal functional of the message ξ and the channel output $Y(\cdot)$. More precisely, the GC (1) is with feedback if the following conditions (a.1) - (a.3) are satisfied:

(a.1) ξ is independent of Z;

(a.2) For each t, $X(t)$ is $\mathcal{F}(\xi) \vee \mathcal{F}_t(Y)$ measurable, where $\mathcal{F}(\xi)$ (resp. $\mathcal{F}_t(Y)$) is the σ-field generated by ξ (resp. $\{Y(u);\ u < t\}$);

(a.3) Stochastic equation (1) has a unique solution in the law sense.

On the other hand, the GC is said to be without feedback if the channel is a functional of the message, namely if the condition (a.1) and

(a.2') $X(t)$ is $\mathcal{F}(\xi)$ measurable,

are satisfied. In this case the input X can be identified with the message ξ and is independent of Z.

We suppose that a constraint given in terms of covariance functions is imposed on the channel input. Let $\mathcal{R}$ be a class of covariance functions on the time interval $[0,T]$, denote by $R_X(s,t)$ $(s,t \in [0,T])$ the covariance function of a process X and define a class $\mathcal{A}^0(\mathcal{R})$ of channel inputs by

$$\mathcal{A}^0(\mathcal{R}) = \{X; X \text{ is independent of } Z \text{ and } R_X(s,t) \in \mathcal{R}\} .$$

The information capacity $C_T^0(\mathcal{R})$ of the GC (2) without feedback is defined by

$$C_T^0(\mathcal{R}) = \sup\{I_T(X^0, Y^0); X^0 \in \mathcal{A}^0(\mathcal{R})\}.$$

We define a class $\mathcal{A}(\mathcal{R})$ of pairs of a message and an input by

$$\mathcal{A}(\mathcal{R}) = \{(\xi, X); (\xi, X) \text{ satisfies (a.1) - (a.3) and } R_X(s,t) \in \mathcal{R}\}.$$

Each $(\xi, X) \in \mathcal{A}(\mathcal{R})$ is an admissible pair to be input. For the GC (1) with feedback, $I_T(X,Y)$ contains not only the information transmitted over the forward channel but also that over the feedback channel. Hence to define the feedback capacity it is mistake to use $I_T(X,Y)$ but we should use $I_T(\xi, Y)$. The feedback capacity $C_T^F(\mathcal{R})$ of the GC (1) is defined by

$$C_T^F(\mathcal{R}) = \sup\{I_T(\xi, Y);\ (\xi, X) \in \mathcal{A}(\mathcal{R})\}$$

Throughout the paper we may assume, without loss of generality, that the means of random variables and stochastic processes are zero.

Every separable Gaussian process has a canonical representation in the sense of Lévy-Hida-Cramér [3]. We are concerned with the case where the canonical representation of the Gaussian process Z has no discrete spectrum, namely we assume that Z is represented canonically in the form

$$Z(t) = \sum_{i=1}^{N} \int_0^t F_i(t,u)\, dB_i(u), \qquad 0 \le t \le T, \tag{5}$$

where $F_i(t,u)$, $i = 1, \dots, N$, are canonical kernels and $dB_i(\cdot)$, $i = 1, \dots, N$, are mutually independent white Gaussian noises with continuous spectral measures $dm_i(t) = E[|dB_i(t)|^2]$ such that $m_i \gg m_{i+1}$. It holds that

$$\mathcal{F}_t(Z) = \mathcal{F}_t(\mathbf{B}), \qquad 0 \le t \le T,$$

where $\mathcal{F}_t(\mathbf{B})$ is the σ-field generated by $\{\mathbf{B}(u) \equiv (B_1(u), \dots, B_N(u)); u < t\}$.

We consider the GC (1) with feedback and the GC (2) without feedback. The main result is stated as follows.

THEOREM 1. *Let the noise Z of the GC's (1) with feedback and (2) without feedback have the canonical representation of the form (5). Assume that the channel input X of the GC (1) and X^0 of the GC (2) have the same covariance function and that X^0 is a Gaussian process. Then*

$$I_T(\xi, Y) \leq 2I_T(X^0, Y^0). \tag{6}$$

As a corollary we can obtain a similar inequality for the channel capacity.

THEOREM 2. *Let $\mathcal{R}$ be a class of covariance functions. If the noise Z of the GC's (1) with feedback and (2) without feedback satisfies the same assumption as in Theorem 1, then*

$$C_T^F(\mathcal{R}) \leq 2C_T^0(\mathcal{R}). \tag{7}$$

One of the most important example of the GC is the white Gaussian channel (WGC) which is presented by

$$Y(t) = \int_0^t x(u)\,du + B(t), \qquad 0 \leq t \leq T, \tag{8}$$

where $B = \{B(t)\}$ is a Brownian motion. We suppose that the average power constraint

$$\int_0^T E[|x(t)|^2]\,dt \leq PT \tag{9}$$

is imposed on the channel inputs, where $P > 0$ is a constant. We denote by $C_T^F(P)$ and $C_T^0(P)$ the feedback capacity and the capacity without feedback, respectively, of the WGC (8) under the constraint (9). It is well known [9] that the capacity of the WGC is not increased by feedback and

$$C_T^F(P) = C_T^0(P) = \frac{1}{2}PT.$$

Another model of interesting GC is given by

$$Y(t) = \int_0^t x(u)\,du + Z(t), \qquad t \geq 0, \tag{10}$$

where the noise Z is a Gaussian process equivalent (or mutually absolutely continuous) to a Brownian motion. It is known [4] that such a process Z is canonically represented as

$$Z(t) = B(t) + \int_0^t \int_0^s f(s,u)\,dB(u)\,ds \equiv \int_0^t F(t,u)\,dB(u), \qquad 0 \leq t \leq T, \tag{11}$$

where $f(s,u) \in L^2[0,T]$ is a Volterra function (i.e., $f(s,u) = 0$ if $s < u$), $F(t,u)$ is the canonical kernel given by .

$$F(t,u) = \begin{cases} 1 + \int_u^t f(s,u)\,ds, & t \geq u \\ 0, & t < u, \end{cases}$$

and $B = \{B(t)\}$ is a Brownian motion. Note that the representation (11) is a special case of (5), so that Theorem 1 and 2 hold for the GC (10). The author[6] has investigated the GC (10) under the constraint

$$\overline{\lim_{T\to\infty}} \frac{1}{T} \int_0^T E[|x(t)|^2]\,dt \leq P. \tag{12}$$

Corresponding to the classes $\mathcal{A}^0(\mathcal{R})$ and $\mathcal{A}(\mathcal{R})$ we define classes $\overline{\mathcal{A}}^0(P)$ and $\overline{\mathcal{A}}(P)$ by

$$\overline{\mathcal{A}}^0(P) = \{X;\ X \text{ is independent of } Z \text{ and satisfies (12)}\}$$

and

$$\overline{\mathcal{A}}(P) = \{(\xi, X);\ (\xi, X) \text{ satisfies (a.1) - (a.3) for each } T > 0 \text{ and } X \text{ satisfies (12)}\},$$

where $X = \{X(t); t > 0\}$, $X(t) = \int_0^t x(u)\,du$. The capacities per unit time $\overline{C}^0(P)$ of the GC (10) without feedback and $\overline{C}^F(P)$ with feedback are defined by

$$\overline{C}^0(P) = \sup\{\overline{I}(X,Y);\ X \in \overline{\mathcal{A}}^0(P)\}$$

and

$$\overline{C}^F(P) = \sup\{\overline{I}(\xi,Y);\ (\xi,X) \in \overline{\mathcal{A}}(P)\},$$

where $\overline{I}(X,Y) = \overline{\lim}_{T\to\infty} \frac{1}{T} I_T(X,Y)$ is the mutual information per unit time.

The capacity $\overline{C}^0(P)$ may be calculated by using the so-called water filling method. On the other hand, no formulas have been known to calculate the feedback capacity $\overline{C}^F(P)$ explicitly. Fortunately, by use of the inequality (7) and a result in [8], we can compute the exact value of $\overline{C}^F(P)$ in a special case.

THEOREM 3. *For the GC (10) with the noise given by (11), it holds that*

$$\overline{C}^F(P) \leq 2\overline{C}^0(P). \tag{13}$$

Especially if the Volterra function $f(s,u)$ is given by

$$f(s,u) = \begin{cases} -e^{u-s}, & s \geq u \\ 0, & s < u, \end{cases} \tag{14}$$

then

$$\overline{C}^F(\frac{1}{2}) = 2\overline{C}^0(\frac{1}{2}) = 1. \tag{15}$$

It should be noted that the equation (15) means that feedback can double the capacity and that the factor 2 in the inequality (6) as well as in (7) can not be replaced by any other constants less than 2.

3. Proof of Main Theorems

Let us consider the GC (1) with the Gaussian noise Z having the canonical representation (5). The reproducing kernel Hilbert space (RKHS) argument is helpful to discuss on the mutual information in the GC [5,6,7]. Let $\mathcal{H} = \mathcal{H}(Z)$ be the RKHS corresponding to $Z = \{Z(t);\ 0 \leq t \leq T\}$. It is known that the RKHS $\mathcal{H}(Z)$ consists of all functions $\Phi(t)$ $(0 \leq t \leq T)$ of the form

$$\Phi(t) = \sum_{i=1}^{N} \int_0^t F_i(t,u)\,\phi_i(u)\,dm_i(u),$$

where $\phi = (\phi_1, ..., \phi_N)$ is an N-dimensional function such that $\sum_{i=1}^{N} \int_0^T |\phi_i(u)|^2\,dm_i(u) < \infty$. The norm $\|\cdot\|_{\mathcal{H}}$ of $\mathcal{H}$ is given by

$$\|\Phi\|_{\mathcal{H}}^2 = \sum_{i=1}^{N} \int_0^T |\phi_i(u)|^2\,dm_i(u).$$

Let $\mathbf{L} = \mathbf{L}(Z)$ be a Hilbert space consisting of N-dimensional stochastic processes $\mathbf{x} = (x_1, ..., x_N) = \{(x_1(t), ..., x_N(t));\ 0 \leq t \leq T\}$ with the norm

$$\|\mathbf{x}\|_{\mathbf{L}}^2 = E\left[\sum_{i=1}^{N} \int_0^T |x_i(t)|^2\,dm_i(t)\right].$$

We denote by $\mathcal{L} = \mathcal{L}(Z)$ the space of all stochastic processes $X = \{X(t);\ 0 \leq t \leq T\}$ of the form

$$X(t) = \sum_{i=1}^{N} \int_0^t F_i(t,u)\,x_i(u)\,dm_i(u), \tag{16}$$

where $\mathbf{x} = (x_1, ..., x_N) \in \mathbf{L}$.

In order to prove our main theorems we need the following four propositions and a theorem (Theorem 4).

PROPOSITION 1 [5]. *Let the channel input X^0 of the GC (2) without feedback be Gaussian. Then $I_T(X^0, Y^0)$ is finite if and only if*

$$X^0(\cdot) \in \mathcal{H}(Z), \tag{17}$$

with probability one.

Then the input X of the GC (1) is of the form (16), to the GC (1) there corresponding an N-dimensional WGC.

$$y_i(t) = \int_0^t x_i(u)\, dm_i(u) + B_i(t), \qquad 0 \le t \le T, \quad i = 1, ..., N, \tag{18}$$

For brevity we write (18) as

$$\mathbf{y}(t) = \int_0^t \mathbf{x}(u)\, d\mathbf{m}(u) + \mathbf{B}(t), \qquad 0 \le t \le T. \tag{19}$$

PROPOSITION 2 [6]. *For the GC (1) with feedback, assume that (ξ, X) satisfies (a.1) - (a.3) and $X \in \mathcal{L}$ is given in the form (16). Then*

$$\mathcal{F}_t(Y) = \mathcal{F}_t(\mathbf{y}), \qquad 0 \le t \le T,$$

where $\mathbf{y} = (y_1, ..., y_N)$ is the output of the WGC (19), and

$$I_T(\xi, Y) = I_T(\xi, \mathbf{y}) = \frac{1}{2} E\left[\sum_{i=1}^{N} \int_0^T |x_i(t) - \widehat{x}_i(t)|^2\, dm_i(t)\right] = \frac{1}{2}\|\mathbf{x} - \widehat{\mathbf{x}}\|_{\mathcal{L}}^2, \tag{20}$$

where $\widehat{\mathbf{x}} = (\widehat{x}_1, ..., \widehat{x}_N)$ and $\widehat{x}_i(t) = E[x_i(t)|\mathcal{F}_t(Y)]$ is the conditional expectation.

PROPOSITION 3 [7]. *For the GC (1) with feedback, assume that $X \in \mathcal{L}$. Then there exist processes $\boldsymbol{\theta} = (\theta_1, ..., \theta_N) \in \mathbf{L}$, $\boldsymbol{\zeta} = (\zeta_1, ..., \zeta_N) \in \mathbf{L}$ and a GC presented by*

$$\widetilde{Y}(t) = \widetilde{X}(t) + Z(t), \qquad 0 \le t \le T, \tag{21}$$

which satisfy the following conditions (i) - (v).

(i) $\widetilde{X}(t) = \sum_{i=1}^N \int_0^t F_i(t,u)\,(\theta_i(u) - \zeta_i(u))\, dm_i(u)$;

(ii) $(\widetilde{X}(\cdot), Z(\cdot))$ *has the same cavariance as* $(X(\cdot), Z(\cdot))$;

(iii) $(\boldsymbol{\theta}(\cdot), \boldsymbol{\zeta}(\cdot), Z(\cdot))$ *forms a Gaussian system* ;

(iv) $\boldsymbol{\theta}$ *is independent of* Z ;

(v) *For each i and t, $\zeta_i(t)$ is $\mathcal{F}_t(\tilde{Y})$ measurable.*
Moreover it holds that

$$I_T(\xi, Y) \le I_T(\boldsymbol{\theta}, \tilde{Y}). \tag{22}$$

To the GC (21) there corresponds the N-dimensional WGC

$$\tilde{\mathbf{y}}(t) = \int_0^t \tilde{\mathbf{x}}(u)\, d\mathbf{m}(u) + \mathbf{B}(t), \qquad 0 \le t \le T, \tag{23}$$

where $\tilde{\mathbf{x}} = (\tilde{x}_1, ..., \tilde{x}_N)$ and $\tilde{x}_i(u) = \theta_i(u) - \zeta_i(u)$. We consider an N-dimensional WGC without feedback given by

$$\mathbf{U}(t) = \int_0^t \boldsymbol{\theta}(u)\, d\mathbf{m}(u) + \mathbf{B}(t), \qquad 0 \le t \le T. \tag{24}$$

PROPOSITION 4 [6,7]. *The σ-fields $\mathcal{F}_t(\mathbf{U}), \mathcal{F}_t(\tilde{\mathbf{y}})$ and $\mathcal{F}_t(\tilde{Y})$ coincide each other, and it holds that*

$$I_T(\boldsymbol{\theta}, \mathbf{U}) = I_T(\boldsymbol{\theta}, \tilde{\mathbf{y}}) = I_T(\boldsymbol{\theta}, \tilde{Y}). \tag{25}$$

Moreover $\boldsymbol{\zeta} = (\zeta_1, ..., \zeta_N)$ of Proposition 3 can be written as

$$\zeta_i(t) = \sum_{j=1}^{N} \int_0^t h_{ij}(t,u)\, dU_j(u), \qquad i = 1, ..., N,$$

where $h_{ij}(t,u), i, j = 1, ..., N$, are Volterra functions such that

$$\sum_{i,j=1}^{N} \int_0^T \int_0^T |h_{ij}(t,u)|^2\, dm_i(t) dm_j(u) < \infty.$$

We can prove that the inequality (6) holds for the WGC (24).

THEOREM 4. *For the WGC (24) we assume that the properties (iii) - (v) are satisfied (replacing Z and $\tilde{Y}$ by $\mathbf{B}$ and $\tilde{\mathbf{y}}$, respectively). Let $\mathbf{x}^0 = (x_1^0, ..., x_N^0)$ be an N-dimensional Gaussian process independent of $\mathbf{B}$ and with the same covariance as $\tilde{\mathbf{x}} = (\tilde{x}_1, ..., \tilde{x}_N)$ and let $\mathbf{y}^0 = (y_1^0, ..., y_N^0)$ be the channel output of a WGC corresponding to the input $\mathbf{x}^0$:*

$$\mathbf{y}^0(t) = \int_0^t \mathbf{x}^0(u)\, d\mathbf{m}(u) + \mathbf{B}(t), \qquad 0 \le t \le T. \tag{26}$$

Then the inequality

$$I_T(\boldsymbol{\theta}, \tilde{\mathbf{y}}) \leq 2I_T(\mathbf{x}^0, \mathbf{y}^0) \tag{27}$$

holds.

Proof of Theorem 4 will be given in the next section.

Based upon Theorem 4 we can prove Theorem 1.

PROOF OF THEOREM 1: Suppose that $X \notin \mathcal{L}(Z)$ so that $X^0 \notin \mathcal{L}(Z)$. Then (17) does not hold,and $I_T(\mathbf{x}^0, \mathbf{y}^0) = I_T(X^0, Y^0)$ is infinite by Proposition 1. Hence the inequality (6) is obvious. Now we suppose that $X \in \mathcal{L}(Z)$. The process X can be written in the form (16). By Proposition 3 and Theorem 4 there exist a GC (21) with properties (i)-(v) of Proposition 3 and a WGC (26) with properties in Theorem 4. It follows from (22), (25) and (27) that

$$I_T(\xi, Y) \leq I_T(\boldsymbol{\theta}, \tilde{Y}) \leq 2I_T(\mathbf{x}^0, \mathbf{y}^0), \tag{28}$$

where $\tilde{\mathbf{y}}$ is the output of the WGC (24). Since X and X^0, and $\mathbf{x} = (x_1, ..., x_N)$ and $\mathbf{x}^0 = (x_1^0, ..., x_N^0)$ have same covariance, we can identify the process X^0 with the process given by

$$\sum_{i=1}^{N} \int_0^t F_i(t, u)\, x_i^0(u)\, dm_i(u).$$

Thus by Proposition 2 we have

$$I_T(\mathbf{x}^0, \mathbf{y}^0) = I_T(X^0, Y^0). \tag{29}$$

The desired inequality (6) follows from (28) and (29). ∎

We now proceed to prove Theorem 2. Let $\mathcal{R}^*$ be the class of all covariance functions of the form

$$R(s, t) = \sum_{i,j=1}^{N} \int_0^s \int_0^t F_i(t, v)\, r_{ij}(u, v)\, dm_i(u) dm_j(v),$$

where $r(u, v) = (r_{ij}(u, v))_{i,j=1,...,N}$ is a symmetric nonnegative function satisfying $\sum_{i=1}^{N} \int_0^T r_{ii}(u, u)\, dm_i(u) < \infty$. Note that a process X belongs to $\mathcal{L}(Z)$ if and only if its covariance function belongs to $\mathcal{R}^*$.

PROOF OF THEOREM 2: If $\mathcal{R} \not\subset \mathcal{R}^*$, then $C_T^0(\mathcal{R}) = C_T^F(\mathcal{R}) = \infty$ ([6]) and (7) is true. Suppose that $\mathcal{R} \subset \mathcal{R}^*$. For any $\epsilon > 0$, there exists an admissible pair $(\xi, X) \in \mathcal{A}(\mathcal{R})$ such that

$$C_T^F(\mathcal{R}) \leq I_T(\xi, Y) + \epsilon, \tag{30}$$

where Y is the channel output corresponding to (ξ, X). By Theorem 1 there exists a Gaussian process $X^0 \in \mathcal{A}^0(\mathcal{R})$ such that

$$I_T(\xi, Y) \leq 2I_T(X^0, Y^0). \tag{31}$$

By definition it is clear that

$$I_T(X^0, Y^0) \leq C_T^0(\mathcal{R}). \tag{32}$$

Since $\epsilon > 0$ is arbitrary, the inequality (7) follows from (30) - (32). ∎

Finally we prove Theorem 3.

PROOF OF THEOREM3: The inequality (13) follows from Theorem 1. For the case where $f(s, u)$ is given by (14), the author [8] has shown that

$$\overline{C}^F\left(\frac{1}{2}\right) \geq 2\overline{C}^0\left(\frac{1}{2}\right) = 1. \tag{33}$$

The equlity (15) follows from (13) and (33). ∎

4. Proof of Theorem 4

In order to prove Theorem 4 we prepare two lemmas. Let $L^2(\mathbf{m})$ be the space of all functions $f(t, u) \equiv (f_{ij}(t, u))_{i,j=1,\dots,N}$ $((t, u) \in [0, T]^2)$ such that

$$\|f\|_2^2 = \sum_{i,j=1}^{N} \int_0^T \int_0^T |f_{ij}(t, u)|^2 \, dm_i(t) dm_j(u) < \infty.$$

Clearly $L^2(\mathbf{m})$ is a Hilbert space with the norm $\|\cdot\|_2$. If each component $f_{ij}(t, u)$ is a Volterra function, $f = (f_{ij})$ is called a Volterra kernel. In the following, for each Volterra kernel $f \in L^2(\mathbf{m})$ and process $\mathbf{x} = (x_1, \dots, x_N) \in \mathbf{L} = \mathbf{L}(Z)$, we denote by $\langle f, d\mathbf{B}\rangle(t)$ a process $\eta(t) = (\eta_1(t), \dots, \eta_N(t))$ given by $\eta_i(t) = \sum_{j=1}^N \int_0^t f_{ij}(t, u)\, dB_j(u)$, and denote by $\langle f, \mathbf{x}, d\mathbf{m}\rangle(t)$ a process $\boldsymbol{\zeta}(t) = (\zeta_1(t), \dots, \zeta_N(t))$, given by $\zeta_i(t) = \sum_{j=1}^N \int_0^t f_{ij}(t, u)\, x_j(u)\, dm_j(u)$.

LEMMA 1. (i) *For each Volterra kernel $f \in L^2(\mathbf{m})$ and process $\mathbf{x} \in \mathbf{L}$, it holds that*

$$\| \langle f, d\mathbf{B}\rangle \|_{\mathbf{L}} = \|f\|_2 \tag{34}$$

and

$$\| \langle f, \mathbf{x}, d\mathbf{m}\rangle \|_{\mathbf{L}} \leq \|f\|_2 \, \|\mathbf{x}\|_{\mathbf{L}}. \tag{35}$$

(ii) *Let* $\mathbf{x}^{(k)} \in \mathbf{L}$ $(k = 1, 2)$ *be a Gaussian process independent of* $\mathbf{B}$, *and* $\mathbf{y}^{(k)}$ *be the channel output of the WGC presented by*

$$\mathbf{y}^{(k)}(t) = \int_0^t \mathbf{x}^{(k)}(u)\, d\mathbf{m}(u) + \mathbf{B}(t), \qquad 0 \le t \le T. \tag{36}$$

Denote by $\hat{\mathbf{x}}^{(k)}(t) = \left\langle f^{(k)}, d\mathbf{y}^{(k)} \right\rangle(t)$ *the conditional expectation of* $\mathbf{x}^{(k)}(t)$ *given* $\mathcal{F}_t(\mathbf{y}^{(k)})$, *where* $f^{(k)} \in L^2(\mathbf{m})$ *is a Volterra kernel. Then*

$$\|\mathbf{x}^{(2)} - \hat{\mathbf{x}}^{(2)}\|_{\mathbf{L}} - \|\mathbf{x}^{(1)} - \hat{\mathbf{x}}^{(1)}\|_{\mathbf{L}} \le \left(1 + \|f^{(1)}\|_2\right) \|\mathbf{x}^{(1)} - \mathbf{x}^{(2)}\|_{\mathbf{L}}. \tag{37}$$

Proof is omitted

LEMMA 2. *Let* $\mathbf{x}^{(n)} \in \mathbf{L}$ $(n = 0, 1, 2, \ldots)$ *be Gaussian process independent of* $\mathbf{B}$, *and* $\mathbf{y}^{(n)}$ *be given by* $\mathbf{y}^{(n)}(t) = \int_0^t \mathbf{x}^{(n)}\, d\mathbf{m}(u) + \mathbf{B}(t)$. *If* $\lim_{n\to\infty} \|\mathbf{x}^{(n)} - \mathbf{x}^{(0)}\|_{\mathbf{L}} = 0$, *then*

$$\lim_{n\to\infty} I_T(\mathbf{x}^{(n)}, \mathbf{y}^{(n)}) = I_T(\mathbf{x}^{(0)}, \mathbf{y}^{(0)}).$$

PROOF: Denote by $\hat{\mathbf{x}}^{(n)} = \left\langle f^{(n)}, d\mathbf{y}^{(n)} \right\rangle$ the conditional expectation of $\mathbf{x}^{(n)}(t)$ given $\mathcal{F}_t(\mathbf{y}^{(k)})$, where $f^{(n)}$ is a Volterra kernel in $L^2(\mathbf{m})$. It is clear that

$$\|f^{(n)}\|_2^2 \le \|f^{(n)}\|_2^2 + \| \left\langle f^{(n)}, \mathbf{x}^{(n)}, d\mathbf{m} \right\rangle \|_{\mathbf{L}}^2 \le \|\hat{\mathbf{x}}^{(n)}\|_{\mathbf{L}}^2 \le \|\mathbf{x}^{(n)}\|_{\mathbf{L}}^2,$$

so that $\{\|f^{(n)}\|_2;\, n = 0, 1, 2, \ldots\}$ is bounded. Using (ii) of Lemma 1, we can easily show that

$$\lim_{n\to\infty} \|\mathbf{x}^{(n)} - \hat{\mathbf{x}}^{(n)}\|_{\mathbf{L}} = \|\mathbf{x}^{(0)} - \hat{\mathbf{x}}^{(0)}\|_{\mathbf{L}}.$$

Therefore it follows from Proposition 2 that

$$\begin{aligned} \lim_{n\to\infty} I_T(\mathbf{x}^{(n)}, \mathbf{y}^{(n)}) &= \frac{1}{2} \lim_{n\to\infty} \|\mathbf{x}^{(n)} - \hat{\mathbf{x}}^{(n)}\|_{\mathbf{L}}^2 \\ &= \frac{1}{2} \|\mathbf{x}^{(0)} - \hat{\mathbf{x}}^{(0)}\|_{\mathbf{L}}^2 = I_T(\mathbf{x}^{(0)}, \mathbf{y}^{(0)}). \ \blacksquare \end{aligned}$$

Now we proceed to prove Theorem 4.

PROOF OF THEOREM 4: (I) At first we assume the following conditions (A.1) and (A.2):

(A.1) The process $\theta_i(\cdot)$ $(i = 1, \ldots, N)$ is mean continuous;

(A.2) The kernel $h_{ij}(t, u)$ is continuous on the set $\{(t, u) \in [0, T]^2; t \neq u\}$.

We denote $\sum_{i=1}^N m_i(\Delta)$ simply by $\mathbf{m}(\Delta)$. We devide the interval $[0, T]$ into 2^n subintervals $\Delta_{n,k} = (t_{n,k-1}, t_{n,k}]$, $k = 1, \ldots, 2^n$, $(t_{n,0} = 0, t_{n,2^n} = T)$, in such a way that

$\mathbf{m}(\Delta_{n,k}) = 2^{-n}\,\mathbf{m}([0,T])$. We define random variables $B^{(n)} = \{B_{i,k}^{(n)}; i = 1,...,N,\ k = 1,...,2^n\}$, $\theta^{(n)} = \{\theta_{i,k}^{(n)}\}$, $U^{(n)} = \{U_{i,k}^{(n)}\}$, $\zeta^{(n)} = \{\zeta_{i,k}^{(n)}\}$, $V^{(n)} = \{V_{i,k}^{(n)}\}$ as follows.

$$
\begin{aligned}
B_{i,k}^{(n)} &= B_i(t_{n,k}) - B_i(t_{n,k-1}),\\
\theta_{i,k}^{(n)} &= \int_{\Delta_{n,k}} \theta_i(u)\, dm_i(u),\\
U_{i,k}^{(n)} &= U_i(t_{n,k}) - U_i(t_{n,k-1}) = \theta_{i,k}^{(n)} + B_{i,k}^{(n)}, && (38)\\
\zeta_{i,k}^{(n)} &= m_i(\Delta_{n,k}) \sum_{j=1}^{N} \sum_{l=1}^{k-1} h_{ij}(t_{n,k}, t_{n,l})\, U_{j,l}^{(n)},\\
V_{i,k}^{(n)} &= \left(\theta_{i,k}^{(n)} - \zeta_{i,k}^{(n)}\right) + B_{i,k}^{(n)} && (39)
\end{aligned}
$$

$(i = 1,...,N,\ k = 1,...,2^n)$. Note that (38) and (39) represent discrete time (N-dimensional) WGC's without feedback and with feedback, respectively. We can easily show that $V_{i,k}^{(n)}$ is a linear combination of $U_{j,l}^{(n)}$, $j = 1,...,N$, $l = 1,...,k$, and conversely $U_{i,k}^{(n)}$ is a linear combination of $V_{j,l}^{(n)}$, $j = 1,...,N$, $l = 1,...,k$. Therefore

$$I(\theta^{(n)}, V^{(n)}) = I(\theta^{(n)}, U^{(n)}) \le I_T(\boldsymbol{\theta}, \mathbf{U}).$$

It can be easily shown that

$$\lim_{n\to\infty} I(\theta^{(n)}, U^{(n)}) = I_T(\boldsymbol{\theta}, \mathbf{U}). \tag{40}$$

Corresponding to the WGC (39) with feedback, we consider a WGC without feedback given by

$$y_{i,k}^{(n)} = x_{i,k}^{(n)} + B_{i,k}^{(n)}, \quad i = 1,...,N,\ k = 1,...,2^n, \tag{41}$$

where $x^{(n)} = \{x_{i,k}^{(n)}\}$ a family of Gaussian random variables with same covariance as $\{\theta_{i,k}^{(n)} - \zeta_{i,k}^{(n)}\}$ and independent of $\mathbf{B}(\cdot)$. We can apply the result due to Cover and Pombra [1] to the discrete time WGC's (39) and (41), and have

$$I(\theta^{(n)}, U^{(n)}) = I(\theta^{(n)}, V^{(n)}) \le 2\, I(x^{(n)}, y^{(n)}), \tag{42}$$

where $y^{(n)} = \{y_{i,k}^{(n)}\}$. We define a Volterra kernel $h^{(n)} = \left(h_{ij}^{(n)}\right) \in L^2(\mathbf{m})$ by

$$h_{ij}^{(n)} = \begin{cases} h_{ij}(t_{n,k}, t_{n,l}), & \text{if} \quad (t,u) \in \Delta_{n,k} \times \Delta_{n,l},\ j < k \\ 0, & \text{otherwise.} \end{cases}$$

We define N-dimensional processes $\boldsymbol{\theta}^{(n)} = \left(\theta_1^{(n)}, \ldots, \theta_N^{(n)}\right)$ and $\boldsymbol{\zeta}^{(n)} = \left(\zeta_1^{(n)}, \ldots, \zeta_N^{(n)}\right)$ by

$$\theta_i^{(n)}(t) = \frac{\theta_{i,k}^{(n)}}{m_i(\Delta_{n,k})} \quad \text{if} \quad t \in \Delta_{n,k}$$

and

$$\zeta_i^{(n)}(t) = \frac{\zeta_{i,k}^{(n)}}{m_i(\Delta_{n,k})} \quad \text{if} \quad t \in \Delta_{n,k}.$$

We can easily see that

$$\boldsymbol{\zeta}^{(n)} = \langle h^{(n)}, d\mathbf{U}\rangle. \tag{43}$$

It is clear from (24), (43) and Lemma 1 that

$$\begin{aligned} \|\boldsymbol{\zeta}^{(n)} - \langle h, d\mathbf{U}\rangle\|_{\mathbf{L}} &\le \|\langle (h^{(n)} - h), \boldsymbol{\theta}, d\mathbf{m}\rangle\|_{\mathbf{L}} + \|\langle (h^{(n)} - h), d\mathbf{B}\rangle\|_{\mathbf{L}} \\ &\le \|h^{(n)} - h\|_2 \left(\|\boldsymbol{\theta}\|_{\mathbf{L}} + 1\right). \end{aligned}$$

By the assumption (A.2) we have $\lim_{n\to\infty} \|h^{(n)} - h\|_2 = 0$. Therefore

$$\lim_{n\to\infty} \|\boldsymbol{\zeta}^{(n)} - \langle h, d\mathbf{U}\rangle\|_{\mathbf{L}} = 0. \tag{44}$$

We define a Gaussian process $\mathbf{x}^{(n)} = \left(x_1^{(n)}, \ldots, x_N^{(n)}\right) \in \mathbf{L}$ by

$$x_i^{(n)}(t) = \frac{x_{i,k}^{(n)}}{m_i(\Delta_{n,k})} \quad \text{if} \quad t \in \Delta_{n,k}.$$

Since $\mathbf{x}^{(n)}$ is independent of $\mathbf{B}$, we see that

$$\mathbf{y}^{(n)}(t) = \int_0^t \mathbf{x}^{(n)}(u)\, d\mathbf{m}(u) + \mathbf{B}(t), \qquad 0 \le t \le T, \tag{45}$$

presents a WGC without feedback. It is clear that

$$I(x^{(n)}, y^{(n)}) \le I_T(\mathbf{x}^{(n)}, \mathbf{y}^{(n)}). \tag{46}$$

By the continuity assumption (A.1) on $\boldsymbol{\theta}$ we can show that

$$\lim_{n\to\infty} \|\boldsymbol{\theta}^{(n)} - \boldsymbol{\theta}\|_{\mathbf{L}} = 0. \tag{47}$$

Since $\mathbf{x}$ and $\boldsymbol{\theta}^{(n)} - \boldsymbol{\zeta}^{(n)}$ have the same covariance, it is true that

$$\begin{aligned}\|\mathbf{x}^{(n)} - \mathbf{x}^0\|_{\mathbf{L}} &= \|(\boldsymbol{\theta}^{(n)} - \boldsymbol{\zeta}^{(n)}) - (\boldsymbol{\theta} - \langle h, d\mathbf{U}\rangle)\|_{\mathbf{L}} \\ &\le \|\boldsymbol{\theta}^{(n)} - \boldsymbol{\theta}\|_{\mathbf{L}} + \|\boldsymbol{\zeta}^{(n)} - \langle h, d\mathbf{U}\rangle\|_{\mathbf{L}}.\end{aligned}$$

Therefore, by (44) and (47), we have

$$\|\mathbf{x}^{(n)} - \mathbf{x}^0\|_{\mathbf{L}} = 0.$$

Applying Lemma 2 we have

$$\lim_{n\to\infty} I_T(\mathbf{x}^{(n)}, \mathbf{y}^{(n)}) = I_T(\mathbf{x}^0, \mathbf{y}^0). \tag{48}$$

Finally the desired inequality (27) follows from (25), (41), (42), (46) and (48).

(II) Secondary we will prove the theorem without assuming (A.1) and (A.2). There exists a Gaussian process $\widetilde{\boldsymbol{\theta}}^n = \left(\widetilde{\theta}_1^n, ..., \widetilde{\theta}_N^n\right) \in \mathbf{L}$ independent of $\mathbf{B}$ satisfying (A.1) and

$$\lim_{n\to\infty} \|\widetilde{\boldsymbol{\theta}}^n - \boldsymbol{\theta}\|_{\mathbf{L}} = 0. \tag{49}$$

There exists a Volterra kernel $\widetilde{h}^n(t,u) = \left(\widetilde{h}_{ij}^n(t,u)\right) \in L^2(\mathbf{m})$ satisfying (A.2) and

$$\lim_{n\to\infty} \|\widetilde{h}^n - h\|_2 = 0. \tag{50}$$

We consider N-dimensional WGC's

$$\widetilde{\mathbf{W}}^n(t) = \int_0^t \widetilde{\boldsymbol{\theta}}^n(u)\, d\mathbf{m}(u) + \mathbf{B}(t), \qquad 0 \le t \le T$$

and

$$\widetilde{\mathbf{U}}^n(t) = \int_0^t (\widetilde{\boldsymbol{\theta}}^n(u) - \widetilde{\boldsymbol{\zeta}}^n(u))\, d\mathbf{m}(u) + \mathbf{B}(t), \qquad 0 \le t \le T,$$

where $\widetilde{\boldsymbol{\zeta}}^n = \langle \widetilde{h}^n, d\widetilde{\mathbf{U}}^n\rangle$. By Proposition 4, we see that $I_T(\widetilde{\boldsymbol{\theta}}^n, \widetilde{\mathbf{U}}^n) = I_T(\widetilde{\boldsymbol{\theta}}^n, \widetilde{\mathbf{W}}^n)$. It follows from (49) and Lemma 2 that

$$\lim_{n\to\infty} I_T(\widetilde{\boldsymbol{\theta}}^n, \widetilde{\mathbf{W}}^n) = \lim_{n\to\infty} I_T(\widetilde{\boldsymbol{\theta}}^n, \widetilde{\mathbf{U}}^n) = I(\boldsymbol{\theta}, \mathbf{U}) = I(\boldsymbol{\theta}, \widetilde{\mathbf{y}}). \tag{51}$$

Let $\widetilde{\mathbf{x}}^n = (\widetilde{x}_1^n, ..., \widetilde{x}_N^n)$ be a Gaussian process with the same distribution as $\widetilde{\boldsymbol{\theta}}^n - \widetilde{\boldsymbol{\zeta}}^n$ and independenat of $\mathbf{B}$, and define a WGC by

$$\widetilde{\mathbf{y}}^n(t) = \int_0^t \widetilde{\mathbf{x}}^n(u)\, d\mathbf{m}(u) + \mathbf{B}(t), \qquad 0 \le t \le T.$$

As we have shown in (I) it is true that

$$I_T(\widetilde{\boldsymbol{\theta}}^n, \widetilde{\mathbf{W}}^n) \leq 2\, I_T(\tilde{\mathbf{x}}^n, \tilde{\mathbf{y}}^n). \tag{52}$$

Using (49), (50), Lemma 1 and Lemma 2, we can show that

$$\lim_{n\to\infty} I_T(\tilde{\mathbf{x}}^n, \tilde{\mathbf{y}}^n) = I_T(\tilde{\mathbf{x}}^0, \tilde{\mathbf{y}}^0). \tag{53}$$

Finally the inequality (27) follows from (51), (52) and (53). ∎

References

[1] Cover,T.M. and Pombra,S.: Gaussian feedback capacitay. IEEE Trans. Inform. Theory, **IT-35** (1989), 37–43.

[2] Ebert,P.M.: The capacity of the Gaussian channel with feedback. Bell System Tech. J., **49** (1970), 8–19.

[3] Hida,T.: Canonical representation of Gaussian processes and their applications. Mem. Coll. Sci. Univ. Kyoto, Ser. A (Math.), **33** (1960), 199–247.

[4] Hitsuda,M.: Representation of Gaussian processes equivalent to Wiener process. Osaka J. Math., **5** (1968), 299–312.

[5] Hitsuda,M.: Mutual information in Gaussian channels. J. Multivariate Anal., **4** (1974), 66–73.

[6] Hitsuda,M. and Ihara,S.: Gaussian channels and the optimal coding. J. Multivariate Anal., **5** (1975), 106–118.

[7] Ihara,S.: On the capacity of the continuous time Gaussian channel with feedback. J. Multivariate Anal., **10** (1980), 319–331.

[8] Ihara,S.: Capacity of mismatched Gaussian channel with and without feedback. Probab. Th. Rel. Fields, **84** (1990), 453–471.

[9] Kadota,T.T., Zakai,M. and Ziv,J.: Mutual information of the white Gaussian channel with and without feedback. IEEE Trans. Inform. Theory, **IT-17** (1971), 368–371.

FOURIER-MEHLER TRANSFORMS IN WHITE NOISE ANALYSIS

Hui-Hsiung Kuo*

Department of Mathematics
Louisiana State University
Baton Rouge, LA 70803, USA

Dedicated to Professor T. Hida on the Occasion of his Retirement

ABSTRACT

The finite-dimensional Fourier-Mehler is generalized to the white noise space. Similar properties as in the finite-dimensional case are established. Its relations with various Laplacians are obtained.

§1. Finite-dimensional Fourier-Mehler Transforms

Consider first the Fourier transform on $\mathbb{R}^k$:

$$\widehat{f}(y) = (2\pi)^{-k/2} \int_{\mathbb{R}^k} e^{-i<x,y>} f(x)dx. \tag{1.1}$$

There are several difficulties to generalize the Fourier transform to infinite dimensional spaces, e.g. the factor $(2\pi)^{-k/2}$ tends to zero as $k \to \infty$, the Lebesgue measure does not exist and the function $e^{-i<x,y>}$ can not be reasonably defined. However, these difficulties can be overcome by rewriting Eq.(1.1) in a dimension free form. Suppose y is an $\mathbb{R}^k$-valued normal random variable with mean zero and covariance operator I. Then for fixed x in $\mathbb{R}^k$,

$$E_y e^{-i<x,y>} = e^{-\frac{1}{2}|x|^2}.$$

Thus if we define the renormalization $: e^{-i<x,y>} :_y$ of $e^{-i<x,y>}$ with respect to the y variable by $: e^{-i<x,y>} :_y = e^{-i<x,y>}/E_y e^{-i<x,y>}$. Then

$$: e^{-i<x,y>} :_y = e^{-i<x,y>+\frac{1}{2}|x|^2}.$$

With this renormalization, we can rewrite Eq.(1.1) as follows:

$$\widehat{f}(y) = \int_{\mathbb{R}^k} : e^{-i<x,y>} :_y f(x)d\mu(x), \tag{1.2}$$

* Research supported by NSF Grant DMS-9001859

where μ is the standard Gaussian measure on $\mathbb{R}^k$. In white noise calculus, the infinite dimensional version of $\mathbb{R}^k$ is the space $\mathcal{S}'(\mathbb{R})$ of tempered distributions. Certainly, the standard Gaussian measure μ on $\mathcal{S}'(\mathbb{R})$ exists and the renormalization makes sense. Thus Eq.(1.2) leads to a definition of Fourier transform on the white noise space $(\mathcal{S}'(\mathbb{R}), \mu)$. This has been done in [Ku 82,Ku 89,Ku 90]. See also [IKT 89,Le 89] for equivalent definitions of Fourier transform.

Next, consider the Fourier-Mehler transform with parameter θ on $\mathbb{R}^k$:

$$(\mathcal{F}_\theta f)(y) = \int_{\mathbb{R}^k} L(x,y) f(x) dx, \tag{1.3}$$

where the kernel function $L(x,y)$ is given by

$$L(x,y) = \left[\pi\left(1 - e^{2i\theta}\right)\right]^{-k/2} \exp\left(\frac{i}{\sin\theta} < x, y > - \frac{i}{2\tan\theta}(|x|^2 + |y|^2)\right).$$

For the case $k = 1$, see [Hi 80]. Just as in the case of Fourier transform (when $\theta = -\frac{\pi}{2}$), there are several difficulties to generalize the Fourier-Mehler transform to infinite-dimensional spaces. However, the same trick as above can be used to rewrite Eq.(1.3) in a dimension free form. Consider the following function

$$g(x,y) = \exp\left(\frac{i}{\sin\theta} < x, y > - \frac{i}{2\tan\theta}|y|^2\right).$$

Supppose y is an $\mathbb{R}^k$-valued normal random variable with mean zero and covariance operator I. Then for fixed x in $\mathbb{R}^k$, the expectation of $g(x,y)$ with respect to y is given by

$$E_y g(x,y) = \left[2^{-1}\left(1 - e^{2i\theta}\right)\right]^{k/2} \exp\left[\left(\frac{i}{2\tan\theta} - \frac{1}{2}\right)|x|^2\right].$$

If we define the multiplicative renormalization of $g(x,y)$ with respect to y by $\mathcal{N}_y g(x,y) \equiv g(x,y)/E_y g(x,y)$, then $\mathcal{N}_y g(x,y)$ is equal to

$$\left[\pi\left(1 - e^{2i\theta}\right)\right]^{-k/2} (2\pi)^{k/2} e^{\frac{1}{2}|x|^2} \exp\left(\frac{i}{\sin\theta} < x, y > - \frac{i}{2\tan\theta}\left(|x|^2 + |y|^2\right)\right).$$

Therefore, we can rewrite the Fourier-Mehler transform with parameter θ in Eq. (1.3) as follows:

$$(\mathcal{F}_\theta f)(y) = \int_{\mathbb{R}^k} f(x)\, \mathcal{N}_y \exp\left(\frac{i}{\sin\theta} < x, y > - \frac{i}{2\tan\theta}|y|^2\right) d\mu(x). \tag{1.4}$$

The purpose of this paper is to generalize the Fourier-Mehler transform to white noise calculus. Although this has been done in [Ku 83b], the definition in [Ku 83b] is quite formal and not well-defined. In this paper we will give a well-defined definition of Fourier-Mehler transform and obtain several new results.

§2. The Space of Hida Distributions

In this section we give a review of the space of Hida distributions [HKPS,HP 90,KT 80a,PS 89,PY 89]. Let $\mathcal{S}(\mathbb{R})$ be the Schwartz space of rapidly decreasing real-valued functions on $\mathbb{R}$. Its dual space $\mathcal{S}'(\mathbb{R})$ consists of tempered distributions. Let A be the following operator

$$A = -(d/dx)^2 + x^2 + 1.$$

Let $|\cdot|_2$ denote the $L^2(\mathbb{R})$-norm. For each $p \in \mathbb{R}$, define $|f|_{2,p} \equiv |A^p f|_2$ and let $\mathcal{S}_p(\mathbb{R})$ be the completion of $\mathcal{S}(\mathbb{R})$ with respect to the norm $|\cdot|_{2,p}$. Then the dual space $\mathcal{S}_p'(\mathbb{R})$ of $\mathcal{S}_p(\mathbb{R})$ is the same as $\mathcal{S}_{-p}(\mathbb{R})$. Moreover, we have the following continuous inclusion maps:

$$\mathcal{S}(\mathbb{R}) \subset \mathcal{S}_p(\mathbb{R}) \subset L^2(\mathbb{R}) \subset \mathcal{S}_p'(\mathbb{R}) \subset \mathcal{S}'(\mathbb{R}), \quad p \geq 0.$$

Let μ be the standard Gaussian measure on the dual space $\mathcal{S}'(\mathbb{R})$ of $\mathcal{S}(\mathbb{R})$, i.e. its characteristic function is given by

$$\int_{\mathcal{S}'(\mathbb{R})} e^{i<x,\xi>} d\mu(x) = e^{-\frac{1}{2}|\xi|_2^2}, \quad \xi \in \mathcal{S}(\mathbb{R}).$$

For brevity, we will use (L^2) to denote $L^2(\mathcal{S}'(\mathbb{R}), \mu)$ with norm $\|\cdot\|_2$. By the Wiener-Itô theorem, (L^2) has the following orthogonal decomposition

$$(L^2) = \bigoplus_{n=0}^{\infty} K_n,$$

where K_n consists of multiple Wiener integrals of order n, i.e. each φ in K_n is of the form $\varphi(x) = <: x^{\otimes n} :, f>$, $f \in \widehat{L}^2(\mathbb{R}^n)$ ($\widehat{}$ denotes the symmetrization). Thus each $\varphi \in (L^2)$ has the following representation

$$\varphi(x) = \sum_{n=0}^{\infty} <: x^{\otimes n} :, f_n > .$$

Moreover, we have

$$\|\varphi\| = \left(\sum_{n=0}^{\infty} n! |f_n|^2_{L^2(\mathbb{R}^n)} \right)^{1/2} .$$

Now, define the *second quantization* $\Gamma(A)$ of A by

$$\Gamma(A) \left(\sum_{n=0}^{\infty} <: x^{\otimes n} :, f_n > \right) = \sum_{n=0}^{\infty} <: x^{\otimes n} :, A^{\otimes n} f_n > .$$

For $p \in \mathbb{R}$, let $\|\varphi\|_{2,p} \equiv \|\Gamma(A)^p \varphi\|_2$. Define

$$(\mathcal{S})_p = \{\varphi \in (L^2); \|\varphi\|_{2,p} < \infty\}, \quad p \geq 0,$$

$$(\mathcal{S})_p = \text{completion of } (L^2) \text{ w.r.t. } \|\cdot\|_{2,p}, \quad p < 0.$$

Obviously, for $p \geq 0$, $(\mathcal{S})_{-p}$ is the dual space $(\mathcal{S})_p^*$ of $(\mathcal{S})_p$. Let $(\mathcal{S})$ be the projective limit of $\{(\mathcal{S})_p; p \geq 0\}$. Then the dual space $(\mathcal{S})^*$ is the inductive limit of $\{(\mathcal{S})_p^*; p > 0\}$. Thus we have the following continuous inclusion maps:

$$(\mathcal{S}) \subset (\mathcal{S})_p \subset (L^2) \subset (\mathcal{S})_p^* \subset (\mathcal{S})^*, \quad p \geq 0.$$

We will call $(\mathcal{S})$ the space of *test functionals* and $(\mathcal{S})^*$ the space of *Hida distributions* (or *generalized Brownian functionals*).

A fundamental tool in white noise analysis is the U-functional map from $(\mathcal{S})^*$ into the space of functions defined on $\mathcal{S}(\mathbb{R})$. First consider the following S-transform defined on (L^2), i.e. for $\varphi \in (L^2)$,

$$(S\varphi)(\xi) = \int_{\mathcal{S}'(\mathbb{R})} \varphi(x+\xi) d\mu(x), \quad \xi \in \mathcal{S}(\mathbb{R}). \tag{2.1}$$

By using the translation formula for μ [Ku 75], we can rewrite the S-transform as follows:

$$(S\varphi)(\xi) = e^{-|\xi|_2^2/2} \int_{\mathcal{S}'(\mathbb{R})} \varphi(x) e^{<x,\xi>} d\mu(x), \quad \xi \in \mathcal{S}(\mathbb{R}).$$

Note that for any ξ in $\mathcal{S}(\mathbb{R})$, the function $e^{<\cdot,\xi>}$ is in $(\mathcal{S})$. Therefore, the S-transform can be extended to the space $(\mathcal{S})^*$ of Hida distributions by using the pairing $\langle\langle \cdot, \cdot \rangle\rangle$ of $(\mathcal{S})^*$ and $(\mathcal{S})$. This extension is called the *U-functional map*, i.e. for $\Phi \in (\mathcal{S})^*$, its U-functional is defined to be

$$U[\Phi](\xi) = e^{-|\xi|_2^2/2} \langle\langle \Phi, e^{<\cdot,\xi>} \rangle\rangle, \quad \xi \in \mathcal{S}(\mathbb{R}). \tag{2.2}$$

Note that $Ee^{<\cdot,\xi>} = e^{\frac{1}{2}|\xi|_2^2}$. Hence if we define $: e^{<\cdot,\xi>} :\equiv e^{<\cdot,\xi>}/Ee^{<\cdot,\xi>}$, then $: e^{<\cdot,\xi>} : = e^{<\cdot,\xi>-\frac{1}{2}|\xi|_2^2}$ and we can write the U-functional of Φ in another form:

$$U[\Phi](\xi) = \langle\!\langle \Phi, : e^{<\cdot,\xi>} : \rangle\!\rangle, \quad \xi \in \mathcal{S}(\mathbb{R}).$$

In §5 we will need the Wick product of two Hida distributions. It is well-known that the product of two U-functionals is also a U-functional. Thus for any two Hida distributions Φ and Ψ, we can define their *Wick product* $\Phi \diamond \Psi$ to be the Hida distribution with U-functional

$$U[\Phi \diamond \Psi](\xi) = U[\Phi](\xi)\, U[\Psi](\xi), \quad \xi \in \mathcal{S}(\mathbb{R}).$$

When Hida introduced the theory of generalized Brownian functionals in [Hi 75], he used the following $\mathcal{T}$-transform:

$$(\mathcal{T}\Phi)(\xi) = \langle\!\langle \Phi, e^{i<\cdot,\xi>} \rangle\!\rangle, \quad \xi \in \mathcal{S}(\mathbb{R}).$$

The U-functional and the $\mathcal{T}$-transform are related by

$$(\mathcal{T}\Phi)(\xi) = e^{-\frac{1}{2}|\xi|_2^2} U[\Phi](i\xi), \quad \xi \in \mathcal{S}(\mathbb{R}),$$

$$U[\Phi](\xi) = e^{-\frac{1}{2}|\xi|_2^2} (\mathcal{T}\Phi)(-i\xi), \quad \xi \in \mathcal{S}(\mathbb{R}).$$

For the definition of Fourier-Mehler transform, we need the multiplicative renormalization of the following formal expression

$$\exp(a < w, x > + b\,|x|_2^2), \quad x \in \mathcal{S}'(\mathbb{R}),$$

where $w \in \mathcal{S}'(\mathbb{R})$, $a, b \in \mathbb{C}$ and $b \neq \frac{1}{2}$. Let e_n be the Hermite function of order n and let P_n be defined by

$$P_n x = \sum_{k=0}^{n-1} < x, e_k > e_k, \quad x \in \mathcal{S}'(\mathbb{R}).$$

Assume for the moment that b is real and $b < \frac{1}{2}$. Let

$$\varphi_n(x) = \exp(a < w, P_n x > + b\,|P_n x|_2^2), \quad x \in \mathcal{S}'(\mathbb{R}).$$

Then $\varphi_n \in (L^2)$ and its expectation is given by

$$E\varphi_n = (1 - 2b)^{-n/2} \exp\left(\frac{a^2}{2(1-2b)} |P_n w|_2^2 \right).$$

Define $\psi_n \equiv \varphi_n / E\varphi_n$. Then $\psi_n \in (L^2)$ and its S-transform from Eq.(2.1) is

$$(S\psi_n)(\xi) = \exp\left(\frac{a}{1-2b} < w, P_n\xi > + \frac{b}{1-2b}|P_n\xi|_2^2\right), \quad \xi \in \mathcal{S}(\mathbb{R}).$$

Now, for each $\xi \in \mathcal{S}(\mathbb{R})$, we have

$$\lim_{n\to\infty}(S\psi_n)(\xi) = \exp\left(\frac{a}{1-2b} < w, \xi > + \frac{b}{1-2b}|\xi|_2^2\right).$$

Moreover, since $|w|_{2,-p} < \infty$ for some $p > 0$, $|P_n\xi|_{2,p} \leq |\xi|_{2,p}$ and $|P_n\xi|_2 \leq |\xi|_2 \leq |\xi|_{2,p}$, there exists some constant c such that for all n

$$|(S\psi_n)(\xi)| \leq \exp(c|\xi|_{2,p}).$$

This implies that the sequence ψ_n converges in $(\mathcal{S})^*$. We will use

$$\mathcal{N}\exp(a < w, x > + b\,|x|_2^2)$$

to denote the limit. Then its U-functional is given by

$$U[\mathcal{N}\exp(a < w, x > + b\,|x|_2^2)](\xi) = \exp\left(\frac{a}{1-2b} < w, \xi > + \frac{b}{1-2b}|\xi|_2^2\right). \tag{2.3}$$

On the other hand, for any $w \in \mathcal{S}'(\mathbb{R})$ and $a, b \in \mathbb{C}, b \neq \frac{1}{2}$, it follows from the Potthoff-Streit characterization theorem [PS 89] that the right hand side of Eq.(2.3) is a U-functional. Thus we have proved the following lemma.

Lemma 2.1. *For any $w \in \mathcal{S}'(\mathbb{R})$ and $a, b \in \mathbb{C}, b \neq \frac{1}{2}$, the multiplicative renormalization*

$$\mathcal{N}\exp(a < w, x > + b\,|x|_2^2)$$

of $\exp(a < w, x > + b\,|x|_2^2)$ *is a Hida distribution. Its U-functional is given by*

$$U[\mathcal{N}\exp(a < w, x > + b\,|x|_2^2)](\xi) = \exp\left(\frac{a}{1-2b} < w, \xi > + \frac{b}{1-2b}|\xi|_2^2\right).$$

Remark. The above Hida distributions are called *white noise Gaussian functions*. We will use the following notation

$$G_{aw,b}(x) \equiv \mathcal{N}\exp(a < w, x > + b\,|x|_2^2),$$

$$G_b(x) \equiv \mathcal{N}\exp(b\,|x|_2^2).$$

§3. Definition and Examples of Fourier-Mehler Transforms

The infinite dimensional analogue of Eq.(1.4) to the space $(\mathcal{S}'(\mathbb{R}), \mu)$ is

$$(\mathcal{F}_\theta \Phi)(y) = \int_{\mathcal{S}'(\mathbb{R})} \Phi(x)\, \mathcal{N}_y \exp\left(\frac{i}{\sin\theta} < x, y > - \frac{i}{2\tan\theta} |y|_2^2 \right) d\mu(x). \tag{3.1}$$

However, we can not use this equation to define the Fourier-Mehler transform because by Lemma 2.1 the above renormalization is a Hida distribution and then this equation makes sense only when Φ is in the space $(\mathcal{S})$ of test functionals. To overcome this severe restriction, we take the U-functional of $\mathcal{F}_\theta \Phi$. By Lemma 2.1 with $x = y$ and $w = x$, we have

$$U\left[\mathcal{N}_y \exp\left(\frac{i}{\sin\theta} < x, y > - \frac{i}{2\tan\theta} |y|_2^2 \right) \right](\xi)$$

$$= \exp\left(e^{i\theta} < x, \xi > - \frac{1}{2} e^{i\theta} \cos\theta\, |\xi|_2^2 \right), \quad \xi \in \mathcal{S}(\mathbb{R}).$$

Therefore, if we take the U-functional of $\mathcal{F}_\theta \Phi$ formally from Eq.(3.1), then we get

$$U[\mathcal{F}_\theta \Phi](\xi) = \langle\!\langle \Phi, \exp\left(e^{i\theta} < \cdot, \xi > - 2^{-1} e^{i\theta} \cos\theta\, |\xi|_2^2 \right) \rangle\!\rangle. \tag{3.2}$$

On the other hand, by using the entire extension $U[\Phi](z\xi)$, $z \in \mathbb{C}$, of the U-functional of Φ, we obtain that

$$U[\Phi](e^{i\theta}\xi) = \langle\!\langle \Phi, \exp(e^{i\theta} < \cdot, \xi >) \rangle\!\rangle \exp\left(-\frac{1}{2} e^{2i\theta} |\xi|_2^2 \right). \tag{3.3}$$

It follows easily from Eqs.(3.2) and (3.3) that

$$U[\mathcal{F}_\theta \Phi](\xi) = U[\Phi]\left(e^{i\theta}\xi\right) \exp\left(\frac{i}{2} e^{i\theta} \sin\theta\, |\xi|_2^2 \right).$$

Observe that, by the Potthoff-Streit characterization theorem [PS 89], the right hand side of the last equation is a U-functional. Therefore, we can use this equation to define the Fourier-Mehler transform.

Definition 3.1. The *Fourier-Mehler transform* of $\Phi \in (\mathcal{S})^*$ with parameter $\theta \in \mathbb{R}$ is the Hida distribution $\mathcal{F}_\theta \Phi$ with U-functional given by

$$U[\mathcal{F}_\theta \Phi](\xi) = U[\Phi]\left(e^{i\theta}\xi\right) \exp\left(\frac{i}{2} e^{i\theta} \sin\theta\, |\xi|_2^2 \right), \quad \xi \in \mathcal{S}(\mathbb{R}).$$

Remark. Note that $\mathcal{F}_0 = I$, $\mathcal{F}_{-\pi/2}$ is the Fourier transform and $\mathcal{F}_{\pi/2}$ is the inverse Fourier transform.

In the following examples, we will use white noise Gaussian function $G_{aw,b}$ given in Lemma 2.1. Note that, in particular, when $a = -2b$

$$U[G_{-2bw,b}](\xi) = \exp\left(\frac{-2b}{1-2b} < w, \xi > + \frac{b}{1-2b} |\xi|_2^2\right), \quad \xi \in \mathcal{S}(\mathbb{R}).$$

When $b \to \infty$, $G_{-2bw,b}$ converges to the *white noise delta function* $\widetilde{\delta}_w$ at $w \in \mathcal{S}'(\mathbb{R})$ whose U-functional is given by

$$U[\widetilde{\delta}_w](\xi) = e^{<w,\xi> - \frac{1}{2}|\xi|_2^2}, \quad \xi \in \mathcal{S}(\mathbb{R}).$$

Example 3.2. $\mathcal{F}_\theta 1 = G_{\frac{i}{2}\tan\theta}$.

Example 3.3. $\mathcal{F}_\theta \widetilde{\delta}_w = G_{i\csc\theta\, w,\, -\frac{i}{2}\cot\theta}$, $w \in \mathcal{S}'(\mathbb{R})$.

Example 3.4. $\mathcal{F}_\theta : e^{a<w,\cdot>} := G_{a\sec\theta\, w,\, \frac{i}{2}\tan\theta}$

Example 3.5. $\mathcal{F}_\theta \mathcal{N}_x \exp(a < w, x > - \frac{1}{2}|x|_2^2) = \mathcal{N}_x \exp\left(ae^{i\theta} < w, x > - \frac{1}{2}|x|_2^2\right)$.

Example 3.6. For the *white noise Hermite function* $\mathcal{H}_n(t) \equiv (\partial_t^*)^n \mathcal{N} e^{-\frac{1}{2}|x|_2^2}$, we have

$$\mathcal{F}_\theta \mathcal{H}_n(t) = e^{in\theta} \mathcal{H}_n(t).$$

Example 3.7. $\mathcal{F}_\theta G_{aw,b} = G_{a(\theta)w,b(\theta)}$, where $a(\theta)$ and $b(\theta)$ are given by

$$a(\theta) = \frac{a}{\cos\theta + 2bi\sin\theta}, \quad b(\theta) = \frac{2b + i\tan\theta}{2(1 + 2bi\tan\theta)}.$$

§4. Elementary Properties of Fourier-Mehler Transforms

In the white noise analysis, the set $\{x(t); t \in \mathbb{R}\}$ is taken as a coordinate system. The Hida (or coordinate) differentiation ∂_t, its adjoint operator ∂_t^*, and the coordinate multiplication by $x(t)$ are defined as follows. We will call a function F on $\mathcal{S}(\mathbb{R})$ *Fréchet differentiable* if there exists a function, denoted by F', from $\mathcal{S}(\mathbb{R})$ into $\mathcal{S}'(\mathbb{R})$ such that

$$F(\xi + \eta) = F(\xi) + < F'(\xi), \eta > + o(\eta), \quad \eta \in \mathcal{S}(\mathbb{R}),$$

where $o(\eta)$ means that there exists $p \in \mathbb{R}$ depending on ξ such that $o(\eta)/|\eta|_{2,p} \to 0$ as $|\eta|_{2,p} \to 0$. Now, let $\Phi \in (\mathcal{S})^*$ and $F = U[\Phi]$. It follows from Eq.(2.2) that F is Fréchet differentiable and F' is given by

$$< F'(\xi), \eta >= - < \xi, \eta > F(\xi) + e^{-\frac{1}{2}|\xi|_2^2} \langle\langle \Phi, < \cdot, \eta > e^{< \cdot, \xi >} \rangle\rangle.$$

Hence for each η in $\mathcal{S}(\mathbb{R})$, $< F'(\cdot), \eta >$ is a U-functional. We define the *Hida derivative* $\partial_t \Phi$ of Φ to be the Hida distribution with the following U-functional:

$$U[\partial_t \Phi](\xi) = F'(\xi; t), \quad \xi \in \mathcal{S}(\mathbb{R}).$$

Note that $\partial_t \Phi$ is a distribution as a function of t. On the other hand, for each fixed $t \in \mathbb{R}$, we define an operator $\widetilde{\partial}_t$ on $(\mathcal{S})$ by

$$\widetilde{\partial}_t \left(\sum_{n=0}^{\infty} <: x^{\otimes n} :, f_n > \right) = \sum_{n=1}^{\infty} n <: x^{\otimes (n-1)} :, f_n(t, \cdot) > .$$

It can be checked that for each t, $\widetilde{\partial}_t$ is a continuous linear map from $(\mathcal{S})$ into itself [HKPS]. Moreover, ∂_t is an extension of $\widetilde{\partial}_t$ to the space $(\mathcal{S})^*$ of Hida distributions. We will use the same notation ∂_t whether it acts on $(\mathcal{S})$ or on $(\mathcal{S})^*$.

The *adjoint operator* ∂_t^* is defined on $(\mathcal{S})^*$ by the duality, i.e. for $\Phi \in (\mathcal{S})^*$,

$$\langle\langle \partial_t^* \Phi, \varphi \rangle\rangle = \langle\langle \Phi, \partial_t \varphi \rangle\rangle, \quad \varphi \in (\mathcal{S}).$$

It can be checked that ∂_t^* is a continuous linear map from $(\mathcal{S})^*$ into itself [HKPS]. The *coordinate multiplication by* $x(t)$ is defined by

$$x(t)\Phi = (\partial_t + \partial_t^*)\Phi.$$

Note that the operator $x(t)\cdot$ is a distribution as a function of t.

Theorem 4.1 (Relation between $\mathcal{F}_\theta$ and ∂_t). *For all* $\Phi \in (\mathcal{S})^*$,

$$\mathcal{F}_\theta(\partial_t \Phi) = e^{-i\theta} \partial_t(\mathcal{F}_\theta \Phi) - i \sin\theta \, \partial_t^*(\mathcal{F}_\theta \Phi).$$

Proof. Let $F = U[\Phi]$ and $G = U[\mathcal{F}_\theta \Phi]$. Then by the definition of Fourier-Mehler transform,

$$U[\mathcal{F}_\theta(\partial_t \Phi)](\xi) = F'(e^{i\theta}\xi; t) \exp\left(\frac{i}{2} e^{i\theta} \sin\theta \, |\xi|_2^2 \right). \tag{4.1}$$

On the other hand, since $U[\partial_t(\mathcal{F}_\theta\Phi)](\xi) = G'(\xi;t)$, we have

$$U[\partial_t(\mathcal{F}_\theta\Phi)](\xi) = e^{i\theta}F'(e^{i\theta}\xi;t)\exp\left(\frac{i}{2}e^{i\theta}\sin\theta\,|\xi|_2^2\right)+ie^{i\theta}\sin\theta\,\xi(t)U[\mathcal{F}_\theta\Phi](\xi). \quad (4.2)$$

It follows from Eqs.(4.1) and (4.2) that

$$\partial_t(\mathcal{F}_\theta\Phi) = e^{i\theta}\mathcal{F}_\theta(\partial_t\Phi) + ie^{i\theta}\sin\theta\,\partial_t^*(\mathcal{F}_\theta\Phi).$$

This yields the equation in the theorem immediately. ■

Theorem 4.2 (Relation between $\mathcal{F}_\theta$ and ∂_t^*). *For all* $\Phi \in (\mathcal{S})^*$,

$$\mathcal{F}_\theta(\partial_t^*\Phi) = e^{i\theta}\partial_t^*(\mathcal{F}_\theta\Phi).$$

Proof. This follows easily from the definition of Fourier-Mehler transform and the fact that $U[\partial_t^*\Phi](\xi) = \xi(t)U[\Phi](\xi)$. ■

Theorem 4.3 (Relation between $\mathcal{F}_\theta$ and $x(t)\cdot$). *For all* $\Phi \in (\mathcal{S})^*$,

$$\mathcal{F}_\theta(x(t)\Phi) = e^{-i\theta}\partial_t(\mathcal{F}_\theta\Phi) + \cos\theta\,\partial_t^*(\mathcal{F}_\theta\Phi).$$

Proof. Apply Theorems 4.1 and 4.2. ■

Theorem 4.4 (Relation between $\mathcal{F}_\theta$ and the translation). *Let* Φ *be in* $(\mathcal{S})^*$ *and* Φ_a *the translation of* Φ *by* $a \in \mathcal{S}(\mathbb{R})$, *i.e.* $\Phi_a(x) = \Phi(x-a)$. *Then*

$$\mathcal{F}_\theta\Phi_a = (\mathcal{F}_\theta\Phi)_{a\cos\theta}\,\exp\left(i\sin\theta < \cdot, a > -\frac{i}{2}\sin\theta\cos\theta\,|a|_2^2\right).$$

Proof. Use the definition of Fourier-Mehler transform and the following facts that for any $\Phi \in (\mathcal{S})^*$, $z \in \mathbb{C}$, and $a \in \mathcal{S}(\mathbb{R})$,

$$U[\Phi_a](\xi) = U[\Phi](\xi - a), \quad \xi \in \mathcal{S}(\mathbb{R}),$$

$$U[e^{z<\cdot,a>}\Phi](\xi) = U[\Phi](\xi + za)\exp\left(z < a, \xi > +\frac{1}{2}z^2|a|_2^2\right), \quad \xi \in \mathcal{S}(\mathbb{R}). \quad ■$$

Theorem 4.5 (Relation between $\mathcal{F}_\theta$ and the $\mathcal{T}$-transform). *For any* $\Phi \in (\mathcal{S})^*$,

$$U[\mathcal{F}_\theta\Phi](\xi) = (\mathcal{T}\Phi)\left(-ie^{i\theta}\xi\right)\exp\left(-\frac{1}{2}e^{i\theta}\cos\theta\,|\xi|_2^2\right).$$

Proof. Use the relation between the U-functional and the $\mathcal{T}$-transform in §2. ■

§5. Further Properties of Fourier-Mehler Transforms

In this section we will show some further properties of Fourier-Mehler transforms. For some of the corresponding results for the Fourier transform, see [Ku 89,Ku 90].

Theorem 5.1. *$\{\mathcal{F}_\theta; \theta \in \mathbb{R}\}$ is a one-parameter group acting on $(\mathcal{S})^*$. Its infinitesimal generator is given by*

$$\mathcal{G} = i\int_{\mathbb{R}} \partial_t^* \partial_t dt + \frac{i}{2}\int_{\mathbb{R}} (\partial_t^*)^2 dt.$$

Remark. In the one-dimensional case, $\mathcal{G} = -\frac{i}{2}(d^2/dx^2 - x^2 + 1)$ [Hi 80]. Also, note that the number operator $N = \int_{\mathbb{R}} \partial_t^* \partial_t dt$ and the adjoint of Gross Laplacian $\Delta_G^* = \int_{\mathbb{R}} (\partial_t^*)^2 dt$. Thus $\mathcal{G} = iN + \frac{i}{2}\Delta_G^*$.

Proof. Direct calculation by using the definition of Fourier-Mehler transform shows that for any $\theta_1, \theta_2 \in \mathbb{R}$, and $\Phi \in (\mathcal{S})^*$,

$$U[\mathcal{F}_{\theta_1}\mathcal{F}_{\theta_2}\Phi](\xi) = U[\mathcal{F}_{\theta_1+\theta_2}\Phi](\xi), \quad \xi \in \mathcal{S}(\mathbb{R}).$$

Therefore, $\mathcal{F}_{\theta_1}\mathcal{F}_{\theta_2} = \mathcal{F}_{\theta_1+\theta_2}$ and so $\{\mathcal{F}_\theta; \theta \in \mathbb{R}\}$ is a one-parameter group acting on $(\mathcal{S})^*$. To find the infinitesimal generator, let $F = U[\Phi]$ and $F_\theta = U[\mathcal{F}_\theta\Phi]$. Then

$$F_\theta(\xi) = F(e^{i\theta}\xi)\exp\left(\frac{i}{2}e^{i\theta}\sin\theta\,|\xi|_2^2\right).$$

Therefore,

$$\frac{d}{d\theta}F_\theta(\xi)\bigg|_{\theta=0} = i < F'(\xi), \xi > + \frac{i}{2}\int_{\mathbb{R}} \xi(t)^2 F(\xi)dt.$$

The theorem follows since

$$U\left[i\int_{\mathbb{R}} \partial_t^*\partial_t\Phi dt + \frac{i}{2}\int_{\mathbb{R}} (\partial_t^*)^2\Phi dt\right](\xi) = i < F'(\xi), \xi > + \frac{i}{2}\int_{\mathbb{R}} \xi(t)^2 F(\xi)dt. \quad \blacksquare$$

Lemma 5.2. *For $\zeta \in \mathcal{S}(\mathbb{R})$, let $T_\zeta\Phi(x) = \Phi(x-\zeta)\exp\left(\frac{1}{2} < x, \zeta > - \frac{1}{4}|\zeta|_2^2\right)$. Then for any $\zeta, \eta \in \mathcal{S}(\mathbb{R})$, $T_\zeta T_\eta = T_{\zeta+\eta}$ acting on $(\mathcal{S})^*$.*

Proof. By direct calculation. ■

Theorem 5.3. *Let ζ be in $\mathcal{S}(\mathbb{R})$. Then $\{T_{t\zeta};\ t \in \mathbb{R}\}$ is a one-parameter group acting on $(\mathcal{S})^*$. Its infinitesimal generator is given by*

$$\kappa_\zeta = -\frac{1}{2}\int_{\mathbb{R}} \zeta(t)\partial_t dt + \frac{1}{2}\int_{\mathbb{R}} \zeta(t)\partial_t^* dt.$$

Proof. By Lemma 5.2, $\{T_{t\zeta};\ t \in \mathbb{R}\}$ is a one-parameter group acting on $(\mathcal{S})^*$. To get the infinitesimal generator, use the second identity in the proof of Theorem 4.4. ■

Theorem 5.4. *The commutation relation for κ_ζ and $\mathcal{G}$ is given by*

$$[\kappa_\zeta, \mathcal{G}] = -\frac{i}{2}\int_{\mathbb{R}} \zeta(t)\partial_t dt - i\int_{\mathbb{R}} \zeta(t)\partial_t^* dt.$$

Proof. Use the following commutation relations

$$[\partial_t, \partial_s] = 0,\ [\partial_t^*, \partial_s^*] = 0,\ [\partial_t, \partial_s^*] = \delta_s(t)I. \quad ■$$

Theorem 5.5. *Let J_θ be the map from $(\mathcal{S})^*$ into itself given by $U[J_\theta\Phi](\xi) = U[\Phi](e^{i\theta}\xi)$. Then $\{J_\theta; \theta \in \mathbb{R}\}$ is a one-parameter group acting on $(\mathcal{S})^*$. Its infinitesimal generator $\mathcal{J}$ is given by*

$$\mathcal{J} = iN = i\int_{\mathbb{R}} \partial_t^*\partial_t dt.$$

Moreover, its commutation relations with $\mathcal{G}$ and κ_ζ are given by

$$[\mathcal{J}, \mathcal{G}] = -\int_{\mathbb{R}} (\partial_t^*)^2 dt,$$

$$[\mathcal{J}, \kappa_\zeta] = \frac{i}{2}\left(\int_{\mathbb{R}} \zeta(t)\partial_t dt + \int_{\mathbb{R}} \zeta(t)\partial_t^* dt\right).$$

Proof. It is obvious that $\{J_\theta; \theta \in \mathbb{R}\}$ is a one-parameter group acting on $(\mathcal{S})^*$. To find the infinitesimal generator, simply use the U-functional map. To find $[\mathcal{J}, \mathcal{G}]$ and $[\mathcal{J}, \kappa_\zeta]$, use the commutation relations for ∂_t and ∂_t^* given in the proof of Theorem 5.4. ■

Theorem 5.6. *For all Φ in $(\mathcal{S})^*$ and any real number p, the following equality holds*

$$\|(\mathcal{F}_\theta\Phi) \diamond G_{\beta(\theta)}\|_{2,p} = \|\Phi\|_{2,p},$$

where $\diamond$ denotes the Wick product and $G_{\beta(\theta)}$ is the white noise Gaussian function with parameter $\beta(\theta) = \frac{i\sin\theta}{2(2i\sin\theta - \cos\theta)}$.

Remark. The equality with $p = 0$ reduces to the Plancherel theorem in the finite dimensional spaces.

Proof. From the definition of Fourier-Mehler transform, we have

$$U[\mathcal{F}_\theta\Phi](\xi)\exp\left(-\frac{i}{2}e^{i\theta}\sin\theta\,|\xi|_2^2\right) = U[\Phi](e^{i\theta}\xi), \quad \xi\in\mathcal{S}(\mathbb{R}).$$

Note that for $\beta(\theta)$ given as in the theorem,

$$U[G_{\beta(\theta)}](\xi) = \exp\left(-\frac{i}{2}e^{i\theta}\sin\theta\,|\xi|_2^2\right).$$

Therefore,

$$U[\mathcal{F}_\theta\Phi)\diamond G_{\beta(\theta)}](\xi) = U[\Phi](e^{i\theta}\xi).$$

This implies that if $\Phi(x) = \sum_{n=0}^{\infty} <: x^{\otimes n} :, f_n >$, then

$$\left((\mathcal{F}_\theta\Phi)\diamond G_{\beta(\theta)}\right)(x) = \sum_{n=0}^{\infty} e^{in\theta} <: x^{\otimes n} :, f_n > .$$

In view of the definition of $\|\cdot\|_{2,p}$ in §2, the assertion follows right away. ■

Theorem 5.7. *For any Φ in $(\mathcal{S})^*$ and any ξ in $\mathcal{S}(\mathbb{R})$,*

$$\langle\langle\mathcal{F}_\theta\Phi, e^{<\cdot,\xi>}\rangle\rangle = \langle\langle\Phi, \mathcal{F}_\theta G_{a(\theta)\xi,b(\theta)}\rangle\rangle\exp\left(\frac{1}{2}e^{i\theta}(e^{i\theta}+\cos\theta)|\xi|_2^2\right),$$

where $G_{a(\theta)\xi,b(\theta)}$ is the white noise Gaussian function with $a(\theta) = e^{i\theta}\sec\theta$ and $b(\theta) = -\frac{i}{2}\tan\theta$.

Remark. In the case of Fourier transform ($\theta = -\frac{\pi}{2}$), the above identity reduces to the following

$$\langle\langle\widehat{\Phi}, e^{<\cdot,\xi>}\rangle\rangle = \langle\langle\Phi, (\widetilde{\delta_\xi})^\frown\rangle\rangle e^{-\frac{1}{2}|\xi|_2^2}.$$

Proof. Use the definition of Fourier-Mehler transform and the relation between the U-functional and the $\mathcal{T}$-transform given in §2. ■

Finally, we give the relations between the Fourier-Mehler transform and the Laplacian operators. Recall that the adjoint of Gross Laplacian Δ_G, the number operator N, the Lévy Laplacian Δ_L and the Volterra Laplacian Δ_V are defined by:

$$\Delta_G^*\Phi = \int_{\mathbb{R}} (\partial_t^*)^2\Phi\,dt,$$

$$N\Phi = \int_{\mathbb{R}} \partial_t^*\partial_t\Phi\,dt,$$

$$\Delta_L \Phi = U^{-1}\left(\int_T F''_L(\cdot;t)dt\right),$$

$$\Delta_V \Phi = U^{-1}\left(\text{trace} F''_V(\cdot)\right),$$

where T is an interval in $\mathbb{R}$ and F''_L and F''_V are the Lévy part and the Volterra part of the U-functional F of Φ, respectively. Note that Δ_G^* and N are defined on the whole space $(\mathcal{S})^*$ of Hida distributions, while Δ_L and Δ_V are defined on their respective domains $\mathcal{D}(\Delta_L)$ and $\mathcal{D}(\Delta_V)$.

The relations between the Fourier transform and the Laplacian operators have been obtained in [HS 88,Ku 89,Ku 90]. In the following we state similar relations for the Fourier-Mehler transform with parameter θ. The proofs require only direct calculation.

Theorem 5.8. *For all Φ in $(\mathcal{S})^*$,*

$$\mathcal{F}_\theta(\Delta_G^*\Phi) = e^{2i\theta}\Delta_G^*(\mathcal{F}_\theta\Phi).$$

Theorem 5.9. *For all Φ in $(\mathcal{S})^*$,*

$$\mathcal{F}_\theta(N\Phi) = N(\mathcal{F}_\theta\Phi) - ie^{i\theta}\sin\theta\,\Delta_G^*(\mathcal{F}_\theta\Phi).$$

Theorem 5.10. *Let T be a finite interval with length $|T|$ and $\Phi \in \mathcal{D}(\Delta_L)$. Then*

$$\mathcal{F}_\theta(\Delta_L\Phi) = e^{-2i\theta}\Delta_L(\mathcal{F}_\theta\Phi) - ie^{-i\theta}\sin\theta\,|T|\,\mathcal{F}_\theta\Phi.$$

Theorem 5.11. *For all Φ in $\mathcal{D}(\Delta_V)$,*

$$\mathcal{F}_\theta(\Delta_V\Phi) = e^{-2i\theta}\Delta_V(\mathcal{F}_\theta\Phi) - 2ie^{-i\theta}\sin\theta\,N(\mathcal{F}_\theta\Phi) - \sin^2\theta\,\Delta_G^*(\mathcal{F}_\theta\Phi).$$

References

[Hi 75] Hida, T.: *Analysis of Brownian Functionals.* Carleton Mathematical Lecture Notes, no. **13** (1975)

[Hi 80] Hida, T.: *Brownian Motion.* Berlin, Heidelberg, New York: Springer (1980)

[Hi 85] Hida, T.: Brownian motion and its functionals; *Ricerche di Matematica* **34** (1985) 183–222

[HKPS] Hida, T., Kuo, H.-H., Potthoff, J. and Streit, L.: *White Noise: An Infinite Dimensional Calculus.* Monograph in preparation

[HP 90] Hida, T. and Potthoff, J.: White noise analysis–an overview;in: *White Noise Analysis–Mathematics and Applications*, T. Hida, etc. (eds.), (1990) 140-165, World Scientific

[HPS 89] Hida, T., Potthoff, J., Streit, L.: White noise analysis and applications; in: *Mathematics + Physics* **3**, L. Streit (ed.). Singapore: World Scientific (1989)

[HS 88] Hida, T. and Saitô, K.: White noise analysis and the Lévy Laplacian;in: *Stochastic Processes in Physics and Engineering*, S. Albeverio etc. (eds.), (1988) 177–184

[IKT 89] Ito, Y., Kubo, I. and Takenaka, S.: Calculus on Gaussian white noise and Kuo's Fourier transformation, *Preprint* (1989)

[KT 80a] Kubo, I. and Takenaka, S.: Calculus on Gaussian white noise I; *Proc. Japan Acad.* **56A** (1980) 376–380

[KT 80b] Kubo, I. and Takenaka, S.: Calculus on Gaussian white noise II; *Proc. Japan Acad.* **56A** (1980) 411–416

[KT 81] Kubo, I. and Takenaka, S.: Calculus on Gaussian white noise III; *Proc. Japan Acad.* **57A** (1981) 433–437

[KT 82] Kubo, I. and Takenaka, S.: Calculus on Gaussian white noise IV; *Proc. Japan Acad.* **58A** (1982) 186–189

[Ku 75] Kuo, H.-H.: *Gaussian Measures in Banach Spaces.* Lecture Notes in Math., Vol. 463. Berlin, Heidelberg, New York: Springer-Verlag (1975)

[Ku 82] Kuo, H.-H.: On Fourier transform of generalized Brownian functionals; *J. Multivariate Anal.* **12** (1982) 415–431

[Ku 83a] Kuo, H.-H.: Brownian functionals and applications; *Acta Appl. Math.* **1** (1983) 175-188

[Ku 83b] Kuo, H.-H.: Fourier-Mehler transforms of generalized Brownian functionals; *Proc. Japan Acad.* **59A** (1983) 312–314

[Ku 89] Kuo, H.-H.: The Fourier transform in white noise calculus; *J. Multivariate Analysis* **31** (1989) 311-327

[Ku 90] Kuo, H.-H.: Lectures on white noise analysis;*Preprint* (1990)

[Le 89] Lee, Y.–J.: Analytic version of test functionals, Fourier transform and a characterization of measures in white noise calculus; *Preprint* (1989), to appear in *J. Funct. Anal.*

[PS 89] Potthoff, J and Streit, L.: A characterization of Hida distributions;*Preprint* (1989), to appear in *J. Funct. Anal.*

[PY 89] Potthoff, J. and Yan J.A.: Some results about test and generalized functionals of white noise; to appear in: *Proc. Singapore Probab. Conf.* (1989), L.Y. Chen (ed.)

A characterization of generalized functions on infinite dimensional spaces and Bargman–Segal analytic functions

Yuh–Jia Lee
Department of Mathematics
National Cheng–Kung University
Tainan, Taiwan, ROC.

ABSTRACT

In this note the U–functionals of generalized functions are characterized in terms of the Bargman–Segal analytic functions on infinite dimensional spaces. The locality of generalized functions are also studied.

1. Introduction.

Let E_0 be a real separable Hilbert space with norm $|\cdot|_0 = \sqrt{<\cdot,\cdot>}$ and $A : \mathscr{D}(A) \longrightarrow H$ a linear operator satisfying the following conditions : (i) $\overline{\mathscr{D}(A)} = H$; (ii) A is positive and self–adjoint; (iii) the inverse A^{-1} exists and is of Hilbert–Schmidt type on H; (iv) the Hilbert–Schmidt norm of A^{-1} is less than 1.

For any $p \in \mathbb{N}$, let E_p denote the completion of $\mathscr{D}(A^p)$ with respect to $|\cdot|_p = |A^p x|_0$ $(x \in \mathscr{D}(A^p))$ and E_{-p} the completion of E_0 with respect to the $|\cdot|_{-p}$–norm $|x|_{-p} = |A^{-p} x|_0$ $(x \in E_0)$. Then E_p is a real separable Hilbert space with dual E_{-p} for each $p \in \mathbb{Z}$. Set $E_\infty = \bigcap_{p\in\mathbb{N}} E_p$ and $E_{-\infty} = \bigcup_{p\in\mathbb{N}} E_{-p}$. Then we have

$$E_\infty \subset E_p \subset E_q \subset E_0 \subset E_{-q} \subset E_{-p} \subset E_{-\infty}$$

for $p \geq q$ and $E_\infty \subset E_0 \subset E_{-\infty}$ forms a Gel'fand friplet [4]. $E_{-\infty}$ will serve as the domain of the generalized functins $(E_{-\infty})$ described as follows.

To start with, let μ denote the Gaussian measure on $E_{-\infty}$ with variance parameter 1. Then μ coincides with the Wiener measure on the abstract Wiener spaces (E_0, E_{-p}) for $p \geq 1$.

It follows from [15] that any $f \in L^2(E_{-\infty}, \mu)$ admits the Wiener–Ito decomposition

$$f = \sum_{n=0}^{\infty} \frac{1}{n!} \Lambda^n(D^n \mu f(0)), \tag{1}$$

where $\mu f = \mu * f$, "D" denotes the E_0–derivative (see [5]) and

$$\Lambda^n(D^n \mu f(0))(x) = \int_{E_{-\infty}} D^n \mu f(0)(x+iy)^n \mu(dy)$$

which exists almost everywhere [12]. Moreover

$$\|f\|^2_{L^2(E_{-\infty},\mu)} = \sum_{n=0}^{\infty} \frac{1}{n!} \|D^n \mu f(0)\|^2_{\mathscr{L}_2^n(E_0)},$$

where $\mathscr{L}_2^n(K)$ denotes the Hilbert–Schmidt n–linear operator over the Hilbert space K.

For $p \in \mathbb{N}_0$, denote by (E_p) the collection of function $f \in L^2(E_{-\infty},\mu)$ such that the number

$$\|f\|^2_{2,p} = \sum_{n=0}^{\infty} \frac{1}{n!} \|D^n \mu f(0)\|^2_{\mathscr{L}_2^n(E_{-p})} \tag{2}$$

is finite. It is easy to see tht $(E_p) = \mathscr{D}(\Gamma(A^p))$ and $(E_p) \subset (E_q)$ for $p > q$, where $\Gamma(A) = \bigoplus_{n=0}^{\infty} A^{\oplus n}$. Set $(E_\infty) = \bigcap_{p \in \mathbb{N}} (E_p)$ and endow it with the projective limit topology. (E_∞) becomes a topological linear space and it will serves as the space of test functions. Denote the dual of (E_∞) by $(E_\infty)^*$. A member of $(E_\infty)^*$ will be called a generalized function on $E_{-\infty}$. When $E_p = \mathscr{S}_p$ and $E_{-\infty} = \mathscr{S}^*$, members of $(E_\infty)^*$ are known as generalized white noise functionals or Hida distributions (see for example, [6,7,11,16,17,18,22]). Denote the $(E_\infty)^*$ –(E_∞) pairing by $\langle\langle\cdot,\cdot\rangle\rangle$. We then extend the convolution $\mu F = \mu * F$ to a function $F \in (E_\infty)^*$ by defining

$$U_F(\xi) = \mu F(\xi) := e^{-\frac{1}{2}|\xi|_0^2} \langle\langle F, e^{<\cdot,\xi>}\rangle\rangle \tag{3}$$

for $\xi \in E_\infty$, where $(\cdot,\cdot)$ denotes the $E_{-\infty}$–E_∞ paining. U_F is also called the U–functional of F. Employing the definition (3) and using the same arguments as the proof of Corollary 2.5 in [16], one see easily that $(E_p)^*$ may be identified as the space (E_{-p}) which is defined as the completion of $(E_0) := L^2[E_{-\infty},\mu]$ with respect to the norm $\|\cdot\|_{2,-p}$ defined by

$$\|f\|_{2,-p}^2 = \sum_{n=0}^{\infty} \frac{1}{n!} \|D^n \mu f(0)\|^2_{\mathscr{L}_2^n(E_{-p})}.$$

Clearly, $(E_\infty)^* = \bigcup_{p\in\mathbb{N}} (E_{-p})$ and

$$(E_\infty) \subset (E_p) \subset (E_0) \subset (E_{-p}) \subset (E_\infty)^*.$$

Moreover, it can be shown that

$$(E_\infty) \subset (E_0) \subset (E_\infty)^*$$

is also a Gel'fand triplet.

The main goal of this paper is to characterize the generalized functions in terms of their U–functionals. It is shown that the U–functionals are nothing but Bargman–Segal analytic functions on some $\mathscr{C}E_p$. Locality of generalized functions are also obtained in this note. When $E_0 = L^2(\mathbb{R}^1)$, $A = -\frac{d^2}{dt^2} + (1+t^2)$, some other results were also obtained in [16,17,19].

Notations

$\mathbb{N}$: natural numbers

$\mathbb{N}_0$: $\mathbb{N} \cup \{0\}$

$\mathbb{Z}$: integers.

$\mathscr{L}^n[X]$: the space of all n–linear operators on a normed linear space X.

$\mathscr{L}_2^n[H]$: the space of all n–linear Hilbert–Schmidt operators on a Hilbert space H.

$\mathscr{C}X$: the complexification of a real normed linear space X, with the Euclidean norm induced by the X–norm, i.e. $\|z\|^2_{\mathscr{C}X} = \|x\|^2_X + \|y\|^2_X$ for $z = x + iy$ $(x,y \in X)$.

Tx^n := $T(x,...,x)$ (n copies), where $T \in \mathscr{L}^n[X]$ and $x \in X$.

$Tx_1...x_n$:= $T(x_1,...,x_n)$ for $x_1,...,x_n \in X$ and $T \in \mathscr{L}^n[X]$.

Df(a) : For notational convenience, this notation denotes the Frechet derivative at a in the direction of any subspace of E_{-p} if it exists.

$(\cdot,\cdot)$: $E_{-\infty}$–E_∞ or E_{-p}–E_p pairing.

$<\cdot,\cdot>_p$: inner product in E_p.

$\tilde{\xi}(x)$:= (x,ξ) for $\xi \in E_p(E)$ and $x \in E_{-p}(E^*)$.

$\langle\langle\cdot,\cdot\rangle\rangle$: $(E)^*$–(E) or (E_{-p})–(E_p) pairing.

2. A connection between Bargman–Segal analytic functions and U–functionals.

In this section, we shall show that (E_p) may be characterized by the space of Bargman–Segal analytic functions on $\mathscr{C}E_{-p}$, $p \in \mathbb{Z}$.

Let H be a real separable Hilbert space with norm $|\cdot| = \sqrt{<\cdot,\cdot>}$ and n_t the Gauss cylinder set measure with variance parameter $t > 0$.

2.1. Definition (Bargman–Segal analytic function) [18,19].

A single–valued function f defined on $\mathscr{C}H$ is called a Bargman–Segal analytic (or entire) function if it satisfies the following conditions :

(i) f is analytic in $\mathscr{C}H$ (see [16]);

(ii) $\sup_P \int_H\int_H |f(Px+iPy)|^2 n_t(dx)n_t(dy) < \infty$,

where the supremum is taken over the set of all finite rank orthogonal projections on H.

Denote the class of Bargman–Segal analytic functions defined above by $K[H,n_t]$. Then $K[H,n_t]$ is a separable Hilbert space with inner product given by

$$\{f,g\}_{H,t} = \lim_{n\to\infty} \int_H\int_H f(P_nx+iP_ny)\overline{g(P_nx+iP_ny)}n_t(dx)n_t(dy),$$

where $\{P_n\}$ is a sequence of finite rank orthogonal projections such that P_n converges strongly to the identify operator on H. The Definition 2.1 is essentially due to Segal [17,18]. When H is of finite dimension, $K[H,\mu_t]$ reduces to the Bargman analytic functions introduced in [1].

We shall show that $K[H,n_t]$ is a Foch–type space given as follows.

2.2. Definition [2]. For $\lambda > 0$, denote by $F^\lambda(H)$ the class of analytic functions on $\mathscr{C}H$ such that

$$\|f\|_{F^\lambda(H)} = \left\{\sum_{k=0}^{\infty} \frac{\lambda^k}{k!} \|D^k f(0)\|^2_{\mathscr{L}_2^n[H]}\right\}^{\frac{1}{2}} < \infty.$$

Suppose $f \in K[H,n_t]$ and P a finite rank orthogonal projection on H. Then it follows immediately from [19; Corollary 3.2] that we have

$$\int_H\int_H |f(Px+iPy)|^2 n_t(dx)n_t(dy)$$

$$= \sum_{k=0}^{\infty} \frac{(2t)^k}{k!}\left[\sum_{i_1,\ldots,i_k=1}^{N} (D^k f(0)e_{i_1}\ldots e_{i_k})^2\right], \tag{4}$$

where $\{e_1,\ldots,e_N)$ is an ONB in P(H).

As an immediate consequence of (4), we have

2.3. Theorem. $K[H,n_t] = F^{2t}(H)$ and $\|f\|^2_{F^{2t}(H)} = \{f,f\}_{H,t}$ for $f \in K[H,n_t]$.

2.4. Corollary. If $f \in (E_P)$, then $U_f \in F^1(E_{-P})$. Conversely, if $U \in F^1(E_{-P})$, then there exists a unique $f \in (E_P)$ such that $U_f = U$.

Proof. The first assertion follows by Theorem 2.3. Conversely, suppose that U is an arbitrary element in $F^1(E_{-p})$, we define

$$f(\varphi) = \sum_{n=0}^{\infty} \frac{1}{n!} \left\{ \sum_{i_1,\ldots,i_n=0}^{\infty} (D^n U(0) e_{i_1} \ldots e_{i_n})(D^n \mu\varphi(0) e_{i_1} \ldots e_{i_n}) \right\}, \tag{5}$$

for $\varphi \in (E_\infty)$, where $\{e_i\}$ is an ONB of H. Then $|f(\varphi)| \leq \|U\|_{F^1(E_{-p})} \|\varphi\|_{2,-p}$ so that $f \in (E_p)$. Clearly $U_f(\xi) = U(\xi)$ for all $\xi \in \mathscr{S}$. ///

From the proof of corollary 2.4, we also have

2.5. Corollary. For $U \in F'(E_{-p})$, $\langle\langle S^{-1}(U), \varphi \rangle\rangle$ is defined by the right hard side of (5).

3. A characterization of U–functionals.

For $p \in \mathbb{Z}$, denote by $\mathscr{A}_p$ the class of analytic function f defined on $\mathscr{C}E_{-p}$ satisfying the condition :

$$\|f\|_{\mathscr{A}_p} = \sup_{z \in \mathscr{C}E_{-p}} \{ |f(z)| e^{-\frac{1}{2}|z|_{-p}^2} \} < \infty$$

(cf. [8]).

Then $\mathscr{A}_p$ is a Banach space and, by restriction, $\mathscr{A}_p$ is continuous embedded in $\mathscr{A}_q$ for $p > q$. Set $\mathscr{A}_\infty = \bigcap_{p \in \mathbb{Z}} \mathscr{A}_p$ and endow it with the projective limit topology. $\mathscr{A}_\infty$ is a topological linear space.

3.1. Proposition.

(a) For $f \in F^1[E_{-p}]$, we have

$$\|f\|_{\mathscr{A}_p} \leq \|f\|_{F^1[E_{-p}]}. \tag{6}$$

(b) Let $\gamma = \sup_{x \in E_0} \langle A^{-1}x, x \rangle$ and let s be the smallest integer such that

$$\gamma^{2s} e^2 \|A^{-1}\|_{HS}^2 < 1. \tag{7}$$

Then there exists an $\alpha_p > 0$ such that

$$\alpha_p \|f\|_{F^1[E_{-p+s+1}]} \leq \|f\|_{\mathscr{A}_p}. \tag{8}$$

Proof.

(a) For $f \in F^1[E_{-p}]$, write $f(z) = \sum_{n=0}^{\infty} \frac{1}{n!} D^n f(0) z^n$, for $z \in \mathscr{C} E_{-p}$. Then we have

$$\begin{aligned} |f(z)| &\leq \sum_{n=0}^{\infty} \frac{1}{n!} \|D^n f(0)\|_{\mathscr{L}^n[E_{-p}]} |z|_{-p}^n \\ &\leq \sum_{n=0}^{\infty} \left(\frac{1}{\sqrt{n!}} \|D^n f(0)\|_{\mathscr{L}_2^n[E_{-p}]}\right) \left(\frac{1}{\sqrt{n!}} |z|_{-p}^n\right) \\ &\leq \|f\|_{F^1[E_{-p}]} \exp\left(\tfrac{1}{2} |z|_{-p}^2\right). \end{aligned} \tag{9}$$

Now (a) follows immediately from the estimation (9).

(b) The inequality (9) will be proved in two steps.

step 1 : Claim that

$$\|D^n f(0)\|^2_{\mathscr{L}^n[E_{-p+s}]} \leq (\gamma^{2s}n)^n e^n \|f\|^2_{\mathscr{A}_p}. \tag{10}$$

In fact, by the Cauchy formula, we may write

$$D^n f(0)h_1h_2\cdots h_n = (\frac{1}{2\pi i})^n \int_{|\lambda_j|=r_j}\cdots\int \frac{f(\lambda_1h_1+\lambda_2h_2+\cdots+\lambda_nh_n)}{\lambda_1^2\lambda_2^2\cdots\lambda_n^2} d\lambda_1 d\lambda_2\cdots d\lambda_n,$$

for $h_1, h_2, \cdots, h_n \in E_{-p+s} \subset E_{-p}$. Consequently,

$$|D^n f(0)h_1h_2\cdots h_n| \leq (r_1r_2\cdots r_n)^{-1}\|f\|_{\mathscr{A}_p} \exp(\frac{1}{2}\sum_{j=1}^n r_j|h_j|_{-p})^2$$

$$\leq (r_1r_2\cdots r_n)^{-1}\|f\|_{\mathscr{A}_p} \exp[\frac{1}{2}\gamma^{2s}(\sum_{j=1}^n r_j|h_j|_{-p+s})^2].$$

So that

$$\|D^n f(0)\|_{\mathscr{L}^n(E_{-p+s})} \leq (r_1r_2\cdots r_n)^{-1} \exp[\frac{1}{2}\gamma^{2s}(\sum_{j=1}^n r_j)^2 \|f\|_{\mathscr{A}_p}. \tag{11}$$

Choose $r_j = \frac{1}{\sqrt{n}}\gamma^{-s}$, one obtain the inequality (10).

step 2 : Proof of the inequality (8).

Let $\{h_j\}$ be an CONS in H_{-p+s+1}. The we have

$$\|f\|^2_{F^1[E_{-p+s+1}]} = \sum_{n=0}^{\infty} \frac{1}{n!}(\sum_{i_1\cdots i_n=1}^{\infty} |D^n f(0)h_{i_1}h_{i_2}\cdots h_{i_n}|^2)$$

$$\leq \sum_{n=0}^{\infty} \frac{1}{n!}\|D^n f(0)\|^2_{\mathscr{L}^n[E_{-p+s}]} \|A^{-1}\|^{2n}_{HS},$$

$$\leq \left\{\sum_{n=0}^{\infty} \frac{1}{n!}(\gamma^{2s}n)^n e^n \|A^{-1}\|^{2n}_{HS}\right\} \|f\|^2_{\mathscr{A}_p}, \tag{12}$$

by (step 1).

Applying the ratio test and employing the assumption (7), the series in $\{\cdots\}$ in equality (12) converges to a finite number, denoted by α_p^{-2}. Then the inequality (8) follows from (12). ///

3.2. Corollary. There exists a constant α_p such that

$$\alpha_p\|f\|_{2,p-s-1} = \alpha_p\|U_f\|_{F^1(E_{-p+s+1})} \leq \|U_f\|_{\mathscr{A}_p} \leq \|U_f\|_{F^1(E_{-p})} = \|f\|_{2,p}.$$

Consequenty $\mathscr{A}_\infty = \bigcap_{p\in\mathbb{Z}} F^1[E_{-p}]$.

3.3. Corollary.

In the case that $E_p = \mathscr{S}_p$, we may choose $s = 1$ in (7) so that

$$F^1[\mathscr{S}_{-p}] \subset \mathscr{A}_p \subset F^1[\mathscr{S}_{-p+2}].$$

Now, it follows from Corollary 2.4 and Proposition 3.1 we reach the following characterization of U–functionals.

3.4. Theorem. In order that a functionals φ defined on E_∞ is a U–functional of some function in $(E_\infty)^*$ if and only if there exists a $p \in \mathbb{Z}$ such that the following conditions are satisfied :

(i) φ is extendable to $\mathscr{C}E_p$ and analytic there.

(ii) $\varphi \in \mathscr{A}_{-p}$.

In other words the collection of U–functionals of $(E_\infty)^*$ is given by

$$\bigcup_{p\in\mathbb{Z}} \mathscr{A}_p = \bigcup_{p\in\mathbb{Z}} F^1[E_{-p}].$$

Proof. In fact, if $\varphi = U_F$ for some $F \in (E_p)$, then φ clearly satisfies (i) and (ii). Conversely, if φ satisfies (i) and (ii), then there exists an r such that $\varphi \in F^1(E_{p+r})$. It follows that there exists an $F \in (E_{-(p+r)})$ such that $\varphi = U_F$.

///

3.5. Corollary. Let $\{\Phi_n, \Phi_0\}$ be a sequence of functions in $(E_\infty)^*$ and $\{f_n, f_0\}$ their U–functionals. Then the following statements are equivalent :

(i) Φ_n converges to Φ_0 in $(E_\infty)^*$.

(ii) There exists some $p \in \mathbb{Z}$ such that f_n converges to f_0 in $F^1(E_p)$.

(iii) There exists some $q \in \mathbb{Z}$ such that f_n converges to f_0 in $\mathscr{A}_q$.

(iv) There exists some $r \in \mathbb{Z}$ and some constant C_r such that for $\xi \in E_r$

$$|f_n(\xi)| \leq C_p\, e^{\frac{1}{2}|\xi|_r^2} \text{ for all } n \in \mathbb{N}_0$$

and

$$f_n(\xi) \longrightarrow f_0(\xi) \text{ pointwise}$$

(cf. [17]).

Proof. It is easy to see that (i) $\Longrightarrow$ (ii) (by corollary 3.2), (ii) $\Longrightarrow$ (i) (by corollary 3.2) and (iii) $\Longrightarrow$ (iv). The implication "(iv) $\Longrightarrow$ (i)" follows by a similar argument as given in [17].

Acknowledgement

The author wishes to express his sincere gratitude to Professor T. Hida for his hospitality during visit Nagoya University in August 1990. The author is also grateful to National Science Council for financial support.

References

1. V. Bargman, On a Hilbert space of analytic functions and an associated integral transform, Part I, Commun, Pure Appl. Math. (1961), 187–214.

2. T. Dwyer, Partial differential equations in Fischer–Fock Spaces for the Hilbert–Schmidt Holomorpht Type, Bull. Amer. Math. Soc. 77(1971), 725–730.

3. I.M. Gel'fand and G.E. Shilov, Generalized Functions, Vol.2, Academic Press (1968).

4. I.M. Gel'fand and N.Ya. Vilenkin, Generalized Functions, Vol.4, Academic Press (1964).

5. L. Gross, Potential theory on Hilbert space, J. Funct. Anal. 1(1967), 123–181.

6. T. Hida, Analysis of Brownian functionals, Carleton Mathematical Lecture Notes No. 13(1975).

7. T. Hida, J. Potthoff and L. Streit, Dirichlet forms and white noise analysis, Commun. Math. Phys. 116(1988), 235–245.

8. Ju. G. Kondrat'ev, Nuclear spaces of entire functions in problems of infinite–dimensional analysis, Soviet Math. Dokl. Vol.22(1980), No.2, 588–592.

9. P. Krée, Solutions faibles d'équations aux dérivées functionnelles I, In Séminaire Pierre Lelong (analyse), Année 1972/73, 142–181, Lecture Notes in Math. V. 410, Springe–Verlag, Berlin, New York, 1974.

10. P. Krée, Solutions faibles d'équations aux dérivées functionnelles II, In "Séminaire Pierre Lelong (analyse)", Année 1973/74, 16–47, Lecture Notes in Math. V. 474, Springer–Verlag, Berlin, New York, 1975.

11. H.–H. Kuo, Brownian functionals and applications, Acta Appl. Math. 1(1983), 175–188.

12. Y.–J. Lee, Integral transform of analytic functions on abstract Wiener space, J. Funct. Anal. 47(1982), 153–164.

13. Y.–J. Lee, Sharp inequalities and regularity of heat semigroup on infinite dimensional space, J. Funct. Anal. 71(1987), 69–87.

14. Y.–J. Lee, Generalized functions on infinite dimensional spaces and its application to white noise calculus, J. Funct. Anal. 82(1989), 429–464.

15. Y.–J. Lee, On the convergence of Wiener–Ito decomposition, Bull. Inst. Math. Acad. Sinica, 17(1989), 305–312.

16. Y.–J. Lee, Analytic version of test functionals, Fourier transform and a characterization of measures in white noise claculus, to appear in J. Func. Anal.

17. J. Potthoff and L. Streit, A characterization of Hida distributions, preprint (Spring, 1990).

18. J. Potthoff and J.A. Yan, Some results about test and generalized functions of white noise, preprint, 1989.

19. I.E. Segal, Mathematical characterization of the physical vacum for a linear Bose–Einstein field. Illinois J. Math. 6(1962), 500–523.

20. I.E. Segal, The complex–wave representation of the free Boson field, In "Topics in Functional Analysis", 321–343, Advances in Mathematics Supplementary Studies, Vol. 3, Academic Press, New York, San Francisco, London, 1978.

21. J.A. Yan, A characterization of white noise functionals, preprint, 1990.

22. Y. Yokoi, Positive generalized Brownian functionals, preprint, 1989.

De Rham-Kodaira Decomposition and Fundamental Spaces of Wiener Functionals

Itaru Mitoma
Department of Mathematics, Saga University
Saga, 840, Japan

ABSTRACT

De Rham-Hodge-Kodaira decomposition theorems are proven on suitable infinite dimensional spaces of differential forms. As applications, we propose Sobolev spaces as the spaces of differential 0-forms.

1. Introduction

Recently differential forms in an infinite dimensional setting have been attracted by several authors[1,2,3,5,15] in connection with the index theory of infinite dimensional Dirac operators and the theory towards the infinite dimensional manifold. We have already proved de Rham-Hodge-Kodaira decomposition theorem on spaces of differential forms in L^2-sense. [1].

The purpose of this paper is to show that de Rham-Hodge-Kodaira decomposition holds on two types of spaces of differential n-forms in L^p-sense. The first space of differential 0-forms yields a generalization of the Sobolev space in the Malliavin calculus[17] and the second one gives the space proposed by Hida[7].

Let B be a real locally convex space, H a separable Hilbert space densely and continuously embedded in B and μ the standard Gaussian measure on B. Let K be another real separable Hilbert space, $\bigwedge^n(\mathrm{K})$ the n-fold anti-symmetric tensor product of K and

$$\bigwedge^n(B,K) = L^2(B,d\mu)\otimes\bigwedge^n(K).$$

Let A be a densely defined closed linear operator from H to K, A^* the adjoint and $\mathcal{P}(\bigwedge^n(K))$ a collection of all $\bigwedge^n(K)$-valued polynomials. We define a densely defined closable linear operator d_n from $\bigwedge^n(B,K)$ to $\bigwedge^{n+1}(B,K)$ by

$$d_n = (n+1)A_{n+1}(A \otimes I \otimes \ldots \otimes I\dot{D})$$

where A_n is the anti-symmetrization operator of degree n and D denotes the H-derivative .

Now we define a densely defined, non-negative self-adjoint operator Δ_n in $\bigwedge^n(B,K)$ such that

$$\Delta_n = d_n^* d_n + d_{n-1} d_{n-1}^*.$$

For $\omega \in \mathcal{P}(\bigwedge^n(K))$, we consider semi-norms $||\omega||_{p,s}, 1 < p < \infty, s = 0,1,2,\cdots$, such that

$$||\omega||_{p,s} = [\int_B ||(I+\Delta_n)^s \omega(x)||^p_{\bigwedge^n(K)} d\mu(x)]^{1/p}.$$
$$= ||(I+\Delta_n)^s \omega||_{L^p}.$$

From now on we use the following convention of notation ;

$$||\cdot||_{L^p} = [\int_B ||\cdot||^p_{\bigwedge^n(K)} d\mu(x)]^{1/p}.$$

Denote by $W_{p,s}(\bigwedge^n(K))$ the completion of $\mathcal{P}(\bigwedge^n(K))$ with respect to the semi-norm $||\cdot||_{p,s}$ and set

$$W^\infty(\bigwedge^n(K)) = \bigcap_{1<p<\infty} \bigcap_{s=0}^{\infty} W_{p,s}(\bigwedge^n(K)),$$

which is called the space of differential n-forms.

For a linear operator S on a Hilbert space, $\sigma(S)$ denotes the spectrum of S. Define $\mathrm{Im}\Delta_n = \Delta_n(W^\infty(\bigwedge^n(K)))$ and $\mathrm{Ker}\Delta_n = \{\omega \in W^\infty(\bigwedge^n(K));\ \ \Delta_n\omega = 0\}$.

Theorem 1. If $\inf \sigma(A^*A)\backslash\{0\} > 0$, then

$$W^\infty(\bigwedge^n(K)) = \mathrm{Im}\Delta_n \oplus \mathrm{Ker}\Delta_n.$$

Now we define another system of semi-norms whose type is different from the above one. It is well known [16] that there exists a unique non-negative self-adjoint operator $\Gamma(A^*A)$ in $L^2(B, d\mu)$ such that

$$\Gamma(A^*A)1 = 1$$

and

$$\Gamma(A^*A) :< x, \xi_1 >< x, \xi_2 > \cdots < x, \xi_n >:$$
$$=:< x, A^*A\xi_1 >< x, A^*A\xi_2 > \cdots < x, A^*A\xi_n >:,$$
$$\xi_j \in \mathfrak{D}(A^*A), \quad j = 1, 2, \cdots, n, \quad n \geq 1,$$

where $< x, \xi >$ is the bilinear form on $B \times H$ extended from the canonical bilinear form on $B \times B^*$ as the random variable, $\mathfrak{D}(\cdot)$ denotes the domain of the operator $\cdot$ and $:< x, \xi_1 >< x, \xi_2 > \cdots < x, \xi_n >:$ is the Wick polynomial defined in Section 2. We define the operator $\Gamma_n(AA^*)$ in $\bigwedge^n(K)$ by

$$\Gamma_n(AA^*) = AA^* \otimes AA^* \otimes \cdots \otimes AA^*,$$

which is non-negative and self-adjoint. Let

$$\Gamma_n = \Gamma(A^*A) \otimes \Gamma_n(AA^*)$$

and define the norm $||| \cdot |||_{p,s}$ on $\mathfrak{D}(\Gamma_n^s)(s \geq 0)$ by

$$||| \omega |||_{p,s} = ||(I + \Gamma_n)^s \omega||_{L^p}.$$

We denote by $V_s^p(\bigwedge^n(K)$ the completion of $\mathfrak{D}(\Gamma_n^s)$ with respect to the norm $||| \cdot |||_{p,s}$ and set

$$H_n(B) = \bigcap_{1<p<\infty} \bigcap_{s=0}^{\infty} V_s^p(\bigwedge^n(K)).$$

Define $\mathrm{Im}\Delta_n = \Delta_n(H_n(B))$ and $\mathrm{Ker}\Delta_n = \{\omega \in H_n(B); \ \ \Delta_n\omega = 0\}$.

THEOREM 2. Suppose that $A^*A \geq 1 + \epsilon$ with a constant $\epsilon > 0$. Then

$$H_n(B) = \mathrm{Im}\Delta_n \oplus \mathrm{Ker}\Delta_n.$$

REMARK. The spaces of differential n-forms $W^\infty(\bigwedge^n(K))$ and $H_n(B)$ are complete countably normed spaces[8].

2. Notations

First of all, we begin by giving general notations. For a real locally convex topological vector space E, we denote by E^* the topological dual space of E and by $< x, \xi >$ $(x \in E, \xi \in E^*)$ the canonical bilinear form on $E \times E^*$. If E is a Hilbert space, we denote by $\|\cdot\|_E$ and $(\cdot,\cdot)_E$ the norm and the inner product of E respectively. Identifying H and the dual space H^*, we have a Gelfand triplet

$$B^* \subset H \subset B.$$

Let μ be a Gaussian measure such that

$$\int_B e^{i<x,\xi>} d\mu(x) = e^{-\frac{\|\xi\|_H^2}{2}}, \quad \xi \in B^*.$$

Identifying $< x, \xi >, \xi \in B^*$, with the Gaussian random variable $< \cdot, \xi >$, we can extend the random variable to any vector $\in H$. We denote the extension by the same symbol $< \cdot, \cdot >$.

Let K be a real separable Hilbert space. For $f_i \in K, i = 1, 2, \cdots, n$, we denote the tensor product of them by

$$f_1 \otimes f_2 \otimes \cdots \otimes f_n.$$

We define the anti-symmetrization operator A_n and the n-exterior product by

$$\begin{aligned} A_n f_1 \otimes f_2 \otimes \cdots \otimes f_n &= \frac{1}{n!} \sum_{\sigma \in \mathfrak{S}_n} (\text{sign } \sigma) f_{\sigma(1)} \otimes f_{\sigma(2)} \otimes \cdots \otimes f_{\sigma(n)} \\ &= f_1 \wedge f_2 \wedge \cdots \wedge f_n \end{aligned}$$

where $\mathfrak{S}_n$ is a symmetric group of degree n.

We denote by $\bigwedge^n(\mathrm{K})$ the completion of the set of all finite linear combinations of such n-exterior products in the natural inner product [14] and set

$\wedge^0(K) = R.$

We introduce the Hilbert space

$$\bigwedge^n(B,K) = L^2(B,d\mu) \otimes \bigwedge^n(K)$$
$$= L^2(B,d\mu;\bigwedge^n(K)),$$

where $L^2(B,d\mu)$ is the Hilbert space of all square integrable functionals on B with respect to the measure μ. We regard it as a space of n-forms on (B,μ) in the L^2 sense.

We denote by $\mathfrak{D}$(T) the domain of operator T. The domain of the C^∞-vectors for T is defined by

$$C^\infty(T) = \bigcap_{n\in N} \mathfrak{D}(T^n).$$

Let A be a densely defined closed linear operator from H to K. We define $\mathcal{P}$ to be a collection of all functionals f on B of the form

$$f(x) = P(<x,\xi_1>,<x,\xi_2>,\cdots,<x,\xi_n>),$$
$$\xi_j \in C^\infty(A^*A), j = 1,2,\cdots,n,$$

where $P(t_1,t_2,\cdots,t_n)$ is a polynomial of n variables . The subspace $\mathcal{P}$ is dense in $L^2(B,d\mu)$. Let $\mathcal{P}(\wedge^n(K))$ be the subspace in $\wedge^n(B,K)$ spanned by $\wedge^n(K)$-valued functionals ω on B of the form

$$\omega(x) = \sum_{j=1}^{m} f_j(x)k_j$$

with $f_j \in \mathcal{P}, j = 1,2,\cdots,m$, and $k_j \in A_n(C^\infty(AA^*)\hat{\otimes}\cdots\hat{\otimes}C^\infty(AA^*))$ $\subset \wedge^n(K), j = 1,2,\cdots,m$, where $\hat{\otimes}$ denotes algebraic tensor product.

We define a densely defined closable linear operator d_n from $\wedge^n(B,K)$ to $\wedge^{n+1}(B,K)$ such that

$$d_n = (n+1)A_{n+1}(A\otimes I\otimes\ldots\otimes I\dot{D})$$

where D denotes the Fréchet derivative on H. Since $\mathfrak{D}(d_n)$ is dense

$\bigwedge^n(B,K)$, the adjoint d_n^* is also closable linear operator from $\bigwedge^{n+1}(B,K)$ to $\bigwedge^n(B,K)$. Henceforth we denote by the same symbols d_n and d_n^* the closed extensions of them.

Now we define a densely defined, non-negative self-adjoint operator Δ_n in $\bigwedge^n(B,K)$ such that

$$\Delta_n = d_n^* d_n + d_{n-1} d_{n-1}^*.$$

For the case $n = 0$, we define

$$\Delta_0 = d_0^* d_0.$$

Before proceeding to prove theorems, we review the concrete form of Δ_n [1]. We take a vector $\omega \in \mathcal{P}(\bigwedge^n(K))$ of the form

$$\omega(x) = P(< x, e_1 >, < x, e_2 >, \cdots, < x, e_m >) g_1 \wedge g_2 \wedge \cdots \wedge g_n,$$

$$e_j \in C^\infty(A^*A), \quad g_k \in C^\infty(AA^*), \ j = 1, 2, \cdots, m, \ k = 1, 2, \cdots, n.$$

Then $\Delta_n \omega(x)$ has the following concrete expression.

$$\Delta_n \omega(x) = \{\sum_{i=1}^{m} < x, A^*Ae_i > \frac{\partial P}{\partial t_i}(< x, e_1 >, < x, e_2 >, \cdots, < x, e_m >)$$

$$- \sum_{i,j=1}^{m} (A^*Ae_i, e_j)_H \frac{\partial^2 P}{\partial t_i \partial t_j}(< x, e_1 >, < x, e_2 >, \cdots, < x, e_m >)\} g_1 \wedge g_2 \wedge \cdots \wedge g_n$$

$$+ \sum_{k=1}^{n} P(< x, e_1 >, < x, e_2 >, \cdots, < x, e_m >) g_1 \wedge g_2 \wedge \cdots \wedge AA^* g_k \wedge \cdots \wedge g_n.$$

Also we have obtained another expression of Δ_n by making use of second quantization operator. We first recall that 'Wick products' $:< x, e_1 >< x, e_2 > \cdots < x, e_m >:$ of random variables $< x, e_i >, x \in B, e_i \in B^*, i = 1, 2, \cdots, m$, with respect to the measure μ are defined by the following recursion relation[10,16] :

$$:< x, e_1 >: = < x, e_1 >,$$

$$:< x, e_1 >< x, e_2 > \cdots < x, e_m >:=< x, e_1 >:< x, e_2 > \cdots < x, e_m >:$$
$$- \sum_{i=2}^{m} E[< x, e_1 >< x, e_i >] :< x, e_2 > \cdots < x, \check{e}_i > \cdots < x, e_m >:, m \geq 2,$$

where E denotes the expectation with respect to the measure μ. It is well known that $L^2(B, d\mu)$ admits the orthogonal decomposition(Fock-Wiener-Itô decomposition)

$$L^2(B, d\mu) = \oplus_{m=0}^{\infty} \mathfrak{L}_m(H),$$

where $\mathfrak{L}_0(H) = R$ and $\mathfrak{L}_m(H)$ is the closed subspace spanned by vectors of the forms $:< x, e_1 > \cdots < x, e_m >:, e_i \in B^*, i = 1, 2, \cdots, m$. The second quantization operator $d\Gamma(S)$ for a self-adjoint operator S in H is defined by the unique self-adjoint operator acting in $L^2(B, d\mu)$ such that for all $e_i \in \mathfrak{D}(S), i = 1, 2, \cdots, m, \ m \geq 1$,

$$d\Gamma(S) :< x, e_1 >< x, e_2 > \cdots < x, e_m >:$$
$$= \sum_{i=1}^{m} :< x, e_1 > \cdots < x, Se_i > \cdots < x, e_m >:$$

and

$$d\Gamma(S)1 = 0.$$

For the vector $\omega(x) =:< x, e_1 >< x, e_2 > \cdots < x, e_m >: g_1 \wedge g_2 \wedge \cdots \wedge g_n$ with $e_i \in C^\infty(A^*A)$, i = 1,2,$\cdots$,m, and $g_k \in C^\infty(AA^*), k = 1, 2, \cdots, n$,

$$\Delta_n \omega(x) = \sum_{i=1}^{m} :< x, e_1 > \cdots < x, A^*Ae_i > \cdots < x, e_m >: g_1 \wedge g_2 \wedge \cdots \wedge g_n$$
$$+ :< x, e_1 >< x, e_2 > \cdots < x, e_m >: \{\sum_{k=1}^{n} g_1 \wedge \cdots \wedge AA^*g_k \wedge \cdots \wedge g_n\}.$$

We define the operator $d\Gamma_n(AA^*)$ acting in $\wedge^n(K)$ by

$$d\Gamma_n(AA^*) = AA^* \otimes I \otimes \cdots \otimes I + I \otimes AA^* \otimes I \otimes \cdots \otimes I + \cdots + I \otimes I \otimes \cdots \otimes AA^*,$$

which is non-negative and self-adjoint[14]. It follows from the above argument that $\Delta_n = d\Gamma(A^*A) \otimes I + I \otimes d\Gamma_n(AA^*)$ on $\mathcal{P}(\wedge^n(K))$.

3. Proof of Theorem 1

We use the heat equation method[13] for the proof. Since Δ_n is non-negative self-adjoint, $-\Delta_n$ generates a strongly continuous semi-group $T_t = e^{-t\Delta_n}$ on $\bigwedge^n(B, K)$.

Define for $1 < p < \infty$,

$$\bigwedge\nolimits_p^n(B, K) = \{\omega; \|\omega\|_{L^p}^p = \int_B \|\omega(x)\|_{\bigwedge^n(K)}^p d\mu(x) < \infty\}.$$

First of all, we show the L^p-continuity of T_t for $p \geq 2$.

Lemma 1. For $p \geq 2, T_t$ is a bounded linear operator from $\bigwedge_p^n(B, K)$ to $\bigwedge_p^n(B, K)$. Further

$$T_t W^\infty(\bigwedge\nolimits^n(K)) \subset W^\infty(\bigwedge\nolimits^n(K)).$$

Proof. Take a vector $\omega(x) = \sum_{j=1}^m f_j(x) k_j$ with $f_j \in \mathcal{P}, j = 1, 2, \cdots, m$, and $k_j \in A_n(C^\infty(AA^*)\hat{\otimes} \cdots \hat{\otimes} C^\infty(AA^*)) \subset \bigwedge^n(K), j = 1, 2, \cdots, m$, where $\hat{\otimes}$ denotes algebraic tensor product. Since $\Delta_n = d\Gamma(A^*A) \otimes I + I \otimes d\Gamma_n(AA^*)$,

$$T_t = Q_t \otimes V_t,$$

where Q_t and V_t are contraction semi-groups with generators $-d\Gamma(A^*A)$ and $-d\Gamma_n(AA^*)$ on $L^2(B, d\mu)$ and $\bigwedge^n(K)$ respectively.

Taking the concrete expression of Δ_n and a manner of constructing the Ornstein-Uhlenbeck semi-group in [17] into account, we have

$$Q_t f_j(x) = \int_B f_j(e^{-tA^*A}x + \sqrt{1 - e^{-2tA^*A}}y) d\mu(y).$$

Hence for $F(x)$ of a finite linear combination of $\{f_j(x)\}$, we get

$$| Q_t F(x) |^p \leq Q_t | F |^p (x). \tag{3.1}$$

Let $\{g_i\}_{i=0}^\infty$ be a C.O.N.S.(Complete Orthonormal System) of $\bigwedge^n(K)$. Noticing (3.1) and

$$T_t \omega(x) = \sum_{j=1}^m (Q_t f_j)(x) \cdot V_t k_j,$$

we have

$$
\begin{aligned}
&\|T_t\omega\|_{L^p} \\
&= [\int_B (T_t\omega(x), T_t\omega(x))^{p/2}_{\wedge^n(K)} d\mu(x)]^{1/p} \\
&= [\int_B (\sum_{i=1}^{\infty} | Q_t(\sum_{j=1}^{m} f_j(V_t k_j, g_i)_{\wedge^n(K)})(x) |^2)^{p/2} d\mu(x)]^{1/p} \\
&\le [\int_B (\sum_{i=1}^{\infty}(Q_t | \sum_{j=1}^{m} f_j(V_t k_j, g_i)_{\wedge^n(K)} |^2)(x))^{p/2} d\mu(x)]^{1/p}.
\end{aligned}
$$

By making use of the integral expression of Q_t and the monotone convergence theorem, we estimate the right hand side of the above inequality by

$$
\begin{aligned}
&[\int_B \{(Q_t \sum_{i=1}^{\infty} | \sum_{j=1}^{m} f_j(V_t k_j, g_i)_{\wedge^n(K)} |^2)(x)\}^{p/2} d\mu(x)]^{1/p} \\
&\le [\int_B (Q_t(\sum_{i=1}^{\infty} | \sum_{j=1}^{m} f_j(V_t k_j, g_i)_{\wedge^n(K)} |^2)^{p/2})(x) d\mu(x)]^{1/p}. \qquad (3.2)
\end{aligned}
$$

Since Q_t is self-adjoint on $L^2(B, d\mu)$ and $Q_t 1 = 1$, (3.2) is equal to

$$
\begin{aligned}
&[\int_B (\sum_{i=1}^{\infty} | \sum_{j=1}^{m} f_j(x)(V_t k_j, g_i)_{\wedge^n(K)} |^2)^{p/2} d\mu(x)]^{1/p} \\
&= [\int_B (\sum_{j=1}^{m} f_j(x) V_t k_j, \sum_{j=1}^{m} f_j(x) V_t k_j)^{p/2}_{\wedge^n(K)} d\mu(x)]^{1/p}. \qquad (3.3)
\end{aligned}
$$

By the contractivity of V_t, (3.3) is dominated by

$$
[\int_B (\omega(x), \omega(x))^{p/2}_{\wedge^n(K)} d\mu(x)]^{1/p} = \|\omega\|_{L^p},
$$

which, together with the fact that

$$
\|T_t\omega\|_{L^p} \le \|T_t\omega\|_{L^2} \qquad (3.4)
$$

for $1 < p \le 2$, completes the proof of Lemma 1.

Now we return to the proof of Theorem 1. By Lemma 1, we have for $p \ge 2$ and $1/p + 1/q = 1$,

$$
\|T_t\omega\|_{L^p}
$$

$$\leq [(\int_B ||T_t\omega(x)||^2_{\bigwedge^n(K)} d\mu(x))^{1/2} (\int_B ||T_t\omega(x)||^{2(p-1)}_{\bigwedge^n(K)} d\mu(x))^{1/2}]^{1/p}$$
$$\leq [||T_t\omega||_{L^2} ||T_t\omega||^{p-1}_{L^{2(p-1)}}]^{1/p}$$
$$\leq ||T_t\omega||^{1/p}_{L^2} ||\omega||^{1/q}_{L^{2(p-1)}}. \tag{3.5}$$

Lemma 1, together with (3.5) and the commutativity of $(I+\Delta_n)$ and T_t, gives that

$$||(I+\Delta_n)^k(T_t\omega - T_s\omega)||_{L^p}$$
$$\leq C_{1,p}||(I+\Delta_n)^k(T_t\omega - T_s\omega)||^{1/p}_{L^2}||(I+\Delta_n)^k\omega||^{1/q}_{L^{2(p-1)}}. \tag{3.6}$$

Here and in the sequel, $C.$ denotes the constant.
Since we have already proved that $\{T_t\omega\}$ forms a Cauchy sequence in $W_2^\infty(\bigwedge^n(K))$
$= \bigcap_{s=0}^\infty W_{2,s}(\bigwedge^n(K))$ [1], the above inequalities (3.4) and (3.6) implies that $\{T_t\omega\}$ forms a Cauchy sequence in $W^\infty(\bigwedge^n(K))$. Hence the strong limit

$$\hat{\omega} = \lim_{t\to\infty} T_t\omega$$

exists in $W^\infty(\bigwedge^n(K))$. By Lemma 1 and the semi-group property of T_t, we get

$$T_t\hat{\omega} = \hat{\omega}.$$

Therefore $\hat{\omega} \in \operatorname{Ker}\Delta_n$.

Again by (3.5), we have for $p \geq 2$,

$$||(I+\Delta_n)^k(T_t\omega - \hat{\omega})||_{L^p}$$
$$\leq ||(I+\Delta_n)^k(T_t\omega - \hat{\omega})||^{1/p}_{L^2}||(I+\Delta_n)^k(\omega - \hat{\omega})||^{1/q}_{L^{2(p-1)}}$$
$$\leq e^{-t\frac{\delta}{p}}||(I+\Delta_n)^k\omega||^{1/p}_{L^2}||(I+\Delta_n)^k(\omega - \hat{\omega})||^{1/q}_{L^{2(p-1)}},$$

where $\delta = \inf \sigma(\Delta_n)\backslash\{0\} > 0$ guaranteed by the condition of Theorem 1 [4].
This implies for $p \geq 2$,

$$\int_0^\infty ||T_t\omega - \hat{\omega}||_{p,k}dt$$
$$\leq C_{2,p}||\omega||_{2(p-1),k}. \tag{3.7}$$

For $1 < p \leq 2$, (3.4) and the condition of Theorem 1 also give

$$\int_0^\infty \|T_t\omega - \hat{\omega}\|_{p,k} dt$$
$$\leq C_{3,p}\|\omega\|_{2,k}. \tag{3.8}$$

The inequalities (3.7) and (3.8) yield that for $\omega \in W^\infty(\wedge^n(K))$, the strong integral

$$G\omega = \int_0^\infty (T_t\omega - \hat{\omega})dt$$

exists in $W^\infty(\wedge^n(K))$. Since Δ_n is continuous on $W^\infty(\wedge^n(K))$, we have

$$\begin{aligned}\Delta_n G\omega &= \int_0^\infty \Delta_n(T_t\omega - \hat{\omega})dt \\ &= \int_0^\infty -\frac{\partial}{\partial t}T_t\omega dt \\ &= \omega - \hat{\omega}.\end{aligned}$$

Thus the proof Theorem 1 is complete.

4. Proof of Theorem 2

Since we have already proved that de Rham-Hodge-Kodaira decomposition holds for the space

$$H_{n,2}(B) = \bigcap\nolimits_{s=0}^{\infty} V_s^2(\bigwedge\nolimits^n(K)),$$

it is sufficient to prove $H_n(B) = H_{n,2}(B)$. The system of semi-norms

$$|||\ \omega\ |||_{p,s} = \|(I + \Gamma_n)^s\omega\|_{L^p}$$

is clearly equivalent to the system of aemi-norms

$$||||\omega||||_{p,s} = \|\Gamma_n^s\omega\|_{L^p}.$$

Take a vector $\omega(x) = \sum_{j=1}^{m} f_j(x)k_j$ as in the proof of Lemma 1. Let $\{\gamma_i\}$ be an independent sequence of coin tossing with values +1 and −1. By the Khinchin inequality[17], we have

$$\|\Gamma_n^s \omega(x)\|^p_{\bigwedge^n(K)}$$

$$= (\sum_{i=1}^{\infty} | \sum_{j=1}^{m} \Gamma(A^*A)^s f_j(x)(\Gamma_n(AA^*)^s k_j, g_i)_{\bigwedge^n(K)} |^2)^{p/2}$$

$$\leq C_p E[| \sum_{i=1}^{\infty} \gamma_i (\sum_{j=1}^{m} \Gamma(A^*A)^s f_j(x)(\Gamma_n(AA^*)^s k_j, g_i)_{\bigwedge^n(K)}) |^p]. \tag{4.1}$$

Let denote the Ornstein-Uhlenbeck semigroup by S_t. By making use of (4.1), the hypercontractivity of S_t and assumption $A^*A \geq 1 + \epsilon$ with a constant $\epsilon > 0$. and following [11], we have some $s' > s$ and another constant c_p coming from the Khinchin inequality such that

$$\int_B \|\Gamma_n^s \omega(x)\|^p_{\bigwedge^n(K)} d\mu(x)$$

$$\leq C_p E[\int_B | \sum_{i=1}^{\infty} \gamma_i (\sum_{j=1}^{m} \Gamma(A^*A)^s f_j(x)(\Gamma_n(AA^*)^s k_j, g_i)_{\bigwedge^n(K)}) |^p \, d\mu(x)]$$

$$\leq C_p E[\{\int_B | \sum_{i=1}^{\infty} \gamma_i S_t^{-1} (\sum_{j=1}^{m} \Gamma(A^*A)^s f_j(x)(\Gamma_n(AA^*)^s k_j, g_i)_{\bigwedge^n(K)}) |^2 \, d\mu(x)\}^{\frac{p}{2}}]$$

$$\leq C_p E[(\sum_{i=1}^{\infty} \gamma_i \{\int_B | S_t^{-1} (\sum_{j=1}^{m} \Gamma(A^*A)^s f_j(x)(\Gamma_n(AA^*)^s k_j, g_i)_{\bigwedge^n(K)}) |^2 \, d\mu(x)\}^{\frac{1}{2}})^p]$$

$$\leq c_p C_p (\sum_{i=1}^{\infty} \int_B | S_t^{-1} (\sum_{j=1}^{m} \Gamma(A^*A)^s f_j(x)(\Gamma_n(AA^*)^s k_j, g_i)_{\bigwedge^n(K)}) |^2 \, d\mu(x))^{\frac{p}{2}}$$

$$\leq c_p C_p (\int_B \sum_{i=1}^{\infty} | \sum_{j=1}^{m} \Gamma(A^*A)^{s'} f_j(x)(\Gamma_n(AA^*)^s k_j, g_i)_{\bigwedge^n(K)} |^2 \, d\mu(x))^{\frac{p}{2}}$$

$$= c_p C_p \|\Gamma(A^*A)^{s'} \otimes \Gamma_n(AA^*)^s \omega\|^p_{L^2}.$$

Again by assumption $A^*A \geq 1 + \epsilon$,

the right hand side of the above inequality $\leq c_p C_p \|\Gamma(A^*A)^{s'} \otimes \Gamma_n(AA^*)^{s'} \omega\|^p_{L^2}$

$$= c_p C_p \|\Gamma_n^{s'} \omega\|^p_{L^2}.$$

Since the system of seminorms $\|\Gamma_n^s \omega\|_{L^2}$ is equivalent to the system of seminorms $||| \, \omega \, |||_{2,s}$, which implies

$$H_{n,2}(B) \subset H_n(B).$$

The opposite inclusion is clear, which completes the proof.

REFERENCES

1. A.Arai and I.Mitoma, De Rham-Hodge-Kodaira decomposition in ∞-dimensions, (Submitted, 1990).

2. A.Arai, A general class of infinite dimensional Dirac operators and path integral representation of their index, (Hokkaido Univ. Preprint ser. in Math. No.61 1989).

3. A.Arai, Path integral representation of the index of Kahler-Dirac operators on an infinite dimensional manifold, *J.Funct.Anal.* 82 (1989) p.330-369.

4. P.A.Deift, Applications of commutation formula, *Duke Math.J.* 45 (1978) p.267-310.

5. A.Jaffe, A.Lesniewski and J.Weitsman, Index of a family of Dirac operators on loop space, *Commun.Math.Physics* 112 (1987) p.75-88.

6. T.Hida, *Brownian motion*, (Springer, Berlin Heidelberg New York, 1980).

7. T.Hida, J.Potthoff and L.Streit, Dirichlet forms and white noise analysis, *Commun.Math.Physics* 116 (1988) p.235-245.

8. I.M.Gelfand and G.E.Shilov, *Generalized functions* 2, (Academic Press, New York London, 1964).

9. I.M.Gelfand and N.Ya.Vilenkin, *Generalized functions* 4, (Academic Press, New York London, 1964).

10. J.Glimm and A.Jaffe, *Quantum Physics*, (Springer, New York,1981).

11. H.Korezlioglu and A.S.Ustunel, A new class of distributions on Wiener spaces, (Preprint 1990).

12. P.Malliavin, Stochastic calculus of variation and hypoelliptic operators, *Proc.International symp.S.D.E.Kyoto*, (Kinokuniya, Tokyo1978).

13. S.Nishikawa, N.Ramachandran and P.Tondeur, The heat equation for Riemanian foliations, *Trans.Amer.Math.Soc.* 319 (1990) p.619-630.

14. M.Reed and B.Simon, *Methods of Modern Mathematical Physics*, Vol.1 (Academic Press, San Diego New York Berkeley Boston London Sydney Tokyo Toronto 1972).

15. I.Shigekawa, DeRham-Hodge-Kodaira's decomposition on an abstract Wiener space, *J.Math.Kyoto Univ.* 26 (1986) p.191-202

16. B.Simon, *The $P(\phi)_2$ Euclidean (Quantum) Field Theory*, (Princeton Univ. Press, Princeton, NJ 1974).

17. S.Watanabe, *Lectures on stochastic differential equations and Malliavin's calculus*, (Springer, Berlin Heidelberg New York Tokyo 1984).

ON THE EXISTENCE OF OPTIMAL RELAXED CONTROL FOR STOCHASTIC DIFFERENTIAL EQUATIONS

MAKIKO NISIO
Department of Mathematics, Kobe University
Rokko, Kobe, 657, Japan

ABSTRACT

This paper deals with optimal controls for a system governed by stochastic partial differential equation. We use an admissible control depending on time and space variable and a Wiener process valued in Hilbert space with nuclear covariance operator. Applying various evaluations, we can prove that the solution depends on relaxed control continuously. This fact derives the existence of an optimal relaxed control.

1. Introduction

In this article we are concerned with control problems of system governed by the following stochastic partial differential equation(SPDE)

$$dq(t,x) = \{\sum_{ij=0}^{d} \frac{\partial}{\partial x_i}(a^{ij}(x,U(t,x))\frac{\partial}{\partial x_j}q(t,x) + f^i(x,U(t,x))\}dt + (\sum_{i=0}^{d} b^i(x)\frac{\partial}{\partial x_i}q(t,x) + g(x))dW(t),$$

$$0 < t \le T, \ x \in R^d, \quad (1.1)$$

where $\frac{\partial}{\partial x_0}$ = identity, U = an admissible control and W is a Wiener process in Hilbert space $\mathfrak{E}$ with nuclear covariance operator. The problem is to minimize a given criterion by choose a suitable control. The main aim of this paper is to show the existence of an optimal relaxed control, under the conditions of ellipticity and regularity(Theorom 4.1). This result is a generalization of previous one,[6] where we used space homogeneous controls and a finite dimensional Wiener process. For our control U, we assume Lipschitz condition in x, with a given Lipschitz constant(see (2,1)). So, we need careful evaluations, although we can apply the same arguments as 6.

Section 2 is preliminaries and we introduce a relaxed system. In section 3, we will prove the continuous dependence of the solution on the relaxed system(Theorem 3.1). This continuity derives the existence of an optimal one(Theorem 4.1).

2. Preliminaries

Let us set

$\mathfrak{U} = \{ u;\ R^d \to \Gamma \ \text{ such that } \ |u(x) - u(y)| \leq K|x-y| \ \}$
where Γ is a given convex compact subset of R^L and K a given constant. Endowing with the compact uniform topology, $\mathfrak{U}$ becomes a convex compact metric space. $\sigma(\mathfrak{U})$ denotes the topological σ-field on $\mathfrak{U}$. Λ denotes the set of all Borel measures λ on $[0,T]\times\mathfrak{U}$, such that

$$\lambda([0,t]\times\mathfrak{U}) = t, \qquad \text{for } t \leq T, \tag{2.1}$$

carrying the weak convergence topology. For such λ, there is a kernel λ', such that $\lambda([0,t]\times A) = \int_0^t \lambda'(s,A)ds$, $\lambda'(t,\ )$ is a probability measure on $\sigma(\mathfrak{U})$ for each t and $\lambda'(\ ,A)$ is Borel measurable for each A. Since $[0,T]\times\mathfrak{U}$ is compact, Λ is also a compact metric space. Let $\mathfrak{P} = \mathfrak{P}(\Lambda)$ be the space of all probabilities on $(\Lambda,\sigma(\Lambda))$, carrying the weak convergence topology. Then Prohorov's theorem asserts that $\mathfrak{P}$ is a compact metric space. Moreover, noting

$$|\int_0^T\int_{\mathfrak{U}} a(u)b(t)\lambda'(t,du)dt| \le \sup_u|a(u)|\,\|b\|_{L^1[0,T]} \quad (2.2)$$

we can easily see that, for $A \in C(\mathfrak{U})$ and $B \in L^2[0,T]$,

$$\int A(u)B(t)\lambda_n(dt,du) \to \int A(u)B(t)\lambda(dt,du) \quad (2.3)$$

whenever $\lambda_n \to \lambda$ weakly.

Let $\mathfrak{E}$ be a separable real Hilbert space and Q a non-negative self-adjoint nuclear operator on $\mathfrak{E}$. Now we introduce a relaxed system,[1,6]

Definition 2.1. $R = (\Omega,F,F_t,P,W,\mu)$ is called a relaxed system, if

(i) (Ω,F,F_t,P) is a probability space

(ii) W is an F_t-Wiener process in $\mathfrak{E}$ with covariance operator Q

and

(iii) μ is an F_t-adapted Λ-valued random variable (Λ-r.v. in short).

In particular, if $\mu'(t,du) = \delta_{U(t)}(du)$, where δ_r = delta measure at r, then (Ω,F,F_t,P,W,U) is called an admissible system, and denoted by $A = (\Omega,F,F_t,P,W,U)$.

Let $\pi(R)$ be the image measure of (W,μ) on $C([0,T];\mathfrak{E})\times\Lambda$. Again endowing with the weak convergence topology on the space $\Pi = \{ \pi(R), R\in\mathfrak{R} \}$, where $\mathfrak{R}$ denotes the set of all relaxed controls, Prohorov's theorem asserts the following proposition.

Proposition 2.1. Π is a compact metric space.

For simplicity, we say R_n converges to R, if $\pi(R_n) \to$

$\pi(R)$ weakly.

Hereafter we always assume the following (A1)~(A5).

(A1) $a^{ij} : R^d \times \Gamma \to R^1$, $i,j = 0,\cdots,d$, are bounded and Lipschitz continuous

(A2) $b^i : R^d \to \mathbb{C}$, $i = 0,\cdots,d$, are bounded and differentiable in $\mathbb{C}$ and their derivatives $\frac{\partial b^i}{\partial x_k} : R^d \to \mathbb{C}$, are bounded and uniformly continuous

(A3) $g : R^d \to \mathbb{C}$, $f^i : R^d \times \Gamma \to R^1$, $i = 0,\cdots,d$, are bounded and uniformly continuous

(A4) there is a positive p, such that $\|g(x)\|_Q$ and $\bar{f}$ are in Ψ_p $\left(= L^2(R^d, (1+|x^2|)^p dx)\right)$, where $\bar{f} = \sup_{\gamma\in\Gamma} |f^i(\;,\gamma)|$

(A5) ellipticity; there is a positive m_0, such that

$$2\sum_{i,j=1}^{d} a^{ij}(x,\gamma)y_i y_j - \left\| \sum_{i=1}^{d} b^i(x)y_i \right\|_Q^2 \geq m_0|y|^2,$$

$$x \in R^d,\ \gamma \in \Gamma,\ y = (y_1,\cdots,y_d) \in R^d.$$

For simplicity, we say

$$|a^{ij}(x,\gamma)|,\ |\bar{f}(x)|,\ |\bar{f}|_p,\ \|b^i\|_{\mathbb{C}},\ \left\|\frac{\partial b^i}{\partial x_k}\right\|_{\mathbb{C}},\ |\|g(x)\|_Q|_p \leq K_0,$$
$$|a^{ij}(x,\gamma) - a^{ij}(x',\gamma')| \leq c_0(|x - x'| - |\gamma - \gamma'|),$$

where $|\ |_p$ = norm in Ψ_p, and by structure constants we mean T, K, m_0, c_0, K_0 and Trace Q.

Now we are concerned with a Gelfand triple $(H^1, L^2(R^2), H^{-1})$ and put $H = L^2(R^d)$, $\|\ \| = \|\ \|_H$, $\|\ \|_1 = \|\ \|_{H^1}$ and $\|\ \|_* = \|\ \|_{H^{-1}}$.

Let us consider SPDE for $R = (\Omega, F, F_t, P, W, \mu)$,

$$\begin{aligned} dq(t) &= \tilde{L}(t;\mu)q(t)dt + Mq(t)dW(t),\ 0 < t \leq T, \\ q(0) &= \varphi\ (\in \Psi_p), \end{aligned} \tag{2.4}$$

where

$$\tilde{L}(t,\mu)\psi = \sum_{i,j=0}^{d} \left(\frac{\partial}{\partial x_i}(\tilde{a}^{ij}(t,x;\mu)\frac{\partial\psi}{\partial x_j} + \tilde{f}^i(t,x;\mu)\right)$$

and

$$M\psi = \sum_{i,j=0}^{d} b^i(x)\frac{\partial\psi}{\partial x_i} + g(x)$$

noting that $\bar{h}(t,x;\mu) = \int_{\mathcal{U}} h(x,u(x))\mu'(t,du)$ is determined in $L^2(dt\times dP)$.

First we will recall the existence theorem in our convenient way.

Theorem[3,5] There exists a unique solution $q = q(\ ,\varphi,R) \in L^2(\Omega\times[0,T];H^1)\cap L^2(\Omega;C([0T];\Psi_p))$. Moreover, there is a positive N, depending only on structurc constants, such that

$$E(\sup_{t\leq T} |q(t)|_p^2) \leq N(|\varphi|_p^2 + 1) \tag{2.5}$$

$$E\int_0^T \|q(t)\|_1^2 dt \leq N(|\varphi|_p^2 + 1) \tag{2.6}$$

and

$$E(|q(t)|_p^4) \leq N(|\varphi|_p^4 + 1) \tag{2.7}$$

These evaluations still hold for p=0.

We will derive the modulus of continuity of q(t) uniformly in R. Hereafter K_i stands for a constant depending only on structure constants.

Proposition 2.2

$$E\ \|q(t) - q(s)\|_*^2 \leq K_1(\ 1+\|\varphi\|^2)|t-s| \tag{2.8}$$

Proof Let $\eta_j \in C_0^\infty$(smooth function with compact support) and $\eta_j, j=1,\cdots$ be an orthonormal base of H^1. Then we have

$$\|q(t) - q(s)\|_*^2 = \sum_{j=1}^\infty (q(t)-q(s),\eta_j)^2$$

$$\leq 2(t-s) \int_s^t \|\mathcal{L}(\theta;\mu)q(\theta)\|_*^2 d\theta$$

$$+ 2\sum_{j=1}^\infty \left(\int_s^t (Mq(\theta),\eta_j)dW(\theta)\right)^2 \tag{2.9}$$

$$\left\| \frac{\partial}{\partial x_i}\left(\tilde{a}^{ij}(\theta;\mu)\frac{\partial q(\theta)}{\partial x_j}\right)\right\|_* \leq \left\|\tilde{a}^{ij}(\theta;\mu)\frac{\partial q(\theta)}{\partial x_j}\right\|$$

$$\leq K_2\left\|\frac{\partial q(\theta)}{\partial x_j}\right\| \leq K_2\|q(\theta)\|_1. \tag{2.10}$$

From (2.6) and (2.10), we see

$$E\int_s^t \|L(\theta,\mu)q(\theta)\|_*^2 d\theta \leq K_3(\|\varphi\|^2 + 1). \tag{2.11}$$

Taking the base e_k of $\mathfrak{E}$, such that $Qe_k = \zeta_k e_k$, we have

$$E\sum_{j=1}^{\infty}\left(\int_s^t (Mq(\theta),\ \eta_j)dW(\theta)\right)^2$$

$$= \sum_{j=1}^{\infty} E\int_s^t \|(Mq(\theta),\ \eta_j)\|_Q^2\, d\theta$$

$$= \sum \zeta_k \int_s^t E\left\|\sum_{i=1}^{d} b^i{}_k \frac{\partial q(\theta)}{\partial x_i} + g_k\right\|_*^2 d\theta \tag{2.12}$$

where $c_k(x) = (c(x), e_k)_{\mathfrak{E}}$ for $c = b^i,\ g$. From (A2), we see

$$\left\|b^i{}_k \frac{\partial q(\theta)}{\partial x_i}\right\|_* \leq K_4\left\|\frac{\partial q(\theta)}{\partial x_i}\right\|_* \leq K_4\|q(\theta)\|. \tag{2.13}$$

Appealing to (2.5), we can conclude (2.8) by (2.9) and (2.11) ~ (2.13).

3. Continuous dependence of $q(\ ,\varphi,R)$ on R

Putting $\hat{\psi}(\tau) = \int_{-\infty}^{\infty} e^{-2\pi i\tau t}\psi(t)dt$, for H^1-function ψ on R^1, and

$$\|\psi\|^2_{\mathscr{H}_r(D)} = \int_{-\infty}^{\infty}\|\psi(t)\|^2_{H^1(D)}dt + \int_{-\infty}^{\infty}|\tau|^{2r}\|\hat{\psi}(\tau)\|^2_{H^{-1}(D)}d\tau,$$

we will define $\mathscr{H}_r(D)$ and $\mathscr{H}_r(T,D)$ as follows,

$$\mathscr{H}_r(D) = \{\ \psi\ ;\ R^1 \to H^1 \text{ such that } \|\psi\|_{\mathscr{H}_r(D)} < \infty\ \}$$

and

$$\mathcal{K}_r(T,D) = \{\psi = \text{restriction of } \varphi \text{ on } [0,T], \varphi\in\mathcal{K}_r(D)\}$$

with the norm

$$\|\psi\|_{\mathcal{K}_r(T,D)} = \inf\{\|\psi\|_{\mathcal{K}_r(D)}; \varphi = \psi \text{ a.e. on } [0,T]\}.$$

Let D be the centered open sphere with radius ℓ. Then the following evaluation holds, for $r=2^{-3}$

$$E\|q(\ ,\varphi,R)\|^2_{\mathcal{K}_r(T,D)} \le K_5(1+\|\varphi\|^2). \tag{3.1}$$

Since the imbedding $\mathcal{K}_r(T,D) \to L([0,T]\times D)$ is compact[4], (3.1) concludes that $\{q(\ ,\varphi,R),\ R\in\mathfrak{R}\}$ is totally bounded in law as $L^2([0,T];\ H(D))$-r.v. On the other hand (2.5) yields

$$E\int_{|x|\ge\ell} |q(t,x)|^2 dx \le K_6(\ 1 + |\varphi|_p^{\ 2})\ (1+\ell)^{-p}.$$

Hence, we have

Lemma 3.1. $\{ q(\ ,\varphi,R); R\in\mathfrak{R}\}$ is totally bounded in law as $L^2([0,T];\ H)$-r.v.

Now we will prove

Theorem 3.1 As $R_n \to R$,

$$q(\ ,\varphi,R_n) \to q(\ ,\varphi,R) \text{ in law as } L^2([0,T];H)\text{-r.v.} \tag{3.2}$$

Proof. Put $R_n = (W_n,\mu_n)$, $q_n = q(\ ,R_n)$, $R = (W,\mu)$ and $q = q(\ ,R)$. Appealing to Lemma 3.1, we can choose a subsequence n_j, such that $(W_{n_j},\mu_{n_j},q_{n_j})$ converges in law as $C([0,T];\mathfrak{C})\times\Lambda\times L^2([0,T];H)$-r.v.. Therefore Skorohod's theorem asserts that there exist $(\hat{W}_{n_j},\hat{\mu}_{n_j},\hat{q}_{n_j})$ and $(\hat{W},\hat{\mu},\hat{q})$ on a probability space $(\Omega,\hat{\mathcal{F}},\hat{P})$ such that

$$(\hat{W}_{n_j},\hat{\mu}_{n_j},\hat{q}_{n_j}) \text{ has the same law as } (W,\mu,q) \tag{3.3}$$

and, with probabilty 1,

$$\|\hat{W}_{n_j}(t) - \hat{W}(t)\|_{\mathfrak{C}} \to 0 \quad \text{uniformly on } [0,T] \tag{3.4}$$

$$\hat{\mu}_{n_j} \to \hat{\mu} \quad \text{weakly} \tag{3.5}$$

$$\hat{q}_{n_j} \to \hat{q} \quad \text{in } L^2([0,T];H). \tag{3.6}$$

Hence $(\hat{W},\hat{\mu})$ has the same law as (W,μ). Moreover we may assume that $\hat{q}_{n_j}$ converges to $\hat{q}$ in $L^2([0,T]\times\Omega;H^1)$ weakly by (2.6). So, $\hat{q} \in L^2([0,T]\times\Omega;H^1)$.

Now we will show that $\hat{q}$ is a solution of Ex.2.4 for $(\hat{W},\hat{\mu})$. By (A1) and the definition of $\mathfrak{U}$, $\tilde{a}^{ik}(t,x;\mu_{n_j})$ satisfies Lipschitz condition in x and

$$\tilde{a}^{ik}(t;\mu_{n_j}) \in H^1_{loc}, \qquad \text{w.p.1}, \tag{3.7}$$

So, we get, for $\eta \in C_0^\infty$

$$< \mathfrak{L}(s;\hat{\mu}_{n_j})\hat{q}_{n_j}(s),\eta > = (\hat{q}_{n_j}(s),\mathfrak{L}^*(s;\hat{\mu}_{n_j})\eta) \tag{3.8}$$

where $\mathfrak{L}^*$= dual operator of $\mathfrak{L}$, and

$$|(\hat{q}_{n_j}(s),\mathfrak{L}^*(s;\hat{\mu}_{n_j})\eta) - (\hat{q}(s),\mathfrak{L}^*(s;\hat{\mu}_{n_j})\eta)|$$

$$\le K_7 \|\eta\|_{H^2} \|\hat{q}_{n_j}(s) - \hat{q}(s)\|. \tag{3.9}$$

On the other hand, (2.3) derives

$$\int_0^t(\hat{q}(s),\mathfrak{L}^*(s;\hat{\mu}_{n_j})\eta)ds \to \int_0^t(\hat{q}(s),\mathfrak{L}^*(s;\hat{\mu})\eta)ds \quad \text{w.p.1.}$$

Recalling (3.8) and (3.9), we get

$$\int_0^t<\mathfrak{L}(s;\hat{\mu}_{n_j})\hat{q}_{n_j}(s),\eta>ds \to \int_0^t<\mathfrak{L}(s;\hat{\mu})\hat{q}(s),\eta>ds \quad \text{w.p.1}$$

since $\hat{q} \in L^2([0,T]\times\Omega;H^1)$.

Next we will treat stochastic integrals. From (A2) and (2.8), we see

$$E|\int_s^t (b^i(x)(\partial_i \hat{q}_{n_j}(\theta) - \partial_i \hat{q}_{n_j}(s)),\eta)d\hat{W}_{n_j}(\theta)|^2$$

$$= \int_s^t E\|(\hat{q}_{n_j}(\theta) - \hat{q}_{n_j}(s),\partial_i(b^i(x)\eta)\|_Q^2 d\theta$$

$$\leq K_8(1 + \|\varphi\|^2)\|\eta\|_H^2 2(t-s)^2,$$

where $\partial_i = \frac{\partial}{\partial x_i}$. Hence, using the routine, we get

$$\int_0^t (M\hat{q}_{n_j}(\theta),\eta)d\hat{W}_{n_j}(s) \to \int_0^t (M\hat{q}(s),\eta)d\hat{W}(s) \quad \text{in } L^2(\Omega).$$

This fact concludes that $\hat{q}$ is a solution for $(\hat{\mu},\hat{W})$. Since the solution of Eq.1.1 is uniquue, Lemma 3.1 completes the proof of Theorem.

4. **Optimal relaxed system**

Let F, G ; $H \to R^1$ be quadratic growth and satisfy (4.2) and (4.3) respectively,

$$|F(\varphi)|,\ |G(\varphi)| \leq c(1+\|\varphi\|^2) \tag{4.1}$$

$$|F(\varphi) - F(\psi)\| \leq c(1+\|\varphi\|+\|\psi\|)\|\varphi-\psi\| \tag{4.2}$$

$$|G(\varphi) - G(\psi)\| \leq c(1+\|\varphi\|+\|\psi\|)\|\varphi-\psi\|_* \tag{4.3}$$

with a positive constant c. For $R = (W,\mu)$, the criterion J and the value function V are defined by

$$J(\varphi,R) = E\int_0^T F(q(t,\varphi,R))dt + G(q(T,\varphi,R))$$

and

$$V(\varphi) = \inf_{R\in\mathfrak{R}} J(\varphi,R)$$

respectively. Appealing to Proposition 2.1 and Theorem 3.1, we can see the existence of an optimal control.

Theorem 4.1 For $\varphi \in \Psi_p$, there is an optimal relaxed system $R^* = R^*(\varphi)$, namely $J(\varphi,R^*) = V(\varphi)$.

Proof. By virtue of Proposition 2.1, it is enough to show that $J(\varphi,R)$ is continuous in R.

Suppose $R_n \to R$. Again appealing to Skorohod's theorem, we may assume, by (3.6),

$$q_n \to q \quad \text{in } L^2([0,T];H), \quad \text{w.p.1.} \tag{4.4}$$

Since (2.7) yields the uniform integrability, we have

$$E \int_0^T F(q_n(t))dt \to E \int_0^T F(q(t))dt. \tag{4.5}$$

On the other hand, (4.4) derives that there is a subsequence n_j such that, for a.a.t,

$$q_{n_j}(t) \to q(t) \quad \text{in } H, \quad \text{w.p.1.}$$

Now, combining with Proposition 2.2, we see

$$E\, G(q_{n_j}(T)) \to E\, G(q(T)), \tag{4.6}$$

Since $\{ E\, G(q_n(T)),\ n = 1,2,\cdots\}$ is bounded in R^1 by (2.7), we can conclude that $E\, G(q_n(T))$ tends to $E\, G(q(T))$. Thus we complete the proof.

Since a relaxed control turns out to be an admissible control under the Roxin condition, we can get an optimal admissible control. Put $c(x,u) = (a^{ij}(x,u), f^i(x,u),$ $ij=0,1,\cdots d)$ and $C(\mathfrak{U}) = \{c(\ ,u);\ u\in\mathfrak{U}\}$.

Roxin condition (convexity condition). $C(\mathfrak{U})$ is a convex subset of $C(R^d;R^{(d+1)(d+2)})$.

For any probability measure v on $\mathfrak{U}$, let us set

$$\mathfrak{U}(v) = \{ u \in \mathfrak{U} \ ; \ c(\ ,u) = \tilde{c}(\ ,v) \}$$

Then, Roxin condition yields that $\mathfrak{U}(v)$ is non-empty and compact. Hence we can take a Borel selector $\tilde{S}$ of $\mathfrak{U}(v)$ by the routine.[7] Setting $U(t) = \tilde{S}(\mu'(t))$ for $R = (W,\mu)$, the solution $q(\ ,\varphi,R)$ turns out to be the solution for an admissible system $A = (W,U)$. Now, Theorem 4.1 asserts the following theorem.

Theorem 4.2 Under the Roxin condition, there is an optimal admissible system $A^* = A^*(\varphi)$, for $\varphi \in \Psi_p$, namely

$$V(\varphi) = \inf_R J(\varphi,R) = \inf_A J(\varphi,A) = J(\varphi,A^*)$$

References

1. W.H.Fleming and M.Nisio, *On stochastic relaxed control for partially observed diffusions*, *Nagoya Math. J.* **93**(1984)pp.71-108.
2. K.Itô, *Foundations of Stochastic Differential Equations in Infinite Dimensional Spaces*,(SIAM, Philadelphia,1984).
3. N.V.Krylov and B.L.Rozovskii, *Stochastic partial differential equations and diffusion processes*, *Russian Math. Survey*, **37**(1982)pp.81-105.
4. J.L.Lions, *Equations Differentialles Opérationnelles et Problèmes aus Limites*,(Springer,Berlin,1961).
5. Metivier and T.Pellaumail, *Stochastic Integrals*(A.P. New York,1980).
6. N.Nagase and M.Nisio, *Optimal controls for stochastic partial differential equations*, *SIAM J. Control Optim.* **28**(1990)pp.186-213.
7. D.W.Stroock and S.R.S.Varadhan, *Multidimensional Diffusion Processes*(Springer,Berlin,1979).

Lévy's Brownian Motion and Stochastic Variational Equation

Dedicated to Professor Takeyuki Hida

Akio Noda

Department of Mathematics
Aichi University of Education
Kariya 448,Japan

Abstract

Towards a stochastic variational analysis of Lévy's Brownian motion, we present partial results on the variation $\delta X_C(t)$ taken along a curve C and well described as Lévy's stochastic infinitesimal eqation, as well as on another type of variation arising from the infinitesimal change of the curve in the parameter space.

1. Introduction.

Let $\{X(x); x \in \mathbf{R}^2\}$ be Lévy's Brownian motion with two-dimensional parameter space. The aim is to investigate the stochastic behavior of the random field $X(x)$ observed when the parameter x moves over the plane $\mathbf{R}^2$. To this end, we first take an arbitrary continuous curve $C = \{x(t) \in \mathbf{R}^2; t_0 \leq t < t_1\}$ to consider the Gaussian process $X_C(t) = X(x(t))$, obtained by restricting the parameter to the curve C.

Then the canonical representation

$$(1) \qquad \bar{X}_C(t) = X_C(t) - X_C(t_0) = \int_{t_0}^{t} F_C(t,u)dB_C(u) , \qquad t_0 \leq t < t_1,$$

is known to be a basic object expressing an aspect of the stochastic movement of $X(x)$ along the curve C (cf. [2]). In particular, the variation $\delta X_C(t) = X_C(t+dt) - X_C(t)$ on the infinitesimal piece of C between $x(t)$ and $x(t+dt)$ can be analyzed from (1) and the innovation part of $\delta X_C(t)$, independent of the past σ-field $\mathcal{F}_t(\bar{X}_C) = \sigma\{\bar{X}_C(u); t_0 < u \leq t\}$, is given by the infinitesimal increment $dB_C(t)$ of a standard Brownian motion.

We should here take into account another movement of $X(x)$ pointing in the direction of the unit normal $n(t)$ at each point $x(t)$. In order to study this aspect of the behavior exhibited by the random field , we take the variation δC of the curve to get a new curve $C + \delta C = \{x(t) + \delta C(t)n(t); t_0 \leq t < t_1\}$, and observe what sort of changes take place between the original process $X_C(t)$ and the infinitesimal transformed process $X_{C+\delta C}(t)$. Among others, we wish to solve the variational problem about the change of the canonical kernel : What form does the variation

$$\delta F_C(t,u) = F_{C+\delta C}(t,u) - F_C(t,u) \tag{2}$$

take for every value (t,u) in $D = \{t_0 < u < t < t_1\}$?

This is our variational approach to the study of Gaussian random fields including Lévy's Brownian motion as a prototype , and the author owes his variational point of view to T.Hida (cf. [3] and [10].)

The purpose of this paper is to present partial results towards the variational problem mentioned above . For general curve C it would be difficult to obtain the exact form of the canonical kernel $F_C(t,u)$. We therefore restrict, as a first step , the range of all transformed curves to a special tractable class of curves $C_a = \{(t, a|t|)\}, a \in R$. (For the reason of this choice , see Remark in Section 2.)

The problem of determining the canonical kernels $F_{C_a}(t,u)$ with the present class of C_a is by no means trivial . In our process of reaching the canonical kernel for a Gaussian process $Y(t)$ in general non-stationary setting , the key step we have to achieve is to derive Lévy's stochastic infinitesimal equation (cf. [4] , [5]) from the covariance function $\Gamma_Y(s,t)$ of $Y(t)$. In most familiar cases , it takes the normal form (i.e. Ito's stochastic differential equation)

$$\delta Y(t) = dt \int_{t_0}^{t} g(t,u)dY(u) + \sigma(t)dB(t), \tag{3}$$

where $m(t) = \int_{t_0}^{t} g(t,u)dY(u) \in L^2(\Omega, \mathcal{F}, P)$ and $\sigma(t) > 0$.

In Section 2 we establish the normal form for our principal object $X_{C_a}(t), X(x)$ being Lévy's Brownian motion . But there are various Gaussian processes that do not admit the normal form at all. In fact,we have encountered singular types of stochastic infinitesimal equations in the study of $X_C(t)$ derived from some Gaussian random fields $X(x)$ (see Remark) . In the final section, we discuss fractional Brownian motions (cf. [6]) as a typical example of non-normal forms.

2. Lévy's Brownian motion.

This section is devoted to the study of the Gaussian process $X_{C_a}(t) = X_a(t)$ mentioned in the previous section. For each $k > 0$ fixed , set

$$\bar{X}_a(t) = X_a(t) - X_a(-k) = X((t, a|t|)) - X((-k, ak)) , \qquad -k \le t \le k, \tag{1'}$$

where $X(x), x \in \mathbf{R}^2$, is Lévy's Brownian motion having the covariance function

$$(4) \qquad \Gamma_1(x,y) = \{|x| + |y| - |x-y|\}/2,$$

$|x|$ being the Euclidean norm. The main feature of such a random field $X(x)$ lies in the fact that every process $X_{L_a^{\pm}}(t)/(1+a^2)^{1/4}$ is a standard Brownian motion, where $L_a^+ = \{(t, at)\ ;\ t \geq 0\}$ $\left(L_a^- = \{(t, a|t|); t \leq 0\}\right)$, half-line emanating from the origin . Thus the present process (1') is a conjunction of two Brownian motions $X_{L_a^-}(t),\ t \leq 0$, and $X_{L_a^+}(t), t \geq 0$, which are mutually dependent for any $a \neq 0$.

We therefore begin with computing the covariance of $X_{L_a^-}(-r)$ and $X_{L_a^+}(s)$:

$$(5) \qquad K(r,s) := E[X_{L_a^-}(-r)X_{L_a^+}(s)]/\sqrt{1+a^2} = \left\{r + s - \sqrt{r^2+s^2+2\alpha rs}\right\}/2 ,$$

where we put $\alpha := (1-a^2)/(1+a^2) \in (-1,1)$. This is a homogeneous positive definite kernel and has the derivative

$$(6) \qquad L(r,s) := \frac{\partial^2}{\partial r \partial s}K(r,s) = (1-\alpha^2)rs(r^2+s^2+2\alpha rs)^{-3/2}/2 , \quad r,s > 0 .$$

Associated with this positive definite symmetric kernel $L(r,s)$, we define , for each $k_1, k_2 > 0$, the integral operator L_{k_1,k_2} by

$$(7) \qquad (L_{k_1,k_2}\phi)(r) := \int_0^{k_1} L(r,s)\phi(s)ds, \qquad 0 < r < k_2 ,$$

where $\phi(s)$ is any L^1 -function on $(0,k_1)$. Since $L(r,s)$ is homogeneous of degree -1 , we have $\int_0^{k_2}\int_0^{k_1} L^2(r,s)drds = \infty$, and hence we cannot appeal to the usual L^2-kernel theory as in [2], Chapter 6.

LEMMA 1. *The operator* L_{k_1,k_2} *induces a bounded operator from* $L^\infty((0,k_1))$ *to* $L^\infty((0,k_2))$ *with norm*

$$(8) \qquad \|L_{k_1,k_2}\|_{\infty,\infty} := \sup_{\|\phi\|_{L^\infty((0,k_1))}\leq 1} \left\| \int_0^{k_1} L(r,s)\phi(s)ds\right\|_{L^\infty((0,k_2))} = \frac{a^2}{1+a^2} < 1 .$$

PROOF: The estimate (8) follows from (5) and (6) ; we can set $\phi(s) \equiv 1$ in (8) to get

$$\begin{aligned}
\|L_{k_1,k_2}\|_{\infty,\infty} &= \sup_{0<r<k_2} \int_0^{k_1} L(r,s)ds = \sup_{0<r<k_2} \left[\frac{\partial}{\partial r} K(r,s)\right]_{s=0}^{s=k_1} \\
&= \{1 - \inf_{0<r<k_2} (r + \alpha k_1)/\sqrt{r^2 + k_1^2 + 2\alpha k_1 r}\}/2 \\
&= \{1 - \lim_{t\downarrow 0} (t+\alpha)/\sqrt{1 + 2\alpha t + t^2}\}/2 \\
&= (1-\alpha)/2 = a^2/(1+a^2).
\end{aligned}$$

We are now in a position to derive Lévy's stochastic infinitesimal equation for the Gaussian process $\bar{X}_a(t), -k \le t \le k$. On the interval $[-k, 0)$, there is no problem to write

$$\delta\bar{X}_a(t) = \bar{X}_a(t+dt) - \bar{X}_a(t) = (1+a^2)^{1/4} dB(t).$$

For each $t \in (0,k)$, we have to compute the conditional expectation

$$\mu_t(dt) := E[\delta\bar{X}_a(t) \mid \mathcal{F}_t(\bar{X}_a)] \tag{9}$$

with $\mathcal{F}_t(\bar{X}_a) := \sigma\{\bar{X}_a(u); -k < u \le t\}$. In view of the sufficient smoothness of $K(r,s)$, this $\mu_t(dt)$ is expected to be of the normal form $m(t)dt$. In fact , we can write

$$m(t) = \int_{-k}^{0} g_t(-r) d\bar{X}_a(r) - \int_0^t f_t(s) d\bar{X}_a(s). \tag{10}$$

with kernels g_t and f_t satisfying the equation

$$\begin{aligned}
E[m(t)\bar{X}_a(u)] &= \frac{\partial}{\partial t} E[\bar{X}_a(t)\bar{X}_a(u)] \\
&= \sqrt{1+a^2}\ \{ \chi_{[-k,0]}(u) \frac{\partial}{\partial t} K(-u,t) - \frac{\partial}{\partial t} K(k,t)\}
\end{aligned}$$

for all $u \in [-k, t]$. Differentiating it with respect to the variable u , we find that the pair g_t and f_t must be a solution of the following system of integral equations:

$$\begin{cases} \int_{-k}^{0} g_t(-r) L(-r,u) dr - f_t(u) = 0, & 0 < u < t, \\ g_t(-u) - \int_0^t f_t(s) L(-u,s) ds = L(-u,t), & -k < u < 0. \end{cases} \tag{11}$$

Using the bounded operator L_{k_1,k_2} , we rewrite (11) into the form of Fredholm equation of the second kind :

$$(I - L_{t,k}L_{k,t})g_t(r) = L(r,t), \qquad 0 < r < k , \tag{12}$$

and

$$(I - L_{k,t}L_{t,k})f_t(s) = \int_0^k L(u,t)L(u,s)du, \qquad 0 < s < t . \tag{12'}$$

Now , Lemma 1 enables us to form Neumann's series to get the desired solutions

$$\begin{aligned} g_t(r) &= \sum_{j=0}^{\infty}(L_{t,k}L_{k,t})^j L(r,t) \\ &= \sum_{j=0}^{\infty} L_{2j+1}(t,r;k,t) \in L^{\infty}((0,k)) , \end{aligned} \tag{13}$$

and

$$\begin{aligned} f_t(s) &= \sum_{j=0}^{\infty}(L_{k,t}L_{t,k})^j \int_0^k L(t,u)L(u,s)du \\ &= \sum_{j=1}^{\infty} L_{2j}(t,s;k,t) \in L^{\infty}((0,t)) , \end{aligned} \tag{13'}$$

where the L_m , $m \geq 1$, denote the iterated kernels given by

$$L_{2j+1}(r,s;k,t) = \int_{(0,k)^j}\int_{(0,t)^j} L(r,u_1)L(u_1,v_1)L(v_1,u_2)\cdots \cdots L(u_j,v_j)L(v_j,s)du_1\cdots du_j dv_1\cdots dv_j,$$

and

$$L_{2j}(r,s;k,t) = \int_{(0,k)^j}\int_{(0,t)^{j-1}} L(r,u_1)L(u_1,v_1)L(v_1,u_2)\cdots \cdots L(v_{j-1},u_j)L(u_j,s)du_1\cdots du_j dv_1\cdots dv_{j-1},$$

for any $r,s,k,t > 0$.

The innovation part $\delta\bar{X}_a(t) - m(t)dt$, independent of $\mathcal{F}_t(\bar{X}_a)$, has the variance $\sqrt{1+a^2}dt$ and is expressible as $(1+a^2)^{1/4}dB(t)$ for each $t \in (0,k)$. We thus obtain the following result.

Theorem 2. *For Lévy's Brownian motion $X(x)$, the variation $\delta\bar{X}_a(t)$ of the process $\bar{X}_a(t)$ along the curve C_a takes this normal form:*

$$(14)\quad \delta\bar{X}_a(t)=\chi_{(0,k)}(t)\{\int_{-k}^{0} g_t(-u)d\bar{X}_a(u)-\int_0^t f_t(u)d\bar{X}_a(u)\}dt+(1+a^2)^{1/4}dB(t),$$

$$-k<t<k, t\neq 0,$$

where g_t and f_t are given by (13) and (13'), respectively.

Associated with the Volterra kernel $f_t(s), 0<s<t<k$, we define

$$(15)\quad R_t(s)=-f_t(s)+\sum_{j=0}^{\infty}(-1)^{j-1}\int\cdots\int_{\{t>u_1\cdots>u_j>s\}} f_t(u_1)f_{u_1}(u_2)\cdots$$
$$\cdots f_{u_j}(s)du_1\cdots du_j,$$

where the series converges absolutely for each (t,s) in $\{0<s<t<k\}$. This Volterra kernel $R_t(s)$ satisfies the resolvent equation

$$R_t(s)+f_t(s)+\int_s^t R_t(u)f_u(s)du=0,$$

and the known procedure for the normal form of $\delta\bar{X}_a(t)$ ([2],[4]) leads us to another expression of $m(t)$, written below in terms of the innovation $dB(u)$:

$$(10')\quad m(t)=(1+a^2)^{1/4}\left[\int_{-k}^{0}\{g_s(-u)+\int_0^t R_t(s)g_s(-u)ds\}dB(u)+\int_0^t R_t(u)dB(u)\right],$$

$$0<t<k.$$

Integrating the stochastic infinitesimal equation (14) with this (10'), we get the desired canonical representation of $\bar{X}_a(t)$.

Theorem 3. *The canonical kernel $F_a(t,u)$ of $\bar{X}_a(t)$, $-k\le t\le k$, is given by*

$$(16)\quad F_a(t,u)=\begin{cases}(1+a^2)^{1/4} & (-k\le u\le 0),\\ (1+a^2)^{1/4}\left\{1+\int_u^t R_s(u)ds\right\} & (0<u\le t\le k),\\ (1+a^2)^{1/4}\left\{1+\int_0^t g_s(-u)ds+\iint_{\{0<s<r<t\}} R_r(s)g_s(-u)dsdr\right\} & \\ \qquad (-k\le u<0<t\le k).\end{cases}$$

We are ready to observe one aspect of changes from the Brownian motion $\bar{X}_0(t)$ to the transformed process $\bar{X}_{da}(t)$ corresponding to the variation $\delta C(t) = |t|da$ of the straight line C_0. Namely, we estimate the variation of the canonical kernel

$$\delta F_0(t,u) = F_{da}(t,u) - F_0(t,u) = F_{da}(t,u) - 1 \tag{17}$$

for each (t,u) in D.

COROLLARY 4. *For infinitesimal da , we have*

$$\delta F_0(t,u) \sim \begin{cases} (da)^2/4 & \text{if} \quad -k \leq u < t < 0, \\ & \text{or if} \quad 0 < u < t \leq k, \\ \left\{\frac{5}{4} + \frac{u(2t-u)}{(t-u)^2}\right\}(da)^2 & \text{if} \quad -k \leq u < 0 < t \leq k. \end{cases} \tag{18}$$

PROOF: It suffices to note that

$$L(t,s) \sim g_t(s) \sim 2ts(t+s)^{-3}(da)^2 + O((da)^4) \ ,$$

and

$$f_t(s) \sim -R_t(s) \sim \left\{4ts \int_0^k (t+u)^{-3}(u+s)^{-3}u^2 du\right\}(da)^4 + O((da)^6) \ ,$$

which yield (18) as a consequence of (16).

REMARK. A Gaussian random field $X(x)$ is called *linearly additive* if every process $X_L(t)$, L being an arbitrary straight line , has independent increments. Such a Gaussian random field can be characterized in terms of the structure function $d_X(x,y) = E[(X(x)-X(y))^2]$. Indeed, this $d_X(x,y)$ turns out to be an L^1-embeddable distance on $\mathbf{R}^2$, and expressible in the form

$$d_X(x,y) = \mu(A_x \ominus A_y) \ , A_x = \{h \in H \ ; \ x \in h\} \ , \tag{19}$$

where μ is a measure on the set H of all half-spaces h not containing the origin (cf. [8]~[10]). For Lévy's Brownian motion , we have $d_X(x,y) = |x-y|$ and the corresponding invariant measure μ_0 on H .

Taking the variation δC of L , we wish to observe what sort of changes take place between the additive process $X_L(t)$ and the infinitesimal transformed process $X_{L+\delta C}(t)$. In this paper, we chose a special deformation of the form $\delta C(t) = a|t|$ and studied the process $X_{C_a}(t)$ only for Lévy's Brownian motion . As a generalization of Theorem 2, we consider linearly additive Gaussian random fields associated with non-invariant measures μ on H ,and find interesting examples of non-normal forms for the variation $\delta X_{C_a}(t)$; we will discuss this kind of generalization in a separate paper.

3. Fractional Brownian motions.

In this section we discuss, from the variational point of view, the fractional Brownian motion $X(x)$ having the covariance function

$$(4') \qquad \Gamma_\alpha(x,y) = \{|x|^\alpha + |y|^\alpha - |x-y|^\alpha\}/2\,, \quad \alpha \neq 1\,,\ 0<\alpha<2\,.$$

In particular, the process $X_L(t)$, L being an arbitrary half-line emanating from the origin, does not admit the normal form, and we are going to show what kind of singularity actually arises in the study of the variation $\delta X_L(t)$, $t>0$.

We start wth the canonical representation of $X_L(t)$:

$$(20) \qquad X_L(t) = \int_0^t F_\alpha(t,u)dB(u)\,, \qquad t \geq 0,$$

where the canonical kernel is given below.

PROPOSITION 5. *([7]) The canonical kernel $F_\alpha(t,u)$ of the fractional Brownian motion $X_L(t)$ is of the form*

$$F_\alpha(t,u) = c_\alpha f_\alpha(u/t) u^{\frac{\alpha-1}{2}}, c_\alpha^2 = \sin\frac{\pi\alpha}{2}\Gamma(\alpha+1)/\Gamma\left(\frac{\alpha-1}{2}\right)^2,$$

where for $0 < s < 1$

$$(21) \qquad f_\alpha(s) = \begin{cases} \displaystyle\int_0^s u^{-\alpha}(1-u)^{\frac{\alpha-1}{2}}du + B\left(1-\alpha, \frac{\alpha+1}{2}\right) & (0<\alpha<1), \\ \displaystyle\int_s^1 u^{-\alpha}(1-u)^{\frac{\alpha-3}{2}}du & (1<\alpha<2). \end{cases}$$

Our idea is this : Take the generalized derivative D_β of fractional order $\beta = (\alpha-1)/2$ to study the Gaussian process $Y(t) = (D_\beta X_L)(t)$ instead of $X_L(t)$ itself. Then the variational structure of $Y(t)$ will be analyzed more effectively. Explicitly, we define

$$(22) \qquad Y(t) = \begin{cases} (I_{-\beta}X_L)(t) & (-1/2<\beta<0), \\ \displaystyle\frac{d}{dt}\left(I_{1-\beta}X_L\right)(t) & (0<\beta<1/2), \end{cases}$$

where $\quad (I_\gamma\varphi)(t) = \int_0^t \frac{(t-s)^{\gamma-1}}{\Gamma(\gamma)}\varphi(s)ds$, fractional integral of order $\gamma > 0$.

The well-definedness of the process $Y(t)$, $t \geq 0$, follows from the result in [11] , or directly from the following canonical representation of $Y(t)$.

Proposition 6. *The canonical kernel $F_Y(t,u)$ of (22) is of the form $c_\alpha k_\beta(u/t)/\Gamma(1-\beta)$ with*

$$k_\beta(s) = \begin{cases} \int_s^1 \left\{(s^{-1}-1)^{-\beta} - (s^{-1}-u^{-1})^{-\beta}\right\} u^{-2\beta-1}(1-u)^{\beta-1}du \\ \quad + (s^{-1}-1)^{-\beta}\left\{\int_0^s u^{-2\beta-1}(1-u)^{\beta-1}du + B(-2\beta,\beta+1)\right\} \\ \qquad\qquad\qquad\qquad\qquad\qquad (-1/2<\beta<0), \\ \int_s^1 (u/s-1)^{-\beta}u^{-\beta-1}(1-u)^{\beta-1}du \qquad (0<\beta<1/2). \end{cases} \tag{23}$$

Moreover, we have

$$\lim_{s\uparrow 1} k_\beta(s) = k_\beta(1-) = \begin{cases} |\beta|/(\beta+1) + 1/|\beta| & (-1/2<\beta<0), \\ B(1-\beta,\beta) & (0<\beta<1/2), \end{cases}$$

$$k_\beta(s) = O(s^{-|\beta|}) \qquad \text{as } s \downarrow 0 ,$$

and

$$\frac{d}{ds}k_\beta(s) = \begin{cases} O((1-s)^{-1}) & \text{as } s \uparrow 1 , \\ O(s^{-1+\beta\wedge 0}) & \text{as } s \downarrow 0 . \end{cases}$$

Proof: The proof is given only in the case $0<\alpha<1$, i.e. $-1/2<\beta<0$. Writing $c_\alpha = c$, $f_\alpha = f$ and $B(1-\alpha, \frac{\alpha+1}{2}) = b$ for notational simplicity, we get by Proposition 5

$$Y(t) = \frac{c}{\Gamma(1-\beta)}\int_0^t u^\beta dB(u)\int_u^t (-\beta)(t-s)^{-\beta-1}f(u/s)ds .$$

The last integral $J := \int_u^t (-\beta)(t-s)^{-\beta-1} f(u/s)ds$ is computed as follows:

$$
\begin{aligned}
J &= \int_u^t (-(t-s)^{-\beta})' ds \int_0^{u/s} v^{-2\beta-1}(1-v)^{\beta-1} dv + b\int_u^t (-(t-s)^{-\beta})' ds \\
&= \left[(t-s)^{-\beta}\right]_{s=t}^{s=u} \int_0^{u/t} v^{-2\beta-1}(1-v)^{\beta-1} dv \\
&\qquad + \int_{u/t}^1 \left[(t-s)^{-\beta}\right]_{s=u/v}^{s=u} v^{-2\beta-1}(1-v)^{\beta-1} dv + b(t-u)^{-\beta} \\
&= (t-u)^{-\beta} \left[\int_0^{u/t} v^{-2\beta-1}(1-v)^{\beta-1} dv \right. \\
&\qquad \left. + b + \int_{u/t}^1 \left\{1 - \left(\frac{t-u/v}{t-u}\right)^{-\beta}\right\} v^{-2\beta-1}(1-v)^{\beta-1} dv\right] ,
\end{aligned}
$$

which yields the desired expression

$$
Y(t) = \frac{c}{\Gamma(1-\beta)} \int_0^t k_\beta(u/t) dB(u) .
$$

Having obtained the exact form of $k_\beta(s)$, $0 < s < 1$, we can easily show the stated properties of this function . We finally note that the above Wiener integral is well defined , and that the canonical property for $k_\beta(u/t)$ follows from that for $f_\alpha(u/t)$ via the invertible integral operator $I_{-\beta}$.

We are ready to drive the stochastic infinitesimal equation of Y(t) :

$$
\delta Y(t) = \int_0^t \{F_Y(t+dt,u) - F_Y(t,u)\} dB(u) + \int_t^{t+dt} F_Y(t+dt,u) dB(u) .
$$

By proposition 6, we obtain

$$
\delta Y(t) = \frac{c_\alpha}{\Gamma(1-\beta)} \left[dt \int_0^t k_\beta'(u/t)(-u/t^2) dB(u) + k_\beta(1-) dB(t)\right] . \tag{24}
$$

Here are notable facts about this non-normal form in order .

i) The innovation part is of the form $\sigma dB(t)$ with $\sigma := c_\alpha k_\beta(1-)/\Gamma(1-\beta) > 0$, and there is no problem ;

ii) The $\mathcal{F}_t(Y)$-measurable term

$$
m(t) = -\frac{c_\alpha}{\Gamma(1-\beta)t} \int_0^t \left[s k_\beta'(s)\right]_{s=u/t} dB(u) \tag{25}
$$

does not belong to $L^2(\Omega, \mathcal{F}, P)$, because the kernel function $sk'_\beta(s)$ diverges like $(1-s)^{-1}$ as s approaches 1. This causes us a serious problem; by using the framework of T.Hida ([1]) based upon $L^2([0,t))$ with each finite intervel $[0,t)$ instead of the whole interval $[0,\infty)$, we would like to interprete the above $m(t)$ as being a *generalized Brownian linear functional.*

In view of the trouble arising in the simplest case of half-line L , the variational study of $X_{C_a}(t)$ would be a challenging problem for the fractional Brownian motions.

References

[1] T.Hida, *Brownian motion and its functionals*, Ricerche di Matematica **34** (1985), 183-222.

[2] T.Hida and M.Hitsuda, "Gaussian processes," (in Japanese) , Kinokuniya, Tokyo, 1976.

[3] T.Hida and SiSi, *Variational calculus for Gaussian random fields*, Proc.IFIP Warsaw (1988), to appear.

[4] P.Lévy, *Random functions : General theory with special reference to Laplacian random functions*, Univ.California Publications in Statistics I **12** (1953), 331-390.

[5] P.Lévy, *A special problem of Brownian motion, and a general theory of Gaussian random functions*, Proc . of the Third Berkeley Symp. on Math. Statist. and Probability (1955), 133-175.

[6] B.B.Mandelbrot, "The fractional geometry of nature," W.H.Freeman and Company, San Francisco, 1977.

[7] G.M.Molchan and Ju.I.Golosov, *Gaussian stationary processes with asymptotic power spectrum*, Soviet Math.Dokl. **10** (1969), 134-137.

[8] T.Mori, *Representation of linearly additive random fields*, preprint.

[9] A.Noda, *Generaleized Radon transform and Lévy's Brownian motion , I and II*, Nagoya Math.J. **105** (1987), 71-87 and 89-107.

[10] A.Noda, *White noise representations of some Gaussian Random fields*, Bulletin of Aichi Univ. of Education **38** (1989), 35-45.

[11] A.M.Sekretarev, *On the existence of mean-square derivatives of fractional order for Hilbert random functions*, Theory of Probab. and Math. Statist. **27** (1983), 141-145.

Generalized Radon – Nikodym Derivatives and Cameron – Martin Theory

Dedicated to Professor T. Hida

on the occasion of his retirement from Nagoya University

J. Potthoff†

BiBoS, Universität Bielefeld, Bielefeld, FRG
Department of Mathematics, Lousiana St. University, Baton Rouge, USA

and

L. Streit

BiBoS, Universität Bielefeld, Bielefeld, FRG
Area de Matemática, Universidade do Minho, Braga, Portugal

Abstract. It is shown that those measures on the Schwartz space of tempered distributions which are represented by a positive Hida distribution in the space $(\mathcal{S})^*$ admit under strongly continuous affine transformations a generalized Radon – Nikodym derivative (with respect to the white noise measure) in $(\mathcal{S})^*$.

1. Introduction.

In the article [PY 89] it was shown that translations $\tau_y : x \mapsto x + y$ and scalings $\sigma_\lambda : x \mapsto \lambda x$, $x, y \in \mathcal{S}'(\mathbb{R})$, $\lambda \in \mathbb{R}$, on the Schwartz space of tempered distributions $\mathcal{S}'(\mathbb{R})$ induce continuous transformations on the space $(\mathcal{S})$ of white noise test functionals (cf. [HPS 88, Ko 80, KT 80, KY 89, KPS 90, PS 89, PY 89,Yo 90] and below for a construction of $(\mathcal{S})$). As a consequence one finds that those measures on $(\mathcal{S}'(\mathbb{R}), \mathcal{B})$, $\mathcal{B}$ being the weak Borel algebra on $\mathcal{S}'(\mathbb{R})$, which are represented by an element in the dual $(\mathcal{S})^*$ of $(\mathcal{S})$ (cf. [Yo 90]) have a *generalized*

† Partially supported by the Council on Research, Louisiana State University, and the National Science Foundation under grant DMS – 9001859

Radon – Nikodym derivative in $(\mathcal{S})^*$ under these transformations. (Moreover, in [PY 89] explicit formulae were given for these generalized Radon – Nikodym derivatives.) Let us explain this in more detail by the simplest example: the white noise measure μ itself.

Let $(\mathcal{S}'(\mathbb{R}), \mathcal{B}, \mu)$ be the white noise probability space, i.e. μ is the centered Gaussian measure on $(\mathcal{S}'(\mathbb{R}), \mathcal{B})$ whose covariance is determined by the scalar product of $L^2(\mathbb{R})$ (with Lebesgue measure). Since $1 \in (\mathcal{S})^*$ we have for all $\varphi \in (\mathcal{S})$

$$\int \varphi(x) d\mu(x) = < 1, \varphi >,$$

where $< \cdot, \cdot >$ denotes dual pairing. Thus we may consider μ as being represented by $1 \in (\mathcal{S})^*$. For $\varphi \in (\mathcal{S})$, $y \in \mathcal{S}'(\mathbb{R})$ denote by φ_y the element $\varphi(\cdot + y)$ in $(\mathcal{S})$ (which is first defined by using the pointwise defined, strongly continuous version $\widetilde{\varphi}$ of φ [AHP 90, Yo 90]). Since the mapping $\widehat{\tau}_y : \varphi \mapsto \varphi_y$ is continuous on $(\mathcal{S})$ [PY 89], $\varphi \mapsto < 1, \varphi_y >$ is a linear continuous functional on $(\mathcal{S})$. Therefore there exists $E_y \in (\mathcal{S})^*$ so that

$$\begin{aligned} < 1, \varphi_y > &= \int \widetilde{\varphi}_y(x) d\mu(x) \\ &= \int \widetilde{\varphi}(x) d\mu_y(x) \\ &= < E_y, \varphi > \end{aligned}$$

for all $\varphi \in (\mathcal{S})$. Here $\mu_y(E) = \mu(E - y)$, $E \in \mathcal{B}$, and we may say that μ_y is represented by $E_y \in (\mathcal{S})^*$.

If $y \in L^2(\mathbb{R})$ it is well-known (and not hard to check) that

$$\int \widetilde{\varphi}(x) d\mu_y(x) = \int \widetilde{\varphi}(x) e^{<x,y> - \frac{1}{2}|y|_2^2} d\mu(x),$$

where $| \cdot |_2$ is the norm of $L^2(\mathbb{R})$. In this case we have

$$E_y(x) = \frac{d\mu_y}{d\mu}(x) =: e^{<x,y>} : \in L^p(\mathcal{S}'(\mathbb{R}), \mu),\ p \geq 1,$$

with the notation $: \exp < \cdot, y >: \equiv \exp\left(< \cdot, y > -1/2|y|_2^2\right)$. On the other hand, it is easy to see that for $y \in \mathcal{S}'(\mathbb{R})$, $: \exp < \cdot, y >:$ still makes sense as an element in $(\mathcal{S})^*$. Thus we find that $E_y =: exp < \cdot, y >:$ and we can interprete E_y as a generalized Radon – Nikodym derivative: $E_y = d\mu_y / d\mu \in (\mathcal{S})^*$.

Similarly one can discuss the scaling transformation σ_λ (see above) and more general measures represented by an element in $(\mathcal{S})^*$ [PY 89].

Throughout we shall denote by $d\nu/d\mu$ the element in $(\mathcal{S})^*$ which represents a measure ν on $(\mathcal{S}'(\mathbb{R}), \mathcal{B})$ whenever such a representing Hida distribution exists.

The purpose of this article is to generalize the above mentioned results in [PY 89] to arbitrary strongly continuous affine transformations on $\mathcal{S}'(\mathbb{R})$. This will be possible due to a theorem in [PS 89] which characterizes the space $(\mathcal{S})^*$ of Hida distributions through the $\mathcal{S}-$ and $\mathcal{T}-$transforms of its elements. In Section 2 we shall sketch this and a related result for $(\mathcal{S})$. In Section 3 we describe the generalized Cameron – Martin theory of affine transformations on $\mathcal{S}'(\mathbb{R})$.

2. Characterization of Hida Distributions.

Consider the space $(L^2) \equiv L^2(\mathcal{S}'(\mathbb{R}), \mu)$ of square integrable functions on $(\mathcal{S}'(\mathbb{R}), \mathcal{B}, \mu)$. According to the Wiener – Itô decomposition theorem (e.g. [Hi 80]) every $\varphi \in (L^2)$ is in one-to-one correspondence with a sequence $(f^{(n)}; n \in \mathbb{N}_0)$ of elements $f^{(n)}$ in $\widehat{L^2(\mathbb{R}^n)}$, where $\widehat{\cdot}$ stands for symmetrization, and $f^{(0)} \in \mathbb{C}$. Moreover

$$\|\varphi\|_2^2 = \sum_{n=o}^{\infty} n! |f^{(n)}|^2_{L^2(\mathbb{R}^n)}, \tag{2.1}$$

$\|\cdot\|_2$ denoting the norm of (L^2). Assume that A is a self-adjoint operator on $L^2(\mathbb{R})$ with $\operatorname{infspec} A \geq 1$. If we define for suitable φ, $\Gamma(A)\varphi$ as corresponding to the sequence $(A^{\otimes n} f^{(n)}; n \in \mathbb{N}_0)$ then we may consider the system of norms given by

$$\|\varphi\|_{2,p} := \|\Gamma(A)^p \varphi\|_2, \quad p \in \mathbb{R}_+.$$

From now on we make the choice $A = -\frac{d^2}{du^2} + u^2 + 1$. We denote by $(\mathcal{S})_p$, $p \in \mathbb{R}_+$, the completed subspace of (L^2) with elements φ so that $\|\varphi\|_{2,p} < \infty$. $(\mathcal{S})$ denotes the projective limit of the family $\{(\mathcal{S})_p; p \in \mathbb{N}_0\}$, $(\mathcal{S})^*$ its dual. Elements in $(\mathcal{S})^*$ will be called *Hida distributions*.

$(\mathcal{S})$ and $(\mathcal{S})^*$ have been investigated in a number of articles. We refer the interested reader to [HPS 88, KPS 90, PS 89, PY 89] and the references quoted there. Apart from the properties of these two spaces which where mentioned in Section 1 we want to recall that $(\mathcal{S})$ is a nuclear Fréchet algebra, and that $\Phi \in (\mathcal{S})^*$ if and only if there exists $p \in \mathbb{N}_0$ so that $\Gamma(A)^{-p}\Phi \in (L^2)$, i.e. if

$$\|\Phi\|^2_{2,-p} := \sum_{n=0}^{\infty} n! |(A^{\otimes n})^{-p} F^{(n)}|_2^2 \tag{2.2}$$

is finite. Here $(F^{(n)};\ n \in \mathbb{N}_0)$ is the sequence of symmetric tempered distributions $F^{(n)}$ over $\mathbb{R}^n$, which is in one-to-one correspondence with Φ. (The chaos expansion of $(\mathcal{S})$ – considered as a subspace of (L^2) – induces a chaos expansion of $(\mathcal{S})^*$.)

It is easy to see that for all $\lambda \in \mathbb{C}$, $\xi \in \mathcal{S}(\mathbb{R})$ the function $\exp \lambda(< \cdot, \xi >)$ belongs to $(\mathcal{S})$. Define for $\Phi \in (\mathcal{S})^*$

$$\mathcal{T}\Phi(\xi) := < \Phi, e^{i<\cdot,\xi>} >, \quad \xi \in \mathcal{S}(\mathbb{R}) \tag{2.3}$$

and

$$\mathcal{S}\Phi(\lambda\xi) := < \Phi, :e^{\lambda<\cdot,\xi>}: >, \quad \xi \in \mathcal{S}(\mathbb{R}),\ \lambda \in \mathbb{C}, \tag{2.4}$$

where we have set

$$:e^{\lambda<x,\xi>}: := e^{\lambda<x,\xi> - \lambda^2/2|\xi|_2^2}.$$

Clearly we have the formula

$$\mathcal{T}\Phi(\xi) = e^{-\frac{1}{2}|\xi|_2^2}\mathcal{S}\Phi(i\xi). \tag{2.5}$$

Moreover, if $\Phi \in (\mathcal{S})^*$ corresponds to $(F^{(n)}; n \in \mathbb{N}_0)$ then

$$\mathcal{S}\Phi(\xi) = \sum_{n=0}^{\infty} < F^{(n)}, \xi^{\otimes n} >, \tag{2.6}$$

and in particular for $\varphi \in (L^2)$ with decomposition given by $(f^{(n)}; n \in \mathbb{N}_0)$ we have

$$\mathcal{S}\varphi(\xi) = \sum_{n=0}^{\infty} \int_{\mathbb{R}^n} f^{(n)}(u)\xi^{\otimes n}(u)du. \tag{2.6$'$}$$

Let us put $|f|_{2,p} := |A^p f|_2$, $p \in \mathbb{R}_+$.

Definition 2.1. Let F be a complex valued function on $\mathcal{S}(\mathbb{R})$. We call F a *U–functional* if and only if the following two conditions are satisfied.

(i) F has a *ray entire* extension on $\mathcal{S}(\mathbb{R})$, i.e. for all $\eta, \xi \in \mathcal{S}(\mathbb{R})$ the function $\lambda \mapsto F(\eta + \lambda\xi)$, $\lambda \in \mathbb{R}$, has an entire analytic extension (denoted by $z \mapsto F(\eta + z\xi)$, $z \in \mathbb{C}$).

(ii) There exists $p \in \mathbb{N}_0$ and constants $K_1, K_2 > 0$ so that for all $z \in \mathbb{C}$ and all $\xi \in \mathcal{S}(\mathbb{R})$

$$|F(z\xi)| \le K_1 e^{K_2|z|^2|\xi|_{2,p}^2}. \tag{2.7}$$

Remark. The name "*U*–functional" has traditional reasons: it has been used in the classical paper [Hi 75] and subsequent articles for the functional $\mathcal{S}\Phi(i\xi)$, where Φ is a Hida distribution.

The main result in [PS 89] is the following.

Theorem 2.2. If $\Phi \in (\mathcal{S})^*$ then $\mathcal{S}\Phi$ and $\mathcal{T}\Phi$ are U–functionals. Conversely, if F is a U–functional then there exist $\Phi_{\mathcal{T}}$ and $\Phi_{\mathcal{S}}$ in $(\mathcal{S})^*$ so that $\mathcal{T}\Phi_{\mathcal{T}} = \mathcal{S}\Phi_{\mathcal{S}} = F$.

Sketch of the proof. The proof of the first statement is easy. The fact that $\mathcal{S}\Phi$ and $\mathcal{T}\Phi$ are ray entire on $\mathcal{S}(I\!R)$ follows directly from (2.6), (2.2) and (2.5). Since $\Phi \in (\mathcal{S})^*$ implies that $\Gamma(A)^{-p}\Phi \in (L^2)$ for some $p \in I\!N_0$, we may estimate as follows

$$\begin{aligned}|\mathcal{S}\Phi(z\xi)| &= |(\Gamma(A)^{-p}\Phi, \Gamma(A)^p : e^{z<\cdot,\xi>} :)| \\ &= |(\Gamma(A)^{-p}\Phi, : e^{z<\cdot,A^p\xi>} :)| \\ &\leq \|\Gamma(A)^{-p}\Phi\|_2 \| : e^{z<\cdot,A^p\xi>} : \|_2.\end{aligned}$$

The last norm can be easily computed and is equal to $\exp(\frac{1}{2}|z|^2|A^p\xi|_2^2)$. Therefore (2.7) holds for $\mathcal{S}\Phi$ and by (2.5) also for $\mathcal{T}\Phi$.

Now assume that F is a U–functional, and consider for $\xi \in \mathcal{S}(I\!R)$ the expansion of $F(z\xi)$, $z \in \mathbb{C}$, at zero:

$$F(z\xi) = \sum_{n=0}^{\infty} z^n F^{(n)}(\xi). \tag{2.8}$$

Cauchy's theorem and (2.7) show that for all $n \in I\!N$ we have the estimate

$$|F^{(n)}(\xi)| \leq (n!)^{-\frac{1}{2}} K^n |\xi|_{2,p}^n.$$

Here and in the following K denotes a constant depending only on K_1, K_2 and p, and which we may change from step to step below. Now one can show (cf. [PS 89]) that $F^{(n)}$ (which is homogeneous of degree n in ξ) extends by polarization to an n–multilinear form (denoted again by $F^{(n)}$) on $\mathcal{S}(I\!R)$ and the above bound yields

$$|F^{(n)}(\xi_1, \ldots, \xi_n)| \leq (n!)^{-\frac{1}{2}} K^n \prod_{k=1}^{n} |\xi_k|_{2,p},$$

for all $\xi_1, \ldots, \xi_n \in \mathcal{S}(I\!R)$. From the last estimate we may conclude that there exist $q \in I\!N$ and $\rho \in (0,1)$ so that for all $n \in I\!N$ we have

$$|(A^{\otimes n})^{-q} F^{(n)}|_2 \leq (n!)^{-\frac{1}{2}} \rho^n.$$

This implies that $(F^{(n)};\, n \in I\!N_0)$ is a sequence so that the sum on the right hand side of (2.2) converges (with p replaced by q). Consequently the sequence of continuous n–multilinear forms $(F^{(n)};\, n \in I\!N_0)$ defines a unique element $\Phi_{\mathcal{S}}$ in $(\mathcal{S})^*$ so that $\mathcal{S}\Phi_{\mathcal{S}} = F$ (cf. (2.6)). $\Phi_{\mathcal{T}}$ can be constructed as above by using the U–functional $\exp(-\frac{1}{2}|\xi|_2^2)F(i\xi)$ (cf. (2.5)). □

Remarks. Lee has proved in [Le 89] another characterization of $(\mathcal{S})^*$ (in fact of $(\mathcal{S})_p$, $p \in \mathbb{R}$) in terms of analytic properties of the S – transform. He used Bargmann – Segal spaces in his approach. However, for practical purposes we find our result more convenient.

Lee [Le 90] and Yan [Ya 90] have localized Theorem 2.2 in the sense that they give an estimate on q in the above proof.

Theorem 2.2 and its proof are readily generalized to white noise with d–dimensional "time" parameter.

A first (more or less trivial) consequence of Theorem 2.2 is the characteriztion of all measures on $(\mathcal{S}'(\mathbb{R}^d), \mathcal{B})$ which are represented by a Hida distribution.

Corollary 2.3. Let ν be a measure on $(\mathcal{S}'(\mathbb{R}^d), \mathcal{B})$. Then $d\nu/d\mu \in (\mathcal{S})^*$ if and only if the characteristic function of ν is a U–functional.

In [PS 89] a number of examples were given to illustrate the use of Theorem 2.2 and Corollary 2.3. Here is another example from quantum field theory.

Example (Free Massless Bosons). The free massless Boson field over $\mathbb{R}^d$, $d \geq 2$, is given by the centered Gaussian measure ν_0 on $\mathcal{S}'(\mathbb{R}^d)$ with covariance operator $(-\Delta)^{-1}$, i.e. with characteristic function given by

$$\int e^{i<x,\xi>} d\nu_0(x) = e^{-\frac{1}{2}|(-\Delta)^{-\frac{1}{4}}\xi|_2^2}, \quad \xi \in \mathcal{S}(\mathbb{R}^d).$$

In order to show that $d\nu_0/d\mu \in (\mathcal{S})^*$ we apply Corollary 2.3, and only have to prove that for some $p \in \mathbb{N}$

$$|(-\Delta)^{-\frac{1}{4}}\xi|_2 \leq \text{const.}|\xi|_{2,p}.$$

By the Plancherel theorem we have

$$\begin{aligned} |(-\Delta)^{-\frac{1}{4}}\xi|_2^2 &= (2\pi)^{-d/2} \int_{\mathbb{R}^d} |k|^{-1} |\tilde{\xi}(k)|^2 dk \\ &\leq \text{const.} \sup_{k \in \mathbb{R}^d} |(1+|k|^2)^q \tilde{\xi}(k)|, \end{aligned}$$

with $q > (d-1)/2$ and where $\tilde{}$ denotes Fourier transformation. Since the system of Schwartz space norms is equivalent to the system $\{|\cdot|_{2,p};\ p \in \mathbb{N}_0\}$ (cf. e.g. [Si 74]) we have that for some $p \in \mathbb{N}_0$

$$\begin{aligned} |(-\Delta)^{-\frac{1}{4}}\xi|_2^2 &\leq \text{const.}|\tilde{\xi}|_{2,p} \\ &= \text{const.}|\xi|_{2,p}, \end{aligned}$$

where the last equality follows from the fact that Fourier transformation leaves A invariant. Consequently, the free massless Boson field (for $d \geq 2$) has a representation in terms of white noise: $d\nu_0/d\mu \in (\mathcal{S})^*$.

Introducing vector–valued fields this example shows that also the measure of the free electromagnetic field (cf. e.g. [Sch 66]) in dimensions greater than 2 has a generalized Radon – Nikodym derivative with respect to the white noise measure which is a Hida distribution.

The space $(\mathcal{S})$ of test functionals can also be characterized in terms of analytic properties of the S–transform of its elements. The following result which was proved in [KPS 90], has also been indicated independently in [Ko 80].

Let us formulate the following two conditions for a complex valued function F on $\mathcal{S}(\mathbb{R})$.

(C.1) F is ray entire on $\mathcal{S}(\mathbb{R})$.

(C.2) For every $p \in \mathbb{N}_0$, $\varepsilon > 0$, and all $\xi \in \mathcal{S}(\mathbb{R})$

$$\lim_{R\to\infty} M(R,\xi)e^{-\varepsilon R^2} = 0, \tag{2.9}$$

uniformly on the set $B_p^* = \{\xi \in \mathcal{S}(\mathbb{R}); |\xi|_{2,-p} = 1\}$, where

$$M(R,\xi) = \sup_{z\in\mathbb{C},|z|=R} |F(z\xi)|.$$

Theorem 2.4. Assume that $\varphi \in (\mathcal{S})$ then $F := S\varphi$ admits conditions C.1 and C.2. Conversely, if F satisfies conditions C.1 and C.2 then there exists $\varphi \in (\mathcal{S})$ whose S–transform is F.

Sketch of the proof. The strategy of the proof is very similar to the one applied in the proof of Theorem 2.2. We merely indicate the essential differences.

Consider the first statement. It is easy to establish C.1 for $S\varphi$. Since $\varphi \in (\mathcal{S})_q$ for all $q \in \mathbb{N}_0$, we have

$$\begin{aligned} |S\varphi(z\xi)| &= |(\varphi, :e^{z<\cdot,\xi>}:)| \\ &\leq \|\varphi\|_{2,q} e^{\frac{1}{2}|z|^2|\xi|^2_{2,q}}, \end{aligned}$$

for all $q \in \mathbb{N}_0$, $z \in \mathbb{C}$. Given $p \in \mathbb{N}_0$ and $\varepsilon > 0$ choose q large enough so that $2^{2(p-q)} < \varepsilon$. If $|\xi|_{2,-p} \leq 1$ then

$$M(R,\xi)e^{-\varepsilon R^2} \leq \|\varphi\|_{2,q} e^{-\frac{1}{2}\varepsilon R^2},$$

where $M(R,\xi) = \sup_{|z|=R} |S\varphi(z\xi)|$, because $\|A^{p-q}\| \leq 2^{p-q}\,(p \leq q)$. This finishes the proof of the first part of the theorem.

Now assume that F satisfies C.1 and C.2. Because of (2.9) and Cauchy's theorem, the expansion coefficients in (2.8) admit an estimate of the following form. For every $p \in \mathbb{N}$ and every $K > 0$ there exists $n_0 \in \mathbb{N}$ so that for all $n \in \mathbb{N}$ with $n \geq n_0$, and all $\xi \in \mathcal{S}(\mathbb{R})$ we have

$$|F^{(n)}(\xi)| \leq e^{-Kn}(n!)^{-\frac{1}{2}}|\xi|^n_{2,-p}.$$

As in the proof of Theorem 2.2, $F^{(n)}$ extends by polarization to an n-multilinear form on $\mathcal{S}(\mathbb{R})$, and the above estimate translates into an estimate of the $L^2(\mathbb{R}^n)$-norm of $(A^{\otimes n})^p F^{(n)}$: for every $p \in \mathbb{N}$ and every $K > 0$ there is $n_0 \in \mathbb{N}$ so that for all $n \geq n_0$

$$|(A^{\otimes n})^p F^{(n)}|_2 \leq (n!)^{-\frac{1}{2}} e^{-Kn}.$$

Thus, if we define φ by the chaos expansion corresponding to $(F^{(n)}; n \in \mathbb{N}_0)$, we find that φ has a finite $(\mathcal{S})_p$ norm

$$\|\varphi\|_{2,p} = \Big(\sum_{n=0}^{\infty} n!|(A^{\otimes n})^p F^{(n)}|_2^2\Big)^{\frac{1}{2}},$$

for every $p \in \mathbb{N}_0$, and consequently $\varphi \in (\mathcal{S})$. $\mathcal{S}\varphi = F$ follows from (2.6) and (2.8). □

3. Cameron – Martin Theory of Affine Transformations.

In this section we consider a rather straightforward application of Corollary 2.3 to strongly continuous transformations on $\mathcal{S}'(\mathbb{R}^d)$, which we believe nevertheless to be worth mentioning.

Throughout we let $d = 1$ (without loss of generality) and we shall assume that we are given a measure ν on $(\mathcal{S}'(\mathbb{R}), \mathcal{B})$ which is represented by $d\nu/d\mu \in (\mathcal{S})^*$, i.e. whose characteristic function is U-functional. In particular we have for some $p \in \mathbb{N}_0$, $K_1, K_2 > 0$

$$\begin{aligned} |\int e^{iz<x,\xi>} d\nu(x)| &= |T\frac{d\nu}{d\mu}(z\xi)| \\ &\leq K_1 e^{K_2|z|^2|\xi|^2_{2,p}}, \end{aligned} \tag{3.1}$$

for all $z \in \mathbb{C}$, $\xi \in \mathcal{S}(\mathbb{R})$.

Since $T d\nu/d\mu$ is ray entire, it is easy to see that $T d\nu/d\mu$ is continuous with respect to $|\cdot|_{2,p}$. Thus an application of Minlos' theorem (cf. e.g. [Hi 80]) shows that ν is carried by $\mathcal{S}_{-q}(\mathbb{R})$ (the dual of $\mathcal{S}_q(\mathbb{R})$ which is the completion of $\mathcal{S}(\mathbb{R})$

under $|\cdot|_{2,q}$): $\nu(\mathcal{S}_{-q}(\mathbb{R})) = 1$, for all $q > p + \frac{1}{2}$, because the injection of $\mathcal{S}_p(\mathbb{R})$ into $\mathcal{S}_q(\mathbb{R})$ has Hilbert–Schmidt norm equal to $\|A^{p-q}\|_{H.S.} < \infty$.

In the following we shall identify a measure ω on $(\mathcal{S}'(\mathbb{R}), \mathcal{B})$ such that $\omega(\mathcal{S}_{-r}(\mathbb{R}))$ $= 1$, $r \in \mathbb{R}_+$, with its restriction to $(\mathcal{S}_{-r}(\mathbb{R}), \mathcal{B}_r)$, where $\mathcal{B}_r$ is the Borel algebra of $\mathcal{S}_{-r}(\mathbb{R})$. (This is meaningful since $\mathcal{B} \cap \mathcal{S}_{-r}(\mathbb{R}) = \mathcal{B}_r$.)

Fix $q > p + \frac{1}{2}$, and assume that B is a linear continuous mapping from $\mathcal{S}_r(\mathbb{R})$ into $\mathcal{S}_q(\mathbb{R})$, for some $r \in \mathbb{R}_+$. Therefore there is a constant $c_{q,r} > 0$ so that for all $\xi \in \mathcal{S}(\mathbb{R})$ we have

$$|B\xi|_{2,q} \leq c_{q,r}|\xi|_{2,r}. \tag{3.2}$$

Moreover we have a continuous linear mapping $B^* : \mathcal{S}_{-q}(\mathbb{R}) \to \mathcal{S}_{-r}(\mathbb{R})$ with

$$< B^*x, \xi > = < x, B\xi >,$$

for all $\xi \in \mathcal{S}_r(\mathbb{R})$, $x \in \mathcal{S}_{-q}(\mathbb{R})$. In particular, we may consider B^* as a ν–a.e. defined mapping of $\mathcal{S}'(R)$ into itself. Note that B^* is $\mathcal{B}_q - \mathcal{B}_r$ measurable, and we may define as usual a measure ν_B on $(\mathcal{S}_{-r}(\mathbb{R}), \mathcal{B}_r)$ by

$$\nu_B(E) := \nu(B^{*-1}E), \quad E \in \mathcal{B}_r. \tag{3.3}$$

ν_B will be identified with a measure on $(\mathcal{S}'(\mathbb{R}), \mathcal{B})$ (and denoted by the same symbol). Clearly we have for $\xi \in \mathcal{S}(\mathbb{R})$:

$$\int e^{i<x,\xi>} d\nu_B(x) = \int e^{i<B^*x,\xi>} d\nu(x)$$
$$= \int e^{i<x,B\xi>} d\nu(x).$$

Obviously, it follows from this that also the characteristic function of ν_B is a U–functional. We have proved the following result.

Lemma 3.1. Assume that ν and B are as above. Then ν_B as defined by (3.3) is represented by a Hida distribution $d\nu_B/d\mu \in (\mathcal{S})^*$.

Remark. One can derive an "explicit" formula for $d\nu_B/d\mu$ similarly as was done in [PY 89] for the case where B is multiplication by a non–zero number. The result is the following formula.

$$\frac{d\nu_B}{d\mu} = \frac{d\mu_B}{d\mu} : (\Gamma(B^*)\frac{d\nu}{d\mu}), \tag{3.4}$$

where $\cdot : \cdot$ is the Wick product (e.g. [PY 89]), and $d\mu_B/d\mu$ is described by

$$\mathcal{S}\frac{d\mu_B}{d\mu}(\xi) = \exp(-\frac{1}{2}(\xi + B\xi, \xi - B\xi)).$$

To present the main result of this section we note the following lemma which is a direct consequence of the reflexivity of $\mathcal{S}(\mathbb{R})$ and $\mathcal{S}'(\mathbb{R})$.

Lemma 3.2. If B is a linear, continuous mapping on $\mathcal{S}(\mathbb{R})$, then it has a strongly continuous adjoint B^* on $\mathcal{S}'(\mathbb{R})$: $< B^*x, \xi >=< x, B\xi >$, $x \in \mathcal{S}'(\mathbb{R})$, $\xi \in \mathcal{S}(\mathbb{R})$. Conversely, if B^* is a linear, strongly continuous mapping on $\mathcal{S}'(\mathbb{R})$, then it is the adjoint of a linear, continuous mapping B on $\mathcal{S}(\mathbb{R})$.

Theorem 3.3. Assume that B^* is a linear, strongly continuous mapping on $\mathcal{S}'(\mathbb{R})$, and let $y \in \mathcal{S}'(\mathbb{R})$. Consider a measure ν on $(\mathcal{S}'(\mathbb{R}), \mathcal{B})$, and the measure $\nu_{B,y}$ given by

$$\nu_{B,y}(E) = \nu(B^{*-1}E - y), \quad E \in \mathcal{B}.$$

If ν is such that $d\nu/d\mu \in (\mathcal{S})^*$, then also $d\nu_{B,y}/d\mu \in (\mathcal{S})^*$. Moreover

$$\frac{d\nu_{B,y}}{d\mu} = \frac{d\mu_{B,0}}{d\mu} : \frac{d\mu_{id,y}}{d\mu} : (\Gamma(B^*)\frac{d\nu}{d\mu}), \tag{3.5}$$

where $\mu_{B,y}(E) = \mu(B^{*-1}E - y)$, $E \in \mathcal{B}$.

Proof. If B^* is linear and strongly continuous on $\mathcal{S}'(\mathbb{R})$, it has a continuous adjoint B according to Lemma 3.2. In particular for every $p \in \mathbb{N}_0$ there is a $q \in \mathbb{N}_0$ so that B is continuous from $\mathcal{S}_p(\mathbb{R})$ into $\mathcal{S}_q(\mathbb{R})$. Whenever ν is such that $d\nu/d\mu \in (\mathcal{S})^*$, we may apply Lemma 3.1 to conclude that $\nu_B = \nu(B^{*-1}\cdot)$ is represented in $(\mathcal{S})^*$.

In order to take the translation by y into account one can apply the results in [PY 89]. Alternatively, one can use the following simpler argument. By Theorem 2.2 we have that $\Phi \in (\mathcal{S})^*$ implies that Φ_y, defined by

$$< \Phi_y, \varphi >=< \Phi, \hat{\tau}_y\varphi >, \quad \varphi \in (\mathcal{S}),$$

belongs to $(\mathcal{S})^*$ too:

$$\begin{aligned} \mathcal{T}\Phi_y(\lambda\xi) &=< \Phi, e^{i\lambda<\cdot+y,\xi>} > \\ &= e^{i\lambda<y,\xi>} < \Phi, e^{i\lambda<\cdot,\xi>} > \\ &= e^{i\lambda<y,\xi>}\mathcal{T}\Phi(\lambda\xi), \end{aligned}$$

and if $\mathcal{T}\Phi$ is a U–functional, so is $e^{i<y,\cdot>}\mathcal{T}\Phi$ for every $y \in \mathcal{S}'(\mathbb{R})$. Formula (3.5) follows from (3.4) and a trivial computation. □

We conclude this paper with the following illustration.

Example. Choose $B = (1-\Delta)^\alpha$, $\alpha \in \mathbb{R}$. Let us first show that B is continuous on $\mathcal{S}(\mathbb{R})$. Let $\xi \in \mathcal{S}(\mathbb{R})$, $q \in \mathbb{N}$, and apply Fourier transformation and the Plancherel theorem: for certain $n, n', m, r \in \mathbb{N}$ we have

$$\begin{aligned}
|(1-\Delta)^\alpha \xi|_{2,q} &= |(1+\widehat{k^2})^\alpha \widetilde{\xi}|_{2,q} \\
&\leq \text{const.} \sup_{k\in\mathbb{R}} |(1+k^2)^n D_k^m (1+k^2)^\alpha \widetilde{\xi}(k)| \\
&\leq \text{const.} \sup_{k\in\mathbb{R}} \sum_{l=0}^{m} |(1+k^2)^{n'} D_k^l \widetilde{\xi}(k)| \\
&\leq \text{const.} |\widetilde{\xi}|_{2,r} \\
&= \text{const.} |\xi|_{2,r}.
\end{aligned}$$

Here we denoted $(\widehat{k\psi})(k) = k\psi(k)$ and made again use of the facts that the family of Schwartz space norms is equivalent to the family of $\mathcal{S}_p(\mathbb{R})$ norms, and that A is invariant under Fourier transformation.

Choose $\alpha = -\frac{1}{2}$ and consider μ_B. It follows from Theorem 3.3 that $d\mu_B/d\mu \in (\mathcal{S})^*$. Moreover, we have

$$\mathcal{T}\frac{d\mu_B}{d\mu}(\xi) = e^{-\frac{1}{2}(\xi,(1-\Delta)^{-1})\xi)}.$$

Observe that $(1-\Delta)^{-1}$ has an integral kernel given by $(1-\Delta)^{-1}(t,s) = \frac{1}{2}e^{-|t-s|}$, t, $s \in \mathbb{R}$. Thus we have shown that the measure of the Ornstein – Uhlenbeck process has a generalized Radon – Nikodym derivative with respect to the white noise measure which is a Hida distribution.

Acknowledgement. We are grateful to Professor T. Hida, for his kind invitation to the conference *Gaussian Random Fields*, Nagoya, August 1990, and – of course – for continuous inspiration.

References

[AHP 90] Albeverio, S., Hida, T., Potthoff, J., Röckner, M. and Streit, L.: Dirichlet forms in terms of white noise analysis II – Construction of infinite dimensional diffusions; *Rev. Math. Phys.* **1** (1990) 313–323

[Hi 75] Hida, T.: *Analysis of Brownian Functionals.* Carleton Mathematical Lecture Notes no. **13** (1975)

[Hi 80] Hida, T.: *Brownian Motion.* Berlin, Heidelberg, New York: Springer (1980)

[HKPS] Hida, T., Kuo, H.–H., Potthoff, J. and Streit, L.: *White Noise: An Infinite Dimensional Calculus.* Monograph in preparation

[HPS 88] Hida, T., Potthoff, J. and Streit, L.: Dirichlet forms and white noise analysis; *Commun. Math. Phys.* **116** (1988) 235–245

[Ko 80] Kondrat'ev, Ju.G.: Nuclear spaces of entire functions in problems of infinite -dimensional analysis; *Soviet Math. Dokl.* **22** (1980) 588–592

[KT 80] Kubo, I. and Takenaka, S.: Calculus on Gaussian white noise I; *Proc. Japan Acad.* **56A** (1980) 376–380

[KY 89] Kubo, I. and Yokoi, Y.: A remark on the space of testing random variables in the white noise calculus; *Nagoya Math. J.* **115** (1989) 139–149

[KPS 90] Kuo, H.–H., Potthoff, J. and Streit, L.: A characterization of white noise test functionals; *Preprint* (1990), to appear in *Nagoya Math. J.*

[Le 89] Lee, Y.–J.: Analytic version of test functionals, Fourier transform and a characterization of measures in white noise calculus; *Preprint* (1989), to appear in *J. Funct. Anal.*

[Le 90] Lee, Y.–J.: A characterization of generalized functions on infinite dimensional spaces and Bargmann–Segal analytic functions; contribtuion to these proceedings

[PS 89] Potthoff, J. and Streit, L.: A characterization of Hida distributions, *Preprint* (1989), to appear in *J. Funct. Anal.*

[PY 89] Potthoff, J. and Yan, J.–A.: Some results about test and generalized functionals of white noise; to appear in *Proc. Singapore Probab. Conf.* (1989), L.Y. Chen (ed.)

[Sch 66] Schweber, S.S.: *An Introduction to Relativistic Quantum Field Theory.* New York: Harper and Row (1966)

[Si 71] Simon, B.: Distributions and their Hermite expansions; *J. Math. Physics* **12** (1971) 140–148

[Ya 90] Yan, J.–A.: A characterization of white noise functionals; *Preprint* (1990)

[Yo 87] Yokoi, Y.: Positive generalized functionals; *Hiroshima Math. J.* **20** (1990) 137–157

On determinism of symmetric α-stable processes of generalized Chentsov type

Yumiko SATO

Department of General Education, Aichi Institute of Technology
Yakusa-cho, Toyota 470-03 JAPAN

Shigeo TAKENAKA

Department of Mathematics, Faculty of Science, Hiroshima University
Higashisenda-machi, Naka-ku, Hiroshima 730 JAPAN

February 18, 1991

ABSTRACT. Stochastic processes called symmetric α-stable processes of generalized Chentsov type are constructed by integral geometric manners. As Gaussian processes are determined by their 2-dimensional marginals, these processes are completely determined by their n-dimensional marginal distributions for some n. We illustrate a concrete example of symmetric α stable process which is determined by its 4-dimensional marginal distributions.

INTRODUCTION.

DEFINITION 1. *A stochastic process $\{X(t); t \in \mathbf{R}\}$ is called an H-self-similar $S\alpha S$ (Symmetric Stable of index α) process with stationary increments, $0 < \alpha \leq 2$, (an (α,H)-process in abbreviation) if it satisfies the following conditions:*

i) *Any finite linear combination $\sum_{k=1}^{n} a_k X(t_k)$ is subject to an $S\alpha S$ law, that is,*

$$\mathbf{E}[\exp(iz\sum_{k=1}^{n} a_k X(t_k))] = \exp(-c|z|^{\alpha}),$$

where $z \in \mathbf{R}$ and $c = c(a_1, \cdots, a_n, t_1, \cdots, t_n)$ is a positive constant (under this condition $\{X(t)\}$ is called an $S\alpha S$ family).

Preparing this paper, the authors used a WYSIWYG editing program for mathematical writing named WPMP, a spelling checker for WPMP and a TEXconverter named CVWT. The authors thank Professor I. Kubo, Hiroshima Univ., and Professor H. Yaguchi, Mie Univ., who wrote these programs, for their allowance to use them. We also thank Professor H. Naito, Nagoya Univ., for his TEXnical help.

ii) *For any positive r, the process* $X_r(t) \equiv X(rt)$ *shares all finite dimensional joint distribution functions with* $r^H X(t)$ *(equivalent each other as stochastic processes), that is,*

$$\mathbf{E}[\exp(i \sum_{k=1}^{n} z_k X(rt_k))] = \mathbf{E}[\exp(i \sum_{k=1}^{n} z_k r^H X(t_k))]$$

for any $t_1, \cdots, t_n$ *and* $z_1, \cdots, z_n \in \mathbf{R}$, $n \in \mathbf{N}$ *(self-similarity of exponent H).*

iii) *For any* $h \in \mathbf{R}$, *the shifted process*

$$\{X^h(t) \equiv X(t+h) - X(h); t \in \mathbf{R}\}$$

is equivalent to $\{X(t)\}$ *(the difference is stationary under the parallel transforms).*

Note that, if we take $h = 0$ and $t = 0$ then $X^0(0) \equiv 0$. This means $X(0) = 0$ a.s.

Using a stable version of Wiener integral, the following example of (α,H)-process is known ([1],[3]).

Example 1. For $0 < \alpha \leq 2$, $0 < H < 1$ and $H \neq \frac{1}{\alpha}$, the process

$$\Delta_H(t) \equiv \int_{\mathbf{R}} (|x-t|^{H-\frac{1}{\alpha}} - |x|^{H-\frac{1}{\alpha}}) dZ_\alpha(x), \tag{0.1}$$

where $\{Z_\alpha(x); x \in \mathbf{R}\}$ is an SαS-process with stationary independent increments, is an (α,H)-process.

Another type of (α,H)-process is obtained by an integral geometric method as follows. Before showing new examples, we need to prepare some notations.

DEFINITION 2. *An SαS family* $\mathcal{Y}_\alpha = \{Y(B); B \in \mathcal{B}, \mu(B) < \infty\}$ *with parameter in the Borel field* $\mathcal{B}$ *of a measure space* $(E, \mathcal{B}, \mu)$ *is called an SαS random measure (or SαS white noise) controlled by* $(E, \mathcal{B}, \mu)$ *if it satisfies:*

i) $\mathbf{E}[\exp(izY(B))] = \exp(-\mu(B)|z|^\alpha)$.

ii) $Y(\cup B_k) = \sum Y(B_k)$, *a.s., for any countable disjoint family* $\{B_k\}$, *and the summands of the right-hand side are mutually independent.*

Let $E = \mathbf{R}_+ \times \mathbf{R}$, the half plane. Take a measure $d\mu_\beta(x_0, x) = x_0^{\beta-2} dx_0 dx$ on E. For any $t \in \mathbf{R}$, define a set S_t as

$$S_t \equiv \{(x_0, x) \in E; |x-t| \leq x_0\} \Delta \{(x_0, x) \in E; |x| \leq x_0\}, \tag{0.2}$$

where Δ means the symmetric difference. Then we have another example of (α,H)-process (see [9]),

Example 2. For $0 < \alpha \le 2$, $0 < H < \frac{1}{\alpha}$, the process

$$X_1^\alpha(t) \equiv Y(S_t) \tag{0.3}$$

is an $(\alpha$,H)-process, where $\mathcal{Y}_\alpha = \{Y_\alpha(B)\}$ is an SαS random measure controlled by $d\mu_{\alpha H}(= d\mu_\beta,\ \beta = \alpha H)$ on E.

In the area of parameter (α, H), $0 < \alpha \le 2$, $0 < H < \min\{1, \frac{1}{\alpha}\}$, now we have two examples of $(\alpha$,H)-processes. Calculating their spectrums (the spherical components of Lévy measure) of finite-dimensional distributions, we know that these two processes are not equivalent, that is, Δ_H has continuous spectrums, while X_1^α has pure point spectrums ([7]).

Let us calculate the characteristic function of $(X_1^\alpha(t_1), X_1^\alpha(t_2), ..., X_1^\alpha(t_N))$.

$$\begin{aligned}\mathbf{E}[\exp(i\sum_{j=1}^N z_j X_1^\alpha(t_j))] &= \mathbf{E}[\exp(i\sum_{j=1}^N z_j Y_\alpha(S_{t_j}))] \\ &= \prod_{\{1,...,N\}\supset A\neq\phi} \exp(-\mu_\beta(S_A)|\sum_{j\in A} z_j|^\alpha),\end{aligned} \tag{0.4}$$

where S_A denotes the set $(\bigcap_{j\in A} S_{t_j}) \cap (\bigcap_{k\notin A} \complement S_{t_k})$ and $\complement S_{t_k}$ is the complement of S_{t_k}.

The above equation tells us that the N-dimensional joint distribution function of the process $\{X_1^\alpha\}$ is determined by 2^N-1 scalars $\{\mu_\beta(S_A); A \subset \{1, ..., N\}, A \neq \phi\}$. Using this fact, the first author obtains the following result.

Let μ and μ' be σ-finite measures on $E \equiv \mathbf{R}_+ \times \mathbf{R}$ such that $\mu(S_t) < \infty$, $\mu'(S_t) < \infty$ for any $t \in \mathbf{R}$ (the set S_t is defined by (0.2)). Consider two stochastic processes defined by

$$X(t) \equiv Y_\alpha(S_t) \quad \text{and} \quad X'(t) \equiv Y'_\alpha(S_t),$$

where $\{Y_\alpha\}$ and $\{Y'_\alpha\}$ are SαS random measures controlled by μ and μ' respectively. Then we have

THEOREM 1. (Y.SATO [7]). *If, for any $t, s \in \mathbf{R}$, the distribution functions of random vectors $(X(t), X(s))$ and $(X'(t), X'(s))$ are equal, then the processes $\{X(t)\}$ and $\{X'(t)\}$ are equivalent.*

Let us call the above process with defining set S_t of (0.2) an SαS process of Chentsov type. The above theorem tells us that an SαS process of Chentsov type is determined by its 2-dimensional distributions. Here we recall that a Gaussian process is determined by its covariance functions. More generally, an SαS random field with parameter space $\mathbf{R}^d$ of Chentsov type is determined by its $(d+1)$-dimensional distributions (Y.Sato [8]).

In this paper, we introduce a generalized notion of processes of Chentsov type, and consider their determinism. We also give a new concrete example of $(\alpha$,H)-process which has the same 2-dimensional marginals as X_1 but is different from both Δ_H and X_1 .

§1. Lemmas from set theory. Let $(F, \mathcal{B}, \mu)$ be a σ-finite measure space and fix N elements $B_1, B_2, \ldots, B_N$ of $\mathcal{B}$. Related to our subject, we are concerned with the sets of the form

$$S(\mathbf{e}) = \left(\bigcap_{j:e_j=1} B_j\right) \cap \left(\bigcap_{k:e_k=0} \complement B_k\right), \tag{1.1}$$

where $\mathbf{e} = (e_1, e_2, \cdots, e_N) \in \{0,1\}^N$, and the scalars

$$a(\mathbf{e}) \equiv \mu(S(\mathbf{e})). \tag{1.2}$$

Set

$$\tilde{S}(\mathbf{e}) = \left(\bigcap_{j:e_j=1} B_j\right) \quad \text{and} \quad \tilde{a}(e) = \mu(\tilde{S}(\mathbf{e})). \tag{1.3}$$

We define the length function of $\mathbf{e}$ as

$$\ell(\mathbf{e}) = \#\{j : e_j = 1\}. \tag{1.4}$$

The set $\tilde{S}(\mathbf{e})$ (or $\tilde{a}(\mathbf{e})$) is related to an $\ell(\mathbf{e})$-dimensional distribution function of the process that we are studying(see §2).

Let us introduce two different orders in 2^N=$\{0,1\}^N$:

a) Lexicographic order, that is,

$$\mathbf{e} \leq \mathbf{f} \text{ if and only if } e_{k+1} < f_{k+1} \text{ for } k = \max\{j; e_1 = f_1, e_2 = f_2, \cdots, e_j = f_j\}. \tag{1.5}$$

This order is equivalent to the order as a binary integer

$$\sum_{j=1}^{N} 2^{N-j} e_j. \tag{1.6}$$

b) Lattice (partial) order,

$$\mathbf{e} \preccurlyeq \mathbf{f} \quad \text{if and only if} \quad e_k \leq f_k \quad \text{for any} \quad k = 1, \ldots, N. \tag{1.7}$$

It is easy to see that

$$\mathbf{e} \preccurlyeq \mathbf{f} \quad \text{implies} \quad \mathbf{e} \leq \mathbf{f}. \tag{1.8}$$

Using the order $\preccurlyeq$, we easily obtain the relation of the sets $S(\cdot)$ and $\tilde{S}(\cdot)$ and also the $a(\cdot)$ and $\tilde{a}(\cdot)$.

PROPOSITION 1.1. *For any* $\mathbf{e} \in 2^N$,

$$\tilde{S}(\mathbf{e}) = \bigcup_{\mathbf{e} \preccurlyeq \mathbf{f}} S(\mathbf{f}) \quad \textit{(disjoint union)} \quad \textit{and} \quad \tilde{a}(\mathbf{e}) = \sum_{\mathbf{e} \preccurlyeq \mathbf{f}} a(\mathbf{f}). \tag{1.9}$$

We define two column vectors

$$\mathbf{a} = (a(\mathrm{e}); \mathrm{e} \in 2^N - \{(0,0,\cdots,0)\}) \qquad \text{and} \qquad \tilde{\mathbf{a}} = (\tilde{a}(\mathrm{e})),$$

where we arrange the elements of the vectors in the order of $\leq$. The above proposition tells us that there exists a $(2^N\text{-}1)\times(2^N\text{-}1)$ matrix M, whose elements are 0 or 1, such that

$$\tilde{\mathbf{a}} = M \cdot \mathbf{a}. \tag{1.10}$$

Moreover, by the relation (1.8), M is an upper triangular matrix and all of its diagonal elements are equal to 1. Hence the matrix M is reversible. So, we could deduce all of the information of $\{a(\mathrm{e}); \mathrm{e} \in 2^N\}$ from that of $\{\tilde{a}(\mathrm{e}); e \in 2^N\}$.

Note that, $\mathrm{e}_{\max} \equiv (1,1,1,\cdots,1)$ is the unique element satisfying $\ell(\mathrm{e}) = N$ and hence, for any $\mathrm{e} \neq \mathrm{e}_{\max}$, the scalar $\tilde{a}(\mathrm{e})$ is related to marginal distributions of dimension less than N.

Suppose that we know all of the information of $\tilde{a}(\mathrm{e})$ for $\ell(\mathrm{e}) < N$.

Under what condition can we obtain all the values of $a(\mathrm{e})$ for any $\mathrm{e} \in 2^N$?

Next proposition known as the **principle of inclusion and exclusion** in combinatorics tells us an answer (see [6]).

PROPOSITION 1.2. *If there exists an* $\mathrm{e}_0 \in 2^N$ *such that* $a(\mathrm{e}_0) = 0$, *(or* $S(\mathrm{e}) = \phi$*), then we can calculate all the values of* $a(\mathrm{e})$ *from the information of* $\{\tilde{a}(\mathrm{e}); \ell(\mathrm{e}) < N\}$.

N.B. In the next section we give a probabilistic aspect of this proposition.

PROOF: Let M_{e_0} denote the $(2^N\text{-}2)\times(2^N\text{-}2)$ matrix obtained from the matrix M by removing e_0-th column and $\mathrm{e}_{\max}$-th row.

1) If $\mathrm{e}_0 = \mathrm{e}_{\max}$, M_{e_0} is an upper triangular matrix with diagonal elements 1 like M. So, $\det(M_{\mathrm{e}_{\max}}) = 1$.

2) In case $\mathrm{e}_0 \neq \mathrm{e}_{\max}$, we can assume that

$$\mathrm{e}_0 = (1,1,\cdots,\underset{\substack{\uparrow \\ k-1}}{1},0,\cdots,0).$$

If e_0 has other form, we change the arrangement of the index set $\{t_j; j = 1,2,\cdots,N\}$ so that e_0 has the required form. It is easy to see that the lattice order $\preccurlyeq$ and the relation (1.8) are invariant under this rearrangement. Thus, all arguments on M are valid for this new arrangement.

Denote R_{f} the f-th row vector of the matrix M_{e_0}. Let us calculate the vector

$$\mathbf{L} = \sum_{\mathrm{f} \succcurlyeq \mathrm{e}_0} R_{\mathrm{f}}(-1)^{\ell(\mathrm{f})} - (0,0,0,\cdots,1).$$

It is obvious that, if $\mathbf{e} \not\succcurlyeq \mathbf{e}_0$ then the e-th element $\mathbf{L_e}$ of $\mathbf{L}$ is equal to 0. For $\mathbf{e} \succcurlyeq \mathbf{e}_0$,

$$\mathbf{L_e} = \sum_{f_k, f_{k+1}, \cdots, f_N = 0,1} M_{\mathbf{f},\mathbf{e}} (-1)^{k-1+\#\{j; f_j=1, j=k,\cdots,N\}},$$

here $M_{\mathbf{f},\mathbf{e}}$ means the (f-e)-th element of the matrix $M_{\mathbf{e}_0}$ for $\mathbf{f} \neq \mathbf{e}_{\max}$ and $M_{\mathbf{e}_{\max},\mathbf{e}} = 0$, 1 if $\mathbf{e} \neq \mathbf{e}_{\max}$ and $\mathbf{e} = \mathbf{e}_{\max}$ respectively. It is obvious that if $\mathbf{f} \succcurlyeq \mathbf{e}$ and $f_j = 0$ then $f' = (f_1, f_2, \cdots, f_{j-1}, 1, f_{j+1}, \cdots, f_N) \succcurlyeq \mathbf{e}$. This means that $M_{\mathbf{f},\mathbf{e}} = M_{\mathbf{f}',\mathbf{e}}$. Considering this fact, let us continue the above calculation.

$$\begin{aligned}
\mathbf{L_e} &= \sum_{f_k=0,1} \left(\sum_{f_{k+1},\cdots,f_N=0,1} M_{\mathbf{f},\mathbf{e}} (-1)^{k-1+\#\{j; f_j=1, j=k+1,\cdots,N\}} \right) (-1)^{f_k} \\
&= \sum_{f_k=0, f_{k+1},\cdots,f_N=0,1} M_{\mathbf{f},\mathbf{e}} (-1)^{k-1+\#\{j; f_j=1, j=k+1,\cdots,N\}} \\
&\quad + \sum_{f_k=1, f_{k+1},\cdots,f_N=0,1} M_{\mathbf{f}',\mathbf{e}} (-1)^{k-1+\#\{j; f_j=1, j=k+1,\cdots,N\}} \times (-1) \\
&= 0.
\end{aligned}$$

Thus we have

$$\sum_{\mathbf{f} \succcurlyeq \mathbf{e}_0} M_{\mathbf{f}} (-1)^{\ell(\mathbf{f})} = (0, , 0, \cdots, 1).$$

Changing the order of rows as $\mathbf{e}_0 \to \mathbf{e}_{\max} - 1$, $\mathbf{e}_0 + 1 \to \mathbf{e}_0$, $\mathbf{e}_0 + 2 \to \mathbf{e}_0 + 1$, etc, the matrix $M_{\mathbf{e}_0}$ in question becomes an upper triangular matrix with diagonal elements 1. Thus we have the required result. ∎

Example of M for $N = 3$ and $M_{\mathbf{e}_0}$ in case of $\mathbf{e}_0 = (1, 0, 0)$.

	(0,0,1)	(0,1,0)	(0,1,1)	(1,0,0)	(1,0,1)	(1,1,0)	(1,1,1)	
(0,0,1)	1	0	1	0	1	0	1	
(0,1,0)	0	1	1	0	0	1	1	
(0,1,1)	0	0	1	0	0	0	1	
(1,0,0)	0	0	0	1	1	1	1	−
(1,0,1)	0	0	0	0	1	0	1	+
(1,1,0)	0	0	0	0	0	1	1	+
(1,1,1)	0	0	0	0	0	0	1	−

fig 1

In this case, the above proof means that the weighted sum of the bottom 4 row vectors is equal to (0,0,0,1,0,0,0) and that the determinant of the right-lower 3 × 3 minor matrix is 1.

§2. SαS processes of generalized Chentsov type and their determinism. In this section we generalize the class of SαS processes of Chentsov type, using more general defining sets, and consider their determinism.

Let $E \equiv \mathbf{R}_+ \times \mathbf{R}$. Instead of C_t given in the introduction, let us consider the following set $C_{N,t}$:

$$\begin{aligned} C_{N,t} &\equiv C_{N,t}(a_1, b_1, a_2, b_2, \cdots, a_N, b_N) \\ &= \{(x_0, x) \in E; a_k x_0 < x - t < b_k x_0, k = 1, \cdots, N\}, \end{aligned} \tag{2.1}$$

where we fix $-\infty < a_1 < b_1 < a_2 < \cdots < a_N < b_N < +\infty$. $C_{N,t}$ is the union of N disjoint cones with common vertex $(0, t)$. Let $S_t = C_{N,t} \Delta C_{N,0}$, $\mathcal{B} = \mathcal{B}(E)$ be the topological Borel field and μ be a σ-finite absolutely continuous measure on E with condition $\mu(S_t) < \infty$ for any S_t. Let $\mathcal{Y}_\alpha = \{Y_\alpha(B)\}$ be an SαS random measure controlled by (E, μ).

DEFINITION 3. *The SαS process*

$$X_N^\alpha(t) \equiv Y_\alpha(S_t) \tag{2.2}$$

constructed by above setup is called an SαS process of **generalized Chentsov type.**

N.B. The process which we consider in the introduction is a special case of this type (take C_t as $C_{1,t}(-1, 1)$), so we call this the process of generalized Chentsov type.

Our main result is:

THEOREM 2 (MAIN THEOREM). *Let $\{X_N^\alpha(t)\}$ and $\{X_N^{\alpha\prime}(t)\}$ be two SαS processes of generalized Chentsov type which share the same system of cones $\{C_{N,t}\}$ and are defined by control measures μ and μ', respectively. If, for any $t_1, t_2, \cdots, t_{2N} \in \mathbf{R}$, two $\mathbf{R}^{2N}$-valued random variables*

$(X_N^\alpha(t_1), X_N^\alpha(t_2), \cdots, X_N^\alpha(t_{2N}))$ and $(X_N^{\alpha\prime}(t_1), X_N^{\alpha\prime}(t_2), \cdots, X_N^{\alpha\prime}(t_{2N}))$

are subject to a common SαS law, then the two processes $\{X_N^\alpha(t)\}$ and $\{X_N^{\alpha\prime}(t)\}$ are equivalent to each other as processes.

The above theorem tells us that we have examples of SαS processes of generalized Chentsov type which are determined by their $2N$-dimensional marginal distributions.

Before proving the Theorem 2, let us prepare two lemmas.

LEMMA 2.1. *If, for any choice of $t_1, t_2, \cdots, t_{n+1}$, there exists an $\mathrm{e}_0 \in \{0,1\}^{n+1}$ such that*

$$S(\mathrm{e}_0) = \Big(\bigcap_{j:e_j=1} S_{t_j}\Big) \cap \Big(\bigcap_{k:e_k=0} S_{t_k}\Big) = \phi \quad (\textit{ or } \quad \mu(S(\mathrm{e}_0)) = 0),$$

then the corresponding SαS process

$$X^{\alpha}(t) = Y_{\alpha}(S_t)$$

is determined by its n-dimensional marginals.

PROOF: Fix $m \geq n$, because $S(\mathrm{e}_0) = \phi$, $S((\mathrm{e}_0, \mathrm{e}_1)) = \phi$ for any $\mathrm{e}_1 \in \{0,1\}^{m-n}$. This means that we can calculate the weights of spectrums of m-dimensional joint distribution of $X(t)$ from its (m-1)-dimensional marginals. Continuing this procedure, we obtain the proof of the lemma. ∎

Using the above lemma, it is enough to prove the main theorem to find an $\mathrm{e}_0 \in \{0,1\}^{2N+1}$ such that $S(\mathrm{e}_0) = \phi$. The next lemma tells us the relation between intersection of the cones and intersection of their symmetric differences.

To avoid confusion, denote

$$C(\mathrm{e}) = \left(\bigcap_{j:e_j=1} C_{t_j}\right) \cap \left(\bigcap_{k:e_k=0} \complement C_{t_k}\right) \tag{2.3}$$

instead of $S(\mathrm{e})$ for the case of cones C_t or $C_{N,t}$.

LEMMA 2.2. *$S(\mathrm{e}) = \phi$ for an $\mathrm{e} \in \{0,1\}^n$ for the time points $\{t_1, t_2, \cdots, t_n\}$ if and only if $C(\tilde{\mathrm{e}}) = C(\tilde{\mathrm{e}}^*) = \phi$, where $\tilde{\mathrm{e}} = (0, \mathrm{e}) \in \{0,1\}^{n+1}$, $(\tilde{\mathrm{e}}^*)_k = 1 - \tilde{\mathrm{e}}_k$ and the time points are taken as $\{0, t_1, t_2, \cdots, t_n\}$.*

PROOF: The above fact is easily obtained by the relation

$$\bigcap(A_k \Delta B) = \bigcap\{(A_k \cap \complement B) \cup (\complement A_k \cap B)\} = \{\complement B \cap (\bigcap A_k)\} \cup \{B \cap (\bigcap \complement A_k)\}.$$

Consider $A_k = C_{t_k}$ or $\complement C_{t_k}$ and $B = C_0$, then $A_k \Delta B = S_{t_k}$ or $\complement S_{t_k}$ respectively. And the union of the left hand side of above equation is a disjoint union. ∎

PROOF OF THE THEOREM 2: Here $S_t = C_{N,t} \Delta C_{N,0}$.

1) For $N = 1$, take 4 time points $0 < t_1 < t_2 < t_3$ and take $e = (0,1,0,1)$. The element (x_0, x) of the set $C(\mathrm{e})$, if it exists, has to satisfy following four inequalities:

(0) $a_1 x_0 \geq x$ or $b_1 x_0 \leq x$,

(1) $a_1 x_0 < x - t_1 < b_1 x_0$

(2) $a_1 x_0 \geq x - t_2$ or $b_1 x_0 \geq x - t_2$

and

(3) $\quad a_1x_0 < x - t_3 < b_1x_0$.

From (1) and (3) we have $a_1x_0 + t_3 < x < b_1x_0 + t_1$. This inequality contradicts (2). Thus $C(\mathrm{e}) = \phi$. In the same manner we have $C(\mathrm{e}^*) = \phi$. This proves the theorem for $N = 1$.

2) For $N > 1$, take $2N + 2$ time points $0 = t_0 < t_1 < t_2 < \cdots < t_{2N+1}$. Let us show that $C(\mathrm{e}) = C(\mathrm{e}^*) = \phi$ for $\mathrm{e} = (0,1,0,\cdots,0,1) \in \{0,1\}^{2N+2}$ by induction.

Let us divide $C_{N,t}$ into two disjoint sets A_t and B_t. A_t is a single cone defined by $\{a_1, b_1\}$ and B_t is the union of the remaining (N-1) cones. The assumptions of our induction are:

i) $\quad A_{s_1} \cap \complement A_{s_2} \cap A_{s_3} = \phi$ for any $s_1 < s_2 < s_3$

and

ii) $\quad B_{s_1} \cap \complement B_{s_2} \cap B_{s_3} \cap \cdots \cap \complement B_{s_{2N-2}} \cap B_{s_{2N-1}} = \phi$ for any $s_1 < s_2 < \cdots < s_{2N-1}$.

Our goal is to show

$$C_{N,s_1} \cap \complement C_{N,s_2} \cap C_{N,s_3} \cap \cdots \cap \complement C_{N,s_{2N}} \cap C_{N,s_{2N+1}} = \phi, \tag{2.4}$$

for any $s_1 < s_2 < \cdots < s_{2N+1}$.

Because both $(0,1,0,1,\cdots,1) \in \{0,1\}^{2N+2}$ and its conjugate $(1,0,1,\cdots,0)$ contain $(1,0,1,\cdots,1) \in \{0,1\}^{2N+1}$, the proof of the theorem is obtained directly from (2.4).

Let us start the proof. Note that, if $t < s$ then $A_t \cap B_s = \phi$. We call this property the rule iii).

$$\begin{aligned}
&C_{N,s_1} \cap \complement C_{N,s_2} \cap C_{N,s_3} \cap \cdots \cap \complement C_{N,s_{2N}} \cap C_{N,s_{2N+1}} \\
&\quad = (A_{s_1} \cup B_{s_1}) \cap \complement(A_{s_2} \cup B_{s_2}) \cap (A_{s_3} \cup B_{s_3}) \cap \complement(A_{s_4} \cup B_{s_4}) \cap \cdots \cap (A_{s_{2N+1}} \cup B_{s_{2N+1}}) \\
&\quad = \{[(A_{s_1} \cap \complement A_{s_2} \cap \complement B_{s_2} \cap A_{s_3}) \cup (A_{s_1} \cap \complement A_{s_2} \cap \complement B_{s_2} \cap B_{s_3})] \\
&\qquad\qquad \cap \complement(A_{s_4} \cup B_{s_4}) \cap \cdots \cap (A_{s_{2N+1}} \cup B_{s_{2N+1}})\} \\
&\qquad \cup \{(B_{s_1} \cap \complement A_{s_2} \cap \complement B_{s_2}) \cap (A_{s_3} \cup B_{s_3}) \cap \complement(A_{s_4} \cup B_{s_4}) \cap \cdots \cap (A_{s_{2N+1}} \cup B_{s_{2N+1}})\}
\end{aligned}$$

(the 1st and the 2nd terms disappear by the rules i) and iii))

Repeat this procedure to $(A_{s_3} \cup B_{s_3}) \cap \complement(A_{s_4} \cup B_{s_4}) \cap (A_{s_5} \cup B_{s_5})$, and so on.

$$\begin{aligned}
&= (B_{s_1} \cap \complement A_{s_2} \cap \complement B_{s_2} \cap B_{s_3} \cap \complement A_{s_4} \cap \complement B_{s_4} \cap B_{s_5} \cap \complement A_{s_6} \cap \complement B_{s_6} \cap \cdots \cap (A_{s_{2N+1}} \cup B_{s_{2N+1}})\} \\
&= (\complement A_{s_2} \cap \complement A_{s_4} \cap \complement A_{s_6} \cap \cdots \cap \complement A_{s_{2N}} \cap A_{s_{2N+1}} \cap B_{s_1} \cap \complement B_{s_2} \cap \cdots \cap \complement B_{s_{2N-2}} \cap B_{s_{2N-1}}) \\
&\qquad\qquad \cap \complement B_{s_{2N}} \cap (A_{s_{2N+1}} \cup B_{s_{2N+1}}).
\end{aligned}$$

By the rule ii), we have the required results. ∎

Let us consider a degenerate case $a_1 = -\infty$ (or equivalently $b_N = \infty$). If $N=1$, then the process becomes a process with independent increments and is determined by its

1-dimensional distributions. By the same method, we can prove that the process of generalized Chentsov type related to cone $C_t(-\infty, b_1, a_2, \cdots, b_N)$ is determined by its $(2N\text{-}1)$-dimensional marginals.

§3. Self-similar processes and examples. In the introduction, we have two examples of (α,H)-processes $\{\Delta_H(t)\}$ and $\{X_1^\alpha(t)\}$ for the parameter area $0 < \alpha \leq 2$, $0 < H < \min(1, 1/\alpha)$. The latter is a concrete example of SαS processes of Chentsov type which we discussed in §2. This case, the cone is $C_t(-1,1)$ and the measure is $d\mu_{\alpha H}(x_0, x) = x_0^{\alpha H-2} dx_0 dx$. The self-similarity of $X_1^\alpha(t)$ comes from the following fact.

For any positive r, the n-dimensional joint characteristic function of $\{X_1^\alpha(rt)\}$ is,

$$\begin{aligned} \mathbf{E}[\exp(i\sum_{j=1}^{n} z_j X_1^\alpha(rt_j))] &= \mathbf{E}[\exp(i\sum z_j Y_\alpha(S_{rt_j}))] \\ &= \prod_{\mathrm{e}\in\{0,1\}^n} \exp(-\mu_{\alpha H}(S_r(\mathrm{e}))|\sum_{j:e_j=1} z_j|^\alpha) \end{aligned} \tag{3.1}$$

(here $S_r(\mathrm{e})$ denotes the set $(\bigcap_{j:e_j=1} S_{rt_j}) \cap (\bigcap_{k:e_k=0} \complement S_{rt_k})$)

$$\begin{aligned} &= \exp(-\mu_{\alpha H}(S_1(\mathrm{e}))|\sum_j |r|^{\alpha H/\alpha} z_j|^\alpha) \\ &= \mathbf{E}[\exp(i\sum z_j |r|^H X_1^\alpha(t_j))]. \end{aligned} \tag{3.2}$$

This means that, for any n and n time points $t_1, t_2, \cdots, t_n$, the distribution functions of two random variables

$(X_1^\alpha(rt_1), X_1^\alpha(rt_2), \cdots, X_1^\alpha(rt_n))$ and $(r^H X_1^\alpha(t_1), r^H X_1^\alpha(t_2), \cdots, r^H X_1^\alpha(t_n))$

are equal. Here we use only the fact

$$\mu_{\alpha H}(S_{rt}) = r^{\alpha H} \mu_{\alpha H}(S_t). \tag{3.3}$$

It is easy to see that (3.3) holds also for any set $S_t = C_{N,t} \Delta C_{N,0}$ of the last section for any $C_{N,t}(a_1, b_1, a_2, b_2, \cdots, a_N, b_N)$. Thus, we obtain **a series of examples** of (α,H)-processes which possibly have **different determinism.**

Taking N=2 and a_1, b_1, a_2, b_2 as special values, we give a concrete example $\{X_2^\alpha\}$. As we will see soon, this example has the following properties:

I. This process $\{X_2^\alpha\}$ is determined by its 4-dimensional distributions and not by 3-dimensional distributions. This fact suggests us that the result of Theorem 2 is best possible (see the proof of proposition 3.3.).

Comparison with $\{X_1^\alpha\}$ yields:

II. $\{X_1^\alpha\}$ and $\{X_2^\alpha\}$ have the same 2-dimensional marginals (proposition 3.1) ,

and

III. The spectrum of 3-dimensional distribution of $\{X_2^\alpha\}$ consists of 14 discrete points while that of $\{X_1^\alpha\}$ consists of 12 points. Thus, two processes are different from each other (proposition 3.3.).

Example 3. Choose N=2 and $a_1 = -1, b_1 = -1/3, a_2 = 1/3, b_2 = 1$. Let

$$X_2^\alpha(t) = Y_\alpha(S_t),$$

where $S_t = C_{2,t} \Delta C_{2,0}$ and $\mathcal{Y}_\alpha$ is an SαS random measure controlled by $d\mu_{\alpha H}$.

PROPOSITION 3.1. *There exists a positive constant γ such that $\{X_1^\alpha(t)\}$ and $\{\gamma \cdot X_2^\alpha(t)\}$ share the same two dimensional distributions.*

To show this fact, we need a new concept. In our construction of SαS processes of generalized Chentsov type, we are able to have a set of stochastic processes $\{\{X^\alpha(t) \equiv Y_\alpha(S_t)\}; 0 < \alpha \leq 2\}$ simultaneously with a set up $\{(E, d\mu), S_t\}$ being fixed. Let us call $\{\{X^\alpha(t)\}\}$ a **conjugate** class of generalized Chentsov type, and call the Gaussian process $\{X^{(2)}(t) \equiv X^2(t)\}$ the **Gaussian element** of this class. Then we have

PROPOSITION 3.2. *Let $\{\{X^\alpha(t)\}; 0 < \alpha \leq 2\}$ and $\{\{X^{\alpha'}(t)\}\}$ be two conjugate classes which are constructed by $\{(E, \mu), S_t\}$ and $\{(E', \mu'), S_t'\}$ respectively. If the Gaussian elements $\{X^{(2)}(t)\}$ and $\{X^{(2)'}(t)\}$ are subject to the same law, then, for any $0 < \alpha < 2$, the processes $\{X^\alpha(t)\}$ and $\{X^{\alpha'}(t)\}$ have common 2-dimensional distributions. Moreover, if, for some α and H, the processes $\{X^\alpha(t)\}$ and $\{X^{\alpha'}(t)\}$ are (α, H)-processes of generalized Chentsov type, then, for any $0 < \alpha < 2$, there exists a constant c_α such that $\{X^\alpha(t)\}$ and $\{c_\alpha X^{\alpha'}(t)\}$ have common 2-dimensional distributions.*

PROOF: Fix a conjugate class $\{\{X^\alpha(t)\}; 0 < \alpha \leq 2\}$ and consider 2-dimensional characteristic function

$$\begin{aligned} &\mathbf{E}[\exp i\{X^\alpha(s)z_1 + X^\alpha(t)z_2\}] \\ &\quad = \exp -\{\mu(S_s \cap \complement S_t)|z_1|^\alpha + \mu(\complement S_s \cap S_t)|z_2|^\alpha + \mu(S_s \cap S_t)|z_1 + z_2|^\alpha\}. \end{aligned}$$

The characteristic function is determined by the scalars $\mu(S_s \cap \complement S_t)$, $\mu(\complement S_s \cap S_t)$ and $\mu(S_s \cap S_t)$.

Let us consider 2-dimensional characteristic function of the Gaussian element

$$\mathbf{E}[\exp i\{X^{(2)}(s)z_1 + X^{(2)}(t)z_2\}] = \exp -\{\frac{1}{2}\sigma_s z_1^2 + \sigma_{st} z_1 z_2 + \frac{1}{2}\sigma_t z_2^2\},$$

where σ_s, σ_t and σ_{st} are the variances and the covariance of the process $\{X^{(2)}\}$. On the other hand, the same characteristic function has the following form:

$$\exp -\{\mu(S_s \cap \complement S_t)|z_1|^2 + \mu(\complement S_s \cap S_t)|z_2|^2 + \mu(S_s \cap S_t)|z_1 + z_2|^2\}.$$

The quantities $\mu(S_s \cap \complement S_t)$ etc. are uniquely determined by σ_s, σ_t and σ_{st} as

$$\mu(S_s \cap \complement S_t) = \frac{1}{2}(\sigma_s - \sigma_{st}), \quad \mu(\complement S_s \cap S_t) = \frac{1}{2}(\sigma_s - \sigma_{st}) \text{ and } \mu(S_s \cap S_t) = \frac{1}{2}\sigma_{st}.$$

Thus, we have the proof of the first part.

Suppose that, for some α and H, $\{X^\alpha(t)\}$ and $\{X^{\alpha'}(t)\}$ are (α,H)-processes. Then the Gaussian elements $\{X^{(2)}(t)\}$ and $\{X^{(2)'}(t)\}$ are (2,αH/2)-processes. In general, if $\{B^H(t); t \in R\}$ is a $(2,H)$-process (a fractional Brownian motion) for some $0 < H < 1$, then its variance is $E[B^H(t)]^2 = \gamma|t|^{2H}$ with a positive constant γ and the covariance function is calculated as

$$\begin{aligned}\mathbf{E}[B^H(t)B^H(s)] &= \frac{1}{2}E[B^H(t)^2 + B^H(s)^2 - (B^H(t) - B^H(s))^2] \\ &= \frac{\gamma}{2}(|t|^{2H} + |s|^{2H} - |t-s|^{2H}).\end{aligned}$$

Thus from 1-dimensional distributions the Gaussian process $\{B^H(t)\}$ is completely determined. Hence $\{X^{(2)}(t)\}$ and $\{cX^{(2)'}(t)\}$ are equivalent with some c. Using the first part, this completes the proof of the proposition. ∎

PROPOSITION 3.3. *Two processes $\{X_1^\alpha(t)\}$ and $\{X_2^\alpha(t)\}$ are not equivalent (though their 2-dimensionl distributions are the same).*

PROOF: To prove proposition 3.3., it is enough to give a set of explicit four time points $\{t_1, t_2, t_3, t_4\}$ on which $\{X_2^\alpha\}$ can not be determined by its 3-dimensional marginals. By Lemmas 3.1 and 3.2, this is equivalent to show that there is no pair e ,e* $\in \{0,1\}^5$ such that $C(\mathrm{e}) = C(\mathrm{e}^*) = \phi$ with respect to the five time points $\{0, t_1, t_2, t_3, t_4\}$. Let us start to prove this for explicite points {-2,-1,0,1,2}.

The defining equations of five cones are

$$|x+2| < x_0 < 3\cdot|x+2|, \qquad |x+1| < x_0 < 3\cdot|x+1|,$$
$$|x| < x_0 < 3\cdot|x|, \qquad |x-1| < x_0 < 3\cdot|x-1|,$$
and
$$|x-2| < x_0 < 3\cdot|x-2|.$$

Let us fix $x = -5$. Then according to the value of x_0 the point (x_0, x) moves in the areas $C(\mathrm{e})$ as follows:

x_0	$1 \sim 3$	$3 \sim 4$	$4 \sim 5$	$5 \sim 6$	$6 \sim 7$
e	$(0,0,0,0,0)$	$(1,0,0,0,0)$	$(1,1,0,0,0)$	$(1,1,1,0,0)$	$(1,1,1,1,0)$
val	0	16, 1	24, 3	28, 7	30, 15

$7 \sim 9$	$9 \sim 12$	$12 \sim 15$	$15 \sim 18$	$18 \sim 21$	$21 \sim$
$(1,1,1,1,1)$	$(0,1,1,1,1)$	$(0,0,1,1,1)$	$(0,0,0,1,1)$	$(0,0,0,0,1)$	$(0,0,0,0,0)$
31	15, 30	7, 28	3, 24	1, 16	0

table 1

The 3rd row of the table indicates the binary value of $\mathbf{e} = (e_1, e_2, \cdots, e_5)$ and its reverse $(e_5, e_4, \cdots, e_1)$. From the symmetry, the reverse appears in the same area of x_0 with $x = +5$. In the same manner, let us pick up new values of e for different values of x:

$(x_0, -3)$	$3 \sim 4$	$4 \sim 5$
val	12, 6	14

$(x_0, -2)$	$1 \sim 2$
	8, 2

$(x_0, -1)$	$0 \sim 1$	$1 \sim 2$
	20, 5	22, 13

$(x_0, -0.5)$	$1.5 \sim 2.5$	$2.5 \sim 4.5$
	18, 9	25, 19

$(x_0, -0.2)$	$0.2 \sim 0.6$	$1.8 \sim 2.2$
	4	26, 11

and

$(x_0, 0)$	$1 \sim 2$	$2 \sim 3$	$3 \sim 6$
	10	27	17

table 2

Among 32 5-bits binary numbers, 29 numbers appear in the above tables. The exceptions are $21 = (1,0,1,0,1)$, $23 = (1,0,1,1,1)$ and $29 = (1,1,1,0,1)$. Fortunately, these exceptions contain no pair e and e*. This fact means that the process $\{X_2\}$ is determined by its 4-dimensional marginals but not by 3-dimensional marginals. Thus we have the required result. ∎

In addition, we can easily prove, by direct calculation, that three sets $C((1,0,1,0,1))$, $C((1,0,1,1,1))$ and $C((1,1,1,0,1))$ are all empty.

Finally, let us sum up our examples of (α, H)-processes.

	$\Delta_H(t)$	$X_1^\alpha(t)$	$X_2^\alpha(t)$
spectrum	continuous	discrete	
1-dim. distributions	all are the same		
2-dim. distributions	different from $X_1^\alpha(t)$, $X_2^\alpha(t)$	same	
3-dim. distributions	different each other		

table 3

REFERENCES

1. S. Cambanis and M. Maejima, *Two classes of self-similar stable processes with stationary increment*, Stoch. Proc. Appl. **32** (1989), 305–392.
2. N. N. Chentsov, *Lévy's Brownian motion of several parameters and generalized white noise*, Theory Probab. Appl. **2** (1957), 265–266.
3. Y. Kasahara and M. Maejima, *Weighted sums of i.i.d. random variables attracted to integrals of stable processes*, Probab. Th. Rel. Fields **78** (1988), 75–96.
4. P. Lévy, "Processus stochastiques et mouvement brownien," Gauthier-Villars, Paris, 1937.
5. T. Mori, *Chentsov type representation of linearly additive random fields*, preprint (1990).
6. H. J. Ryser, "Combinatorial Mathematics. Carus Math. Mono. 14.," Math. Assoc. America, 1963.
7. Y. Sato, *Joint distributions of some self-similar stable processes*, preprint (1989).
8. Y. Sato, *Distributions of stable random fields of Chentsov type*, Nagoya M. J. (1991) (to appear).
9. S. Takenaka, *Integral geometric constructions of self-similar stable processes*, Nagoya M. J. (1991) (to appear).

Keywords. stable process, self-similar process, determinism
1980 *Mathematics subject classifications*: 60G99,60E99,60D05,62H05

A W.N.C. Viewpoint on Intersection Local Times

Narn-Rueih Shieh*

Department of Mathematics

National Taiwan University

Taipei, Taiwan

* This work is supported in part by National Science Council, Taipei, Taiwan.

Abstract Consider the formal functional

$$\int_{I_1}\cdots\int_{I_k}\prod_{i=1}^{k-1}u_i(B(t_{i+1})-B(t_i))dt_1\cdots dt_k,$$

where $B(t)$ is planar Brownian motion and $I_1,\cdots,I_k$ are k, $k\geq 2$, disjoint compact intervals in $\{t>0\}$ and $u_1,\cdots,u_k$ are k positive tempered measures on R^2. Using White Noise Calculus, we show that the integrand is a generalized Brownian functional while the integral itself is an ordinary Brownian functional.

§1. Introduction

Let $B(t)$ be planar Brownian motion. Let $k\geq 2$ and $I_1,\cdots,I_k$ be k disjoint compact intervals in $\{t>0\}$, with increasing order. It has been proved by Wolpert [11] and Geman, Horowitz and Rosen [2] that the formal functional

$$\Phi_k=\int_{I_1}\cdots\int_{I_k}\prod_{i=1}^{k-1}\delta(B(t_{i+1})-B(t_i))dt_1\cdots dt_k$$

can be realized as the L^2 limit of $\{\Phi_{k,\varepsilon}\}$ as $\varepsilon\downarrow 0$, where we replace δ by the density of planar Gaussian distribution $N_2(0,\varepsilon^2)$ to obtain $\Phi_{k,\varepsilon}$. The functional Φ_k plays an important role both in multiple points problem [2, 7, 11] and in some physical models arising from quantum fields and polymers; see Wolpert [12], Dynkin [1], Westwater [10] and the references cited there. In this paper, we shall consider the following formal functional

$$\Psi_k = \int_{I_1} \cdots \int_{I_k} \prod_{i=1}^{k-1} u_i(B(t_{t+1}) - B(t_i)) dt_1 \cdots dt_k, \tag{1.1}$$

where u_i are positive tempered measures on R^2. Using White Noise Calculus (Hida Calculus), we shall show that the integrand in (1.1) can be regarded as a generalized Brownian functional while Ψ_k itself is an ordinary Brownian functional. Thus, we not only retrieve the above-mentioned result on Φ_k but also give an interesting viewpoint on the connexion between generalized and ordinary Brownian functionals. The case $k = 2$ has been studied by Kuo and Shieh [5] from the viewpoint of generalized Ito's formulae. However, the technique for general $k \geq 2$ studied here is much more involved.

There exists already a vast literature on the formulation of White Noise Calculus. Here, we merely mention and refer to a latest overview by Hida and Putthoff [3]. We shall adopt the following triples

$$S(R) \hookrightarrow L^2(R) \hookrightarrow S^*(R), \text{ and}$$

$$(S) \hookrightarrow (L^2) \hookrightarrow (S)^*, \text{ where}$$

$(L^2) = L^2(S^*(R), \text{white noise measure})$, as those appeared in [3, pp142-149]. The members in (L^2) and $(S)^*$ are called respectively ordinary and generalized Brownian functionals. Define, for $t \geq 0$ and $x \in S^*(R)$,

$$B_1(t,x) = \langle x, 1_{[0,t]} \rangle, B_2(t,x) = \langle x, 1_{[-t,0]} \rangle, \text{ and}$$

$$B(t,x) = (B_1(t,x), B_2(t,x));$$

then B_i and B are respectively linear and planar Brownian motion. Kallianpur and Kuo [4] showed that, for any tempered distribution u on R, the formal composite

$u(B_1(t))$ belongs to another space $(L^2)^-$ of generalized Brownian functionals and has an orthogonal series expansion in $(L^2)^-$. Our series expansion in Proposition 2.1 below is motivated from them.

§2. The main results

In this section, $H_n(x)$ will denote the n-th Hermite polynomial on R

$$H_n(x) = (-1)^n \exp(-x^2)(\frac{d}{dx})^n \exp(x^2).$$

Let $u_1, \cdots, u_k$ be k, $k \geq 2$, positive tempered measures on R^2. For $i = 1, \cdots, k-1$ and $m_i, n_i \in Z_{\geq 0}$ and $0 < t_i < t_{i+1}$, let $K_{i,m_i,n_i}(t_i, t_{i+1})$ denote the integral

$$\int_{R^2} H_{m_i}\Big(\frac{x_1}{\sqrt{2(t_{i+1}-t_i)}}\Big) H_{n_i}\Big(\frac{x_2}{\sqrt{2(t_{i+1}-t_i)}}\Big) e^{-\frac{x_1^2+x_2^2}{2(t_{i+1}-t_i)}} du_i(x_1, x_2).$$

Moreover, we write, for $\overline{m} = (m_1, \cdots, m_{k-1})$, $|\overline{m}| = m_1 + \cdots + m_{k-1}$ and $\overline{m}! = m_1! \cdots m_{k-1}!$.

Proposition 2.1. For $0 < t_1 < \cdots < t_k < \infty$, the formal functional

$$\Psi_k(t_1, \cdots, t_k) = \prod_{i=1}^{k-1} u_i(B(t_{i+1}) - B(t_i))$$

is in $(S)^*$ and has the following series expansion in $(S)^*$

(2.1) $$\Psi_k(t_1, \cdots, t_k) = (\frac{1}{2\pi})^{k-1} \sum_{\alpha,\beta=0}^{\infty} \psi_{\alpha,\beta}(t_1, \cdots, t_k), \text{ where}$$

$$\psi_{\alpha,\beta}(t_1, \cdots, t_k) = \sum_{\substack{|\overline{m}|=\alpha \\ |\overline{n}|=\beta}} \frac{1}{\overline{m}!\,\overline{n}!\,2^{|\overline{m}|}\,2^{|\overline{n}|}} \prod_{i=1}^{k-1} \Big\{ \frac{K_{i,m_i,n_i}(t_i, t_{i+1})}{t_{i+1} - t_i}$$

$$\cdot H_{m_i}(\frac{B_1(t_{i+1}) - B_1(t_i)}{\sqrt{2(t_{i+1}-t_i)}}) H_{n_i}(\frac{B_2(t_{i+1}) - B_2(t_i)}{\sqrt{2(t_{i+1}-t_i)}}) \Big\}.$$

Proof. Let $\mathcal{H}_\theta(B_i)$ denote the space of θ-th chaos (θ-ple Wiener integrals) with respect to B_i. We observe that

$$\psi_{\alpha,\beta}(t_1,\cdots,t_k)\in\mathcal{H}_\alpha(B_1)\otimes\mathcal{H}_\beta(B_2). \tag{2.2}$$

Using the same arguments as those in the proof of Shieh and Yokoi [8, Theorem 3.1, pp377-378], we can show that the summation in (2.1) converges in $(S)^*_{-p}$ whenever p is large enough. Along arguing, we need the following estimates on $K_{i,m_i,n_i}(t_i,t_{i+1})$. Firstly,

$$|K_{i,m_i,n_i}|\le\Big(\sup_{\substack{t>0\\x_1,x_2}}|e^{-\frac{x_1^2+x_2^2}{4t}}H_{m_i}(\frac{x_1}{\sqrt{2t}})H_{n_i}(\frac{x_2}{\sqrt{2t}})|\Big)$$
$$\cdot\int_{R^2}e^{-\frac{x_1^2+x_2^2}{4(t_{i+1}-t_i)}}du_i(x_1,x_2);$$

note that the positivity of u_i is needed for the validity of the expression. Secondly, according to Szegö [9, (8.91.10)], the supremum in the above parenthesis is $O((m_in_i2^{m_i}2^{n_i})^{1/2}\cdot(m_in_i)^{-1/12})$. This completes the proof. ∎

Theorem 2.2. For any $k\ge2$, Ψ_k defined by (1.1) is an ordinary Brownian functional, i.e. $\Psi_k\in(L^2)$.

Proof. By Proposition 2.1 and the continuity of $(t_1,\cdots,t_k)\to\Psi_k(t_1,\cdots,t_k)$, $t_i\in I_i$, in $(S)^*$, the functional Ψ_k is in $(S)^*$ and has the following series expansion

$$(\frac{1}{2\pi})^{k-1}\sum_{\alpha,\beta=0}^{\infty}\psi_{\alpha,\beta},$$

where

$$\psi_{\alpha,\beta}=\int_{I_1}\cdots\int_{I_k}\psi_{\alpha,\beta}(t_1,\cdots,t_k)dt_1\cdots dt_k.$$

By (2.2),

$$E\psi_{\alpha,\beta}\psi_{\alpha',\beta'}=0,\ \text{whenever}\ (\alpha,\beta)\neq(\alpha',\beta').$$

Therefore, the finiteness of the summuation:

$$\sum_{\alpha,\beta}E\psi^2_{\alpha,\beta}<\infty. \tag{2.3}$$

will ensure that $\Psi_k \in (L^2)$, which is our object. Observe that

$$E\psi^2_{\alpha,\beta} = \int_{t_i, t'_i \in I_i} \cdots \int E\psi_{\alpha,\beta}(t_1, \cdots, t_k)\psi_{\alpha,\beta}(t'_1, \cdots, t'_k) dt_1 dt'_1 \cdots dt_k dt'_k.$$

By the definition of $\psi_{\alpha,\beta}(t_1, \cdots, t_k)$ in (2.1),

$$\begin{aligned} &E\psi_{\alpha,\beta}(t_1, \cdots, t_k)\psi_{\alpha,\beta}(t'_1, \cdots, t'_k) \\ &= \frac{1}{2^{2\alpha}2^{2\beta}} \sum_{\substack{|\overline{m}|=|\overline{m}'|=\alpha \\ |\overline{n}|=|\overline{n}'|=\beta}} \sum \frac{1}{\overline{m}!\,\overline{m}'!\,\overline{n}!\,\overline{n}'!} \Big(\prod_{i=1}^{k-1} \frac{K_{i,m_i,n_i}(t_i, t_{i+1}) K_{i,m'_i,n'_i}(t'_i, t'_{i+1})}{(t_{i+1} - t_i)(t'_{i+1} - t'_i)} \Big) \\ &\cdot E \prod_{i=1}^{k-1} H_{m_i}(\frac{X_i}{\sqrt{2}}) H_{m'_i}(\frac{X'_i}{\sqrt{2}}) \cdot E \prod_{i=1}^{k-1} H_{n_i}(\frac{Y_i}{\sqrt{2}}) H'_{n_i}(\frac{Y'_i}{\sqrt{2}}), \end{aligned}$$

where

$$X_i = \frac{B_1(t_{i+1}) - B_1(t_i)}{\sqrt{t_{i+1} - t_i}}, \quad Y_i = \frac{B_2(t_{i+1}) - B_2(t_i)}{\sqrt{t_{i+1} - t_i}}, \text{ and}$$

X'_i, Y'_i are defined samely for t'_i. Because of the facts that $t_i, t'_i \in I_i$ and that I_i are disjoint,

$$EX_iX'_i = EY_iY'_i = \frac{t_{i+1} \wedge t'_{i+1} - t_i \vee t'_i}{\sqrt{(t_{i+1} - t_i)(t'_{i+1} - t'_i)}}, \qquad \begin{array}{l} a \wedge b = \min(a, b) \\ a \vee b = \max(a, b), \end{array}$$

$$EX_iX'_{i+1} = EY_iY'_{i+1} = \frac{(t_{i+1} - t'_{i+1})^+}{\sqrt{(t_{i+1} - t_i)(t'_{i+2} - t'_{i+1})}}, \quad a^+ = \max(a, 0),$$

$$EX'_iX_{i+1} = EY'_iY_{i+1} = \frac{(t'_{i+1} - t_{i+1})^+}{\sqrt{(t_{i+2} - t_{i+1})(t'_{i+1} - t'_i)}}$$

while all other X_i, X'_j have zero covariances, i.e. $EX_iX'_j = 0$, samely for Y_i, Y'_j.

Now, we appeal to the diagram formula in Major [6, p51] which concerns with the expectation of products of Hermite polynomials. From the covariance relations of X_i, X'_j, Y_i, Y'_j mentioned above, if the product

$$E \prod_{i=1}^{k-1} H_{m_i}(\frac{X_i}{\sqrt{2}}) H_{m'_i}(\frac{X'_i}{\sqrt{2}}) E \prod_{i=1}^{k-1} H_{n_i}(\frac{Y_i}{\sqrt{2}}) H'_{n_i}(\frac{Y'_i}{\sqrt{2}})$$

does not vanish, then it is of the form

$$2^\alpha 2^\beta \frac{\overline{m}!\,\overline{m}'!\,\overline{n}!\,\overline{n}'!}{\alpha!\beta!} \prod_{j=1}^{2k-3} \{[E^{\ell_1}(X_1X_1')E^{\ell_1^*}(Y_1Y_1')]\cdot$$

$$[E^{\ell_2}(X_1'X_2)E^{\ell_2^*}(Y_1'Y_2) \text{ or } E^{\ell_2}(X_1X_2')E^{\ell_2^*}(Y_1Y_2')]\cdot$$

$$[E^{\ell_3}(X_2X_2')E^{\ell_3^*}(Y_2Y_2')]\cdot[\cdots]\cdots\cdot[E^{\ell_{2k-3}}(X_{k-1}X_{k-1}')$$

$$E^{\ell_{2k-3}^*}(Y_{k-1}Y_{k-1}')]\},$$

where $\ell_1+\cdots+\ell_{2k-3}=\alpha$, $\ell_1^*+\cdots+\ell_{2k-3}^*=\beta$. An illustration for $k=4$, $\alpha=10$, $\beta=0$, $\overline{m}=(3,4,3)$, $\overline{m}'=(4,5,1)$, then $\ell_1=3$, $\ell_2=1$, $\ell_3=3$, $\ell_4=2$, $\ell_5=1$. Note that, in case $\overline{m}=\overline{m}'$ and $\overline{n}=\overline{n}'$, then $\ell_{2j-1}=m_j$, $\ell_{2j}=0$, $\ell_{2j-1}^*=n_j$, $\ell_{2j}^*=0$. In view of the estimates on K_{i,m_i,n_i} given in the proof of Proposition 2.1, to prove (2.3) in case $k>2$ we may merely consider those terms whose indices m_i, m_i', n_i, n_i' are large enough, say $\geq N_0$, such that

(i) $K_{i,m_i,n_i} \leq C(m_i!\,n_i!\,2^{m_i}2^{n_i})^{1/2}(m_in_i)^{-1/12}$, and

(ii) $\dfrac{m_1!\cdots m_{k-1}!\,m_1^2\cdots m_{k-1}^2}{(m_1+\cdots+m_{k-1})!}\leq 1, \quad \dfrac{n_1!\cdots n_{k-1}!\,n_1^2\cdots n_{k-1}^2}{(n_1+\cdots+n_{k-1})!}\leq 1;$

and samely for m_i', n_i'. In fact, those terms involving, say, $m_1=0,\cdots,N_0-1$ are reduced to the lower-fold categories. Thus,

$$\sum_{\substack{\alpha,\beta\\ m_i,n_i\geq N_0}} E\psi_{\alpha,\beta}^2 \leq C \sum_{\substack{m_i,n_i\\ m_i'n_i'\geq N_0}} \frac{1}{(\prod_{i=1}^{k-1} m_im_i'n_in_i')^{\frac{13}{12}}}$$

$$\cdot \int\cdots\int_{t_i,t_i'\in I_i} \left\{\frac{E^{\ell_1}(X_1X_1')E^{\ell_1^*}(Y_1Y_1')\cdots}{\prod_{i=1}^{k-1}(t_{i+1}-t_i)(t_{i+1}'-t_i')}\right\} dt_1dt_1'\cdots dt_kdt_k'.$$

Since $\sum_{h\geq N_0}\frac{1}{h^{13/12}}<\infty$, we need only to show that the integral in t_i, t_i' in the above display is uniformly bounded. To see this, we consider, for example, the integration over $t_1\leq t_1'<t_2<t_2'<\cdots<t_k\leq t_k'$. Then, we have

$$\int\cdots\int_{t_i<t_i'} \left\{\frac{(t_2-t_1')^{\ell_1+\ell_1^*}}{[(t_2-t_1)(t_2'-t_1')]^{\frac{\ell_1+\ell_1^*}{2}+1}}\cdot\frac{(t_2'-t_2)^{\ell_2+\ell_2^*}}{[(t_3-t_2)(t_2'-t_1')]^{\ell_2+\ell_2^*}}\cdot\right.$$

$$\left.\frac{(t_3-t_2')^{\ell_3+\ell_3^*}}{[(t_3-t_2)(t_3'-t_2')]^{\frac{\ell_3+\ell_3^*}{2}+1}}\cdots\cdot\frac{(t_k-t_{k-1}')^{\ell_{2k-3}+\ell_{2k-3}^*}}{[(t_k-t_{k-1})(t_k'-t_{k-1}')]^{\frac{\ell_{2k-3}+\ell_{2k-3}^*}{2}+1}}\right\}$$

$$dt_1dt_1'\cdots dt_kdt_k'.$$

We integrate successively in the order of $t'_k, t'_{k-1}, \cdots t'_1$, while in the meantime we drop the terms involving $\frac{t_{i+1}-t'_i}{t_{i+1}-t_i}$, $\frac{t'_i-t_i}{t_{i+1}-t_i}$, $\frac{t'_i-t_i}{t'_i-t'_{i-1}}$. The resultant

$$\leq C\frac{1}{(\ell_{2k-3}+\ell^*_{2k-3}+4)(\ell_{2k-5}+\ell^*_{2k-5}+4)\cdots(\ell_1+\ell^*_1+4)} \leq C.$$

Finally, the case $k = 2$ is much more simple. The summution in (2.3) involves only $\overline{m} = \overline{m}' = m_1$ and $\overline{n} = \overline{n}' = n_1$, and there is no need to concern ℓ_{2j} and ℓ^*_{2j}. In such a case, we use certain similar, yet simpler, calculations as above, where the inequality $a + b \geq 2\sqrt{ab}$ will be used instead of the display (ii) above. This completes the whole proof. ∎

Acknowledgment The author would like to express the gratitude to Professor T. Hida for his kind invitation to attend at White Noise Analysis Conference in Bielefeld W. Germany, July 10-15 1989 and Gaussian Random Fields Conference in Nagoya Japan, August 14-20 1990. The content of this paper was announced in W.N.A. Conference and is corrected and completed between W.N.A. and G.R.F. Conferences.

References

[1] Dynkin, E. B.: Self-intersection guage for random walks and for Brownian motion, Ann. Probab. 16 (1988), 1-57.

[2] Geman, D., Horowitz, J. and Rosen, J.: A local time analysis of intersections of Brownian paths in the plane, Ann. Probab. 12 (1984), 86-107.

[3] Hida, T. and Putthoff, J.: White noise analysis–an overview, White Noise Analysis (1990), 140-165, World Scientific.

[4] Kallianpur, G. and Kuo, H. -H.: Regularity property of Donsker's delta function, Appl. Math. Optim. 12 (1984), 89-95.

[5] Kuo, H. -H. and Shieh, N. R.: A generalized Ito's formula for multi-dimensional Brownian motions and its applications, Chinese J. Math. 15 (1987), 163-174.

[6] Major, P.: Multiple Wiener-Ito Integrals, Lecture Notes in Math. vol. 849 (1981), Springer-Verlag.

[7] Rosen, J.: Self intersections of random fields, Ann. Probab. 12 (1984), 108-119.

[8] Shieh, N. R. and Yokoi, Y.: Positivity of Donsker's delta functions, White Noise Analysis (1990), 374-382, World Scientific.

[9] Szegö, G.: Orthogonal Polynomials, Amer. Math. Soc. Colloq. Publ. vol. XXIII (1939).

[10] Westwater, J.: On Edward's model for long polymer chains, Comm. Math. Phys. 72 (1980), 131-174.

[11] Wolpert, R.: Wiener path intersections and local times, J. Funct. Anal. 30 (1978), 329-340.

[12] Wolpert, R.: Local time and a particle picture for Euclidean field theory. J. Funct. Anal. 30 (1978) 341-357.

SOME PROPERTIES OF SOLUTIONS FOR ONE-DIMENSIONAL SPDE'S ASSOCIATED WITH SPACE-TIME WHITE NOISE

Tokuzo SHIGA

Department of Applied Physics, Tokyo Institute of Technology
Oh-okayama, Meguro, Tokyo 152, Japan

ABSTRACT

We consider one-dimensional stochastic partial differential equations (SPDE) associated with a space time white noise, which are driven as a spatially continuum limit from infinite dimensional stochastic differential equations (SDE) associated with independent Brownian motions indiced by the one-dimensonal lattice space Z. Our main concern is to observe different aspects of the SPDE's from the diffusion equations obtained by neglecting the random term of the SPDE's. In particular, propagation of support compactness, strong comparison of solutions, and large time behaviors of the solutions are discussed.

1. Introduction

Let us first consider the following stochastic discretized diffusion equations over the d-dimensional cubic lattice space Z^d :

$$(1) \quad dX_t(i) = (\Delta X_t(i) + b(X_t(t)))dt + a(X_t(i))dB_t(i) \quad (i \in Z^d),$$

where $a(u)$ and $b(u)$ are real-valued continuous functions on R, Δ is the discrete Laplacian over Z^d, i.e.

$$\Delta X(i) = \sum_{|j-i|=1} (X(j) - X(i)) \quad \text{for } i \in Z^d,$$

and $\{B_t(i) : i \in Z^d\}$ is an independent system of one dimensional $(\mathcal{F}_t)$-Brownian motions defined on a probablity space with filtration $(\Omega, \mathcal{F}, \mathcal{F}_t, P)$.

Such a type of the equation (1) appears in various situations of statistical physics and mathematical biology , and it has been extensively investigated.[1,2,3]

Let us next consider a continuum limit of the equation (1) from Z^d to R^d. For $\varepsilon > 0$, we introduce the following rescaled equation :

$$(1)_\varepsilon \quad dX_t(i) = (\varepsilon^{-2}\Delta X_t(i) + b(X_t(i)))dt + \varepsilon^{-d/2}a(X_t(i))dB_t(i), \quad (i \in Z^d).$$

For a solution $X_t(i)$ of $(1)_\varepsilon$, define a two parameter process $X_\varepsilon(t,x)$ $(t \geq 0, x \in R^d)$ as a continuous extension of $X_\varepsilon(t,x) = X_t(i)$ for $x = \varepsilon i$.

Suppose that $X_\varepsilon(0,x)$ converges to a continuous function $X(0,x)$ of $x \in R^d$. Then, under a suitable assumption on $a(u)$ and $b(u)$, one can prove that if $d = 1$, $X_\varepsilon(t,x)$ converges to a two parameter continuous process $X(t,x)$ as $\varepsilon \to 0$ in the sense of probability law, and moreover $X(t,x)$ solves the following SPDE :

(2) $$\frac{\partial X(t,x)}{\partial t} = \Delta X(t,x) + b(X(t,x)) + a(X(t,x))\dot{W}(t,x) \quad (t \geq 0, x \in R),$$

where Δ is the Laplacian over R^d, and $\dot{W}(t,x)$ is a space-time white noise on $[0,\infty) \times R$ which is a centered Gaussian field with the covariance $E(\dot{W}(t,x)\dot{W}(s,y)) = \delta(t-s)\delta(x-y)$.
We notice that the equation (2) should be interpreted in the sense of the Schwartz distributions.

On the other hand, if $d \geq 2$, one can show under a restrictive assumption that $X_\varepsilon(t) = (X_\varepsilon(t,x))$ converges to a limit process $X(t)$ as $\varepsilon \to 0$ in some sense. Moreover in a special case the limit process $X(t)$ solves the following non-random diffusion equation :

(3) $$\frac{\partial X(t,x)}{\partial t} = \Delta X(t,x) + b(X(t,x)) \quad (t \geq 0, x \in R^d).$$

However in other cases the limit process is not function-valued but Schwartz distribution-valued. Thus the SPDE (2) makes sense only in spatially one dimensional case.

The main objective of the present lecture is to discuss the one-dimensional SPDE (2) that is a continuum limit of the SPDE (2).

In order to formulate a solution of (2) more presicely, we introduce C_{tem} (the totality of tempered continuous functions on R) as a subspaces of $C(R)$, the space of continuous functions on R. For $f \in C(R)$, let $|f|_{(p)} = \sup_{x\in R} e^{p|x|}|f(x)|$ for $p \in R$, and set

$$C_{tem}(R) = \{f \in C(R) \mid |f|_{(p)} < \infty \text{ for all } p < 0\}.$$

We will always assume a linear growth condition on $a(u)$ and $b(u)$; there exists a constant $C > 0$ such that

(4) $$|a(u)| + |b(u)| \leq C(1 + |u|) \quad \text{for } u \in R.$$

An $\{\mathcal{F}_t\}$-predictable functional $X(t,\cdot) = (X(t,x,\omega))$ is a C_{tem}- valued solution of (2), if it is a C_{tem}-valued continuous process such that for every C^∞-function φ with compact support,

(5) $$\langle X(t),\varphi\rangle = \langle X(0),\varphi\rangle + \int_0^t (\langle X(s),\Delta\varphi\rangle + \langle b(X(s)),\varphi\rangle)ds + \int_0^t\int_R a(X(s,x))\varphi(x)\dot{W}(s,x)dsdx,$$

where the last term is a stochastic integral in the Ito sense.[4]

Then it is not difficult to see that, under the condition (4), $X(t,\cdot)$ is a C_{tem}-valued solution of (5) if and only if $X(t,\cdot)$ is an $\{\mathcal{F}_t\}$-predictable and C_{tem}-valued continuous process that satisfies the following stochastic integral equation (SIE):[5]

(6) $$X(t,x) = G(t)X(0,x) + \int_0^t\int_R G(t-s,x,y)b(X(s,y))dsdy + \int_0^t\int_R G(t-s,x,y)a(X(s,y))\dot{W}(s,y)dsdy.$$

where $G(t,x,y) = \frac{1}{\sqrt{4\pi t}}\exp(-(y-x)^2/4t)$ and $G(t)X(0,x) = \int_R G(t,x,y)X(0,y)dy$.

The SIE (6) is more tractable, and by a standard method one can discuss existence and uniqueness of solutions as in finite dimensional SDE's.

Suppose that a(u) and b(u) are Lipschitz continuous. Then for every $X(0) \in C_{tem}$, there exists a unique C_{tem}-valued solution $X(t) = (X(t,x))$ for the SPDE (2).[5]

If we only assume that a(u) and b(u) are continuous, instead of Lipschitz continuity, then for every $X(0) \in C_{tem}$, there exist an $\{\mathcal{F}_t\}$-predictable and C_{tem}-valued continuous process $X(t) = ((X(t,x))$ and a space-time white noise $W(t,x)$ defined on a probability space with filtration $(\Omega, \mathcal{F}, \mathcal{F}_t, P)$ such that X(t) is a solution of the SPDE (2) with the white noise W(t,x).

As a next stage, we would like to investigate the properties of the solution for the SPDE (2). In particular, we are interested in different aspects of solutions of the SPDE (2) from those of the diffusion equation, although (2) may be regarded as a random perturbation of the diffusion equation.

The present lecture treats the following two problems. The first is concerned with *positivity of solutions* (more generally, *comparison of solutions* or *a maximum principle*). For example, let

(7) $\quad a(u) = \sqrt{|u|}$ and $b(u) = 0$.

Then the SPDE (2) is associated with a one-dimensional measure-valued branching diffusion process.[6] Accordingly, if $X(0) \geq 0$ is a continuous function with compact support, we have a non- negative solution X(t) , which is uniquely determined in the law sense, but not in the path-wise sense. Furthermore, the solution $X(t) = (X(t,x))$ also has compact support for every $t > 0$.[7] In particular, it is to be noted that the strongly positivity (everywhere positivity) does not hold. We call it *SCP property* (i.e. the support compactness propagates along the time axis t). Of course, this phenomenon does not hold for the diffusion equations, which would be caused by a strong effect of the random term involving a(u).

We will generalize this result. Roughly speaking, supposing $a(0) = 0$, if $|a(u)|$ grows up faster than $u^{1/2}$ at a neighborhood of $u = 0$, the SCP property holds.

On the other hand C. Mueller [8] recently discussed the following case:

(8) $\quad a(u) = |u|^{\alpha}$ $(\alpha \geq 1)$ and $b(u) \equiv 0$.

He proved that if $X(0) \geq 0$ is a continuous function with compact support and $X(0,x) > 0$ for some $x \in R$, then the solution $X(t) = (X(t,x))$ of the SPDE (2) satisfies

(9) $\quad P(X(t,x) > 0 \text{ for every } x \in R \mid t < \sigma_{\infty}) = 1 \quad \text{for } t > 0,$

where $\sigma_{\infty} = \lim_{n \to \infty} \sigma_n$ and $\sigma_n = \inf\{t \geq 0 \mid \sup_{x \in R} X(t,x) \geq n\}$.

This result will be extended to a general Lipschitz continuous function a(u) with $a(0) = 0$, moreover, we will also prove *a strong comparison theorem for two solutions* of the SPDE in a general Lipschizian situation of a(u) and b(u).

The second problem is concerned with a limiting behavior of solutions of the SPDE (2) as $t \to \infty$. Assume $b(u) = 0$. For the heat equation every constant function trivially is a solution. The total mass $<X(t),1>$ is also preserved along the time t, if $<X(0),1> < \infty$. However, these are not true for the SPDE (2). Suppose that $b(u) \equiv 0$, $a(0) = 0$, and a(u) satisfies a reasonable regularlity assumption. Then, it will be shown that $< X(t),1>$ converges to zero as $t \to \infty$ almost surely, if $X(0) \geq 0$ and $<X(0),1> < \infty$. Furthermore it will be also shown that $X(t) = (X(t,x))$ converges weakly to zero locally in $x \in R$ as $t \to \infty$ in the probability sense if $X(0) \geq 0$ has bounded moment in $x \in R$.

2. SCP-property

In this section we discuss the SCP-property of solutions for the SPDE (2). Before proceeding these problems we first present a nonnegativity result and a weak comparison theorem for solutions of (2).

Theorem 2.1 Supposing that a(u) and b(u) are Lipschitz continuous, let $X(t) = (X(t,x))$ and $Y(t) = (Y(t,x))$ be the C_{tem}-valued solutions of the SPDE (2) with initial conditions $X(0)$ and $Y(0) \in C_{tem}$. If $X(0) \geq Y(0)$, then it holds

$$P(X(t) \geq Y(t) \quad \text{for every } t \geq 0) = 1.$$

Furthermore if we add the following condition;

(9) $a(0) = 0$ and $b(0) \geq 0$,

Then for every $X(0) \geq 0 \in C_{tem}$, the C_{tem}-valued solution $X(t)$ of the SPDE (2) satisfies

$$P(X(t) \geq 0 \quad \text{for every } t \geq 0) = 1.$$

Theorem 2.2 Suppose that a(u) and b(u) are continuous with the linear growth condition (4), but not Lipschitz continuous. Further assume the condition (9). Then for every $X(0) \geq 0 \in C_{tem}$, there exist an $\{\mathcal{F}_t\}$-predictable and C_{tem}-valued continuous process $X(t) = ((X(t,x)) \geq 0$ and a space-time white noise $W(t,x)$ defined on a probability space with filtration $(\Omega, \mathcal{F}, \mathcal{F}_t, P)$ such that $X(t)$ solves the SPDE (2) with the $W(t,x)$.

Furthermore, suppose that there exist constants $C > 0$ and $0 < \theta < 1$ such that for $u \in R$

(10) $|a(u)| \leq C(|u| + |u|^{\theta})$

(11) $|b(u)| \leq C|u|$.

Let $X(t) = (X(t,x)) \geq 0$ be a C_{tem}-valued solution of the SPDE (2) with an $X(0) \geq 0 \in C_{rap}$. Then $X(t)$ is a C_{rap}-valued solution, where C_{rap} is the totality of rapidly decreasing continuous functions defined on R, i.e. $C_{rap} = \{f \in C(R) \mid |f|_{(p)} < \infty \quad \text{for every } p > 0\}$.

The proofs of Theorem 2.1 and 2.2 are rather standard, and are essentially reduced to one-dimensional method of SDE's.

Now we proceed to the SCP-property of solutions for the SPDE (2).

Theorem 2.3 Suppose that a(u) and b(u) are continuous functions with the linear growth condition (4). Assume further that $a(0) = 0$, and for each $K > 0$ there exists a constant $c_K > 0$ such that

(12) $|a(u)| \geq c_K u^{1/2}$ for $0 \leq u \leq K$,

and that there exists a $C > 0$ such that

(13) $|b(u)| \leq C|u|$ for $u \in R$.

Then the SCP-property holds for the SPDE (2), that is, if $X(0) \geq 0$ is a continuous function with compact support, it holds that $P(X(t)$ has compact support for every $t > 0) = 1$ for every nonnegative C_{tem}-valued solution $X(t)$ of the SPDE (2) with the initial condition $X(0)$.

We remark that under the assumption of Theorem 2.3 we do not know the uniqueness of solutions for the SPDE (2) since a(u) is not assumed to be Lipschitz continuous. The assumption (12) means that the effect of a(u) is stronger at a neibourhood of $u = 0$ rather than Lipschizian cases. Also, it is clear that $a(u) = |u|^\alpha$ satisfies the condition (12) with $0 < \alpha \le 1/2$.

The proof of Theorem 2.3 is essentially the same as that in the case (7), which applies the following fact on a singular boundary value problem of a one-dimensional ODE.

Lemma 2.4 [7] Let $c > 0$ be a fixed constant.
(i) For $r > 0$, there exists a unique positive solution $v(x) = v(x;r) \in C^2(-r, r)$ of the equation

$$(13) \quad v''(x) = cv(x)^2 \text{ for } |x| < r$$

$$\lim_{|x| \to r-} v(x) = \infty.$$

Moreover, it holds that $v(x;r) \to 0$ as $r \to \infty$ uniformly on each compact interval.
(ii) For any continuous function $h \ge 0$ with compact support, consider the following equation:

$$(14) \quad v''(x) = cv(x)^2 - h(x) \quad \text{for } x \in R.$$

Then there exists a unique positive solution $v(x) = v_h(x)$ that is a C^2-function vanishing at infinity. Moreover, if h vanishes in $(-r, r)$, it holds that

$$(15) \quad v_h(x) \le v(x;r) \quad \text{for } |x| < r.$$

In the case of (7), by making use of Lemma 2.4 one can see that for every nonnegative continuous function φ with compact support, we have

$$(16) \quad E(\exp(-\theta\int_0^\infty \langle X(s),\varphi\rangle ds)) = \exp - \langle X(0),v(\cdot,\theta\varphi)\rangle .$$

Here $v(\cdot,\theta\varphi)$ is the solution of (14) with $c = 1/2$ and $h = \theta\varphi$. Combining the relation (16) with Lemma 2.4 and letting $\theta \to \infty$, one can conclude Theorem 2.3 in this case. The general case can also be reduced to this case with some modifications.

The method to use Lemma 2.4 is applicable to another class of one-dimensional SPDE's including a Fleming-Viot diffusion process. Let us introduce the following SPDE:

$$(17) \quad \frac{\partial X(t,x)}{\partial t} = \Delta X(t.x) + s(x,X(t,x)) - X(t,x)\int_R s(y,X(t,y))dy + \sqrt{\alpha(x)X(t,x)}\ \dot{W}(t,x)$$

$$- X(t,x)\int_R \sqrt{\alpha(y)X(t,y)}\ \dot{W}(t,y)dy \quad (t \ge 0,\ x \in R).$$

where $\alpha(x) : R \to R$ is a uniformly positive and bounded continuous function and $s(x,u) : R \times R \to R$ is a continuous function satisfying that for some constant $C > 0$,

$$(18) \quad |s(x,u)| \le C\,|u| \quad \text{for } x \in R \text{ and } u \in R.$$

Then the SPDE (17) is rewritten in the following way : for every C^∞ function φ with compact support,

$$(19) \quad \langle X(t),\varphi\rangle = \langle X(0),\varphi\rangle + \int_0^t (\langle X(s),\Delta\varphi\rangle + \int_R s(x,X(s,x))\varphi(x)dx - \langle X(s),\varphi\rangle \int_R s(x,X(s,x))dx)ds$$

$+ M_t(\varphi)$,

where $M_t(\varphi)$ is an $\{\mathcal{F}_t\}$-martingale with quadratic variation process

(20) $$<M(\varphi)>_t = \int_0^t (<X(s),\alpha\varphi^2> - 2<X(s),\alpha\varphi><X(s),\varphi> + <X(s),\alpha><X(s),\varphi>^2)ds.$$

If $\alpha(x) = \alpha$ is constant and $s(x,u) = s(x)u$, (18) turns to a standard form of Fleming-Viot diffusion process incorpolation a haploid selection. Then it is known that for every C_{tem} probability density function $X(0) = (X(0,x))$ on R, there exists a unique C_{tem} probability density -valued solution $X(t) = (X(t,x))$ of the SPDE (18) in the law sense.[6] Indeed, the SPDE (18) is an infinite-dimensional version of Gillespie and Sato's diffusion model in population genetics.[9] In this case the uniqueness problem remains open. However one can discuss the SCP-property for the SPDE (18).

Theorem 2.5 Let $X(0)$ be a C_{tem} probability density function on R. Then for every C_{tem} probability density-vallued solution $X(t) = (X(t,x))$ of the SPDE (18) with the initial condition $X(0)$, it holds that $P(X(t)$ has compact support for every $t > 0) = 1$.

3. Strong positivity and strong comparison

In this section we establish a strong comparison theorem in a general Lipschizian situation, which extends Mueller's result [8].

Theorem 3.1 Suppose that $a(u)$ and $b(u)$ are Lipschitz continuous. Let $X(t) = (X(t,x))$ and $Y(t) = (Y(t,x))$ be the C_{tem}-valued solutions of the SPDE (2) with the initial conditions $X(0) = ((X(0,x)) \in C_{tem}$ and $Y(0) = ((Y(0,x)) \in C_{tem}$ respectively. If $X(0) \geq Y(0)$ and $X(0,x) > Y(0,x)$ for some $x \in R$, then $P(X(t,x) > Y(t,x)$ for every $t > 0$ and $x \in R) = 1$.

As a corollary of Theorem 3.1 we obtain

Theorem 3.2 Assume further that $a(0) = 0$ and $b(0) \geq 0$ in addition to the Lipschitz continuity on $a(u)$ and $b(u)$. Let $X(t) = (X(t,x))$ be the C_{tem}-valued solution of the SPDE (2) with the initial condition $X(0) = (X(0,x)) \in C_{tem}$. If $X(0) \geq 0$ and $X(0,x) > 0$ for some $x \in R$, then $P(X(t,x) > 0$ for every $t > 0$ and $x \in R) = 1$.

We will outline a direct proof of Theorem 3.2 assuming $b(u) \equiv 0$. Note that the solution $X(t)$ is represented by the following SIE:

(21) $$X(t,x) = G(t)X(0)(x) + \int_0^t\int_R G(t-s,x,y)a(X(s,y))\dot{W}(s,y)dsdy.$$

The first term is clearly positive everywhere for $t > 0$. Accordingly, the problem is how to control the second term. To do this Mueller [8] uses the following Gaussian estimate.

Lemma 3.3 Let $K > 0$ be fixed and let $b(t,y,\omega)$ be an $\{\mathcal{F}_t\}$-predictable functional such that $|b(t,x,\omega)| \leq K e^{-(T-t)|x|}$ for every $0 \leq t \leq T/2$ and every $x \in R$ almost surely. Then there exist $c_1 > 0$ and $c_2 > 0$ depending on K such that for every $0 < \varepsilon < 1$ and every $0 < T < 1$

$$P\left(\left|\int_0^t\int_R G(t-s,x,y)b(s,y)\dot{W}(s,y)dsdy\right| > \varepsilon e^{-(T-t)|x|} \text{ for some } 0 < t < T/2 \text{ and } x \in R\right)$$

$$\leq c_1 T \varepsilon^{-24} \exp(-c_2 \varepsilon^2 T^{-1/4}).$$

We note that if $b(t,x,\omega)$ is non-random, the stochastic integral defines a two parameter Gaussian fields, so the proof of Lemma 3.3 is quite standard. But even for a random $b(t,x,\omega)$, a similar calculation is possible by a time change argument.[7]

Let

$$(22) \quad N(t,x) = \int_0^t\int_R G(t-s,x,y)a(X(s,y))\dot{W}(s,y)dsdy.$$

Using Lemma 3.3 one can estimate $N(t,x)$ as follows.

Lemma 3.4 Let $M > 0$ be fixed. Then there exist $c_1 > 0$ and $c_2 > 0$, depending on M and L, such that if $f(x) \leq \beta I_{[-M,M]}(x)$ for every $x \in R$ with a $\beta > 0$, then for every $0 < \varepsilon < 1$ and $0 < T < 1$

$$P(|N(t,x)| > \varepsilon\beta e^{-(T-t)|x|} \text{ for some } 0 < t < T/2 \text{ and some } x \in R) \leq c_1 \varepsilon^{-24} \exp(-c_2 \varepsilon^2 T^{-1/4}).$$

Outline of the proof of Theorem 3.2

Choose $a < b$ and $\beta > 0$ such that $f(x) > \beta I_{(a,b)}(x)$ for every $x \in R$. Fix an arbitrary $M > 0$ such as $[-M/2, M/2] \supseteq (a,b)$ and $t > 0$.

Step 1. As easily checked, there is an $m_0 = m_0(t,a,b,M)$ such that if $m \geq m_0$ and $(c,d) \supseteq (a,b)$

$$(23) \quad G(s)I_{(c,d)}(x) \geq I_{(c-M/m,\, d+M/m)}(x)/4 \quad \text{for every } t/2m \leq s \leq t/m \text{ and } x \in R.$$

Step 2. Using Lemma 3.4 together with (23) we can show that if $[-M,M] \supseteq (c,d) \supseteq (a,b)$ and $f(x) \geq \alpha I_{(c,d)}(x)$ with an $\alpha > 0$, then for every $m \geq m_0$

$$(24) \quad P(X(s,x) \geq \alpha I_{(c-M/m,\, d+M/m)}(x) \text{ for every } t/2m \leq s \leq t/m \text{ and } x \in R)$$

$$\geq 1 - c_1 16^{24} \exp(-c_2 16^{-2} t^{-1/4} m^{1/4}),$$

where c_1 and c_2 are the constants in Lemma 3.4, which depend on M and L.

Step 3. Define the events A_k and B_k:

$$A_k = [X(t,x) \geq \beta 8^{-k} I_{(a-Mk/m,\, b+Mk/m)}(x) \text{ for every } s \in [t/2m, t/m] + kt/m \text{ and } x \in R]$$

and

$$B_k = [X(t,x) \geq \beta 8^{-k} I_{(a-Mk/m,\, b+Mk/m)}(x) \text{ for every } s \in [0, t/2m] + kt/m \text{ and } x \in R].$$

Since $Y(t,x) = X(t + t_0,x)$ is a unique solution of the SIE (21) with the initial condition $Y(0) = X(t_0)$, by the Markov property, (24), and Theorem 2.1, we see that if $[-M,M] \supseteq (a - M(k-1)/m,$

b + M(k-1)/m),

$$P(A_k | A_{k-1} \cap \cdots \cap A_0) = P(A_k | X(kt/m,x) \geq b8^{-k+1} I_{(a - M(k-1)/m,\, b + M(k-1)/m)}(x))$$

$$\geq 1 - c(m),$$

and

$$P(B_k | B_{k-1} \cap \cdots \cap B_0) = P(B_k | X(kt/m,x) \geq b8^{-k+1} I_{(a - M(k-1)/m,\, b + M(k-1)/m)}(x))$$

$$\geq 1 - c(m),$$

where $c(m) = c_1 16^{24} \exp(- c_2 16^{-2} t^{-1/4} m^{1/4})$.
Noting that

$$P(\bigcap_{0\leq k\leq m/2} A_k \cap \bigcap_{0\leq k\leq m/2} B_k) \geq 1 - (1 - P(\bigcap_{0\leq k\leq m/2} A_k)) - (1 - P(\bigcap_{0\leq k\leq m/2} B_k))$$

$$\geq (1 - c(m))^{m/2} P(A_0) + (1 - c(m))^{m/2} P(B_0) - 1.$$

and

$$\lim_{m\to\infty} P(A_0) = \lim_{m\to\infty} P(B_0) = 1 \text{ and } \lim_{m\to\infty} (1 - c(m))^{m/2} = 1,$$

we obtain for every $t > 0$ and $M > 0$

$$P(X(s,x) > 0 \text{ for every } t/4 \leq s \leq t/2 \text{ and every } a - M/4 \leq x \leq b + M/4)$$

$$\geq \lim_{m\to\infty} P(\bigcap_{0\leq k\leq m/2} A_k \cap \bigcap_{0\leq k\leq m/2} B_k)$$

$$= 1,$$

which completes the proof of Theorem 3.2 in the case $b(u) \equiv 0$.

Theorem 3.1 can also proved by a similar argument to the proof of Theorem 3.2.

4. A limiting behavior of the solution

In this section we discuss a large time behavior of the solution of the SPDE (2) under the following situation :

(25) $b(u) \equiv 0$, and $a(u)$ is a continuous function satisfying $a(0) = 0$ with the linear growth condition (4) .

Then for every nonnegative $X(0) = (X(0,x)) \in C_{tem}$, there exists a C_{tem}-valued nonnegative solution $X(t) = (X(t,x))$ of the SPDE with an $\{\mathcal{F}_t\}$-space-time white noise $W(t,x)$ defined on a suitable probability space with filtration $(\Omega, \mathcal{F}, \mathcal{F}_t, P)$. Since the uniqueness of solutions are not guaranteed, we henceforth fix a nonnegative C_{tem}-valued solution $X(t) = (X(t,x))$ for the SPDE (2). Then we have the following results.

Theorem 4.1 In addition to (25) assume that for every $M > 0$ there are a constants $c = c_M > 0$ such that

(26) $|a(u)| \geq c\,|u|$ for $0 \leq u \leq M$.

Then if $\langle X(0),1\rangle < \infty$, $P(\lim_{t\to\infty} \langle X(t),1\rangle = 0) = 1$.

Theorem 4.2 Assume that there is a constant $c > 0$ such that either
(i) $a(u) = cu$, or
(ii) $|a(u)| \geq c\, u^{1/2}$ for $u \geq 0$.
If $X(0) \geq 0$, $\in C_{tem}$, and $E(X(0,x))$ is bounded in $x \in R$, then $\sup_{x \in J} X(t,x)$ converges to zero as $t \to \infty$ in the probability sense for every finite interval J.

Theorem 4.1 means that for one-dimensional SPDE's the total mass process vanishes as $t \to \infty$ almost surely, if the noise term is not negligible. As for Theorem 4.2 we can treat only very special cases, but the phenomenon itself would still hold in general one dimensional cases. Furthermore it would be more important problem to investigate the decay order of the solution $X(t,x)$ as $t \to \infty$.

We will first outline the proof of Theorem 4.1.

1°. Notice that $\lim_{t\to\infty} \langle X(t),1\rangle$ exists almost surely, since $\langle X(t),1\rangle$ is a nonnegative martingale.

2°. Set $Z = \sup_{0\leq t<\infty} \langle X(t),1\rangle$. By 1°, to complete the proof of Theorem 4.1 it suffice to show

$$(27) \quad \lim_{t\to\infty} E(\langle X(t),1\rangle^{1/2} : Z \leq M) = 0 \text{ for every } M > 0.$$

3°. Let $\zeta_M = \inf\{t \geq 0 \mid \langle X(t),1\rangle \geq M\}$ for $M > 0$. Noting that

$$(28) \quad \langle X(t),1\rangle = \langle X(0),1\rangle + \int_0^t\int_R a(X(s,x))\dot{W}(s,x)\,ds\,dx,$$

apply the Ito formula to (28) together with (26) and the following inequality [10];
for every interval J,

$$(29) \quad \langle X(s)^2,1\rangle \geq |J|^{-1}(\langle X(s),1\rangle^2 - 2\langle X(s),1 - I_J\rangle^{1/2}\langle X(s),1\rangle^{3/2},$$

where I_J is an indicator function of the interval J and $|J|$ stands for the length of J, so that for any intervals $J(s)$ we get

$$(30) \quad E(\langle X(t\wedge\zeta_M),1\rangle^{1/2}) - \langle X(0),1\rangle^{1/2}$$
$$\leq -c^2/8\, E\Big(\int_0^{t\wedge\zeta_M} \langle X(s)^2,1\rangle\, \langle X(s),1\rangle^{-3/2} ds\Big)$$
$$\leq -c^2/8\, E\Big(\int_0^{t\wedge\zeta_M} |J(s)|^{-1}(\langle X(s),1\rangle^{1/2} - 2\langle X(s),1 - I_{J(s)}\rangle^{1/2})ds\Big)$$
$$\leq -c^2/8 \int_0^t |J(s)|^{-1}(E(\langle X(s),1\rangle^{1/2}: s < \zeta_M) - 2\langle G(s)X(0),1 - I_{J(s)}\rangle^{1/2} ds$$

4°. Suppose that (27) fails, then there is a $\alpha > 0$ such that for sufficiently large s

$$(31) \quad E(\langle X(s),1\rangle^{1/2} : s < \zeta_M) \geq E(\langle X(s),1\rangle^{1/2} : Z \leq M) \geq \alpha.$$

Here we take $J(s) = [-s^\beta, s^\beta]$ for a $1/2 < \beta < 1$. Then

$$\lim_{s\to\infty} \langle G(s)X(0), 1 - I_{J(s)}\rangle = 0,$$

Hence the last term of (30) is minus infinity, which leads to a contradiction. Therefore we have shown (27), completing the proof of Theorem 4.1.

Now we proceed to the proof of Theorem 4.2. In the case (ii) it can be compared with the case (7), for which the result is known.[11] In the case (i) there is a nice duality between two solutions due to the linearity of a(u). That is, let X(t) and Y(t) be two nonnegative C_{tem}-valued solutions of (2) with the initial conditions $X(0) \geq 0 \in C_{tem}$ and $Y(0) \geq 0 \in C_{tem}$. Then

Lemma 4.3 Two one dimensional stochastic processes $\langle X(t),Y(0)\rangle$ and $\langle Y(t),X(0)\rangle$ have the same probability laws, if $\langle X(0),Y(0)\rangle < \infty$.

By this duality the proof of Theorem 4.2 is reduced to Theorem 4.1 in the case (ii).

Finally we refer the forthcoming paper[12] for the details of the section 2 and 3.

References

1. G. Royer, *J. Math. pure et appl.* **56** (1977) 455-478.
2. T. Shiga and A. Shimizu, *J. Math. Kyoto Univ.* **20** (1980) 395-418.
3. T. Shiga, *J. Math. Kyoto Univ.* **20** (1980) 213-242.
4. J.B. Walsh, *Lecture Notes in Math.* **1180** (Springer-Verlag, 1986) 265-439.
5. I. Iwata, *P.T.R.F.* **74** (1987) 141-159.
6. N. Konno and T. Shiga, *P.T.R.F.* **79** (1988) 201-225.
7. I. Iscoe, *Ann. Probab.* **16** (1988) 200-221.
8. C. Mueller (preprint, 1990).
9. T. Shiga, *Lecture Notes in Biomath.* **70** (Springer-Verlag, 1985) 87-99.
10. T.M. Liggett, *Interacting Particle systems* (Springer-Verlag, 1985), p.435.
11. D.A. Dawson, *Z.Wahr. Theorie verw. Geb.* **40** (1977) 125-145.
12. T. Shiga, (preprint, 1990).

Variational Calculus for Lévy's Brownian motion and a generalization

Si Si

Department of Mathematics

University of Yangon

Yangon, Myanmar

Dedicated to Professor Takeyuki Hida

§1. Introduction.

Let $X = \{ X(a);\ a \in R^d \}$ be a d-dimensional parameter Lévy's Brownian motion, and let **C** be a certain collection of $C^\infty -$ manifolds C in R^d with co-dimension 1, being homeomorphic to the unit sphere S^{d-1}. We are interested in a random field $Y = \{ Y(C)\ ;\ C \in \mathbf{C} \}$ derived from X in such a way that

$$Y(C) = \int_C f(C,u)X(u)\ d\sigma(u), \tag{1.1}$$

where $d\sigma$ is the surface element over C, and also we have particular interest in a random field $\{ Y(C)\ ;\ C \in \mathbf{C} \}$ expressed in the form

$$Y(C) = \int_{[C]} g(a,u)x(u)\ dv(u), \quad [C] \text{ the domain with boundary } C, \tag{1.2}$$

where a is a fixed point in R^d, dv is the Lebesgue measure and where x is the white noise which is assumed to be the innovation of the Brownian motion $\{X(a),\ a \in R^d \}$ (see Si Si [13]). More generally, we shall discuss such $Y(C)$ as

$$Y(C) = \Phi(\int h(C, x, u)dv(u)), \tag{1.3}$$

where Φ is a non-random function.

Since the $Y(C)$ is a random function depending on a manifold C, it seems to be natural to discuss the variation $\delta Y(C)$ of $Y(C)$ when C runs through **C**. For this purpose we shall first use the integral representation (for definition, see [2] as well as [3]) of the $Y(C)$ which is viewed as a functional of white noise, then we can appeal to the classical theory of the variational calculus. There, as was expected, we can see a sort of generalization of a stochastic functional differential equation (in terms of P. Lévy, such a generalization is to be a *stochastic infinitesimal eaquation* that was proposed in [8]). With this motivation we are led to discuss how to solve the equation which comes from the variation of $Y(C)$ of the form (1.3).

Having clarified this idea in §4, we come to a general theory of *stochastic variational equation* for $Y(C)$. Assuming that $Y(C)$ is a random function living in the space $(S)^*$ of generalized white noise functionals, we apply the $\mathcal{S}$-transform to have a representation of $Y(C)$ in terms of the U-functional with variable $\xi \in$ E ; let it be denoted by $U(C, \xi)$. Thus the infinitesimal equation for $Y(C)$ may be expressed in terms of an ordinary functional differential equation, and a sufficient condition for the equation to have the unique solution will be given in §4.

The theory discussed in this note is far from a general theory, however there are many good examples of its application as will be seen in the last part of this paper.

§2. Background.

Let E $\subset L^2(R^d) \subset E^*$ be a Gel'fand triple, and let $(E^*, \mathbf{B}, \mu)$ be a white noise, that is a measure space which defines a generalized Gaussian process with the characteristic functional

$$C(\xi) = \int \exp[-\tfrac{1}{2}\|\xi\|^2], \ \|\cdot\| \text{ the } L^2(R^d)\text{-norm.}$$

The space $(S)^*$ of generalized white noise functional and the space (S) of test functionals are introduced in the ususal manner in such a way that

$$(S) \subset L^2(E^*, \mu) \equiv (L^2) \subset (S)^* \tag{2.1}$$

is a Gel'fand triple (see [3]).

The $\mathcal{S}$-transform of $(S)^*$-functional $\varphi(x)$ is defined by

(2.2) $$S : \varphi \longrightarrow (S\varphi)(\xi) \equiv U(\xi) = \int_{E^*} \varphi(x+\xi)d\mu(x),$$

and it gives the U-functional $U(\xi)$ associated with φ.

To fix the idea we take the basic nuclear space E to be the Schwartz space $S = S(R^d)$. Then it is proved that any U-functional $U(\xi)$ is Fréchet differentiable on S (see [3]), and its functional derivative is denoted by $U'(\xi, t)$. The variation $\delta U(\xi)$ of $U(\xi)$ is therefore expressed in the form

(2.3) $$\delta U(\xi) = < U'(\xi, \cdot), \delta\xi(\cdot) >, \quad \delta\xi \in S,$$

or formally

(2.3') $$\delta U(\xi) = \int_{R^d} U'(\xi, t)\delta\xi(t)dv(t),$$

where $\delta\xi$ is still in S and where the kernel $U'(\xi, \cdot)$, that is the functional derivative, may be a generalized function. Such a variational calculus is useful to get innovation of random fields.

We then come to the restriction of parameter t. It is allowed to restrict the parameter t to a C^∞–manifold C in R^d. There is established a Gel'fand triple $E(C) \subset L^2(C, d\sigma(s)) \subset E(C),^*$ $d\sigma(s)$ being the surface element over C, so that the white noise measure μ_C and the induced white noise x_C can be defined, and the space $(S(C))^*$ of generalized white noise functionals on the measure space ($E(C)^*, \mu_C$) is formed. There is a natural restriction of μ to have μ_C (we refer to [6] for details) and accordingly we should have the following injection

$$(S)^* \longrightarrow (S(C))^*$$

which is also naturally defined.

In terms of the U-functional, the above mapping is simply to be the restriction of the variable t of ξ to the manifold C :

$$U(\xi) \longrightarrow U(\xi_C), \xi_C(u) = \xi(u) \mid_C \text{ being a member of } S(\mathrm{C}).$$

§3. Random fields depending on manifold.

We are interested in a class of random variables $Y(C)$ depending on a smooth manifold

C in R^d. We assume that the C runs through the class $\mathbf{C}$ given by

(A.1) $$\mathbf{C} = \{C\,;\ C^\infty\text{−manifold homeomorphic to } S^{d-1},\ \text{convex}\},$$

and that

(A.2) each $Y(C)$ is a member of $(S)^*$, where the basic nuclear space E is taken to be $S(R^d)$.

The assumption (A.2) implies that there is a U-functional $U(\xi, C)$ on $S(R^d)$ which is associated with $Y(C)$ and is Fréchet differentiable in ξ. Using the function-type representaton of C we are able to introduce the C^∞−topology to $\mathbf{C}$ in the usual manner.

We then come to a representation of $U(\xi, C)$ by expressing C in terms of C^∞−function. By assumption, C has a representation by an R^d-valued C^∞−function $\eta(s)$ defined on the unit sphere S^{d-1}, where s is a spherical parameter. Hence $U(\xi, C)$ can be written in the form $U(\xi,\eta)$. It is noted that the expression $U(\xi,\eta)$ is not unique since there is freedom in choosing a parameter s which is taken to be the variable of the function η. Any deformation of a manifold C within the class $\mathbf{C}$ may also be expressed by a C^∞−function. Thus, if C is fixed and is represented by $\eta(s)$, $s \in S^{d-1}$, then a variation δC of C can uniquely be expressed by another infinitesimal C^∞−function $\delta\eta(s)$ that denotes the shift at the point $\eta(s)$ along the normal direction towards the exterior of the domain $[C]$ enclosed by C.

With the expression established above any variation of $Y(C)$ can be represented by that of $U(\xi,\eta)$. Let C be fixed, therefore η be fixed. Then, a variation $\delta Y(C)$ of $Y(C)$ for a deformation $C \longrightarrow C + \delta C$ corresponds to that of $U(\xi,\eta)$ for $\eta \longrightarrow \eta + \delta\eta$, where ξ is a fixed variable and is unchanged. Such a correspondence may easily define the variation of $Y(C)$ in such a way that

(3.1) $\delta Y(C) \longleftrightarrow \delta U(\xi,\eta)$, where δU can be defined as in (2.3) or (2.3') by taking η to be the variable.

Thus we have an analytic formulation for variations of a random field $Y(C)$ indexed by a manifold C and are ready to apply the classical theory of variational calculus as will be illustrated in the next section. Here we note again that the variation $\delta U(\xi,\eta)$ is independent of the choice of the paraneter s.

§4. Variational calculus for $U(\xi, C)$ and for $Y(C)$.

We start with the fundamental lemma due to P. Lévy [7]. Let $U(\xi,\eta)$ be a U-functional introduced in the last section. To make the notation for U-functionals simple we omit ξ which is always fixed.

Lemma. (P. Lévy) *Assume that the variation $\delta U(\eta)$ is expressed in the form of a variational equation*

(4.1) $$\delta U(\eta) = \int_{S^{d-1}} f(\eta;\, U,\, s)\, \delta\eta(s) d\sigma(s),\ d\sigma : \text{surface element},$$

with the initial condition $U_0 = U(\eta_0)$.

Assume that f is continuous in three variables and satisfies the following condition : Let U_0 and η_0 be fixed, and let ζ be also fixed arbitrarily. For any U, V *and η such that*

$$| U - U_0 | < c,\ | V - U_0 | < c,\ \eta = \eta_0 + \lambda\zeta$$

the following inequalities

$$\int_{S^{d-1}} f(\eta,\, U,\, s)^2 d\sigma(s) \le M^2 \text{ and}$$

$$\int_{S^{d-1}} \{f(\eta,\, U,\, s) - f(\eta,\, V,\, s)\}^2 d\sigma(s) \le K^2 \,| U - V |$$

hold with absolute constants M and K.

Then the equation (4.1) *has the unique solution $U(\eta)$.*

Proof. Set

$$\varphi_0(\lambda) = U_0.$$

Define $\varphi_p(\lambda)$, $p = 1, 2, \cdots$, by

$$\varphi_p(\lambda) = U_0 + \int_0^\lambda d\mu \int_{S^{d-1}} f(\eta_0 + \mu\zeta, \varphi_{p-1}(\mu), s)\zeta(s) d\sigma(s),$$

inductively. Then

$$\varphi_p(\lambda) - \varphi_{p-1}(\lambda)$$

$$= \int_0^\lambda d\mu \int_{S^{d-1}} \{f(\eta_0 + \mu\zeta,\ \varphi_{p-1}, s) \ - \ f(\eta_0 + \mu\zeta,\ \varphi_{p-2}, s)\}\zeta(s) d\sigma(s),$$

holds. Therefore, provided that

(4.2) $\quad | \varphi_i(\mu) - U_0 | \leq c$, for $i = 1, 2, \cdots, p\text{-}1$, and for $\mu < \lambda$,

we have

$$| \varphi_p(\lambda) - \varphi_{p-1}(\lambda) | \leq \int_0^\lambda d\mu K \, | \varphi_{p-1}(\mu) - \varphi_{p-2}(\mu) | \cdot \|\zeta\| \, ,$$

where $\|\|$ is the $L^2(S^{d-1}, d\sigma) - norm$. Hence

$$| \varphi_1(\lambda) - U_0 | \leq M\lambda \, ,$$

$$| \varphi_2(\lambda) - \varphi_1(\lambda) | \leq \tfrac{1}{2} MK\lambda^2 \, ,$$

and generally

$$| \varphi_p(\lambda) - \varphi_{p-1}(\lambda) | \leq \tfrac{1}{p} MK^{p-1}\lambda^p \, , \; p = 3, 4, \cdots,$$

from which we have

$$| \varphi_p(\lambda) - U_0 | \leq -\tfrac{M}{K} \textstyle\sum_{j=1}^{p} \frac{(K\lambda)^p}{p} \leq -\frac{M}{K}\{e^{K\lambda} - 1\} \, .$$

The inequality (4.2) has therefore been proved for all $p \geq 1$ and for all λ less than min $\{R, -\frac{1}{K}\log(1 + -\frac{cK}{M})\}$. Thus the convergence of the $\varphi_p(\lambda)$ has been established and U $=$ $\varphi(\lambda)$ can therefore be defined as the limit of the $\varphi_p(\lambda)$. Obviously $U |_{\lambda=0} = U_0$ holds, and with the choice $\zeta = \delta\eta$ we have

(4.3) $\quad \frac{d}{d\lambda} U |_{\lambda=0} = \delta U.$

Thus the lemma is proved. (Q.E.D.)

Our main theorem is now almost straight forward from the above Lemma, if we paraphrase the variation of the stochastic variational equation for $Y(C)$ in terms of the U-functional which is the S-transform of $Y(C)$.

Let the stochastic variational equation for $Y(C)$ is expressed in the form

$$\delta Y(C) = \int_C f(C, Y, s) \, x_C(\eta(s))\delta n(s) d\sigma(s),$$

$$Y(C_0) = Y_0,$$

where C is represented by η and x_C is the induced white noise. We are interested in the existence and the uniqueness of the solution to the equation (4.4). Here is a basic theorem.

Theorem. *Given a stochastic variational equation* (4.4). *Let C have an analytic expression* *in terms of $C^\infty -$ function η and assume that the function f in (4.4) admits an expression of the form $f(\eta, Y, s)$. If this function f satisfies the condition in the Lemma above, then in the neighbourhood of C_0 the equation (4.4) has the unique solution that lives in the space $(S)^*$ and the S-transform of the solution is given by the solution to the equation (4.1).*

Proof follows immediately from the Lemma, if we put $\xi(s)\delta n(s) = \zeta(s)$.

Example 1. (An additive field ; see [4]) Consider the following stochastic variational equation

$$\delta Y(C) = \int_C g(\eta(s)) x_C(\eta(s)) \delta n(s) d\sigma(s), \quad C \in \mathbf{C}. \tag{4.5}$$

The unique solution to this equation gives a Gaussian random field expressed in the form

$$Y(C) = \int_{[C]} g(u) x(u)\, du + Y_0$$

with a constant random variable Y_0, where $[C]$ denotes the domain in R^d enclosed by C.

Example 2. To fix the idea d is taken to be 2. We choose a subclass $\mathbf{C}_0$ of $\mathbf{C}$ such that

$$\mathbf{C}_0 = \{\ C \in \mathbf{C}\ ;\ C \text{ runs through the origin } \mathbf{o} \in R^2\}$$

and define

$$Y(C) = \int_{[C]} |\ u\ |^{-1/2}\, x(u)\ dv(u). \tag{4.6}$$

If, in particular C is taken to be $C_a = \{$ circle with diameter $\bar{oa}$ $\}$, then $\{\ Y(C_a),\ a \in R^2\ \}$ is a Lévy's Brownian motion.

This $Y(C)$ is a solution to the stochastic variational equation

$$\delta Y(C) = \int_C |\ \eta(s)\ |^{-1/2}\, x_C(\eta(s)) \delta n(s) d\sigma(s) \tag{4.7}$$

as in Example 1. In order to be given an ordinary random variable δn has to be chosen so that its support excludes the origin $\mathbf{o}$. Even with such a choice we still have rich enough δn's to squeeze out the white noise x_C from the variations $\delta Y(C)$, where C is fixed and δC runs through $\mathbf{C}_0$.

The system $\{\ x_C\ ;\ C \in \mathbf{C}_0\ \}$ determines consistently the original white noise x that has defined the field $Y(C)$. Thus, if the curve, that is the parameter of $Y(C)$, moves within $\mathbf{C}_0$, we can recover the white noise x, which is now thought of as the innovation of $Y(C)$. It is therefore interesting to compare this fact to the result in the paper [11] where a method of getting innovation is presented.

Example 3. Assume that f in (4.4) is linear in Y; for instance let the variation of the associated U-functional be given by

$$\delta U(C) = \int_C U(C)\, f(\eta(s))\xi(\eta(s))\delta n(s)d\sigma(s). \tag{4.8}$$

Then we are given the solution

$$U(C) = A\ \exp[\int_{[C]} f(u)\xi(u)du],\ A = A(\xi) \text{ being a constant.}$$

Hence, the random field $Y(C)$ is

$$Y(C) = A'\ \exp[\int_{[C]} f(u)x(u)du\ \text{-}\ -\tfrac{1}{2}\int_{[C]} f(u)^2 du], \tag{4.9}$$

where A' is a constant random variable determined by the initial condition. Thus we are given a field of martingale type indexed by a manifold C, $C \in \mathbf{C}$.

Remark.1. It is interesting to consider the case where the deformations of C in $\mathbf{C}$ are to be made only by conformal transformation acting on R^d, in particular the case $d = 2$ where the theory of holomorphic functions of a complex variable can be used. (We refer to e.g. [1].)

Remark.2. If we consider a random field $Y(C)$ given by the integral over a curve C :

$$Y(C) = \int_C g(x(s))d\sigma(s),$$

then the variational equation is of the form

$$\delta Y(C) = \int_C\{(\tfrac{\partial}{\partial n}g)(s)n(s)d\sigma(s) + g(x(s))\delta(d\sigma(s))\}.$$

We shall discuss how to solve such a variational equation in the forthcoming paper.

Acknowledgement. The author is grateful to Dr. K. Saitô who helped in typing the final manuscript of this paper.

[References]

[1] B.A. Dubrovin, A.T. Fomenko and S.P. Novikov, Modern Geometry - Methods and Applications. Part 1. (English Translation), Springer-Verlag 1984. Graduate Texts in Math. 93.

[2] T. Hida, Brownian motion. Springer-Verlag 1980, Applications of Mathematics 11.

[3] T. Hida, H.-H. Kuo, J. Potthoff and L. Streit, White noise - An infinite dimensional calculus. Monograph, to appear.

[4] T. Hida, K.-S. Lee and Si Si, Multidimensional parameter white noise and Gaussian random fields. in *White Noise Theory*, the Balakrishnan volume, 1987, 177 - 183.

[5] T. Hida and J. Potthoff, Whit noise analysis - An Overview. in *White noise analysis. Mathematics and Applications.* World Scientific, 1990, ed. T. Hida, H.-H. Kuo, J. Potthoff and L. Streit, 140 -165.

[6] T. Hida and Si Si, Variational calculus for Gaussian random fields. Proc. IFIP Warsaw, 1988. Springer-Verlag, Lecture Notes in Control and Information Sciences, no.136, ed. J. Zabczyk. 86 - 97. (Correction : The second integral of the formula (5.1) should be deleted.)

[7] P. Lévy, Problèmes concrets d'analyse fonctionnelle. Gauthier- Villars. 1951. (in particular, Deuxième Partie.)

[8] ——, Random functions: General theory with special reference to Laplacian random functions. Univ. of California publications in statistics, I, no.12, (1953) 331 - 388.

[9] H.P. McKean, Brownian motion with several-dimensional time. Theory of Probabilty and its Appl. 8 (1963), 335 - 354.

[10] Si Si, A note on Lévy's Brownian motion. Nagoya Math. J. 108 (1987), 121 - 130;

[11] —, A note on Lévy's Brownian motion II. Nagoya Math. J. 114 (1989),165-172.

[12] —, Gaussian processes and conditional expectations. BiBoS Notes 292/87. Universität Bielefeld, Bielefeld, B.R. Deutschland, 1987.

[13] —, Innovation approach to Lévy's Brownian motion. Proc. of Preseminar for Inter-

national Conference on Gaussian random fields held at Nagoya, 1990, ed. T. Hida and K. Saitô, 1991, Part 2, 27-33.

On the maximum Markovian self-adjoint extensions of one-dimensional diffusion operators

M. TAKEDA

Department of Mathematics, Himeji Institute of Technology

Shosha 2167, Himeji 671-22, Japan

Abstract. We characterize the maximum Markovian self-adjoint extension of a bdx-symmetric operator, $\mathcal{L}\phi = \frac{1}{b}(a\phi')'$, $\phi \in C_0^\infty(r_1, r_2)$ under some conditions on functions a, b. As an application we remark that $\mathcal{L}$ has a unique Markovian self-adjoint extension if and only if both boundaries are not regular in Feller's sense.

1. Introduction.

In this note we consider one-dimensional diffusion operators of the type

$$(1.1) \qquad \mathcal{L}\phi = \frac{1}{b}(a\phi')', \qquad \phi \in C_0^\infty(\Omega),$$

where Ω is an interval in R^1. Under some conditions (cf. (2.1) below) the operator $\mathcal{L}$ becomes a symmetric operator on $L^2(\Omega, bdx)$. Let $\mathcal{A}_{\mathcal{M}}(\mathcal{L})$ be the family of all self-adjoint extensions of $\mathcal{L}$ which generate Markovian semigroups. Then the semi-order on $\mathcal{A}_{\mathcal{M}}(\mathcal{L})$ can be defined (cf. (2.3) below) because each element of $\mathcal{A}_{\mathcal{M}}(\mathcal{L})$ is non-positive self-adjoint operator. It is known in [4] that the Friedrichs extension of $\mathcal{L}$ belongs to $\mathcal{A}_{\mathcal{M}}(\mathcal{L})$ and it is minimum one. Our aim is to characterize the maximum element of $\mathcal{A}_{\mathcal{M}}(\mathcal{L})$ (Theorem 1). As an application of Theorem 1, we can say that the operator $\mathcal{L}$ has a unique Markovian self-adjoint extension, i.e. $\sharp(\mathcal{A}_{\mathcal{M}}(\mathcal{L})) = 1$, if we can identify the maximum one with the Friedrichs extension.

In the case that $a = b$, the operater $\mathcal{L}$ is called *"generalied Schrödinger operator"* and above problom was investigated in [1],[4],[8] and [9]. In [10], Wielens studied the relation between the essential self-adjointness and the boundary classification of Feller. The essential self-adjointness is of course a stronger notion than the uniqueness of Markovian self-adjoint extension, and in [9] we gave several examples of generalized Schrödinger operators which have a unique Markovian extension but are not essentially self-adjoint.

In final section, we remark that if the function a is strictly positive everywhere, the uniqueness of Markovian self-adjoint extension of $\mathcal{L}$ is equivalent with that both boundaries of Ω are not regular.

2. The maximum Markovian extension.

Let Ω be an open interval (r_1, r_2), where $-\infty \leq r_1 < r_2 \leq \infty$. Let a and b be functions on Ω such that

$$
(2.1) \qquad \begin{aligned}
&i)\ a > 0,\ dx\text{-a.e.},\ a \in L^1_{loc}(\Omega,\ dx),\\
&ii)\ b > 0,\ dx\text{-a.e.},\ b \in L^1_{loc}(\Omega,\ dx),\\
&iii)\ \frac{a}{\sqrt{b}},\ \frac{a'}{\sqrt{b}} \in L^2_{loc}(\Omega,\ dx)
\end{aligned}
$$

where a' means the derivative in the distribution sense. Then we consider the diffusion operator

$$
(2.2) \qquad \mathcal{L}\phi = \frac{1}{b}(a\phi')', \qquad \phi \in C_0^\infty(\Omega).
$$

Under the condition (2.1) iii), the operator $\mathcal{L}$ becomes a symmetric operator on $L^2(\Omega,\ bdx)$. Let us denote by $\mathcal{A}_{\mathcal{M}}(\mathcal{L})$ the family of all self-adjoint extensions of $\mathcal{L}$ that generates Markovian semi-groups;

$$
\mathcal{A}_{\mathcal{M}}(\mathcal{L}) = \left\{ A : \begin{array}{c} A \text{ is a self-adjoint extension of } \mathcal{L} \text{ such that} \\ 0 \leq e^{tA} f \leq 1,\ bdx\text{-a.e. if } 0 \leq f \leq 1,\ bdx\text{-a.e.} \end{array} \right\}.
$$

We call an element of $\mathcal{A}_{\mathcal{M}}(\mathcal{L})$ *Markovian extension* of $\mathcal{L}$. Recall that semi-order "$\prec$" on $\mathcal{A}_{\mathcal{M}}(\mathcal{L})$ is defined as

$$\text{(2.3)} \quad A_1 \prec A_2 \quad \text{if} \quad \mathcal{D}[\sqrt{-A_1}] \subset \mathcal{D}[\sqrt{-A_2}] \quad \text{and}$$
$$(\sqrt{-A_1}u, \sqrt{-A_1}u)_{bdx} \geq (\sqrt{-A_2}u, \sqrt{-A_2}u)_{bdx}, \text{ for any } u \in \mathcal{D}[\sqrt{-A_1}].$$

Then, the Friedrichs extension of $\mathcal{L}$ belongs to $\mathcal{A}_{\mathcal{M}}(\mathcal{L})$(cf. [Theorem 2.11;4]) and it is the minimum one of $\mathcal{A}_{\mathcal{M}}(\mathcal{L})$ with respect to above semi-order.

Let us define a linear space $\mathcal{F}^+$ by

$$\mathcal{F}^+ = \left\{ u \in L^2(\Omega, bdx); \begin{array}{l} \text{there exists a function } g \in L^2(\Omega, adx) \text{ such that} \\ (u, -\frac{1}{b}(a\phi)')_{bdx} = (g, \phi)_{adx} \text{ for any } \phi \in C_0^\infty(\Omega) \end{array} \right\}$$

We denote by Du the function g appearing in the definition of $\mathcal{F}^+$ and define the bilinear form $\mathcal{E}^+$ by

$$\mathcal{E}^+(u,v) = \int_{r_1}^{r_2} a(x)Du(x)Dv(x)dx, \quad u, v \in \mathcal{F}^+.$$

Then $(\mathcal{E}^+, \mathcal{F}^+)$ is a closed form on $L^2(\Omega, bdx)$. Moreover we can show that $(\mathcal{E}^+, \mathcal{F}^+)$ becomes a Dirichlet space. In fact, set

$$\overline{\mathcal{F}} = \left\{ u \in L^2(\Omega, bdx); \begin{array}{l} u \text{ has a } dx\text{-version } \widetilde{u} \text{ that is absolutely} \\ \text{continuous on } \{x \in \Omega; \widetilde{a}(x) > 0\} \text{ and} \\ \dfrac{d\widetilde{u}}{dx} \in L^2(\Omega, adx) \end{array} \right\},$$

and denote $\frac{d\widetilde{u}}{dx}$ by ∇u. Here $\widetilde{a}$ is an absolutely continuous version of a. Then we have

Lemma 1. *The bilinear form* $\overline{\mathcal{F}}(u,v) = \int_{r_1}^{r_2} a(x)\nabla u(x)\nabla v(x)dx$, $u,v \in \overline{\mathcal{F}}$, *is a Dirichlet space on* $L^2(\Omega, bdx)$.

Proof. Note that by Proposition 2.4 in [2] the set $\{x \in \Omega; \widetilde{a}(x) > 0\}$ is equal to the regular set $R(a)(= \{x \in \Omega; \int_{x-\epsilon}^{x+\epsilon} a^{-1}(t)dt < \infty$ for some $\epsilon > 0\})$. Then the closedness of $(\overline{\mathcal{E}}, \overline{\mathcal{F}})$ can be proved by the same argument as in Theorem 2.2 in [3]. The Markov property of $(\overline{\mathcal{E}}, \overline{\mathcal{F}})$ is clear by the definition.

If the function ua is absolutely continuous, the function u has a version $\widetilde{u}$ that is absolutely continuous on $\{x \in \Omega; \widetilde{a}(x) > 0\}$. Thus let us define the space $\widehat{\mathcal{F}}$ by

$$\widehat{\mathcal{F}} = \left\{ u \in L^2(\Omega, bdx); \begin{array}{c} ua \text{ has an absolutely continuous} \\ \text{version on } \Omega \text{ and } \dfrac{d\widetilde{u}}{dx} \in L^2(\Omega, adx) \end{array} \right\}.$$

Then we have

Lemma 2. *It holds that* $\widehat{\mathcal{F}} = \overline{\mathcal{F}}$.

Proof. By the definition of $\widehat{\mathcal{F}}$, $\widehat{\mathcal{F}} \subset \overline{\mathcal{F}}$. First we prove that $\widehat{\mathcal{F}}$ is a closed subspace of $\overline{\mathcal{F}}$ with respect to $\overline{\mathcal{E}}_1$. Let $u_n \subset \widehat{\mathcal{F}}$ such that $\overline{\mathcal{E}}_1(u_n - u_m, u_n - u_m) \to 0\,(n, m \to \infty)$. Then there exists $u \in \overline{\mathcal{F}}$ such that $u_n \to u$ in $L^2(\Omega, bdx)$ and $\nabla u_n \to \nabla u$ in $L^2(\Omega, adx)$ by Lemma 1. Since for any $\phi \in C_0^\infty(\Omega)$

$$\begin{aligned} |(ua, -\phi')_{dx} - (u_n a, -\phi')_{dx}| &\le \left(\int_{r_1}^{r_2} (u_n - u)^2 bdx\right)^{\frac{1}{2}} \cdot \left(\int_{r_1}^{r_2} \frac{a^2}{b} \phi'^2 dx\right)^{\frac{1}{2}} \\ &\to 0\,(n \to \infty), \end{aligned}$$

and

$$\begin{aligned} &|(u_n a, -\phi')_{dx} - (\nabla u a + u a', \phi)_{dx}| \\ \le &\left(\int_{r_1}^{r_2} (\nabla u_n - \nabla u)^2 adx\right)^{\frac{1}{2}} \cdot \left(\int_{r_1}^{r_2} a\phi^2 dx\right)^{\frac{1}{2}} \\ &+ \left(\int_{r_1}^{r_2} (u_n - u)^2 bdx\right)^{\frac{1}{2}} \cdot \left(\int_{r_1}^{r_2} \frac{a'^2}{b} \phi^2 dx\right)^{\frac{1}{2}} \\ &\to 0\,(n \to \infty), \end{aligned}$$

it follows that $(ua, -\phi')_{dx} = (\nabla ua + ua', \phi)_{dx}$. Hence $u \in \widehat{\mathcal{F}}$ by noting that ua and $\nabla ua + ua' \in L^1_{loc}(\Omega, dx)$.

For $u \in \overline{\mathcal{F}}$ let $u^{(n)} = ((-n) \vee u) \wedge n$. Let (α, β) be a connected component of $\{\tilde{a} > 0\}$ such that $[\alpha, \beta] \subset \Omega$. Then for any $\phi \in C_0^\infty(\Omega)$.

$$\begin{aligned}
\int_\alpha^\beta u^{(n)} a\phi' ax &= \lim_{\epsilon \to 0} \int_{\alpha+\epsilon}^{\beta-\epsilon} u^{(n)} a\phi' dx \\
&= \lim_{\epsilon \to 0} \left([\tilde{u}^{(n)} \tilde{a}\phi]_{\alpha+\epsilon}^{\beta-\epsilon} - \int_{\alpha+\epsilon}^{\beta-\epsilon} (u^{(n)} a)' \phi dx \right) \\
&= - \int_\alpha^\beta (\nabla u^{(n)} a + u^{(n)} a') \phi dx.
\end{aligned}$$

Hence $\int_{r_1}^{r_2} u^{(n)} a\phi' dx = -\int_{r_1}^{r_2} (\nabla u^{(n)} a + u^{(n)} a') \phi dx$ for any $\phi \in C_0^\infty(\Omega)$ and which implies that $u^{(n)} \in \widehat{\mathcal{F}}$. Therefore $u \in \widehat{\mathcal{F}}$ by Lemma 1.

We can easily show that $\mathcal{F}^+ = \widehat{\mathcal{F}}$ *and* $D = \nabla$. Therefore we can attain the next lemma by Lemma 1 and Lemma 2.

Lemma 3. $(\mathcal{E}^+, \mathcal{F}^+)$ *is a Dirichlet space on* $L^2(\Omega, bdx)$.

Now we state our main theorem.

Theorem 1. *The self-adjoint operator* A^+ *corresponding to the Dirichlet space* $(\mathcal{E}^+, \mathcal{F}^+)$ *is the maximum element of* $\mathcal{A}_{\mathcal{M}}(\mathcal{L})$.

(Sketch of proof). The assertion that $A^+ \in \mathcal{A}_{\mathcal{M}}(\mathcal{L})$ can be proved by the same manner as in [Lemma 7;9]. For $A \in \mathcal{A}_{\mathcal{M}}(\mathcal{L})$ denote $\mathcal{D}[\sqrt{-A}]$ and $(\sqrt{-A}\cdot, \sqrt{-A}\cdot)_{bdx}$ by $\mathcal{F}_A$ and $\mathcal{E}_A(\cdot, \cdot)$ respectively. Define the operator $T^{[u]}$ by

$$T^{[u]}(\phi) = (u, -\frac{1}{b}(a\phi'))_{bdx}, \quad \phi \in C_0^\infty(\Omega).$$

Then we attain the theorem by Riesz theorem if we can prove the next inequality :

$$(2.4) \qquad T^{[u]}(\phi) \le \mathcal{E}_A(u,v)^{\frac{1}{2}} \cdot \|\phi\|_{L^2(adx)}, \quad for \quad \phi \in C_0^\infty(\Omega).$$

Take $\phi_1, \phi_2 \in C_0^\infty(\Omega)$ such that

$$i)\ \phi_1 = 1 \quad on \quad \mathrm{supp}[\phi]$$

$$ii)\ \phi_2 = 1 \quad on \quad \mathrm{supp}[\phi_1].$$

Then

$$\begin{aligned} T^{[u]} &= (u\phi_1, -\frac{1}{b}(a\phi'))_{bdx} \\ &= (u\phi_1, -\frac{1}{b}(a(\phi_2 \int_c^x \phi(\tau)d\tau))')')_{bdx} \quad (r_1 < c < r_2) \\ &= (u\phi_1, -\mathcal{L}(\phi_2 \int_c^x \phi(\tau)d\tau))_{bdx}. \end{aligned}$$

Since A is a self-adjoint extension of $\mathcal{L}$, the right hand side is equal to $\mathcal{E}_A(u\phi_1, \phi_2 \int_c^x \phi(\tau)d\tau)$. Let $(\widetilde{\Omega}, \widetilde{m}, \widetilde{\mathcal{E}}_A, \widetilde{\mathcal{F}}_A, \Phi)$ be a regular representation of $(\mathcal{E}_A, \mathcal{F}_A)$ (cf. [5]). Then by Lemma 3 in [9]

$$\begin{aligned} \mathcal{E}_A(u\phi_1, \phi_2 \int_c^x \phi(\tau)d\tau) &= \widetilde{\mathcal{E}}_A(\Phi(u)\Phi(\phi_1), \Phi(\phi_2 \int_c^x \phi(\tau)d\tau)) \\ &= \widetilde{\mathcal{E}}_A^c(\Phi(u)\Phi(\phi_1), \Phi(\phi_2 \int_c^x \phi(\tau)d\tau)), \end{aligned}$$

where $\widetilde{\mathcal{E}}_A^c$ is a local part of $\widetilde{\mathcal{E}}_A$ in the Beuling-Deny formula. Let $\widetilde{\mathcal{M}} = (\widetilde{P}_x, \widetilde{X}_t)$ be a Hunt process on $\widetilde{\Omega}$ corresponding to $(\widetilde{\mathcal{E}}_A, \widetilde{\mathcal{F}}_A)$. Let $\widetilde{u}$ be a quasi- continuous version of u. Then it was proved in [4] that the additive functional $\widetilde{A}_t^{[u]} = \widetilde{u}(X_t) - \widetilde{u}(X_0), u \in \widetilde{\mathcal{F}}_A$, can be decomposed as

$$\widetilde{A}_t^{[u]} = \widetilde{M}_t^{[u]} + \widetilde{N}_t^{[u]},$$

where $\widetilde{M}_t^{[u]}$ is a martingale additive functional of finite energy and $\widetilde{N}_t^{[u]}$ is a continuous additive functional of zero energy. Denote by $\widetilde{M}^{c[u]}$ the continuous part of

$\widetilde{M}_t^{[u]}$ and $\widetilde{\mu}^c_{<u,v>}$ the smooth measure corresponding to $< \widetilde{M}^{c[u]}, \widetilde{M}^{c[v]} >$ (cf. [4]). Then by the derivation property of $\widetilde{\mu}^c_{<u,v>}$

$$\begin{aligned}
&\widetilde{\mathcal{E}}^c_A(\Phi(u)\Phi(\phi_1), \Phi(\phi_2 \int_c^x \phi(\tau)d\tau)) \\
=&\frac{1}{2}\int_{\widetilde{\Omega}} d\widetilde{\mu}^c_{<\Phi(u)\Phi(\phi_1),\Phi(\phi_2 \int_c^x \phi(\tau)d\tau)>} \\
=&\frac{1}{2}\int_{\widetilde{\Omega}} \widetilde{\Phi(u)}d\widetilde{\mu}^c_{<\Phi(\phi_1),\Phi(\phi_2 \int_c^x \phi(\tau)d\tau)>} \\
&+\frac{1}{2}\int_{\widetilde{\Omega}} \Phi(\phi_1)d\widetilde{\mu}^c_{<\Phi(u),\Phi(\phi_2 \int_c^x \phi(\tau)d\tau)>}.
\end{aligned}$$

On the other hand, by Lemma 1 and Lemma 4 in [9] we have

$$\begin{aligned}
\widetilde{\mu}^c_{<\Phi(\phi_1),\Phi(\phi_2 \int_c^x \phi(\tau)d\tau)>} &= \Phi(\Gamma(\phi_1, \phi_2 \int_c^x \phi(\tau)d\tau))d\widetilde{m} \\
&= 0.
\end{aligned}$$

Here $\Gamma(\phi,\psi) = 2\frac{a}{b}\phi'\psi'$. Therefore

$$\begin{aligned}
T^{[u]}(\phi) &\le \left(\frac{1}{2}\int d\widetilde{\mu}^c_{<\Phi(u)>}\right)^{\frac{1}{2}} \cdot \left(\frac{1}{2}\int \Phi(\phi_1)^2 d\widetilde{\mu}^c_{<\Phi(\phi_2 \int_c^x \phi(\tau)d\tau)>}\right)^{\frac{1}{2}} \\
&\le \mathcal{E}_A(u,u)^{\frac{1}{2}} \cdot \left(\frac{1}{2}\int \phi_1^2 \cdot \Gamma(\phi_2 \int_c^x \phi(\tau)d\tau, \phi_2 \int_c^x \phi(\tau)bdx\right)^{\frac{1}{2}} \\
&\le \mathcal{E}_A(u,u)^{\frac{1}{2}} \cdot \|\phi\|_{L^2(adx)}.
\end{aligned}$$

Let $\mathcal{F}^0$ be the closure of $C_0^\infty(\Omega)$ with respect to $\mathcal{E}_1^+ (= \mathcal{E}^+ + (,)_{bdx})$. Then we obtain

Corollary 1. *If $\mathcal{F}^+$ is identified with $\mathcal{F}^0$, then $\mathcal{L}$ has a unique Markovian extension .*

3. The uniqueness of Markovian extension.

By the general theory (cf. [4]) the space $\mathcal{F}^+$ can be orthogonally decomposed as

$$\mathcal{F}^+ = \mathcal{F}^0 \oplus (\mathcal{F}^+ \cap \mathrm{Ker}(1 - \mathcal{L}^*)) \tag{3.1}$$

with respect to the inner product $\mathcal{E}_1^+$. Here $\mathcal{L}^*$ is the adjoint operator of $\mathcal{L}$. Therefore, $\mathcal{L}$ has a unique Markovian extension if and only if $\mathcal{F}^+ \cap \mathrm{Ker}(1 - \mathcal{L}^*) = \{0\}$.

Remark 1. The essential self-adjointness of $\mathcal{L}$ is equivalent with the statement that $\mathrm{Ker}(1 - \mathcal{L}^*) = \{0\}$(cf. [7]).

We suppose another condition on the function a:

$$a \quad \text{is strictly positive everywhere on} \quad \Omega. \tag{3.2}$$

Then we can easily show that the domain of the adjoint operator $\mathcal{L}^*$ is given by

$$\mathcal{D}(\mathcal{L}^*) = \left\{ u \in L^2(\Omega, bdx); \begin{array}{l} u \text{ is continuously differentiable on } \Omega \\ \text{and } u' \text{ is absolutely continuous and} \\ \frac{1}{b}(au')' \in L^2(\Omega, bdx) \end{array} \right\}$$

Set $s(x) = \int_c^x \frac{1}{a(t)} dt$, $m(x) = \int_c^x b(t)dt$. Then $\mathcal{L}\phi = \frac{d}{dm}(\frac{d}{ds}\phi)$. By the same argument as in [Example 1.2.2;4], we can show that $\mathcal{F}^+ \cap \mathrm{Ker}(1 - \mathcal{L}^*) = \{0\}$ if and only if both r_1 and r_2 are not regular. Therefore we obtain the following theorem.

Theorem 2. *Under condition (2.1) and (3.2) the operator $\mathcal{L}$ has a unique Markovian extension if and only if both boundaries are not regular.*

Remark 2. In Wielens [10], he investigated the relation between the essential self-adjointness and boundary classification of Feller. He devided the entrance boundary to two classes :

$$r \quad \text{is called a } \textit{strong entrance boundary} \text{ if } \quad \left| \int_c^r s(x)^2 dm(x) \right| = \infty$$

and

$$r \quad \text{is called a } \textit{weak entrance boundary} \text{ if } \quad |\int_c^r s(x)^2 dm(x)| < \infty,$$

and proved that natural, exit, and strong entrance boundaries are corresponding to the limit point in the sence of Wyel, and regular and weak entrance boundaries are corresponding to the limit circle (He treated only generalized Schrödinger operators, but his argument can be extended to our case.). By combining this fact and Theorem 2, we can say that the difference between the essntial self-adjointness and the uniqueness of Markovian extension appears if r_1 or r_2 is a weak entrance boundary.

Remark 3. In the case that $\Omega = R^1$ and $a = b$, Röchner and Sheng [8] has proved the uniqueness of Markovian extension without the condition (3.2). Thus we think that it is possible for the operator $\mathcal{L}$ to have a unique Markovian extension even if the function a has zero points.

References

[1] S. Albeverio and S. Kusuoka, *Maximality of infinite dimensional Dirichlet forms and Høegh-Krohn's model of quantum fields.* to appear.

[2] S. Albeverio, S. Kusuoka and M. Röchner, *On partial integration in infinite dimensional space and applications to Dirichlet forms.* to appear in J. London Math. Soc.

[3] S. Albeverio and M. Röchner, *Classical Dirichlet forms on topological vector spaces - closability and a Cameron-Martin formula*, J. Funct. Anal. **88** (1990), 395–436.

[4] M. Fukushima, "Dirichlet Forms and Markov Processes," North- Holland, Kodansha, 1980.

[5] M. Fukushima, *Regular representations of Dirichlet spaces*, Trans. Amer. Math. Soc. **155** (1971), 455–473.

[6] K. Ito and H. P. Mckean, "Diffusion processes and their sample paths," Springer-Verlag, 1965.

[7] M. Reed and B. Simon, "Methods of Modern Mathematical Physics II," Academic Press, 1975.

[8] M. Röchner and Z. T. Sheng, *On uniqeness of generalized Schrödinger operators.* preprint.

[9] M. Takeda, *The maximum Markovian self-adjoint extensions of generalized Schrödinger operators.* preprint.

[10] N. Wielens, *The essential self-adjointness of generalized Schrödinger operators,* J. Funct. Anal. **61** (1985), 98–115.

THE LAW OF THE ITERATED LOGARITHM FOR LOCAL TIME OF A LÉVY PROCESS

IN-SUK WEE

Department of Mathematics, Korea University
1, Anam-dong, Seoul, 136-701, Korea

Abstract

Let X_t be a one-dimensional Lévy process with local time $L(t,x)$ and $L^*(t) = \sup\{L(t,x) : x \in \Re\}$. Under the assumption which is more general than being a symmetric stable process with index $\alpha > 1$, we obtain the law of iterated logarithm for $L^*(t)$.

1 Introduction

Let X_t, $t \geq 0$ be a one dimensional Lévy process whose characteristic function can be represented as follows ;

$$E\exp(iuX_t) = \exp(t\psi(u))$$

where

$$\psi(u) = ibu + \int \left(e^{iux} - 1 - iux(1+x^2)^{-1}\right) \nu(dx)\,.$$

Here ν is a measure on $\Re - \{0\}$ satisfying $\int(1 \wedge x^2)\nu(dx) < \infty$. Note that we don't include the Gaussian component in $\psi(u)$ since the behavior of B.M. is well-known.

For $x > 0$, define

$$G(x) = \int_{|y|>x} \nu(dy) ,$$
$$K(x) = x^{-2} \int_{|y|\le x} y^2 \nu(dy) .$$

We assume that

$$\limsup \frac{G(x)}{K(x)} < 1 \tag{1.1}$$

as x tends both to 0 and ∞ , and

$$EX_1 = 0 . \tag{1.2}$$

For α-stable processes, $\lim G(x)/K(x) = (2-\alpha)/\alpha$ as $x \to 0$ and $x \to \infty$, so that our assumption is more general than being a symmetric α-stable process with $\alpha > 1$. We also note that (1.1) implies that $E\,|X_1| < \infty$. Assumption (1.2) is necessary to guarantee recurrence of the process. In fact, under (1.1) and (1.2), X_t is point-recurrent since the following criteria for point-recurrence are satisfied[10] ; for any $\lambda > 0$,

$$\int_{-\infty}^{\infty} Re \frac{1}{\lambda - \psi(u)} du < \infty ,$$

$$\int_{|u|<1} Re \left(- \frac{1}{\psi(u)} \right) du = \infty .$$

The purpose of this work is mainly concerned with the asymptotic growth rate of local time of X. Under (1.1) and (1.2), Kesten and Bretagnolle's conditions[5,6,10] for existence of a continuous version of $L(t,x)$ as a function of t are satisfied which are as follows ;

$$\int_{-\infty}^{\infty} Re \frac{1}{1 - \psi(u)} du < \infty , \tag{1.3}$$

$$\int (1 \wedge |x|)\, \nu(dx) = \infty . \tag{1.4}$$

Moreover, it is not hard to show that a jointly continuous version of $L(t,x)$ exists by checking the results obtained by Barlow and Hawkes[3] and Barlow[1].

They improved the result of Getoor and Kesten[6], and proved that under (1.3) and (1.4),

$$\int_{|x|<1/e} \frac{\bar{\varphi}(x)}{x(\log 1/x)^{1/2}}\, dx < \infty \tag{1.5}$$

is necessary and sufficient for the existence of jointly continuous version of local time where

$$\varphi^2(y) = \int (1 - \cos uy)\, Re \frac{1}{1 - \psi(u)}\, du \ ,$$

and $\bar{\varphi}$ denotes the monotone rearrangement of φ. When a jointly continuous $L(t,x)$ exists, it is clear that we can think of $L(t,x)$ as

$$L(t,x) = \lim_{\varepsilon \to 0} \frac{1}{2\varepsilon} \int_0^t \chi_{(x-\varepsilon, x+\varepsilon)}(X(s))\, ds \ .$$

This work is motivated by a result of Griffin[7] which described the asymptotic growth of the local time of a symmetric α-stable process with $\alpha > 1$.

To describe his results, let

$$\begin{aligned} L^*(t) &= \sup\{L(t,x) : x \in \Re^1\} \\ R(t) &= \{X(s) : 0 \le s \le t\} \end{aligned}$$

and denote the Lebesgue measure of $R(t)$ by $m(R(t))$. Griffin showed that for some c, $C \in (0, \infty)$,

$$\limsup_{t \to \infty} t^{-1/\alpha} (llt)^{-(1-1/\alpha)} m(R(t)) = c \ \text{ a.s.} \tag{1.6}$$

$$\liminf_{t \to \infty} t^{-(1-1/\alpha)} (llt)^{1-1/\alpha} L^*(t) = C \ \text{ a.s.} \tag{1.7}$$

where llt denotes $\log\log t$. In the case of Brownian Motion, the law of iterated logarithm implies that the result of type of (1.6) holds even with $\sup_{s \le t} |X_s|$ replacing $m(R(t))$, and the result of type of (1.7) was obtained by Kesten[9]. For symmetric α-stable processes, it is well-known that (1.6) is no more valid with $\sup_{s \le t} | X_s |$ instead of $m(R(t))$. We will now be concerned with the same question for a class of Lévy processes satisfying (1.1) and (1.2). For $x > 0$, define

$$f(x) = G(x) + K(x) \ .$$

It is easy to see that $f^{-1}(1/y)$ is well-defined for y large since f is continuous and strictly decreasing once it reaches the support of ν. In this work, we prove that assuming (1.1) and (1.2),

$$\liminf_{t\to\infty} L^*(t)\, f^{-1}(llt/t)\, llt/t = C \text{ a.s.}\,.$$

Furthermore, under the extra condition

$$\liminf_{x\to\infty} \frac{G(x)}{K(x)} > 0\,, \tag{1.8}$$

it can be shown that

$$\limsup_{t\to\infty} \frac{m(R(t))}{f^{-1}(llt/t)llt} = C \text{ a.s.}$$

where the proof is not contained here. We remark that under (1.8), it is known that there exists no nondecreasing function $a(t)$ such that for $c \in (0, \infty)$

$$\limsup_{t\to\infty} \frac{|X_t|}{a(t)} = c \text{ a.s.}\,.$$

There are very important estimates used to derive necessary probability estimates, which assert that for some constant c, C and positive integer m,

$$E(L(t,x)) \le \frac{ct}{f^{-1}(1/t)} \tag{1.9}$$

$$E(L(t,x) - L(t,x+y))^{2m} \le (2m)! \left(\frac{Ct}{f^{-1}(1/t)}\right)^m \left(\frac{1}{|y|\, K(|y|)}\right)^m\,. \tag{1.10}$$

The proof involved the method employed previously by Jain and Pruitt[8] for the case of a integer valued random walk, which turned out to be very useful in our situation. The analogous estimate to (1.10) for Brownian local time was obtained by Borodin[4] using the Ray's formula[11] which is not valid for general Lévy processes.

In section 2, we will briefly review the basic facts and obtain the important estimates on the local time. We will state and prove the main theorem about the local time in Section 3. We will denote a finite positive constant by C whose value may be distinct from line to line. And whenever necessary, we will number the constants.

2 Basic Estimates

In this section, we will assume throughout that (1.1) and (1.2) hold. We start with some useful facts about $f(x)$. First note that, in general, $x^2 f(x)$ is continuous and strictly increasing once it reaches the support of Lévy measure ν. Also it will be frequently used that from (1.1), there exist $0 < \varepsilon_0 < 1$, a_0, A_0 such that for $x \geq A_0$ or $x \leq a_0$,

$$G(x) \leq (1-\varepsilon_0)K(x)$$

hence there exists $1 < \delta_0 < 2$ such that $x^{\delta_0} f(x)$ is strictly decreasing on $(0, a_0] \cup [A_0, \infty)$. Also we may assume that for any x,

$$G(x) \leq C\, K(x)$$

for some C, which implies that there exists $0 < \delta_1 \leq \delta_0$ such that $x^{\delta_1} f(x)$ is strictly decreasing for any x (See Wee[12]). By using these facts we observe that for $M > 1$, small and large values of x and y,

$$M^{-2} f(x) \leq f(Mx) \leq M^{-\delta_0} f(x) \tag{2.1}$$

$$M^{-1/\delta_0} f^{-1}(y) \leq f^{-1}(My) \leq M^{-1/2} f^{-1}(y) \tag{2.2}$$

It then trivially follows that

$$2^{-1} M^{-2} K(x) \leq K(Mx) \leq 2M^{-\delta_0} K(x) \tag{2.3}$$

Now it is convenient to introduce more notation though they are not used so frequently as G and K. Define for $x > 0$,

$$\alpha(x) = \int_{|y|>x} y(1+y^2)^{-1}\, \nu(dy)\,,$$

$$\beta(x) = \int_{|y|\leq x} y^3(1+y^2)^{-1}\, \nu(dy)\,.$$

We may write X_t as the sum of two independent Lévy processes $X_t^1(a)$, $X_t^2(a)$ where

$$E\exp\left(iuX_t^1(a)\right)$$

$$= \exp\left\{itu(b-\alpha(a)) + t\int_{|x|\leq a}\left(\exp(iux) - 1 - iux(1+x^2)^{-1}\right)\, \nu(dx)\right\}$$

$$E\exp\left(iuX_t^2(a)\right) = \exp\left\{t\int_{|x|>a} (\exp(iux) - 1)\, \nu(dx)\right\}\,.$$

$X_t^2(a)$ is a compound Poisson process of parameter $tG(a)$ only with jumps of size greater than a up to time t. That is,

$$X_t^2(a) = \sum_{s \le t} (X_s - X_{s-}) \, \chi_{[-a,a]^c} (X_s - X_{s-}) \, .$$

By differentiating, it is easy to see that

$$\begin{aligned} EX_t^1(a) &= t\,(b + \beta(a) - \alpha(a)) \\ VarX_t^1(a) &= ta^2 K(a) \, . \end{aligned}$$

Now we will derive two basic estimates (1.9) and (1.10) on the local time which are interesting themselves. The techniques used here heavily rely on the method used by Jain and Pruitt[8] in the case of a integer valued random walk under analogous assumption to ours. The argument starts from the well-known inversion formula to obtain the necessary probability estimate. We will state a series of lemmas whose proofs are contained in Wee[13].

Lemma 2.1 *There exists C_1 such that for any t,*

$$\int | e^{t\psi(u)} | \, du \le \frac{C_1}{f^{-1}(1/t)} \, .$$

Lemma 2.2 *There exists C_2 such that for any x, $\varepsilon > 0$, t,*

$$P\,(|X_t - x| < \varepsilon) \le \frac{C_2 \varepsilon}{f^{-1}(1/t)} \, .$$

Lemma 2.3 *There exists C_3 such that for any x, $\varepsilon > 0$, and sufficiently large t,*

$$\int_0^t P\,(|X_s - x| < \varepsilon) \; ds \le \frac{C_3 \varepsilon t}{f^{-1}(1/t)} \, ,$$

hence

$$EL(t,x) \le \frac{C_3 t}{f^{-1}(1/t)} \, .$$

Lemma 2.4 *There exists C_4 such that for any x, y and $\varepsilon > 0$,*

$$\int_0^\infty | \, P\,(|X_s - x| < \varepsilon) - P\,(|X_s - (x+y)| < \varepsilon) \, | \; ds \le \frac{C_4 \varepsilon}{| \, y \, | \, K(| \, y \, |)} \, .$$

Lemma 2.5 *There exists C_5 such that for any positive integer m, and t large,*

$$E\left(L(t,x)-L(t,x+y)\right)^{2m} \leq (2m)!\left(\frac{C_5 t}{f^{-1}(1/t)}\right)^m \left(\frac{1}{|\,y\,|\,K(|\,y\,|)}\right)^m .$$

Sketch of the proof. Fix x and y and let

$$I_{z,\varepsilon} = (z-\varepsilon, z+\varepsilon)\,, \qquad \Phi_\varepsilon = \chi_{I_{x,\varepsilon}} - \chi_{I_{x+y,\varepsilon}} .$$

It suffices to show that

$$E\left(\int_0^t \Phi_\varepsilon(X_s)\,ds\right)^{2m} \leq (2m)!\,C_5^m \left(\frac{\varepsilon t}{f^{-1}(1/t)}\right)^m \left(\frac{\varepsilon}{|\,y\,|\,K(|\,y\,|)}\right)^m .$$

Furthermore we may restrict ourselves to the case when each X_t has density $p(t,x)$. Write for $N > 2m$,

$$\begin{aligned} E\left(\int_0^t \Phi_\varepsilon(X_s)\,ds\right)^{2m} &= E\left(S_1+S_2+\cdots+S_N\right)^{2m} \\ &= \sum_Q E\left(S_1^{i_1} S_2^{i_2}\cdots S_N^{i_N}\right) \end{aligned}$$

where

$$J_k = [(k-1)t/N, kt/N)\,, \qquad S_k = \int_{J_k} \Phi_\varepsilon(X_s)\,ds\,,$$

$$Q = \{(i_1,\cdots,i_N) : 0 \leq i_1,\cdots,i_N \leq 2m,\ \ i_1+i_2+\cdots+i_N = 2m\}\,.$$

Let

$$\begin{aligned} Q_1 &= \{(i_1,\cdots,i_N)\in Q : i_k = 0 \text{ or } 1 \text{ for any } k\}\,, \\ Q_2 &= Q - Q_1\,. \end{aligned}$$

By calculating the cardinal number of Q_2, we observe that for N sufficiently large,

$$\begin{aligned} E\left(\int_0^t \Phi_\varepsilon(X_s)\,ds\right)^{2m} &\sim \sum_{Q_1} E\left(S_1^{i_1} S_2^{i_2}\cdots S_N^{i_N}\right) \\ &= (2m)! \sum_{T_{2m}(1,N)} E\left(S_{i_1} S_{i_2}\cdots S_{i_{2m}}\right) \end{aligned}$$

where $T_k(i,j) = \{(i_1, \cdots, i_k) : i \le i_1 < i_2 < \cdots < i_k \le j\}$, and $\sim$ denotes the ratio being close to 1 for N large. Now we prove that, by induction on m,

$$| \sum_{T_{2m}(1,N)} E(S_{i_1} S_{i_2} \cdots S_{i_{2m}}) | \le C_5^m \left(\frac{\varepsilon t}{f^{-1}(1/t)}\right)^m \left(\frac{\varepsilon}{|\, y \,|\, K(|\, y \,|)}\right)^m .$$

The essential step in the proof is the following ; for $p(v,w) \ne 0$,

$$\begin{aligned} & E\left(\Phi_\varepsilon(X_u) \mid X_v = w\right) \\ = \; & p(v,w)^{-1} \left\{ \int_{I_{x,\varepsilon}} p(u,z)\, p(v-u, w-z)\, dz \right. \\ & \left. - \int_{I_{x+y,\varepsilon}} p(u,z)\, p(v-u, w-z)\, dz \right\} . \end{aligned}$$

3 LIL for local time

In this section, under the assumptions (1.1) and (1.2), we will prove that

$$\liminf_{t\to\infty} L^*(t) f^{-1}(llt/t)\, llt/t > 0 \text{ a.s.} \tag{3.1}$$

$$\liminf_{t\to\infty} L^*(t) f^{-1}(llt/t)\, llt/t < \infty \text{ a.s.} \tag{3.2}$$

The necessary probability estimates to prove (3.1) is relatively easier to obtain. First we quote the results from Wee[12].

Lemma 3.1 (Theorem 4.6 (2) of Wee[12]) *Under (1.1) and (1.2), there exists $0 < \theta_1 < 1$ such that for t sufficiently large,*

$$P(X_t > 0) \ge \theta_1 , \quad P(X_t < 0) \ge \theta_1 .$$

Lemma 3.2 (Lemma 3.2 (2) of Wee[12]) *Under (1.1) and (1.2), there exist C_6, C_7 such that for a sufficiently large,*

$$P\left(\sup_{s\le t} |\, X_s \,| \le a\right) \ge C_6 \exp\left(-C_7 t f(a)\right) .$$

Lemma 3.3 (Lemma 2.3 of Wee[12]) *For any $d \geq a$, any m,*

$$P\left(|X_t^1(a) - m| \leq d\right) \leq \frac{Cd}{(ta^2K(a))^{1/2}} .$$

Theorem 3.1 *If (1.1) and (1.2) hold, then*

$$\liminf_{t\to\infty} L^*(t) f^{-1}(llt/t) llt/t > 0 \quad a.s. .$$

Proof. Denote $h(t) = t/(llt\, f^{-1}(llt/t))$ and $A_t = \sup_{s\leq t} | X_s |$. Let n be the largest integer not exceeding llt. Using the fact that $t \leq L^*(t)A(t)$ and Lemma 3.2, we observe that for t large, $\zeta < 1$,

$$\begin{aligned} P(L^*(t) \leq \zeta h(t)) &\leq \{P(L^*(llt/t) \leq \zeta h(t))\}^{llt/2} \\ &\leq \left\{P\left(A_{t/llt} \geq t/(\zeta\, h(t)\, llt)\right)\right\}^{llt/2} \\ &\leq \left\{1 - C_6 \exp(-C_7\zeta^{\delta_0})\right\}^{llt/2} \\ &\leq e^{-\xi llt} \end{aligned} \tag{3.3}$$

where

$$e^{-\xi} = \left\{1 - C_6 \exp(-C_7 \zeta^{\delta_0})\right\}^{1/2} .$$

Now setting $t_k = 2^k$, it suffices to show that $\sum P\{L^*(t_k) \leq 2^{-1}\zeta h(t_{k+1})\}$ converges. Using that $h(t_{k+1})/h(t_k) \leq 2$ for k sufficiently large, and (3.3), we have

$$P\left\{L^*(t_k) \leq 2^{-1}\zeta h(t_{k+1})\right\} \leq \exp(-\xi llt_k)$$

whose sum converges if ζ is small enough.

Next we prove the second half of the main result whose proof is much more complicated. We need a lemma to estimate the tail of the distribution of $L^*(t)$, mainly based on the results in section 2. Recall that we are assuming (1.1) and (1.2). The proof is rather complicated and contained in Wee[13].

Lemma 3.4 *For $\eta > 1$, $M > 0$, there exists C_8 such that for t large,*

$$P\left(L^*(\eta t/llt) \geq \frac{Mt}{f^{-1}(llt/t)llt}\right) < C_8\eta^{1/2}/M .$$

Finally, we need an estimate on the distribution of X_t, whose proof is also omitted here.

Lemma 3.5 *There exist* $0 < \theta_2 < 1$, $0 < \rho_1 < 1$, $\rho_2 > 1$, $\gamma > 1$ *such that for* t *sufficiently large,*

$$P\left(\rho_1 f^{-1}(1/\gamma t) \leq X_t \leq \rho_2 f^{-1}(1/\gamma t)\right) > \theta_2 .$$

Now we are ready to finish the proof of main result.

Theorem 3.2

$$\liminf_{t\to\infty} L^*(t) f^{-1}(llt/t) llt/t < \infty \quad a.s.$$

Proof. Let $h(t) = t/(f^{-1}(llt/t)\, llt)$ and $t_k = \exp(k^\lambda)$, $\lambda > 1$. We will use

$$\begin{aligned}\liminf_{t\to\infty} L^*(t)/h(t) \;\leq\; & \limsup_{k\to\infty} L^*(t_k)/h(t_{k+1}) \\ & + \liminf_{k\to\infty}(\sup_x(L(t_{k+1},x) - L(t_k,x)))/h(t_{k+1}) .\end{aligned} \tag{3.4}$$

To prove the lim sup in (3.4) is finite, let $\eta > 1$ be fixed and l be the smallest integer exceeding llt_k/η. Then we have, by Lemma 3.4,

$$\begin{aligned}P\{L^*(t_k) \geq Ch(t_{k+1})\} &\leq lP\{L^*(\eta t_k/llt_k) \geq Ch(t_{k+1})/l\} \\ &\leq C\eta^{1/2} l^2 h(t_k)/h(t_{k+1}) \\ &\leq C(\log k)^2 (t_k/t_{k+1})^{1-1/\delta_0}\end{aligned}$$

whose sum converges if $\lambda > 1$.

To prove that the lim inf in (3.4) is finite, let $s = \eta t/llt$, $\eta > 1$, $a = f^{-1}(1/\gamma s)$, $\rho = \rho_1$, and $A = 2\rho_2\rho_1^{-1} + 2$ where γ, ρ_1, ρ_2 are the constants appearing in Lemma 3.5. Following Griffin's method[7], set

$$\begin{aligned}E_k = \Big\{ & \sup_x (L(ks,x) - L((k-1)s,x)) \leq Mh(t), \\ & \sup_{0\leq u\leq s} |X_{u+(k-1)s} - X_{(k-1)s}| \leq \rho a,\ k\rho a \leq X_{ks} \leq (A+k)\rho a \Big\}\end{aligned}$$

and observe that, for $n = [llt/\eta] + 1$,

$$\bigcap_{k=1}^{n} E_k \subset \{L^*(t) \leq AMh(t)\} .$$

Note that

$$P\left(\bigcap_{k=1}^{n} E_k \mid \mathcal{F}_{(k-1)s}\right) = \prod_{K=1}^{n-1} \chi_{E_k} \cdot P\left\{\sup_x(L(ns,x) - L((n-1)s,x)) \le Mh(t), \sup_{0\le u\le s} |X_{u+(n-1)s} - X_{(n-1)s}| \le \rho a\right\} \cdot P\left(n\rho a \le X_{ns} \le (A+n)\rho a \mid X_{(n-1)s}\right) \text{ a.s.} \tag{3.5}$$

Applying Lemma 3.4 and Lemma 3.2, we have

$$\begin{aligned} & P\left\{\sup_x(L(ns,x) - L((n-1)s,x)) \le Mh(t)\,, \sup_{0\le u\le s} | X_{u+(n-1)s} - X_{(n-1)s} | \le \rho a\right\} \\ \ge\ & C_6 \exp(-C_7\rho^{-\delta}/\gamma) - C_8\eta^{1/2}/M \\ =\ & \theta_3 > 0 \end{aligned}$$

if M is large enough. Also note that by Lemma 3.5, for $x \in [(n-1)\rho a, (n-1+A/2)\rho a]$,

$$P\left(n\rho a \le X_{ns} \le (A+n)\rho a \mid X_{(n-1)s} = x\right) \ge \theta_2\,,$$

and similarly for $x \in [(n-1+A/2)\rho a, (A+n-1)\rho a]$. By taking the iterated conditional expectations of (3.5), we have

$$\begin{aligned} P(L^*(t) \le AMh(t)) &\ge (\theta_2\theta_3)^n \\ &\ge \exp(-2\xi llt/\eta)\,, \end{aligned} \tag{3.6}$$

where $\theta_2\theta_3 = e^{-\xi}$. Applying (3.6) we obtain divergence of sum of

$$P\left(\sup_x(L(t_{k+1},x) - L(t_k,x)) \le AMh(t_{k+1})\right)$$

for appropriate λ.

Acknowledgements

The author would like to thank Professor Hida for his kind invitation to this interesting and stimulating conference on Gaussian random field in Nagoya.

References

[1] M. T. Barlow, Continuity of local times for Lévy processes, *Z. Wahrsch. Verw. Gebiete.* **69**, 1985, 23-35.

[2] M. T. Barlow, Necessary and sufficient condition for the continuity of local times of Lévy processes, *Ann. Probab.* **16**, 1988, 1389-1427.

[3] M. T. Barlow and J. Hawkes, Application de l'entropie métrique a la continuité des temps locaux des processus de Lévy, *C. R. Acad. Sci. Ser. I. Math.* **301**, 1985, 237-239.

[4] A. N. Borodin, On the character of convergence to Brownian local time I, *Probab. Th. Rel. Fields.* **72**, 1986, 231-250.

[5] J. Bretagnolle, Resultats de Kesten sur les processus a accroisements indépendants, *Seminare de Probabilités V, Lecture Notes in Math.* **191** (Springer, Berlin, 1971). 21-36.

[6] R. K. Getoor and H. Kesten, Continuity of local times of Markov processes, *Compositio Math.* **24**, 1972, 277-303.

[7] P. S. Griffin, Laws of the iterated logarithm for symmetric stable processes, *Z. Wahrsch. Verw. Gebiete.* **68**, 1985, 271-285.

[8] N. C. Jain and W. E. Pruitt, Asymptotic behavior of the local time of a recurrent random walk, *Ann. Probab.* **12**, 1984, 64-85.

[9] H. Kesten, An iterated logarithm law for local time, *Duke Math. J.* **32**, 1965, 447-456.

[10] H. Kesten, Hitting probabilities of single points for processes with stationary independent increments, *Mem. Amer. Math. Soc.* **93**, 1969.

[11] D. B. Ray, Sojourn times of a diffusion process, *Ill. J. Math.* **7**, 1963, 615-630.

[12] I. S. Wee, Lower functions for processes with stationary independent increments, *Probab. Th. Rel. Fields.* **77**, 1988, 551-566.

[13] I. S. Wee, The law of the iterated logarithm for local time of a Lévy process. (preprint).

Constructing Kernels via Stochastic Measures *

J. A. Yan

(Institute of Applied Mathematics Academia Sinica, 100080 Beijing, China)

The construction of kernels is a fundamental problem in probability theory and in Markov process theory. Some deep results on this subject have been obtained by Getoor [4], Dellacherie [3] and Albeverio-Ma [1]. In this paper, we investigate this problem in more general case. We shall construct a kernel representing a stochastic Daniell integral on a vector lattice via stochastic measures. In §1 we establish stochastic versions of the Carathéodory theorem and the Daniell-Stone theorem. As an applicaton we obtain a generalization of a theorem of Getoor on the construction of kernels. A notion of Lusin-and Radon-type spaces is introduced in §2. It is shown that this notion is a natural extension of that of Lusin and Radon spaces. The main theorem on the construction of kernels is given in §3. This theorem generalizes the above mentioned theorem of Getoor to Lusin-and Radon-type spaces case.

§1. Stochastic Version of the Daniell-Stone Theorem

In this paper, by a *stochastic space* we mean a triple $(\Omega, \mathcal{F}, \mathcal{N})$, where $(\Omega, \mathcal{F})$ is a measurable space and $\mathcal{N} \subset \mathcal{F}$ is a collection of so-called "negligible sets". Namaly, $\mathcal{N}$ is closed under countable unions and hereditary in the sense that if $A \in \mathcal{N}$, $B \subset A$ and $B \in \mathcal{F}$ then $B \in \mathcal{N}$. In addition, we assume that $\Omega \notin \mathcal{N}$. A typical exemple of $\mathcal{N}$ is of the following form:

$$\mathcal{N} = \{A \in \mathcal{F} : P(A) = 0, \ \forall P \in \mathcal{P}\} \tag{1.1}$$

where $\mathcal{P}$ is a family of probability measures on $(\Omega, \mathcal{F})$.

Let $(\Omega, \mathcal{F}, \mathcal{N})$ be a stochastic space. A property depending on $\omega \in \Omega$ is said to be hold $\mathcal{N}$-a.e. (or $\mathcal{N}$-almost hold) if it holds on Ω except a negligible set. Let $L(\Omega, \mathcal{F})$ (resp. $\bar{L}(\Omega, \mathcal{F})$) denote the set of all real (resp. extended real) valued $\mathcal{F}$-measurable functions on Ω. Let $\mathcal{A}$ be an algerbra on a set E. By a $\mathcal{N}$ - *stochastic measure* ($\mathcal{N}$ - S.M.) on $\mathcal{A}$ we mean a map μ from $\mathcal{A}$ to $\bar{L}_+(\Omega, \mathcal{F})$, the positive part of $\bar{L}(\Omega, \mathcal{F})$, satisfying the following conditions: (i) $\mu(\phi) = 0$,

AMS Subject Classification. 60A10, 62G57

*The project supported by the National Natural Science Foundation of China.

$\mathcal{N}$-a.e.; (ii) if $(A_n) \subset \mathcal{A}$, $A_n \cap A_m = \phi, n \neq m$, and $A = \sum_{n=1}^{\infty} A_n \in \mathcal{A}$, then $\mu(A) = \sum_{n=1}^{\infty} \mu(A_n)$, $\mathcal{N}$-a.e.. If $\mathcal{N}$ is of the form (1.1), we call also a $\mathcal{N}$-S.M. $\mathcal{P}$-S.M.. A $\mathcal{N}$-S.M. μ is said to be *finite,* if $\mu(E) < \infty$, $\mathcal{N}$-a.e.. μ is said to be *σ-finite,* if there exists a countable partition (A_n) of E with $(A_n) \subset \mathcal{A}$ such that $\mu(A_n) < \infty$, $\mathcal{N}$-a.e. for all $n \geq 1$. Two S.M's μ_1 and μ_2 are said to be *$\mathcal{N}$-equivalent,* if $\mu_1(A) = \mu_2(A)$, $\mathcal{N}$-a.e. for each $A \in \mathcal{A}$.

Let $\mathcal{C}$ be a collection of subsets of E. In this paper we denote by $\mathcal{C}_\sigma$ (resp. $\mathcal{C}_\delta$) the collection of all countable unions (resp. countable intersections) of sets in $\mathcal{C}$. We denote by $\mathcal{C}_{\Sigma f}$ the collection of all finite unions of disjoint sets in $\mathcal{C}$.

Let $\mathcal{A}$ be an algebra on a set E and $(\Omega, \mathcal{F}, P)$ a probability space. Recall that any finite P-S.M. μ on $\mathcal{A}$ can be uniquely extended to a P-S. M. on $\sigma(\mathcal{A})$ (see for exemple Morando [5]). The following theorem is a generalization of this result. It can be considered as a stochastic version of the Carathéodory theorem.

Theorem 1.1. *Let $(\Omega, \mathcal{F}, \mathcal{N})$ be a stochastic space, where $\mathcal{N}$ is of the form (1.1). Let $\mathcal{A}$ be an algebra on a set E. Then any σ-finite $\mathcal{N}$-S.M. on $\mathcal{A}$ can be extended to a $\mathcal{N}$-S. M. on $\sigma(\mathcal{A})$. This extension is unique up to $\mathcal{N}$-equivalence.*

Proof. First of all, let μ be a finite $\mathcal{N}$-S. M. on $\mathcal{A}$. Since $\mathcal{N}$ is of the form (1.1), for each $P \in \mathcal{P}$ μ is a P-S.M. on $\mathcal{A}$. We denote by μ_p the unique P-S.M. on $\sigma(\mathcal{A})$ extending the P-S.M. μ on $\mathcal{A}$. For each $B \in \mathcal{A}_\delta$ we choose a fixed but arbitrary sequence $(B_n) \subset \mathcal{A}$ such that $B_n \downarrow B$ and we put

$$\bar{\mu}(B) = \limsup_{n \to \infty} \mu(B_n).$$

Set $\mathcal{B} = \{A \cap B^c : A, B \in \mathcal{A}_\delta\}$ and $\mathcal{A}' = \mathcal{B}_{\Sigma f}$. Then $\mathcal{B}$ is a semi-algebra on E and $\mathcal{A}'$ is an algebra on E. For each $C \in \mathcal{A}'$, we choose a fixed but arbitrary representation of C, say $C = \sum_{i=1}^{n}(A_i \cap B_i^c)$ with $A_i, B_i \in \mathcal{A}_\delta, 1 \leq i \leq n$, and we put

$$\mu'(C) = \sum_{i=1}^{n} \Big[\bar{\mu}(A_i) - \bar{\mu}(A_i \cap B_i)\Big].$$

Then it is easy to see that μ' is a unified version of μ_p on $\mathcal{A}'$, P runing over $\mathcal{P}$. Thus μ' is a $\mathcal{N}$-S. M. on $\mathcal{A}'$ which extends uniquely μ. Let $\mathcal{X} = \{\mathcal{A}_\alpha, \alpha \in \wedge\}$ be the cellection of those algebras $\mathcal{A}_\alpha$ on E such that $\sigma(\mathcal{A}) \supset \mathcal{A}_\alpha \supset \mathcal{A}$ and there exists a unified version μ_α of μ_p on $\mathcal{A}_\alpha$, P runing over $\mathcal{P}$. Then $\mathcal{X}$is a non-empty set. The inclusion relation $\subset$ is a partial order on $\mathcal{X}$. Let $\wedge' \subset \wedge$ be such that $\{\mathcal{A}_\alpha, \alpha \in \wedge'\}$ is linearly ordered. Put $\bar{\mathcal{A}}_\alpha = \bigcup_{\alpha \in \wedge'} \mathcal{A}_\alpha$. For each $A \in \bar{\mathcal{A}}$, we choose arbitrarily an $\mathcal{A}_\alpha$, $\alpha \in \wedge'$, such that $A \in \mathcal{A}_\alpha$ and put $\bar{\mu}(A) = \mu_\alpha(A)$. Then it is easy to see that $\bar{\mu}$ is a unified version of μ_p on $\bar{\mathcal{A}}$, P runing over $\mathcal{P}$. Thus $\bar{\mathcal{A}} \in \mathcal{X}$. By Zorn's lemma $\mathcal{X}$ has a maximal element, say $\mathcal{A}_{\alpha_0}$. We claim that $\mathcal{A}_{\alpha_0} = \sigma(\mathcal{A})$. In fact, assume that

$\mathcal{A}_{\alpha_0} \neq \sigma(\mathcal{A})$. Then $\mathcal{A}_{\alpha_0}$ is not a σ-algebra, because $\sigma(\mathcal{A})$ is the smallest σ-algebra containing $\mathcal{A}$. Set $\mathcal{C} = \{A \cap B^c : A, B \in (\mathcal{A}_{\alpha_0})_\delta\}$, $\mathcal{D} = \mathcal{C}_{\Sigma f}$. Then $\mathcal{D}$ is an algebra and we have $\mathcal{D} \supsetneq \mathcal{A}_{\alpha_0}$. Since μ_{α_0} is a finite $\mathcal{N}$-S.M. on $\mathcal{A}_{\alpha_0}$ and is a unified verion of μ_p on $\mathcal{A}_{\alpha_0}$, P runing over $\mathcal{P}$, there exists a $\mathcal{N}$-S.M. ν on $\mathcal{D}$ which is a unified verion of μ_p on $\mathcal{D}$, $P \in \mathcal{P}$, in view of the first step of the above proof. Therefore, we have $\mathcal{D} \in \mathcal{X}$. This is a contradiction. We have thus proved that $\mathcal{A}_{\alpha_0} = \sigma(\mathcal{A})$. Consequently, μ_{α_0} is a $\mathcal{N}$-S.M. on $\sigma(\mathcal{A})$ which is a unique extension of μ.

Now let μ be a σ-finite $\mathcal{N}$-S.M. on $\mathcal{A}$. Let (A_n) be a countable partition of E with $(A_n) \subset \mathcal{A}$ such that $\mu(A_n) < \infty$, $\mathcal{N}$-a.e. for all $n \geq 1$. Put

$$\bar{\mu}(A) = \sum_{n=1}^{\infty} \frac{1}{2^n} \frac{\mu(A_n \cap A)}{1 + \mu(A_n)}, \qquad A \in \mathcal{A}.$$

Then $\bar{\mu}$ is finite $\mathcal{N}$-S.M. on $\mathcal{A}$. By what proved above, $\bar{\mu}$ can be uniquely extended to a $\mathcal{N}$-S. M. on $\sigma(\mathcal{A})$. We denote it still by $\bar{\mu}$. Put

$$\nu(A) = \sum_{n=1}^{\infty} 2^n \bar{\mu}(A \cap A_n) \mu(A_n), \qquad A \in \sigma(\mathcal{A}).$$

Then ν is a $\mathcal{N}$-S.M. on $\sigma(\mathcal{A})$ extending μ on $\mathcal{A}$. The theorem is proved.

Before going to establish a stochastic version of the Daniell-Stone theorem we introduce some notions. Recall that a *vector latlice* on a set E is a vector space $\mathcal{H}$ of real functions on E with the property that if $f \in \mathcal{H}$ then $|f| \in \mathcal{H}$ and $f \wedge 1 \in \mathcal{H}$. Let $(\Omega, \mathcal{F}, \mathcal{N})$ be a stochastic space. By a *$\mathcal{N}$-stochastic Daniell integral* on $\mathcal{H}$ we mean a map T from $\mathcal{H}$ to $L(\Omega, \mathcal{F})$ satisfying the following conditions: (i) $f \geq 0$, $f \in \mathcal{H}, \Rightarrow T(f) \geq 0$, $\mathcal{N}$-a.e.; (ii) $T(\alpha f + \beta g) = \alpha T(f) + \beta T(g)$, $\mathcal{N}$-a.e., $\forall f, g \in \mathcal{H}$, $\alpha, \beta \in \mathbb{R}$; (iii) $(f_n) \subset \mathcal{H}_+, f_n \downarrow 0 \Rightarrow T(f_n) \to 0$, $\mathcal{N}$-a.e..

Let $(E, \mathcal{E})$ be a measurable space and μ a $\mathcal{N}$-stochastic measure on $\mathcal{E}$. Let f be a positive $\mathcal{E}$-measurable function on E. Set $f_n = \frac{1}{2^n} \sum_{k=0}^{\infty} 1_{[f > k/2^n]}$. Then $f_n \uparrow f$. Put

$$\mu(f) = \lim_{n \to \infty} \frac{1}{2^n} \sum_{k=0}^{\infty} \mu([f > \frac{k}{2^n}]). \tag{1.2}$$

The above limit exists $\mathcal{N}$-a.e., because we have

$$\frac{1}{2^n} \sum_{k=1}^{\infty} \mu([f > \frac{k}{2^n}]) \leq \frac{1}{2^{n+1}} \sum_{k=1}^{\infty} \mu([f > \frac{k}{2^{n+1}}]), \qquad \mathcal{N} - a.e..$$

We call $\mu(f)$ the *integral* of f w.r.t. μ. If f is a $\mathcal{E}$-measurable function such that one of $\mu(f^+)$ and $\mu(f^-)$ is finite $\mathcal{N}$-a.e., we put $\mu(f) = \mu(f^+) - \mu(f^-)$ and call $\mu(f)$ the *integral* of f w.r.t. μ

Let $(\Omega, \mathcal{F}, \mathcal{N})$ be a stochastic space and $\mathcal{H}$ a vector lattice on E. Let T be a $\mathcal{N}$-stochastic Daniell integral on $\mathcal{H}$ and μ a $\mathcal{N}$-S.M. on $\sigma(\mathcal{H})$. If for each $f \in \mathcal{H}$ one has $T(f) = \mu(f)$, $\mathcal{N}$-a.e., then μ is called a $\mathcal{N}$-S.M. *representing* T.

The following theorem is a stochastic version of the Daniell-Stone theorem.

Theorem 1.2. *Let $(\Omega, \mathcal{F}, \mathcal{N})$ be a stochastic space, where $\mathcal{N}$ is of the form (1.1). Let $\mathcal{H}$ be a vector lattice on E. Assume that there exists a sequence $(f_n) \subset \mathcal{H}_+$ such that $f_n \uparrow 1$ on E. then for any $\mathcal{N}$-stochastic Daniell integral T on $\mathcal{H}$ there exists a unique $\mathcal{N}$-S.M. μ on $\sigma(\mathcal{H})$ representing T.*

Proof. For each fixed $P \in \mathcal{P}$, we can regard T as a P-stochastic Daniell integral on $\mathcal{H}$. By using classical method of the measure theory with a minor modification consisting in replacing "inf" and "sup" by "ess.inf" and "ess.sup" we can prove that there exists a unique σ-finite P-S.M. μ_p on $\sigma(\mathcal{H})$ representing P-stochastic Daniell integral T on $\mathcal{H}$. In view of Theorem 1.1, in order to prove the theorem, it suffices to find an algebra $\mathcal{A}$ generating the σ-algebra $\sigma(\mathcal{H})$ and a unified version μ of μ_p on $\mathcal{A}$, P runing over $\mathcal{P}$.

The fellowing proof is inspired by Bauer [2]. Put

$$\mathcal{H}_+^* = \{f : \exists (f_n) \subset \mathcal{H}_+ \text{ such that } f_n \uparrow f\}.$$

For each $f \in \mathcal{H}_+^*$ we choose a fixed but arbitrary sequence $(f_n) \subset \mathcal{H}_+$ such that $f_n \uparrow f$ and we put

$$T^*(f) = \limsup_{n\to\infty} T(f_n).$$

Put

$$\mathcal{D} = \{A \subset E : I_A \in \mathcal{H}_+^*\}, \mathcal{D}_1 = \{A \in \mathcal{D} : T^*(I_A) < \infty, \mathcal{N} - a.e.\}.$$

Then $(\mathcal{D}_1)_\sigma = \mathcal{D}$ and $\sigma(D) = \sigma(\mathcal{H})$ (see Bauer [2]). Set

$$\mathcal{C} = \{A \cap B^c, A, B, \in \mathcal{D}\}, \qquad \mathcal{A} = \mathcal{C}_{\Sigma f}.$$

Then $\mathcal{A}$ is an algebra on E. We put

$$\bar{\mu}(A) = T^*(I_A), \qquad A \in \mathcal{D}.$$

Then it is easily seen that $\bar{\mu}$ is a unified version of μ_p on $\mathcal{D}$, P runing over $\mathcal{P}$. Let $C \in \mathcal{C}$. We choose arbitrarily a representation of C, say $C = A \cap B^c$ with $A, B \in \mathcal{D}$, and then choose a sequence $(A_n) \subset \mathcal{D}_1$, such that $A_n \uparrow A$. We put

$$\mu(C) = \limsup_{n\to\infty} [\bar{\mu}(A_n) - \bar{\mu}(A_n \cap B)].$$

Since for each $P \in \mathcal{P}$, we have obviously

$$\mu_p(C) = \lim_{n\to\infty} \mu_p(A_n \cap B^c) = \lim_{n\to\infty} [\mu_p(A_n) - \mu_p(A_n \cap B^c)], \qquad P - a.e.,$$

μ is a unified version of μ_p on $\mathcal{C}$, P runing over $\mathcal{P}$. Now for each $A \in \mathcal{A}$, we choose arbitrarily a representation of A, say $A = \sum_{i=1}^{n} C_i$ with $(C_i) \subset \mathcal{C}$, $C_i \cap C_j = \phi, i \neq j$. We put

$$\mu(A) = \sum_{i=1}^{n} \mu(C_i).$$

Then μ is a unified versionof μ_p on $\mathcal{A}$, P runing over $\mathcal{P}$. Consequently, μ is a $\mathcal{N}$-stochastic measure on $\mathcal{A}$. μ is obviously σ-finite. Therefore, by Theorem 1.1 μ can be uniquely extended to a $\mathcal{N}$-stochastic measure on $\sigma(\mathcal{A}) = \sigma(\mathcal{H})$. Since $\mu_p(f) = T(f)$, P-a.e., $\forall f \in \mathcal{H}$ and μ is a unified versionof μ_p on $\sigma(\mathcal{A})$, P runing over $\mathcal{P}$, we have $\mu(f) = T(f), \mathcal{N}$-a.e., $\forall f \in \mathcal{H}$. The theorem is proved.

As an application of Theorem 1.2, we obtain the following generalization of a theorem of Getoor on the construction of kernels. Recall that a *kernel* from $(\Omega, \mathcal{F})$ to $(E, \mathcal{E})$ is a positive function $K(\omega, A)$ defined on $\Omega \times \mathcal{E}$ such that $K(\cdot, A)$ is $\mathcal{F}$-measurable for each $A \in \mathcal{E}$ and $K(\omega, \cdot)$ is a measure on $\mathcal{E}$ for each $\omega \in \Omega$.

Theorem 1.3. *Let $(\Omega, \mathcal{F}, \mathcal{N})$ be a stochastic space, where $\mathcal{N}$ is of the form (1.1). Let $\mathcal{H}$ be a vector lattice on E. Assume that there exists a sequence $(f_n) \subset \mathcal{H}_+$ such that $f_n \uparrow 1$ on E. Let T be a $\mathcal{N}$-stochastic Daniell integral on $\mathcal{H}$. If $(E, \sigma(\mathcal{H}))$ is a Radon space, then there exists a kernel K from $(\Omega, \mathcal{F}^u)$ to $(E, \sigma(\mathcal{H}))$ such that for each $f \in \mathcal{H}$ one has $T(f) = K(\cdot, f)$, $\mathcal{N}^u$-a.e., where $\mathcal{F}^u$ stands for the universal completion of $\mathcal{F}$ and $\mathcal{N}^u = \{A \in \mathcal{F}^u : P(A) = 0, \forall P \in \mathcal{P}\}$. If in addition $(E, \sigma(\mathcal{H}))$ is a Lusin space, or if $\mathcal{P}$ is a countable family, then K may be constructed as a kenel from $(\Omega, \mathcal{F})$ to $(E, \sigma(\mathcal{H}))$.*

Proof. According to Theorem 1.2, there exists a σ-finite $\mathcal{N}$-S.M. μ on $\sigma(\mathcal{H})$ representing T. Let (A_n) be a partition of E with $(A_n) \subset \sigma(\mathcal{H})$ such that $\mu(A_n) < \infty$, $\mathcal{N}$-a.e. for all $n \geq 1$. put

$$\bar{\mu}(A) = \sum_{n=1}^{\infty} \frac{1}{2^n} \frac{\mu(A_n \cap A)}{1 + \mu(A_n)}, \qquad A \in \sigma(\mathcal{H}).$$

Then $\bar{\mu}$ is a $\mathcal{N}$-S.M. on $\sigma(\mathcal{H})$ and $\bar{\mu}(E) \leq 1$. Under the assumptions on $(E, \sigma(\mathcal{H}))$, by a theorem of Getoor [4] there exists a kernel $\bar{K}$ representing $\bar{\mu}$ in a self-evident sense. We put

$$K(\omega, A) = \sum_{n=1}^{\infty} 2^n \bar{K}(\omega, A \cap A_n) \mu(A_n)(\omega), \qquad A \in \sigma(\mathcal{H}).$$

Then K is a desired kernel.

§2. Lusin-type and Radon-type Spaces

In order to extending Getoor's results on the construction of kernels to non-separable measurable space case, we propose a notion of Lusin-type and Radon-type spaces as follows.

Definition 2.1. *Let $(E, \mathcal{E})$ be a measurable space. If for each separable sub-σ-algebra $\mathcal{E}_0$ of $\mathcal{E}$ there exists a compact class C on E such that $C \subset \mathcal{E}$ and C_σ contains an algebra (or equivalently, a countable algebra) $\mathcal{A}$ with the property that $\sigma(\mathcal{A}) \supset \mathcal{E}_0$, then we call $(E, \mathcal{E})$ a Lusin-type space.*

Recall that a collection $\mathcal{C}$ of subsets of E is called a compact class if it satisfies the following conditions: $(C_n) \subset \mathcal{C}, \bigcap_{n=1}^{\infty} C_n = \phi \Rightarrow \bigcap_{n=1}^{m} C_n = \phi$ for some integer m.

A measurable space $(E, \mathcal{E})$ is called a *Radon-type space,* if there exists a Lusin-type space $(E', \mathcal{E}')$ and a universally measurable subset E_0 of E' such that $(E, \mathcal{E})$ is isomorphic to $(E_0, E_0 \cap \mathcal{E}')$.

Remark 1. Let $(E, \mathcal{E})$ be a measurable space. If there exists a compact class $\mathcal{C}$ on E such that $\mathcal{C} \subset \mathcal{E}$ and $\mathcal{C}_\sigma$ contains an algebra $\mathcal{A}$ generating $\mathcal{E}$, then $(E, \mathcal{E})$ is obviously a Lusin-type space.

If $(E, \mathcal{E})$ is a separable measurable space, then $(E, \mathcal{E})$ is a Lusin-type space iff it verifies the above mentioned property.

Remark 2. $(R, \mathcal{B}(R))$ is obviously a Lusin-type space. Therefore our notion of Lusin-type (resp. Radon-type) space generalizes that of Lusin (resp. Radon) space.

Lemma 2.1. *Let $(E_n, \mathcal{E}_n)$ be a sequence of Lusin-type (resp. Radon-type) spaces. Then the product measurable space $(\prod_n E_n, \prod_n \mathcal{E}_n)$ is also a Lusin-type (resp. Radon-type) space.*

Proof. We only need to consider the Lusin-type space case. Assume that $\mathcal{F}$ is a separable sub-σ-algebra of $\prod_n \mathcal{E}_n$. It is easy to prove that for each n there exists a separable σ-algebra $\mathcal{F}_n \subset \mathcal{E}_n$ such that $\mathcal{F} \subset \prod_n \mathcal{F}_n$. For each n we take a compact class $\mathcal{C}_n$ on E_n such that $\mathcal{C}_n \subset \mathcal{E}_n$ and $(\mathcal{C}_n)_\sigma$ contains an algebra $\mathcal{A}_n$ with the property that $\sigma(\mathcal{A}_n) \subset \mathcal{F}_n$. We may assume that each $\mathcal{C}_n$ is closed under countable intersetions. Put

$$\mathcal{D} = \{C_n \times \prod_{i \neq n}^{\infty} E_i,\ n \geq 1,\ C_n \in \mathcal{C}_n\}$$

Then $\mathcal{D}$ is a compact class on $\prod_{i=1}^{\infty} E_i$. Put

$$\mathcal{C} = \{\bigcap_{i=1}^{n} D_i,\ n \geq 1,\ D_i \in \mathcal{D}\}.$$

Then $\mathcal{C}$ is a compact class on $\prod_{i=1}^{\infty} E_i$, and we have $\mathcal{C} \subset \prod_{i=1}^{\infty} \mathcal{E}_i$. Moreover, $\mathcal{C}_\sigma$ contains an algebra which generates a σ-algebra containing $\prod_n \mathcal{F}_n$, a priori, containing $\mathcal{F}$. Thus, by Definition 2.1, $(\prod_n E_n, \prod \mathcal{E}_n)$ is a Lusin-type space.

The following lemma gives us a typical example of Lusin-type space.

Lemma 2.2. *Let E be a σ-compact and locally compact Hausdorff space. Then $(E, \mathcal{B}_0(E))$ is a Lusin-type space, where $\mathcal{B}_0(E)$ is the strong Baire σ-algebra on E, which is generated by the collection of all continuous functions with compact support on E.*

Proof. Let $\mathcal{G}$ (resp. $\mathcal{K}$) denote the collection of all open (resp. comact) subsets of E. Set $\mathcal{C} = \mathcal{K} \cap \mathcal{G}_\sigma$ and $\mathcal{D} = \mathcal{G} \cap \mathcal{K}_\sigma$. Then it is well known that one has $\sigma(\mathcal{D}) = \sigma(\mathcal{C}) = \mathcal{B}_0(E)$ and

$\mathcal{D} \subset \mathcal{C}_\sigma$ (See Bauer [2]). Put

$$\mathcal{D}_1 = \{A \cap B^c : A, B \in \mathcal{D}\}, \ \mathcal{A} = \{\bigcup_{i=1}^{n} A_i, \ n \geq 1, \ A_i \in \mathcal{D}_1, \ i \leq i \leq n\}.$$

Assume $D = A \cap B^c \in \mathcal{D}_1$, where $A, B \in \mathcal{D}$. Choose a sequence $(C_n) \subset \mathcal{C}$ and a sequence $(B_n) \subset \mathcal{K}$ such that

$$A = \bigcup_{n=1}^{\infty} C_n, \ B = \bigcup_{n=1}^{\infty} B_n.$$

For each C_n we choose a sequence $(G_{ni, \ i \geq 1}) \subset \mathcal{G}$ such that $C_n = \bigcap_{i=1}^{\infty} G_{ni}$. Then we have

$$A \cap B^c = \bigcup_{n=1}^{\infty} [C_n \cap (\bigcap_{m=1}^{\infty} B_m^c)] = \bigcup_{n=1}^{\infty} [\bigcap_{i,m} (G_{ni} \cap B_m^c)].$$

Since $C_n \cap B^c \in \mathcal{K}, G_{ni} \cap B_m^c \in \mathcal{G}$, we see from the above expression that $C_n \cap B^c \in \mathcal{C}$. Thus, $A \cap B^c \in \mathcal{C}_\sigma$. Consequently, $\mathcal{A} \subset \mathcal{C}_\sigma$. According to the above Remark 1 we conclude that $(E, \mathcal{B}_0(E))$ is a Lusin-type space.

§3 Construction of Kenels Representing $\mathcal{N}$-Stochastic Daniell Integrals

In this section we shall use a compact class argument to construct a kernel representing a $\mathcal{N}$-stochastic Daniell integral on a vector lattice. The key-stone for this construction is Theorem 1.2.

The following lemma is well known. For reader's convenience we give its proof here.

Lemma 3.1. *Let $\mathcal{A}$ and $\mathcal{A}_1$ to two rings on E such that $\mathcal{A}_1 \supset \mathcal{A}$. Let μ be a finite positive function defined on $\mathcal{A}_1$ such that $\mu(\phi) = 0$ and μ is finitely additive in the sense that if $A, B \in \mathcal{A}_1$ and $A \cap B = \phi$ then $\mu(A \cup B) = \mu(A) + \mu(B)$. Assume that there exists a compact class $\mathcal{C}$ on E such that $\mathcal{C} \subset \mathcal{A}_1$ and*

$$\mu(A) = sup\{\mu(C) : C \subset A, C \in \mathcal{C}\}, \ \forall A \in \mathcal{A}.$$

Then the restriction of μ to $\mathcal{A}$ is a measure on $\mathcal{A}$.

Proof. Let $A_n \in \mathcal{A}$ be such that $A_n \downarrow \phi$. It suffices to prove that $\mu(A_n) \to 0$. For given $\mathcal{E} > 0$, there exists for each n a $C_n \in \mathcal{C}$ such that $C_n \subset A_n$ and $\mu(A_n) \leq \mu(C_n) + \mathcal{E}/2^n$. Since $\bigcap_n C_n \subset \bigcap_n A_n = \phi$, there exists an integer m such that $\bigcap_{n=1}^{m} C_n = \phi$. Therefore, for $k \geq m$, we have

$$A_k \subset A_m = \bigcap_{n=1}^{m} A_n = (\bigcap_{n=1}^{m} A_n) \cap (\bigcup_{n=1}^{m} C_n^c) \subset \bigcup_{n=1}^{m} (A_n \setminus C_n),$$

form which it follwos that

$$\mu(A_k) \leq \sum_{k=1}^{m} \mu(A_n \setminus C_n) < \mathcal{E}.$$

The lemma is proved.

Let $(\Omega, \mathcal{F}, \mathcal{N})$ be a stochastic space and $(E, \mathcal{E})$ a measurable space. Two kernels K_1 and K_2 from $(\Omega, \mathcal{F})$ to $(E, \mathcal{E})$ are said to be *$\mathcal{N}$-equivalent* if there exists a set $N \in \mathcal{N}$ such that $K_1(\omega, \cdot) = K_2(\omega, \cdot)$ for each $\omega \in \Omega \setminus N$. Let μ be a $\mathcal{N}$-S. M. on $\mathcal{E}$ and K a kernel from $(\Omega, \mathcal{F})$ to $(E, \mathcal{E})$. If $K(\cdot, A) = \mu(A)$, $\mathcal{N}$-a.e., for each $A \in \mathcal{E}$, we call K a *kernel representing* μ.

The following theorem generalizes a theorem of Getoor [4] to Lusin and Radon-type spaces case.

Theorem 3.1. *Let $(\Omega, \mathcal{F}, \mathcal{N})$ be a stochastic space and $(E, \mathcal{E})$ a measurable space. Let μ be a σ-finite $\mathcal{N}$-stochastic measure on $\mathcal{E}$.*

1) If $(E, \mathcal{E})$ is a Lusin-type space, then for each separable sub-σ-algebra $\mathcal{E}_0$ of $\mathcal{E}$ there exists a kernel K form $(\Omega, \mathcal{F})$ to $(E, \mathcal{E}_0)$ representing μ on $\mathcal{E}_0$. Such a kernel K is unique up to $\mathcal{N}$-equivalence.

2) If $(E, \mathcal{E})$ is a Radon-type space and $\mathcal{N}$ is of the form (1.1), then for each separable sub-σ-algebra $\mathcal{E}_0$ of $\mathcal{E}$ there exists a kernel K from $(\Omega, \mathcal{F}^u)$ to $(E, \mathcal{E}_0)$ such that $\mu(A) = K(\cdot, A)$, N^u-a.e., for each $A \in \mathcal{E}_0$, where $\mathcal{F}^u$ and $\mathcal{N}^u$ are defined in Theorem 1.3. If in addition $\mathcal{P}$ is a countable family, then K may be constructed as a kernel from $(\Omega, \mathcal{F})$ to $(E, \mathcal{E}_0)$, and we have $\mu(A) = K(\cdot, A)$, $\mathcal{N}$-a.e. for each $A \in \mathcal{E}_0$.

Proof. We may assume that μ is a finite $\mathcal{N}$-S. M. (see the proof of Theorem 1.3). To prove 1), we take a compact class $\mathcal{D} \subset \mathcal{E}$ such that $\mathcal{D}_\sigma$ contains a countable algebra $\mathcal{A} = (A_n)$ with the property that $\sigma(\mathcal{A}) \supset \mathcal{E}_0$. For each i, we choose a sequence $(C_{ik}, k \geq 1) \subset \mathcal{D}$ such that $C_{ik} \uparrow A_i, (k \to \infty)$. Let $\mathcal{C} = \{C_{ik,i,k\geq 1}\}$ and let $\mathcal{A}_1$ be the algebra generated by $\mathcal{A} \cup \mathcal{C}$. Then it is easily seen that there exists a set $N \in \mathcal{N}$ such that for each $\omega \in \Omega \setminus N$ one has $\mu(E)(\omega) < \infty$ and $\mu(\cdot)(\omega)$ satisfies the conditions of Lemma 3.1. Thus, for each $\omega \in \Omega \setminus N$, $\mu(\cdot)(\omega)$ is a finite measure on A. Now, for $\omega \in \Omega \setminus N$, let $K(\omega, \cdot)$ denote the unique measure on $\sigma(\mathcal{A})$ extending $\mu(\cdot)(\omega)$ on $\mathcal{A}$, and for $\omega \in N$, put $K(\omega, \cdot) = 0$. Then by the monotone class theorem K is a kernel from $(\Omega, \mathcal{F})$ to $(E, \sigma(\mathcal{A}))$ representing μ on $\sigma(\mathcal{A})$. The restriction of K to $\mathcal{E}_0$ give us a desired kernel. Since $\mathcal{E}_0$ is separable, the uniqueness (up to $\mathcal{N}$-equivalence) of such a kernel is an easy consequence of the monotone class theorem.

Let's now prove 2). We may assume that there exists a Lusin-type space $(G, \mathcal{G})$ such that $E \in \mathcal{G}^u$ and $\mathcal{E} = E \cap \mathcal{G}$. Assume that the sequence (A_n) generates $\mathcal{E}_0$. For each n, we choose a $B_n \in \mathcal{G}$ such that $B_n \cap E = A_n$. Let $\mathcal{G}_0$ denote the σ-algebra on G generated by (B_n). Then we have $\mathcal{E}_0 \subset E \cap \mathcal{G}_0$. Put

$$\bar{\mu}(A) = \mu(A \cap E), \; A \in \mathcal{G}.$$

Then $\bar{\mu}$ is a finite $\mathcal{N}$-stochastic measure on $\mathcal{G}$. According to 1) there exists a kernel $\bar{K}$ from $(\Omega, \mathcal{F})$ to $(G, \mathcal{G}_0)$ representing $\bar{\mu}$. Moreover, for each $\omega \in \Omega$, $\bar{K}(\omega, G) < \infty$. Therefore, $\bar{K}$ can be uniquely extended to a kernel $\bar{K}^u$ from $(\Omega, \mathcal{F}^u)$ to $(G, \mathcal{G}_0^u)$. Set $\Omega_0 = \{\omega \in \Omega : \bar{K}^u(\omega, G \setminus E) =$

$0\}$. Then $P(\Omega_0) = 1, \forall P \in \mathcal{P}$. Put

$$K(\omega, A) = 1_{\Omega_0}(\omega)\bar{K}^u(\omega, A), \qquad A \in \mathcal{E}_0.$$

Then K is a kernel from $(\Omega, \mathcal{F}^u)$ to $(E, \mathcal{E}_0)$ and for each $A \in \mathcal{E}_0$ one has $\mu(A) = K(\cdot, A)$, $\mathcal{N}^u$-a.e. Finally, if $\mathcal{P}$ is a countable family, then $\mathcal{N}$ can be defined by a single probability measure. Thus we may assume that $\mathcal{P}$ consists of a single probability measure P. Put

$$\lambda(A) = \int \frac{\bar{\mu}(A)(\omega)}{1 + \bar{\mu}(G)(\omega)} P(d\omega), \; A \in \mathcal{G},$$

then λ is a finite measure on $(G, \mathcal{G})$. Since $E \in \mathcal{G}^u$, we can find a set $E_0 \in \mathcal{G}$ such that $E_0 \subset E$ and $\lambda(E_0) = \bar{\lambda}(E)$, where $\bar{\lambda}$ stands for the measure on $\mathcal{G}^u$ which extends uniquely the measure λ. In view of Lemma 2.1, $(E_0, E_0 \cap \mathcal{G})$ is a Lusin-type space. Since now $E_0 \in \mathcal{E}$, we can put

$$\mu_0(A) = \mu(A), \; A \in E_0 \cap \mathcal{E}.$$

and we have actually $E_0 \cap \mathcal{G} = E_0 \cap \mathcal{E}$. For $E_0 \cap \mathcal{E}_0$ applying the result of part 1 we get a kernel K_0 from $(\Omega, \mathcal{F})$ to $(E_0, E_0 \cap \mathcal{E}_0)$ representing μ_0 on $E_0 \cap \mathcal{E}_0$. Finally, put

$$K(\omega, A) = K(\omega, A \cap E_0), \; A \in \mathcal{E}_0.$$

Then it is easy to see that K is a kernel from $(\Omega, \mathcal{F})$ to $(E, \mathcal{E}_0)$ which represents μ on $\mathcal{E}_0$. The theorem is proved.

From Theorems 1.2 and 3.1 we get immediately the following theorem, the statement about the uniqueness being a consequence of the monotone class theorem.

Theorem 3.2. *Let $(\Omega, \mathcal{F}, \mathcal{N})$ be a stochastic space, where $\mathcal{N}$ is of the form (1.1). Let $\mathcal{H}$ be a vector lattice on E and let T be a $\mathcal{N}$-stochastic Daniell integral on $\mathcal{H}$. Assume that there exists a sequence $(f_n) \subset \mathcal{H}_+$ such that $f_n \uparrow 1$ on E and that $(E, \sigma(\mathcal{H}))$ is a Radon-type space. Then for each sub vector lattice $\mathcal{H}_0$ of $\mathcal{H}$ such that $\sigma(\mathcal{H}_0)$ is separable, there exists a unique kernel K from $(\Omega, \mathcal{F}^u)$ to $(E, \sigma(\mathcal{H}_0))$ such that for each $f \in \mathcal{H}_0$ one has $T(f) = K(\cdot, f)$, $\mathcal{N}^u$-a.e.. If $(E, \sigma(\mathcal{H}))$ is a Lusin-type space, or if $\mathcal{P}$ is a countable family, then K may be constructed as a kernel from $(\Omega, \mathcal{F})$ to $(E, \sigma(\mathcal{H}_0))$. In this case, for each $f \in \mathcal{H}_0$ one has $T(f) = K(\cdot, f)$, $\mathcal{N}$-a.e..*

Acknowledgement. This work was done during my visit to Institut de Mathematique, Université Louis Pasteur (Strasbourg) in May and June 1990. It was presented at the Nagoya conference on Gaussian random fields in August 1990. I would like to express my deepest gratitude to Professors P. A. Meyer and T.Hida for their kind invitation and hospitality. I am also grateful to Professors Z. M. Ma, P.A. Meyer and G. Mokobdzki for helpful discussions. The financial support by the Kajima Foundation of Japan is also acknowledged.

References

[1] S. Albeverio, Z. M. Ma, A note on quasicontinuous kernels representing quasi-linear positive maps, 1990 (preprint)

[2] H. Bauer, Wasrscheinlichkeits theorie and Grundzüge der Maßtheorie, Walter de Gruyter & Co. Berlin 1968

[3] C. Dellacherie, Sur la construction de noyaux boréliens, Sém. Probab. X, LN in Math. 511, Springer, 1976, 545-577.

[4] R. K. Getoor, On the construction of kernels, Sém. Probab. 1X, LN in Math. 465, Springer, 1975, 443-463.

[5] Ph. Morando: Mesures aléatoires, Sém. Probab. III, LN in Math. 88, Springer, 1969, 190-229.

INFINITE-DIMENSONAL ROTATION GROUP AND BROWNIAN MOTION

HISAAKI YOSHIZAWA

Department of Applied Mathematics
Okayama Sience University
Okayama 700, Japan

Abstract. A continuously parametrized series of representations of the full rotation group of Hilbert space is constructed. A discrete series of representations, realized by homogeneous chaos of N.Wiener is presented for the sake of introduction to the continuous series. Their relations with Brownian motion are also mentioned.

§1. Introduction and Basic Notions

1.1 *Outline and Background*

The purpose of the present report is to consider some types of representations of the rotation group of the Hilbert space, which we define in 1.4 and denote by O_∞, and their connections with the Brownian motion.

A series of irreducible unitary representations of the group O_∞ is connected both with the homogeneous chaos of N. Wiener[1] in 1938 and with Fock representations[2].

In §2 we shall give a brief sketch about a theory of this class of representations. It is known now, but it plays a kind of introduction to constructiong another class of representations in §4. For this purpose, it is necessary and convenient for us to consider some kind of continuous one-parameter subgroups of O_∞ such as the group of shift transformations of the Brownian motion (§3). Thus in this report we do not express O_∞ as a projective limit of rotation groups of finite dimensions. This is a continuation of the viewpoint adopted in the report of 1969[3].

1.2. *Rigged Hilbert space*

Let H be a real Hilbert space with the scalar product $< \varphi, \psi >$ and the norm $| \varphi |$. Let E be a topological vector space satisfying the condition that it is contained in H as a dense subspace and the topology of E is stronger than the norm topology of H, and moreover that the topology of E is given by countable number of Hilbert norms and E is nuclear. Let E^* be the conjugate space (equipped with the strong topology) of E. Then we can assume that

$$E \subset H \subset E^*.$$

We denote elements of E by $\xi, \eta, \cdots$ and those of E^* by $x, y, \cdots$; the canonical bilinear form on $E \times E^*$ is denoted by $< \xi, x >$. Thus it is a *rigged* Hilbert space and the triple E, H, E^* is a *troika* defined by I. Gelfand[4].

For simplicity we confine ourselves to the special case where H is the Hilbert space $L^2(-\infty, \infty)$ over the real line and E is the Schwartz space $\mathcal{S}$. Then the shift

$$\tau(t) : \xi(s) \to \xi(s-t), (-\infty < t < \infty)$$

leaves the space E stable.

1.3. *Brownian Motion*

Using the troika of spaces, we can define Gauss measure on E^* as follows. Put

$$C(\xi) = \exp(- | \xi |^2).$$

Then $C(\xi)$ is a continuous positive-definite function on E; hence, by the theorem of Bochner-Minlos-Sozanov[4], there exists a probability measure μ on the weak Borel field of E^* such that

$$C(\xi) = \int e^{i<\xi,x>}\mu(dx),$$

where the integration is taken over E^*. We call μ the *Gauss measure*, (we do not need the variance parameter).

Now the shift $(\tau(t))$ defines a one-parameter group of measure-preserving transformations on the probability space (E^*, μ). We call it the flow of the Brownian motion or simply *Brownian motion*. Needless to say, it is nothing else but the flow defined by the Wiener's Brownian motion.

1.4. *Rotation Group of Infinite Dimension*

We denote by O_∞ the set of all such operators g on H that are orthogonal and leave E stable. Such g are shown to be homeomorphisms of E. They constitute a transformation group on E and also on E^* canonically. We shall define topology of O_∞ in §3. The group O_∞ is an analogue to the rotation group $SO(3)$ or $O(3)$ in the 3-dimensional space.

Instead of the rotation group, the unitary group U_∞ on the complex Hilbert space is sometimes more convenient and more important theoretically in general, as is the case of finite-dimensionality. But we do not enter this consideration in this report, since the range of our problems here is too restricted to O_∞ to be investigated by the method of U_∞.

§2. Homogeneous Chaos

The homogeneous chaos of Wiener combined with the group O_∞ and the spherical harmonics of Laplace combined with the rotation group show striking resemblance, which we outline in this §. After Wiener[1] several related problems were considered, in particular by S. Kakutani[5,6] and K.Itô[7,8].The following line of description is along the present reporter's research which was inspired by articles[1,5,7,9] and reported to the 1961 meeting of Mathematical Society of Japan.

2.1. *Spheres*

We shall use the following notations: The group O_3 of rotations around the origin of the three-dimensional space E_3; the unit sphere S with the center at the origin, and the uniform probability measure m on S.

Then we have the following dual characterization.

(1° S) *The uniform measure m on S is characterized by O_3-invariance and O_3-ergodicity.*

(1° G) *Gauss measure μ in the space E^* is characterized by O_∞-invariance and O_∞ -ergodicity.*(See Umemura[10])

2.2 *Spherical Harmonics*

The following facts of the classical spherical harmonics are well-known.

(2° S) *Let $f(x)$ be a harmonic polynomial on E_3 with complex coefficients and of ℓ -th degree. Let E_ℓ be the linear set constituted of these polynomials (plus the constant 0) in the Hilbert space $L^2(S,m) = L^2(E,m)$. E_ℓ is the spherical harmonics of order ℓ. We have*

$$L^2(S,m) = \sum_{\ell=0}^{\infty} \oplus E_\ell.$$

(3° S) *For* $\ell = 0, 1, \cdots, E_\ell$ *is the eigenspace of the spherical Laplacian*

$$-\Lambda = \frac{1}{\sin\theta}\frac{\partial}{\partial\theta}(\sin\theta\frac{\partial}{\partial\theta}) + \frac{1}{\sin^2\theta}\frac{\partial^2}{\partial\varphi^2}.$$

The spherical functions of Laplace constitute an orthogonal base of E_ℓ :

$$Y_\ell^m(\theta, \varphi) = e^{im\varphi}P_\ell^m(\cos\theta), (-\ell \leq m \leq \ell).$$

The perfectly analogous propositions hold in the case of (E^*, μ). We define first a polynomial $f(x)$ on E^* as follows:

$$f(x) = P(<\xi_1, x>, \cdots, <\xi_p, x>),$$

where $P(t_1, \cdots, t_p)$ is a polynomial of real variables $t_1, \cdots, t_p$ with complex coefficients and $\xi_1, \cdots, \xi_p \in E$. Let $E_c^* = E^* + iE^*$ be the complexification of E^*. Then every polynomial on E^* is canonically extended to E_c^*. Now denote by $\mathcal{M}_n$ the linear set of homogeneous polynomials of degree n.

(2° G) *Every polynomial belongs to* $L^2(E^*, \mu)$, *and the Wiener transform of* $f(x)$ *of* $\mathcal{M}_n$

$$\tilde{f}(y) = \int f(y + ix)\mu(dx) \ (\textit{integrated over } E^*)$$

belongs to $\mathcal{M}_n, (n = 0, 1, \cdots)$. *Denote by* $\mathcal{H}_n$ *the closure of the Wiener transform of* $\mathcal{M}_n$. Then

$$L^2(E^*, \mu) = \sum_{n=0}^{\infty} \oplus \mathcal{H}_n.$$

The function space $\mathcal{H}_n$ defined above is nothing else but the homogeneous chaos of degree n.

Now we shall consider a form on E, which corresponds to the spherical Laplacian Λ on the sphere S. Let $(\xi_1, \xi_2, \cdots)$ be an orthogonal base of E, and E_n be the subspace spanned by $(\xi_1, \cdots, \xi_n)$. We denote independent variables in E_n also by $\xi_1, \cdots, \xi_n$. Then it is known[11] that there exists a form Δ_∞ on E such that

$$\Delta_\infty = \lim_{n \to \infty} \frac{1}{n} \Lambda_n$$

in a certain sense, where Λ_n are the spherical Laplacians in E_n. Moreover, when Δ_∞ is restricted to a single variable, say ξ_1, we have

$$\Delta_\infty = \frac{d^2}{d\xi_1^2} - \xi_1 \frac{d}{d\xi_1}.$$

We can consider Δ_∞ an operator in $L^2(E^*, \mu)$.

(3° G) *Each $\mathcal{H}_n (n = 0, 1, \cdots)$ is an eigenspace of the operator Δ_∞. In particular, $\mathcal{H}_n$ contains Hermite polynomial $H_n(\xi_1)$, which is Wiener transform of the monomial of degree n.*

Proofs of these propositions are carried out in non-computational way. The above operator Δ_∞ coincides with Laplacian of Lévy[12]. Interesting investigations[13] on these Laplacians are being reported recently.

2.3 *Irreducible Representations*

We shall consider unitary representations of O_3 and O_∞. Let

$$R(h) : f(s) \longrightarrow f(sh), f \in L^2(S, m), h \in O_3;$$

and
$$T(g) : f(x) \longrightarrow f(xg), f \in L^2(E^*, \mu), g \in O_\infty.$$

They are unitary operators and they constitute *unitary representations* of O_3 and O_∞, respectively: $T(g_1 g_2) = T(g_1)T(g_2)$, and the mapping

$$O_\infty \in g \longrightarrow T(g)f$$

is continuous for each f.

(4° S) *The spherical Laplacian* Λ *commutes with every operator* $R(h)$. *Conversely, any (closed) linear operator which commutes with every* $R(h)$ *is expressed by a polynomial, or more generally, by some function of* Λ.

From this fact immediately follow the fact that the restriction $R_\ell = (R_\ell, E_\ell)$ of the representation $(R, L^2(S))$ to the subspace E_ℓ is a representation of O_3, and also the following proposition.

(5° S) *Each representation* $R_\ell = (R_\ell, E_\ell)$ *of* O_3 *is irreducible.*

The identical propositions hold also for O_∞.

(4° G) *The operator* Δ_∞ *on* E^* *commutes with every* $T(g)$. *Conversely, any operator which commutes with all* $T(g)$ *is a function of* Δ_∞.

(5° G) *Each representation* $\mathcal{T}_n = (T_n, \mathcal{H}_n)$ *of* O_∞, *that is the restriction of* $(T(g), L^2(E^*))$ *to* $\mathcal{H}_n$, is irreducible.

As for irreducibility, we mention the following remark. Irreducibility is closely connected with various concepts. They are, besides the commutant property (4°) and among others, ergodicity, tensor products, and being representations of class 1 (see the following). Accordingly we can prove the irreducibility by several different methods.

Each irreducible representation, constructed above, has such a unique function in E_ℓ or $\mathcal{H}_n$ that depends only on a single variable, that is, $P_\ell(\cos\theta)$ in (3° S) or $H_n(x)$ in (3° G). These irreducible representations are called to be of *class* 1. The representations R_ℓ and $\mathcal{T}_n$ exhaust all representations of class 1 of O_3 and O_∞, respectively. Moreover, all irreducible representations are of class 1 for O_3, but not for O_∞; O_∞

admits a larger *discrete* series of representations which includes the above representations[16].

We should like to mention here that problems with which this report is closely related were considered from various viewpoints and contained in articles by several authors, in particular, H.P.McKean[14], T.Hida[15], Okamoto and Sakurai[16].

§3. Geometry of O_∞ and E^*

3.1. *Topology in* O_∞

We descrete the contents of this § according to the report[3]. We define the compact-open topology in the transformation group O_∞, that is, given a compact set F and an open set V in E, a neighborhood of the identity e of O_∞ is defined as the set of all such g that $g(\xi) - \xi \in V$ for every $\xi \in F$. Then O_∞ is a topological group. In fact it is an infinite-dimensional and complete continuous group. Moreover, the manifold O_∞ is a nuclear space, which makes it possible to define differentiability and the like of functions on O_∞.

3.2. *Rotations of finite rank*

(6°) *Every orthogonal operator g of finite rank (namely, such a g that leaves a subspace of finite codimension fixed point-wise) belongs to O_∞. All such g constitute an invariant subgroup N_0 of O_∞, and the closure N of N_0 is the nontrivial closed invariant subgroup of O_∞.*

(7°) *Each shift operator $\tau(t)(t \neq 0)$ does not belong to N.*

We may say this proposition means that the operator $\tau(t)$ is essentially infinite-dimensional.

3.3. *Transversal fields*

We define some subgroups of O_∞, which are used in §4. The following sets Z_+ and Z_- are closed subgroups of O_∞ :

$$Z_\pm = \{g : \tau(-t)g\tau(t) \to e, \text{ as } t \to \pm\infty\}.$$

They are, in some sense, an infinite-dimensional analogue to nilpotent subgroups of semisimple Lie groups.

In the same way we define subgroups $Z_\pm(\gamma)$, for a one-parameter subgroup $\gamma = (\gamma(t))$.

(8°) *If* $(\gamma(t))$ *satisfies the condition that* $<\xi, \gamma(t)\eta> \to 0$ *as* $t \to \infty$ *for any* ξ *and* η, *then the subgroup* N *is contained in both* $Z_+(\gamma)$ *and* $Z_-(\gamma)$.

The shift τ obviously satisfies the above condition. The fields generated by these subgroups in E^* can be regarded transversal to the flow in E^* defined by the given one-parameter subgroup γ. Similarly to the case of classical flows, the present reporter showed that the existence of transversal fields implies such properties of the Brownian motion as ergodicity, strong mixing-up property, and having countable Lebesgue spectrum.

§4. Continuous Series of Representations

4.1. *Analogy to Hyperbolic Group*

In this § we shall briefly describe a class of representations of O_∞, which is different from the series of representations $\mathcal{T}_n$ (constructed in §2), although our research is not yet satisfactory. To explain our motivation, we shall consider here analogy of our representations to the representations of the hyperbolic group (which is 3-dimensional Lorentz group and, at the same time, is locally isomorphic to the group $SL(2,R)$ of real 2×2 matrices with determinant equal to 1), though at the present it is not clear whether this analogy has any substantial meaning beyond its superficiality.

The representations of group $SL(2,R)$ are known[4], and the irreducible unitary representations of class 1 are classified as follows:

the principal continuous series, constituted of representations $D_{i\rho}$, $(-\infty < \rho < \infty)$;

the supplementary series, constituted of $D_s, (-1 < s < 1, s \neq 0)$;

the discrete series, constituted of (reducible and non-unitary) representations $D_n(n = 0, 1, 2, \cdots)$, each of which is decomposed in two irreducible unitary representations D_n^+ and D_n^- and one finite-dimensional nonunitary irreducible representation F_n.

Each of these representations $D_\lambda = (T_\lambda, \mathfrak{H}_\lambda)(\lambda \in \mathbf{C})$ is realized in a nuclear space of smooth functions on the real line and by operators $T_\lambda(g)$ expressed by a common form parametrized by a complex number λ. From this fact it is reasonable to think that the representations $D_s, D_n^\pm$ and F_n are obtained by the analytic prolongation of $D_{i\rho}$ of the principal continuous series.

4.2. *Gauss Decomposition*

Analogously to the subgroups Z_+ and Z_-, we define closed subgroups K_+ and K_- as follows:

$$K_\pm = \{g; \tau(-t)g\tau(t) \text{ converges to some } a, \text{ as } t \to \pm\infty\}.$$

We may consider the manifolds $K_\pm$ an infinite-dimensional analogue to the Grassman manifold. The elements a in the definition of $K_\pm$ constitute a subgroup D of $K_\pm$, and we have the following propositions:

(9°) $Z_\pm$ *are invariant subgroups of* $K_\pm$, *respectively, and*

$$K_+ = Z_+ D, K_- = DZ_-.$$

(10°) $O_\infty = Z_+ D Z_-$. (We call it *Gauss decomposition* of O_∞.)

(11°) $D \supset \tau$.

(12°) $K_- = Z_+ \setminus O_\infty$, (*It does not mean a group-isomorphism.*)

4.3. *Toward Construction of Continuous Series*

We describe briefly a method of defining a class of representations of O_∞, which we will call a *continuous series of representations* of O_∞. Our method is an extension of the method of Gelfand's[4] which is used in the case of $SL(2, \mathbf{C})$ and $SL(2, R)$.

First we define the *quasi-regular representation* as follows. Let $\mathcal{H}$ be the vector space of functions on K_-.

Let $T(g)$ be the following mapping on $\mathcal{H} : f(k) \rightarrow f(k\bar{g})$, where we denote by k and $k\bar{g}$ those cosets in $Z_+ \setminus O_\infty$ that contain k of K_- and kg of O_∞, respectively. Then $T(g)$ are continuous and constitute a representation of O_∞ on $\mathcal{H}$. We shall denote it by $\mathcal{R} = (T, \mathcal{H})$.

(13°) *For* $g \in N, T(g)$ *is the identity operator; that is,* $\mathcal{R}$ *is a representation of the factor group* G/N.

Now we construct representations which constitute a continuous series. Let $\chi(t)$ be a unitary character of the one-parameter subgroup $\tau = \tau(t)$: $\chi(t) = \exp(i\chi t)$, where the same notation χ stands for the parameter too, $-\infty < \chi < \infty$. We call a function $f(k)$ on $K_- = Z_+ \setminus O_\infty$ χ-

homogeneous when it satisfies the condition $f(\tau(t)k) = \chi(t)f(k)$. Note that this concept is well-defined.

(14°) *All χ-homogeneous functions on K_- constitute a subspace $\mathcal{H}_\chi$ of $\mathcal{H}$.*

(15°) *Each operator $T(g)$ on $\mathcal{H}$ leaves $\mathcal{H}_\chi$ stable.*

Hence we denote the restriction of $T(g)$ on $\mathcal{H}_\chi$ by $T_\chi(g)$.

(16°) *For each χ of τ, the pair $\mathcal{R}_\chi = (T_\chi, \mathcal{H}_\chi)$ defines a representation of O_∞; it is a representation of G/N.*

We note that for each $\chi, T_\chi(g)$ and $\mathcal{H}_\chi$can be rewritten in another expression, where $\mathcal{H}_\chi$ is the space of functions on Z_-.

Concluding the report, we mention some problems which are desirable to be settled:

Equivalence and irreducibility of $\mathcal{R}_\chi$; invariant Hermitian forms in $\mathcal{H}$ and $\mathcal{H}_\chi$; irreducible decomposition of the representation $\mathcal{R}$; possible relation between our continuous series and the series of homogeneous chaos $\mathcal{T}_n$; and existence or non-existence of different continuous series.

We hope to discuss on another occasion these problems as well as the details of what we explained only briefly in this §.

References

1. N.Wiener, The homogeneous chaos, *Amer. J. Math.* 60 (1938), 897-936.
2. V.Fock, Konfigrationsraum und zweite Quantelung, *Zeits. Physik* 75 (1932), 622-647
3. H.Yoshizawa, Rotation group of Hilbert space and its application to

Brownian motion, in *Proc. Internat. Conf. Functional Analysis and Related Topics* (Tokyo, 1969), 414-423.

4. I.M.Gelfand, M.I.Graev and N.Ya. Vilenkin, *Generalized Functions*, Vol 5 (in Russian)(Moscow, 1962).
5. S.Kakutani, Determination of the spectrum of the flow of Brownian motion, *Proc. Nat. Acad. Sci. U.S.A.* 36 (1950), 319-323.
6. S.Kakutani, Spectral analysis of stationary Gaussian processes, *Proc. 4th Berkeley Symp. Probability* Vol. 2, (1961), 239-247.
7. K.Itô, Multiple Wiener integral, *J. Math. Soc. Japan* 3 (1951), 157-169.
8. K.Itô, Spectral type of the shift transformation of differential process with stationary increments, *Trans. Amer. Math. Soc.* 81 (1956), 253-263.
9. I.E.Segal, Tensor algebras over Hilbert spaces **I**, *Trans. Amer. Math. Soc.*81 (1956), 106-134.
10. Y.Umemura, Measures on infinite dimensional vecter spaces, *Publ. Res. Inst. Math. Sci. Ser. A*, 1 (1965), 1-47
11. Y.Umemura, On the infinite dimensional Laplacian operator, *J. Math. Kyoto Univ.* 4 (1965), 477-492.
12. P.Lévy, *Problèms concerts d'analyse fonctionnelle, Troisième partie*, (Gauthier-Villars, 1951).
13. N.Obata, A characterization of the Lévy Laplacian in terms of infinite dimensional rotation groups, *Nagoya Math. J.* 118 (1990), 111-132.
14. H.P.McKean, Geometory of differential space, *Ann Probability* 1 (1973), 197-206.

15. T.Hida, *Brownian motion*, (Springer-Verlag, 1980).

16. K.Okamoto and T.Sakurai, On a certain class of irreducible unitary representations of the infinite dimensional rotation group II, *Hiroshima Math. J.* 12 (1982), 385-397.

LAW EQUIVALENCE OF ORNSTEIN–UHLENBECK PROCESSES

J. ZABCZYK
Institute of Mathematics, Polish Academy of Sciences
Śniadeckich 8, 00–950 Warsaw, Poland

ABSTRACT
The paper is concerned with absolute continuity of laws corresponding to Markov–Gaussian processes on Hilbert spaces.

1. Introduction and Preliminaries

Let $(\Omega, \mathcal{F}, \mathsf{P})$ be a probability space with a filtration $(\mathcal{F}_t)_{t\geq 0}$ and $W(t)$, $t \geq 0$ a cylindrical Wiener process on a separable Hilbert space U; we assume that the covariance of $W(1)$ is the identity operator I. Let moreover H be a separable Hilbert space, $S(t)$, $t \geq 0$, a C_0-semigroup of operators on H with the infinitesimal generator A and B a bounded linear operator from U into H. An Ornstein–Uhlenbeck process $X(t)$, $t \geq 0$ is defined by the formula

$$X(t) = \int_0^t S(t-s)B\,dW(s), \qquad t \geq 0. \tag{1}$$

The stochastic integral in (1) is well defined if and only if

$$\int_0^t |S(t)B|^2_{\mathrm{HS}}\,dt < +\infty, \qquad t \geq 0, \tag{2}$$

where $|\cdot|_{\mathrm{HS}}$ denotes the Hilbert–Schmidt norm of an operator. If the condition (2) holds then the process (1) is stochastically continuous and therefore has a predictable version.

It is well known that if $\dim H < +\infty$ then the class of time homogeneous, Markov–Gaussian processes X, $X(0) = 0$, is identical with the class of processes of type (1). Results of Ref. 3 suggest that a similar characterization is valid in the infinite dimensional case with the operator B possibly unbounded.

Let us fix $T > 0$. It follows from (2) that

$$\mathsf{E} \int_0^T |X(S)|^2 \, ds < +\infty. \tag{3}$$

Therefore the law of X is a Gaussian measure $\mathcal{L}(X)$ on the Hilbert space $\mathcal{H} = L^2[0, T; H]$. Let $\mathcal{L}(\tilde{X})$ be the law on $\mathcal{H}$ of another Ornstein–Uhlenbeck process

$$\tilde{X}(t) = \int_0^t \tilde{S}(t-s)\tilde{B}\, d\tilde{W}(s), \qquad t \geq 0 \tag{4}$$

corresponding to a C_0-semigroup $\tilde{S}(t)$, $t \geq 0$, with the infinitesimal generator $\tilde{A}$, to a linear, bounded operator $\tilde{B}$: $\tilde{U} \to H$ and to a cylindrical Wiener process $\tilde{W}$ on $\tilde{U}$.

The paper is concerned with necessary and with sufficient conditions for the absolute continuity of the law $\mathcal{L}(X)$ with respect to $\mathcal{L}(\tilde{X})$. The aim is to obtain explicit conditions in terms of A, $\tilde{A}$ and B, $\tilde{B}$.

The case $B = \tilde{B} = I$ has been treated by several authors under different sets of conditions. S. M. Kozlow[5] obtained necessary and sufficient conditions for equivalence in the case of elliptic operators A, $\tilde{A}$ acting on manifolds without boundaries. The case of diagonal and commuting operators A and $\tilde{A}$ was discussed by T. Koski and W. Loges[7]. S. Peszat[9] considered operators $\tilde{A}$ of the form $\tilde{A} = A + K$ with K from the class of perturbations introduced by N. Dunford and J. T. Schwartz, see Ref. 2. Systems with delay operators and with analytic generators were studied in our paper[10]. For motivation to study absolute continuity of Ornstein–Uhlenbeck processes we refer to Ref. 7 and Ref. 1.

General necessary and sufficient conditions for absolute continuity of Gaussian measures are given by the following theorem in Ref. 8, II, §3.

Theorem [Feldman and Hajek]. *Let μ and $\tilde{\mu}$ be centered Gaussian measures on a separable Hilbert space with the covariance operators Q and $\tilde{Q}$. Then they are either singular or mutually absolutely continuous (equivalent). They are equivalent if and only if*

$$\operatorname{Image} Q^{1/2} = \operatorname{Image} \tilde{Q}^{1/2} \tag{5}$$

and

The operator $(\tilde{Q}^{-1/2}Q^{1/2})(\tilde{Q}^{-1/2}Q^{1/2})^* - I$ is Hilbert-Schmidt with all eigenvalues greater than -1. (6)

In the formulation of (6) one assumes that (5) holds and that $\tilde{Q}^{-1/2}$ is the pseudoinverse of $\tilde{Q}^{1/2}$.

In our situation one has

Proposition 1. *The covariance operator Q of the measure $\mu = \mathcal{L}(X)$ is of integral type*

$$Q\varphi(t) = \int_0^T q(t,s)\varphi(s)\,ds, \qquad t \in [0,T], \qquad \varphi \in \mathcal{H} \tag{7}$$

where

$$q(t,s) = \int_0^{t\wedge s} S(t-r)BB^*S^*(s-r)\,dr, \qquad t,s \in [0,T]. \tag{8}$$

However, even in the simplest one dimensional case, $\dim H = \dim U = 1$, explicit formulae for the operators $Q^{1/2}$ and $\tilde{Q}^{1/2}$ are not available and therefore a direct application of the Feldman–Hajek theorem to our problem is not possible.

In Section 2, see Theorem 1 and Theorem 2, we give more explicit versions of conditions (5) and (6). They are formulated in terms of the associated control systems

$$\dot{y} = Ay + Bu, \qquad y(0) = 0, \qquad u \in L^2[0,T;U] \tag{9}$$

$$\dot{\tilde{y}} = \tilde{A}\tilde{y} + \tilde{B}\tilde{u}, \qquad \tilde{y}(0) = 0, \qquad \tilde{u} \in L^2[0,T;\tilde{U}] \tag{10}$$

In Section 3 we apply them to various classes of processes. Possible extensions are discussed briefly in Section 4.

2. Control Theoretic Version of Equivalence Condition

Let L and $\tilde{L}$ be the "input–output" transformations corresponding to systems (9) and (10). Thus $L\colon \mathcal{U} = L^2[0,T;U] \to \mathcal{H}$, and $\tilde{L}\colon \tilde{\mathcal{U}} = L^2[0,T;\tilde{U}] \to \mathcal{H}$ are given by the formulae:

$$Lu(t) = \int_0^t S(t-s)Bu(s)\,ds, \qquad t \in [0,T], \qquad u(\cdot) \in \mathcal{U} \tag{11}$$

$$\tilde{L}\tilde{u}(t) = \int_0^t \tilde{S}(t-s)\tilde{B}\tilde{u}(s)\,ds, \qquad t \in [0,T], \qquad \tilde{u}(\cdot) \in \tilde{\mathcal{U}}. \tag{12}$$

Without any loss of generality we can assume that

$$\operatorname{Ker} B = \{0\}, \qquad \operatorname{Ker} \tilde{B} = \{0\}. \tag{13}$$

Theorem 1. *Measures $\mathcal{L}(X)$ and $\mathcal{L}(\tilde{X})$ are equivalent if and only if*

$$\operatorname{Image} L = \operatorname{Image} \tilde{L} \tag{14}$$

and the operator

$$(\tilde{L}^{-1}L)(\tilde{L}^{-1}L)^* - I \tag{15}$$

is Hilbert–Schmidt.

Remark. In (15) the symbol $\tilde{L}^{-1}$ stands for the pseudoinverse of $\tilde{L}$. If (14) holds then, by the closed graph theorem, the operator

$$\mathcal{M} = \tilde{L}^{-1}L \tag{16}$$

is linear and bounded.

Proof. We show that (14) is equivalent to (5) and (15) to (6). It is well known, see also Lemma 1, that operators $Q^{1/2}$ and L have the same image if and only if there exist constant k_1, $k_2 > 0$ such that

$$k_1 |Q^{1/2}\varphi| \le |L^*\varphi| \le k_2 |Q^{1/2}\varphi|, \qquad \varphi \in \mathcal{H}.$$

By direct computation $Q = LL^*$. Therefore

$$|Q^{1/2}\varphi|^2_{\mathcal{H}} = \langle Q\varphi, \varphi\rangle_{\mathcal{H}} = \langle LL^*\varphi, \varphi\rangle_{\mathcal{H}} = |L^*\varphi|^2_{\mathcal{U}}. \tag{17}$$

Conversely Image L = Image $Q^{1/2}$ so the first part of the theorem follows. To prove the second part we need the following

Lemma 1. *If H_1, H_2 and H are Hilbert spaces and L_1: $H_1 \to H$, L_2: $H_2 \to H$ are bounded linear operators such that*

$$|L_1^* h| = |L_2^* h|, \qquad \textit{for all } \ h \in H, \tag{18}$$

then Image L_1 = Image L_2 *and*

$$|L_1^{-1}h| = |L_2^{-1}h|, \qquad \textit{for all } \ h \in \text{Image}\, L_1. \tag{19}$$

Proof. By a classical separation theorem for convex sets the condition (18) is equivalent to

$$\overline{\{L_1 u;\ |u| \le 1\}} = \overline{\{L_2 v;\ |v| \le 1\}}. \tag{20}$$

Since the unite balls in Hilbert spaces are weakly compact therefore (20) is equivalent to

$$\{L_1 u;\ |u| \le 1\} = \{L_2 v; |v| \le 1\}. \tag{21}$$

To show (19) one can assume that L_1 and L_2 are invertible operators. Let $e = L_1 h_1 = L_2 h_2 \neq 0$ and, by contradiction, $|h_1| > |h_2| = 1$. Then

$$\frac{e}{|h_2|} = L_2\left(\frac{h_2}{|h_2|}\right) \in \{L_2 v; |v| \le 1\}.$$

But $\frac{e}{|h_2|} = L_1\left(\frac{h_1}{|h_2|}\right)$ and $\left|\frac{h_1}{|h_2|}\right| > 1$, therefore $\frac{e}{|h_2|} \notin \{L_1 u; |u| \le 1\}$, a contradiction.

To complete the proof of Theorem 1 let us notice that from Lemma 1 and identity $LL^* = Q$ one has

$$|Q^{-1/2}h| = |L^{-1}h|, \qquad |\tilde{Q}^{-1/2}h| = |\tilde{L}^{-1}h|, \qquad h \in \operatorname{Image} L.$$

Consequently, the transformations

$$P_1 = Q^{-1/2}L \qquad \text{and} \qquad P_2 = \tilde{Q}^{-1/2}\tilde{L}$$

are isometries of $\mathcal{U}$ and $\tilde{\mathcal{U}}$ onto $\mathcal{H}$. Moreover $\tilde{Q}^{-1/2}Q^{1/2} = P_2\tilde{L}^{-1/2}LP{-}1_1$ and

$$\begin{aligned}(\tilde{Q}^{-1/2}Q^{1/2})(\tilde{Q}^{-1/2}Q^{1/2})^* &= P_2\tilde{L}^{-1}LP_1P_1^{-1}(P_2\tilde{L}^{-1}LP_1^{-1})^* \\ &= P_2\mathcal{M}P_1^{-1}(P_1^{-1})^*\mathcal{M}^*P_2^* = P_2\mathcal{M}\mathcal{M}^*P_2^{-1}.\end{aligned}$$

Consequently operators

$$(\tilde{Q}^{-1/2}Q^{1/2})(\tilde{Q}^{-1/2}Q^{1/2})^* - I \qquad \text{and} \qquad \mathcal{M}\mathcal{M}^* - I$$

are isometrically equivalent. So one of them is Hilbert–Schmidt if and only if the other is. In addition $|\mathcal{M}^*u| \neq 0$ for $u \notin 0$. The proof is complete.

Remark. Condition (14) and the operator $\mathcal{M}$ have clear control-theoretic interpretations. The condition (15) holds if and only if the control systems (9) and (10) generate the same sets of trajectories. Moreover $\mathcal{M}u = \tilde{u}$, if and only if the controls u and $\tilde{u}$ generate the same output y.

The next theorem gives an explicit formula for the operator $\mathcal{M}$.

Theorem 2. *Assume that* (14) *holds. Then for all Lipschitz continuous functions* $u \in \mathcal{U}$ *and almost all* $t \in [0,T]$:

$$\mathcal{M}u(t) = \tilde{B}^{-1}(A - \tilde{A})\int_0^t S(t-s)Bu(s)\,ds + \tilde{B}^{-1}u(t). \tag{22}$$

We need two lemmas.

Lemma 2. *If* $u \in \mathcal{U}$ *and the function* $y = Lu$ *is absolutely continuous then*

(i) $y(t) \in D(A)$ *for almost all* $t \in [0,T]$,

(ii) $Ay(\,\cdot\,)$ *is integrable on* $[0,T]$,

(iii) $y(t) = \int_0^t Ay(s)\,ds + \int_0^t Bu(s)\,ds, \quad t \in [0,T]$.

Proof. Fix $v \in D(A^*)$. Then

$$\langle y(t), v\rangle = \langle \int_0^t y(s)\, ds, a^* v\rangle + \langle \int_0^t Bu(s)\, ds, v\rangle, \qquad t \in [0, T].$$

Consequently, for all $t \in [0, T]$

$$y(t) = \int_0^t Bu(s)\, ds + A\Big(\int_0^t y(s)\, ds\Big).$$

Let $t \in (0, T]$ be a number such that $y(\cdot)$ is differentiable at t and $\frac{d}{dt}\big(\int_0^t Bu(s)\, ds\big) = Bu(t)$. Then, taking $h \downarrow 0$ and using the fact that A is a closed operator one obtains

$$\frac{d}{dt} y(t) = Ay(t) + Bu(t).$$

Lemma 3. *If $u \in \mathcal{U}$ is Lipschitz continuous then also $y = Lu$ is Lipschitz continuous and therefore absolutely continuous.*

Proof. The result follows from the estimate:

$$|y(t+h) - y(t)| \le \int_t^{t+h} |S(t+h-s)Bu(s)|\, ds + \int_0^t |S(r)|\,|B|\,|u(t+h-r) - u(t-r)|\, dr$$

valid for $0 \le t \le t + h \le T$.

Proof of Theorem 2. Assume that a function $u \in \mathcal{U}$ is Lipschitz continuous. By Lemma 2 and Lemma 3 for $y = Lu$, $y(t) \in D(A)$, $\dot{y}(t) = Ay(t) + Bu(t)$, for almost all $t \in (0, T]$. If (14) holds then there exists $\tilde{u} \in \tilde{\mathcal{U}}$ such that $y = \tilde{L}\tilde{u}$. Again by Lemma 3 and Lemma 2

$$y(t) \in D(\tilde{A}), \qquad \dot{y}(t) = \tilde{A}y(t) + \tilde{B}\tilde{u}(t), \qquad \text{for almost all } t \in (0, T).$$

Consequently, for almost all $t \in [0, T]$:

$$\tilde{B}\tilde{u}(t) = (A - \tilde{A})y(t) + Bu(t).$$

However, see Theorem 3, Image $\tilde{B}$ = Image B, and finally

$$\tilde{u}(t) = \tilde{B}^{-1}(A - \tilde{A})y(t) + \tilde{B}^{-1}Bu(t),$$

the required result.

3. Applications

3.1. *Conditions on Operators B and $\tilde{B}$*

Theorem 3. *Assume that laws $\mathcal{L}(X)$ and $\mathcal{L}(\tilde{X})$ are equivalent. Then there exist positive constants k_1, k_2 such that*

$$k_1 BB^* \le \tilde{B}\tilde{B}^* \le k_2 BB^* \tag{23}$$

If, in addition, the set $D(A^*)^2 \cup D(\tilde{A}^*)^2$ is dense in H, then

$$BB^* = \tilde{B}\tilde{B}^* \tag{24}$$

Proof. We show that (23) follows from (14). If follows from (14) that for some $k > 0$ and all $\varphi \in \mathcal{H}$, $|L^*\varphi|^2 \le |\tilde{L}^*\varphi|^2$. Equivalently

$$\int_0^T \left| \int_t^T B^* S(s-t)\varphi(s)\, ds \right|^2 dt \le k \int_0^T \left| \int_t^T \tilde{B}^* \tilde{S}^*(s-t)\varphi(s)\, ds \right|^2 dt. \tag{25}$$

For arbitrary $\delta \in (0, T]$ and all $a \in H$ define

$$\varphi_\delta(s) = \frac{\sqrt{3}}{\delta^{3/2}} I_{[0,\delta]}(s) a, \qquad s \in [0, T]$$

Inserting φ_δ into (25) and passing with $\delta \downarrow 0$ one obtains that

$$|B^* a|^2 \le k |\tilde{B}^* a|^2. \tag{26}$$

It is therefore enough to define $k_1 = k^{-1}$. In a similar way one defines k_2.

To prove the second part of the theorem assume that $v \in D(A^{*2})$. Without any loss of generality one can assume that $\mathbf{P}$-a.s.

$$\langle X(t), v\rangle = \int_0^t \langle X(s), A^* v\rangle\, ds + \langle B^* v, W(t)\rangle, \qquad t \in [0, T]. \tag{27}$$

By the law of the iterated logarithm and (27)

$$\mathbf{P}\left(\overline{\lim_{t\downarrow 0}} \frac{1}{\sqrt{2t \log_2 \frac{1}{t}}} \langle X(t), v\rangle = |B^* v| \right) = 1 \tag{28}$$

Since the laws $\mathcal{L}(X)$ and $\mathcal{L}(\tilde{X})$ are equivalent, (29) holds also for $\tilde{X}$. If, in addition, $v \in D(a^*)^2 \cap D(\tilde{A}^*)^2$, then

$$\mathbf{P}\left(\overline{\lim_{t\downarrow 0}} \frac{1}{\sqrt{2t \log_2 \frac{1}{t}}} \langle \tilde{X}(t), v\rangle = |\tilde{B}^* v| \right) = 1$$

Consequently, for v from a dense subset of H,

$$|B^* v|^2 = \langle BB^* v, v\rangle = |\tilde{B}^* v| = \langle \tilde{B}\tilde{B}^* v, v\rangle.$$

Conjecture 1. We conjecture that the condition (24) is always necessary for the absolute continuity.

Note that replacing in (7)–(8) the operator BB^* by $\tilde{B}\tilde{B}^*$ does not change the law of X therefore, to simplify the formulation of the results, we will assume from now on that

$$\tilde{B} = B, \qquad \operatorname{Ker} B = \{0\}. \tag{29}$$

3.2. *Finite Dimensional and Delay Systems*

Assume that $\dim H = n < +\infty$ and $\dim U = m < +\infty$ and denote by $\{e_k; k = 1, 2, \ldots, m\}$ a basis in U.

The following proposition is an easy corollary of Theorem 1 and Theorem 2, see Ref. 10.

Proposition 1. *Laws $\mathcal{L}(X)$ and $\mathcal{L}(\tilde{X})$ are equivalent if and only if*

(i) *vectors $A^j Be_k$, $j = 0, \ldots, m-1$, $k = 1, \ldots, m$ span the same space H_0 as the vectors $\tilde{A}^j Be_k$, $j = 0, \ldots, n-1$, $k = 1, 2, \ldots, m$,*

(ii) $(A - \tilde{A})(H_0) \subset \operatorname{Image} B$.

Similar characterization can be obtained for delay systems. For simplicity we treat systems in $\mathbf{R}^n$ with one delay only. We assume that

$$\begin{aligned} dX(t) &= \big(A_0 X(t) + A_1 X(t-1)\big)\, dt + B\, dW(t) \\ X(\theta) &= 0, \qquad \text{for } \theta \in [-1, 0], \end{aligned} \tag{30}$$

$$\begin{aligned} d\tilde{X}(t) &= \big(\tilde{A}_0 \tilde{X}(t) + \tilde{A}_1 \tilde{X}(t-1)\big)\, dt + B\, dW(t) \\ \tilde{X}(\theta) &= 0, \qquad \text{for } \theta \in [-1, 0], \end{aligned} \tag{31}$$

where W is an m-dimensional Wiener process.

Solutions to the equations (30) and (31) can be regarded as an infinite-dimensional Ornstein–Uhlenbeck processes, see e.g. Ref. 10, by defining the new state $Z(t)$:

$$Z(t) = (X(t), X_t) \in \mathbf{R}^n \times L^2(-1, 0; \mathbf{R}^n) = H,$$

where $X_t(\theta) = X(t+\theta)$, $t \geq 0$, $\theta \in [-1, 0]$. Theorem 1 implies that if the laws $\mathcal{L}(X)$ and $\mathcal{L}(\tilde{X})$ are absolutely continuous (on $C(0, T; \mathbf{R}^n)$) then the controlled systems

$$\begin{aligned} \dot{y} &= A_0 y(t) + A_1 y(t-1) + Bu(t) \\ y(\theta) &= 0 \qquad \text{for } \theta \in [-1, 0], \quad t \geq 0 \end{aligned} \tag{32}$$

$$\dot{\tilde{y}} = \tilde{A}_0 \tilde{y} \mu t \geq 0 \tag{32}$$

$$\begin{aligned} \dot{\tilde{y}} &= \tilde{A}_0 \tilde{y}(t) + \tilde{A}_1 \tilde{y}(t-1) + Bu(t) \\ \tilde{y}(\theta) &= 0 \qquad \text{for } \theta \in [-1, 0], \quad t \geq 0 \end{aligned} \tag{33}$$

generate the same sets of trajectories.

Denote by H_0 and H_1 the spaces of reachable states respectively for the n-dimensional system

$$\dot{y} = A_0 y(t) + Bu(t) \qquad y(0) = 0 \tag{34}$$

and for the $2n$-dimensional system

$$\dot{y} = A_0 y(t) + Bu(t) \qquad y(0) = 0 \tag{35}$$

$$\dot{z} = A_1 z + A_0 y, \qquad z(0) = 0. \tag{36}$$

They can be easily described algebraically. In a similar way spaces $\tilde{H}_0$ and $\tilde{H}_1$ can be associated with (31). We restrict our consideration to $T \in [0, 2]$. For detailed calculations we refer to Ref. 10.

Proposition 2. *For the delay systems* (30) *and* (31), $\mathcal{L}(X)$ *is equivalent to* $\mathcal{L}(\tilde{X})$ *if and only if:*

$$\text{when } T \leq 1 \text{ and } H_0 = \tilde{H}_0, \ (A_0 - \tilde{A}_0)(H_0) \subset \operatorname{Image} B$$

or

when $T \in (1, 2]$ *and* $H = \tilde{H}_0$, $H_1 = \tilde{H}_1$, $(A_0 - \tilde{A}_0)(H_0) \subset \operatorname{Image} B$, $[A_1 - \tilde{A}_1, A_0 - \tilde{A}_0](H_1) \subset \operatorname{Image} B$.

3.3. The White Noise Case

We assume that

$$\dim U = +\infty \qquad \text{and} \qquad B = I. \tag{37}$$

Note that now

$$\begin{aligned} \mathcal{M}u(t) &= (A - \tilde{A}) \left(\int_0^t S(t - s) u(s) \, ds \right) + u(t) \\ &= \mathcal{K}u(t) + u(t), \qquad t \in [0, T] \end{aligned} \tag{38}$$

Consequently, by Theorem 1 and Theorem 2, the following result holds.

Theorem 4. *If* (37) *holds then* $\mathcal{L}(X)$ *and* $\mathcal{L}(\tilde{X})$ *are equivalent if and only if* $\operatorname{Image} L = \operatorname{Image} \tilde{L}$ *and* $\mathcal{K} + \mathcal{K}^* + \mathcal{K}\mathcal{K}^*$ *is Hilbert–Schmidt.*

We have also the following more specific characterizations.

Theorem 5. *Assume* (37).

(i) *If* Image L = Image $\tilde{L}$ *then* $D(A) = D(\tilde{A})$.

(ii) *If* $D(A) = D(\tilde{A})$ *and the semigroups* $S(t)$, $\tilde{S}(t)$, $t \geq 0$ *are analytic then conversely* Image L = Image $\tilde{L}$.

(iii) *If* Image L = Image $\tilde{L}$ *then the operator* $\mathcal{K}$ *is Hilbert–Schmidt if and only if*

$$\int_0^T \left|(A - \tilde{A})S(t)\right|_{\mathrm{HS}}^2 dt < +\infty \tag{39}$$

Proof. (i) Let $x \in D(A)$, $x \neq 0$ then $y(t) = tx$, $t \geq 0$ is an absolutely continuous solution of the equation

$$\dot{y} = Ay + u, \qquad y(0) = 0 \tag{40}$$

where $u(t) = x - tAx$, $t \geq 0$. Consequently, see Lemma 2, $\dot{y} = \tilde{A}y + v$, $y(0) = 0$ for some $v \in L^2[0,T;H]$. In particular for almost all $t \geq 0$, $tx \in D(\tilde{A})$, so $x \in D(\tilde{A})$.

(ii) It is enough to use the fact, see e.g. Ref. 10, that the set of all solutions y of (40) with $u \in L^2[0,T;H]$ is identical with $W_0^{1,2}(0,T;H) \cap L^2(0,T;D(A))$.

(iii) Let (e_m) be an orthonormal, complete basis in H composed of elements from $D(A)$ and let (f_n) be an orthonormal complete basis in $L^2[0,T;\mathbf{R}^1]$ composed of Lipschitz continuous functions. Functions $F_{n,m}(t) = f_n(t)e_m$, $t \in [0,T]$, $n,m = 1,2,\ldots$ form a complete and orthonormal basis in $\mathcal{H}$. Note that

$$\sum_{n,m} |\mathcal{K}F_{n,m}|_{\mathcal{H}}^2 = \sum_{n,m} \int_0^T \left(\sum_k \langle e_k, \mathcal{K}F_{n,m}(t)\rangle_H^2\right) dt.$$

Moreover

$$\int_0^T \left(\sum_n \langle e_k, \mathcal{K}F_{n,m}(t)\rangle_H^2\right) dt = \sum_n \int_0^T \left(\int_0^t \langle e_k, (A - \tilde{A})S(t-s)e_m\rangle f_n(s)\, ds\right)^2 dt.$$

Since the operators $r_{k,m}$: $L^2[0,T;\mathbf{R}^1] \to L^2[0,T;\mathbf{R}^1]$

$$r_{k,m}f(t) = \int_0^t \langle e_k, S(t-s)e_m f(s)\, ds\rangle, \qquad t \in [0,1]$$

are Hilbert–Schmidt with the Hilbert–Schmidt norms

$$|r_{k,m}|_{\mathrm{HS}}^2 = \int_0^T \int_0^t \langle e_k, (A - \tilde{A})S(t-s)e_m\rangle^2\, dt\, ds,$$

therefore, by Parseval's identity

$$\sum_{n,m} |\mathcal{K}F_{n,m}|^2_{\mathcal{H}} = \int_0^T \int_0^t \sum_m \left|(A-\tilde{A})S(t-s)e_m\right|^2 dt\, ds$$
$$= \int_0^T (T-r)\Big(\sum_m \left|(A-\tilde{A})S(r)e_m\right|^2\Big)\, dr.$$

This way we have shown that the operator $\mathcal{K}$ is Hilbert–Schmidt if and only if

$$\int_0^T (T-r)\Big(\sum_m \left|(A-\tilde{A})S(r)e_m\right|^2\Big)\, dr < +\infty \tag{41}$$

If, for some $r > 0$, $\sum_m \left|(A-\tilde{A})S(r)e_m\right|^2 < +\infty$ then the operator $(A-\tilde{A})S(r)$: $D(A) \to H$ has an extension to a linear bounded operator from H into H, denoted also by $(A-\tilde{A})S(r)$, given by the formula:

$$(A-\tilde{A})S(r)h = \sum_m \langle h, e_m\rangle (A-\tilde{A})S(r)e_m, \qquad h \in H \tag{42}$$

with the series in (42) absolutely convergent:

$$\sum_m \left|\langle h, e_m\rangle (A-\tilde{A})S(r)e_m\right| \le \Big(\sum_m \langle h, e_m\rangle^2\Big)^{1/2} \Big(\sum_m \left|(A-\tilde{A})S(r)e_m\right|^2\Big)^{1/2}.$$

The result now easily follows.

Corollary 1. *If*

$$D(A) = D(\tilde{A}) \tag{43}$$

and

$$\int_0^T \left|(A-\tilde{A})S(r)\right|^2_{\mathrm{HS}}\, dr < +\infty \tag{44}$$

then laws $\mathcal{L}(X)$ and $\mathcal{L}(\tilde{X})$ are absolutely continuous. If the operator A is self-adjoint then condition (44) means that the operator $(A-\tilde{A})(-A)^{-1/2}$ has a Hilbert–Schmidt extension to H.

Conjecture 2. Conditions (43) and (44) are also necessary for absolute continuity of $\mathcal{L}(X)$ and $\mathcal{L}(\tilde{X})$.

4. An Extension

Assume that C and $\tilde{C}$ are linear bounded operators from H into a Hilbert space G. Conditions for absolute continuity of $\mathcal{L}(CX)$ and $\mathcal{L}(\tilde{C}\tilde{X})$ analogical to those formulated in Theorem 1 and Theorem 2 can be given as well. The corresponding control systems are now of the form:

$$\begin{aligned} \dot{y} &= Ay + Bu \\ z &= Cy, \qquad y(0) = 0 \end{aligned} \tag{45}$$

$$\begin{aligned} \dot{\tilde{y}} &= \tilde{A}\tilde{y} + \tilde{B}\tilde{u} \\ \tilde{z} &= \tilde{C}\tilde{y}, \qquad \tilde{y}(0) = 0 \end{aligned} \tag{46}$$

and the operators L and $\tilde{L}$ are given by the formulae:

$$Lu(t) = C\int_0^t S(t-s)Bu(s)\,ds, \qquad u \in \mathcal{U}, t \in [0,T] \tag{47}$$

$$\tilde{L}\tilde{u}(t) = \tilde{C}\int_0^t \tilde{S}(t-s)\tilde{B}\tilde{u}(s)\,ds, \qquad \tilde{u} \in \tilde{\mathcal{U}}, t \in [0,T]. \tag{48}$$

With such changes formulations of Theorem 1 and Theorem 2 remain the same as well as the proofs.

If $\dim H = n < +\infty$, $\dim U = m < +\infty$ and $\dim \tilde{U} = \tilde{m} < +\infty$ then necessary and sufficient conditions for $\operatorname{Image} L = \operatorname{Image} \tilde{L}$ were given by B.Jakubczyk[4]. Namely $\operatorname{Image} L = \operatorname{Image} \tilde{L}$ if and only if the images of linear transformations given by matrices:

$$\begin{bmatrix} CB & 0 & \cdots\cdots & & & \\ CAB & CB & 0 & \cdots\cdots & & \\ \cdots & CAB & CB & 0 & \cdots & \\ \cdots\cdots & & & & & \\ CA^{2n}B & \cdots & \cdots & CAB & CB & \end{bmatrix}, \quad \begin{bmatrix} \tilde{C}\tilde{B} & 0 & \cdots\cdots & & & \\ \tilde{C}\tilde{A}\tilde{B} & \tilde{C}\tilde{B} & 0 & \cdots\cdots & & \\ \cdots & \tilde{C}\tilde{A}\tilde{B} & \tilde{C}\tilde{B} & 0 & \cdots & \\ \cdots\cdots & & & & & \\ \tilde{C}\tilde{A}^{2n}\tilde{B} & \cdots & \cdots & \tilde{C}\tilde{A}\tilde{B} & \tilde{C}\tilde{B} & \end{bmatrix}$$

are the same.

Those are only necessary conditions for the law equivalence of $\mathcal{L}(CX)$ and $\mathcal{L}(\tilde{C}\tilde{X})$. It is an open problem to find conditions which are both necessary and sufficient, see Ref. 11 for a special case.

Acknowledgements

The author thanks A. Chojnowska-Michalik and S. Peszat for helpful discussions which led, in particular, to an improved version of Theorem 5.

The paper is a rewritten version of two earlier preprints Ref. 10 and Ref. 11 and was prepared when the author was visiting the Department of Mathematics, University of Warwick, November/December, 1990.

References

[1] G. Da Prato and J. Zabczyk, *Nonexplosion, boundedness and ergodicity of semilinear stochastic equations.* To appear in J. Diff. Equations.

[2] E. B. Davies, *One-parameter semigroups.* London Mathematical Society Monographs No. 15. London Academic Press, 1980.

[3] K. Ito, *Infinite dimensional Ornstein–Uhlenbeck processes*, Taniguchi Symposium. FA, Katata, 1982, 197–224.

[4] B. Jakubczyk, *Invariants and input-output classifications of linear systems*, PhD dissertation, Institute of Mathematics, Polish Academy of Sciences 1977.

[5] S. M. Kozlow, *Equivalence of measures for linear Ito equations with partial derivatives*, Vest. Moscow University, Ser. Math. and Mech., No. 4, 1977, pp. 47–52 (in Russian).

[6] S. M. Kozlow, *Some questions of stochastic equations with partial derivatives*, Trudy Sem. Petrovskogo, 4 (1978), pp. 147–172 (in Russian).

[7] T. Koski and W. Loges, *Asymptotic statistical inference for stochastic heat flow problem*, Statistics and Probability Letters, Vol. 3, No. 4, 1985, pp. 185–189.

[8] H. H. Kuo, *Gaussian Measures in Banach Spaces*, Lecture Notes in Mathematics 463, Springer-Verlag, Berlin, Heidelberg, New York, 1975.

[9] S. Peszat, *Equivalence of distributions of some Gaussian processes taking values in Hilbert space*, submitted for publication.

[10] J. Zabczyk, *Law equivalence of Ornstein–Uhlenbeck processes and control equivalence of linear systems*, Preprint 457, Institute of Mathematics, Polish Academy of Sciences, June 1989, pp. 1–28.

[11] J. Zabczyk, *Law equivalence of Ornstein–Uhlenbeck processes*, Preprint 476, Institute of Mathematics, Polish Academy of Sciences, September 1990, pp. 1–25.

GAUSSIAN RANDOM FIELDS

Series on Probability and Statistics Vol. 1
The Third Nagoya Lévy Seminar

Nagoya, Japan 15–20 Aug. 1990

edited by K. Itô and T. Hida

ERRATA

Equation (5) on line 4 of page 147 should be

$$\int_{C[a,b]}^{\mathrm{anf}q} \delta F(x|y)dx = (-iq)\int_{C[a,b]}^{\mathrm{anf}q} F(x)\left[\int_a^b y'(t)dx(t)\right] dx \; .$$

Equation on line 5 of page 151 should be

$$\delta F(x|y) = i\int_{[B[a,b]} \exp\left\{ i\int_a^b v(t)dx(t)\right\} d\nu_y(v) \in S^1 \; .$$

World Scientific

www.ingramcontent.com/pod-product-compliance
Ingram Content Group UK Ltd.
Pitfield, Milton Keynes, MK11 3LW, UK
UKHW021850210426
5322IPUK00022B/582